Springer Proceedings in Business and Economics

Springer Proceedings in Business and Economics brings the most current research presented at conferences and workshops to a global readership. The series features volumes (in electronic and print formats) of selected contributions from conferences in all areas of economics, business, management, and finance. In addition to an overall evaluation by the publisher of the topical interest, scientific quality, and timeliness of each volume, each contribution is refereed to standards comparable to those of leading journals, resulting in authoritative contributions to the respective fields. Springer's production and distribution infrastructure ensures rapid publication and wide circulation of the latest developments in the most compelling and promising areas of research today.

The editorial development of volumes may be managed using Springer Nature's innovative EquinOCS, a proven online conference proceedings submission, management and review system. This system is designed to ensure an efficient timeline for your publication, making Springer Proceedings in Business and Economics the premier series to publish your workshop or conference volume.

This book series is indexed in SCOPUS.

Wadim Strielkowski
Editor

Leadership, Entrepreneurship and Sustainable Development Post COVID-19

2022 Prague Institute for Qualification Enhancement (PRIZK) International Leadership Conference, June 24-25, 2022, Prague, Czech Republic

Springer

Editor
Wadim Strielkowski
Prague Institute for Qualification Enhancement
Prague, Czech Republic

ISSN 2198-7246 ISSN 2198-7254 (electronic)
Springer Proceedings in Business and Economics
ISBN 978-3-031-28130-3 ISBN 978-3-031-28131-0 (eBook)
https://doi.org/10.1007/978-3-031-28131-0

This Springer imprint is published by the registered company Springer Nature Switzerland AG
The registered company address is: Gewerbestrasse 11, 6330 Cham, Switzerland

Preface: Leadership, Entrepreneurship and Sustainable Development Post COVID-19

This proceedings volume entitled *Leadership, Entrepreneurship and Sustainable Development Post COVID-19* constitutes an edited volume of original peer-reviewed contributions presented at the 2022 Prague Institute for Qualification Enhancement (PRIZK) International Leadership Conference.

The eighth International Leadership Conference 2022 "Novel Insights in the Leadership in Business and Economics After the COVID-19 Pandemic" (NILBEC2022) was held in Prague, Czech Republic in June 24–25, 2022. The Conference was a unique event organized by the academic and research institution from the Czech Republic and managed to bring together academics, researchers, as well as stakeholders from ten different countries and being proficient in such fields as business, economics, education, social sciences, international relations, energy policy, as well as political sciences responsible for the creation and maintenance of the sets of rules for the future sustainable development of entrepreneurial and academic leadership in the post-COVID era.

This Conference proceedings volume covers a very broad range of topics business, economics, education, social sciences, international relations, sustainability, energy policy, as well as political sciences.

The Conference offered a wide range of keynote speeches, oral presentations and poster presentations, as well as roundtables and workshops. Its main points of discussions and debates centered on the six main parts (alias the conference tracks):

- Part I: Leadership as a social, economic, and cultural phenomenon in the post-COVID era
- Part II: Novel leadership trends in business and economics
- Part III: Leadership in academia and higher education after COVID-19 pandemic
- Part IV: Leadership and the digital surge in business and education
- Part V: Leadership in energy policy: new threats and opportunities
- Part VI: Leadership within the context of Sustainable Development Goals (SDGs)

This proceedings volume will be of interest to a wide audience, including academics, researchers, and policy-makers responsible for the creation and maintenance of the sets of rules for the future sustainable development of entrepreneurial and academic leadership in the post-COVID era.

I sincerely hope that you will find this proceedings volume to be inspiring and enlightening. I also hope that it will ignite many novel ideas on how leadership, entrepreneurship, and sustainable development can be nourished in the post-COVID twenty-first century with all its new threats and challenges.

Wadim Strielkowski
Prague Institute for Qualification Enhancement
Prague, Czech Republic
strielkowski@prizk.cz

Contents

Chapter 1
An Introduction: Pandemic-Inspired Policies for Leadership, Entrepreneurship, and Sustainable Development

Gordon Rausser and Wadim Strielkowski

Abstract In this introduction to the volume of proceedings from the eighth International Leadership Conference (2022) hosted by the Prague Institute for Qualification Enhancement (PRIZK), Czech Republic, in June 2022, we focus on the pandemic-inspired policies for leadership, entrepreneurship, and sustainable development, set within the context of the challenges posed by the recent war in Ukraine. The main focus on the conference was to outline positive changes for the leadership in business and economics based on the lessons learned during the pandemic. It becomes quite apparent to us that the concept of leadership has been deeply affected most recently by the COVID-19 pandemic making the things that seemed impossible just a couple of years ago a reality today. The pandemic caused a "digital surge" in business, economics, and education that is likely to stay here after the pandemic is long gone and forgotten. Due to the COVID-19 pandemic, digital innovations in leadership, entrepreneurship, and sustainable development that would have otherwise taken years to implement were introduced in a matter of days making a perfect example of "creative destruction."

Keywords Pandemic-inspired policies · Leadership · Entrepreneurship · Sustainable development

JEL Classification J18, H12, Q01

G. Rausser
University of California Berkeley, Berkeley, CA, USA
e-mail: rausser@berkeley.edu

W. Strielkowski (✉)
Prague Institute for Qualification Enhancement, Prague, Czech Republic
e-mail: strielkowski@prizk.cz

W. Strielkowski (ed.), *Leadership, Entrepreneurship and Sustainable Development Post COVID-19*, Springer Proceedings in Business and Economics, https://doi.org/10.1007/978-3-031-28131-0_1

1.1 Introduction

The eighth International Leadership Conference (2022) "Novel Insights in the Leadership in Business and Economics After the COVID-19 Pandemic" (NILBEC 2022) held by the Prague Institute for Qualification Enhancement (PRIZK), Czech Republic, in June 24–25, 2022, managed to bring together academics, researchers, and stakeholders from ten different countries and being proficient in such fields as business, economics, education, social sciences, international relations, energy policy, and political sciences responsible for the creation and maintenance of the sets of rules for the future sustainable development of entrepreneurial and academic leadership in the post-COVID era.

The effects of the COVID-19 outbreak on sustainable development and future prospects can be seen through the innovations in development and humanitarian efforts within the pandemic framework (Ranjbari et al., 2021; Rausser et al., 2021). There are some important lessons emerging from the NILBEC 2022 International Conference proceedings on how innovation for development efforts has been successful in the COVID-19 pandemic and how to find the ways that such efforts can be made more effective. The proceedings offer a review of the innovation landscape around COVID-19, first through a global lens and then through an emphasis on investment for development and humanitarian efforts which can be conducted for seeing these efforts and outcomes. Following the examination of the overall role of innovation in COVID-19 responses, the proceedings review the innovation efforts that are under way within international development and humanitarian responses to the pandemic, how these efforts are working, and how they may need to be strengthened in order to meet pressing health, social, and economic challenges and ensure society's long-term resilience. In just over 30 original papers that have been selected, peer-reviewed, and included into this proceeding volume, there are multiple examples of development investments that are using innovation to respond to the COVID-19 response, both addressing direct impacts on the health of populations and indirect impacts on economy and society.

1.2 Impacts of the Pandemic

The impacts of COVID-19 are not limited to public health and sustainable development, because challenges related to managing global trade or energy policy are also present (Rausser et al., 2022; Strielkowski et al., 2022a). There is a high level of uncertainty associated with the COVID-19 pandemic (Xu, 2021). The pandemic could have serious impacts on the world with regard to the quality of life and sustainable development in politics, environment, and economy, as well as on the world economy (Čábelková et al., 2022; Tvaronavičienė et al., 2022). Not only the economic losses but also social parameters such as changes in psychological sustainability have been affected by COVID-19 (Simionescu et al., 2022). The pandemic changes the way humans live on a global scale, by impacting the context of social, economic, and ecological aspects. To rescue economies and the mental health of citizens from the COVID-19 pandemic, many countries developed strategies, but years of struggles are needed to restore economic stability (Strielkowski et al., 2022b; Panneer et al., 2022). A post-pandemic recovery may emerge, but this is likely to hinge on business leaders' ability to cope with the shocks in productivity and growth caused by COVID-19 (Karmaker et al., 2021).

Within the context of COVID-19, governments across the globe are trying to redefine what a post-pandemic world looks like. As war in Ukraine compounds the impact of COVID-19, it is clear that the world faces an uncertain future, one defined by economic and social crises (Pereira et al., 2022). As the effects of the COVID-19 pandemic continue to unfold, Ukraine's war has plunged the world into chaos. The Russian invasion of Ukraine in February 2022 has not only caused massive humanitarian, migration, and refugee crises, but it has added a negative jolt to the world economy, which is still dealing with the COVID-19 pandemic (Chumachenko & Chumachenko, 2022).

1.3 COVID-19 and the Digital Surge

Most recently, there is a significant promise in frontier technology advances, and rapid developments in the vaccine for the COVID-19 pandemic are the best examples. The growth of innovation investments by governments and businesses across much of the world amid the COVID-19 pandemic's enormous human and economic costs, as shown by the Global Innovation Index (2021), exemplifies the growing recognition that new ideas are crucial for surmounting the pandemic and for ensuring post-pandemic economic growth (Shah et al., 2022). The effects of the pandemic-related economic slowdown were very heterogeneous across sectors, according to the new feature in the GII's (Global Innovation Index) Global Innovation Tracker. Countries that had their workers-support services, like job-matching, guidance to unemployed workers, or smart metering in the industry and households (Rausser et al., 2018), digitalized prior to the pandemic were better able to support their workers during the COVID-19 crisis.

Prior to COVID-19, just a few countries worldwide digitalized their state-run job services system allowed for unemployment insurance claims to be processed in under 1 min, and that system was expanded to handle the high volume of claims received during the pandemic at a similar rate. Technology is one of the largest drivers of development in our increasingly digitalized global economy. It is the new technologies, which are mutually dependent, interconnected, and mutually reinforcing, which are aided by digitization and connectivity, and which are multiplied by its effects (Low & Bu, 2022). With the significant usage of technology to access essential needs such as healthcare and education, the effect of digital disruptions on social equity is crucially important to understand.

1.4 COVID-19 and Education

The COVID-19 pandemic opened the door to new technologies that would endure in the classroom after the disaster subsided (Strielkowski 2022). Due to lockdowns and the necessity to conduct distance learning, many institutions started using educational technology driven by artificial intelligence that predict the navigational behavior of learners in an online course from their computer mouses or touchscreens and allow for the real-time identification of pedagogically valuable behaviors. The technology represents an AI-driven online tutor that can assess students' strengths and weaknesses and deliver personalized individual instructions. Thus, it can considerably extend the capacity of a teacher in a classroom full of students and let the teacher to focus on a bigger picture, while the AI would focus on individual student's needs (Kesavan et al., 2022).

Using AI in teaching due to the COVID-19 pandemic is just one but also a very clear example of the Schumpeterian "creative destruction" in making. Back in 1942, economist Joseph Schumpeter formulated a theory of "creative destruction," suggesting that the business cycle operates on a long-term wave of innovation. For Schumpeter, economic development is a natural outcome of forces inherent to markets, created by opportunities for generating profits (Tülüce & Yurtkur, 2015). The key, Schumpeter observed, is that an evolutionary process rewards improvements and innovations, while penalizing less effective ways to organize resources. According to Schumpeter, technological innovations increase economic growth and raise the standard of living. Creative destruction plays a critical role in economic entrepreneurship and development. Creative destruction is the breaking up of long-standing practices in order to make room for innovation, and it is seen as the driving force behind capitalism (Schumpeter, 1942). Essentially, creative destruction theory suggests that long-standing arrangements and assumptions should be destroyed to free resources and energy for use in innovations.

We can now see how digital innovations in higher education that would have normally taken years due to the various contradictory administrative regulations are now introduced promptly in a matter of days.

1.5 Conclusions

In general, it is clear that business leaders around the world must take advantage of the COVID-19 pandemic crisis to rethink their supply networks, take steps to understand their vulnerabilities, and then act to create resilience. From better decision-making to the need for international collaboration, here are seven things we can learn from the pandemic. With the right approach, the COVID-19 crisis could become an opportunity to step up and generate more value and positive social impacts, instead of simply falling back into the status quo. From economic systems to sustainable development, there is a chance that decision-makers will step out from the status quo and create positive changes that will benefit humanity. Public and private sectors must actively participate in forming critical decisions that will create impact, particularly in times of crisis. In a world torn apart by pandemics, climate change, and civil unrest, we need responsible corporate leaders to lead with public interests at heart. Leaders worldwide can form an agenda that addresses interrelationships among climate, health, workforce needs, supply chains, digital, finance, and inequality, as well as economic development.

Within the context of the pathways for the development and growth in the post-COVID era, it can be noted that innovation makes room for new things, by dismantling things that stopped working or that were never good in the first place. Therefore, it is essential that managers enable innovation and risk-taking over and over. The creative destruction process, which is part of an economic transition toward a new strong wave of growth, which is based on fundamentals and improvements in innovation, accounts for a transition between old and new technologies, old and new industries, and old and new allocations of resources, including jobs.

Overall, it becomes apparent that what seemed impossible prior to the COVID-19 pandemic has now become a "new normal." The whole concept of leadership and leaders has undergone profound transformation and changes. COVID-19 pandemic caused a "digital surge" in the ways business and education are perceived or conducted today. Hence, there exists a pressing need for eliciting recent updates in the concepts of leadership in business and economics that would bring new insights into this topic and define the place and the role of leadership within the context of the European Union's Green Deal and the United Nations' Sustainable Development Goals (SGDs).

References

Čábelková, I., Smutka, L., & Strielkowski, W. (2022). Public support for sustainable development and environmental policy: A case of the Czech Republic. *Sustainable Development, 30*(1), 110–126. https://doi.org/10.1002/sd.2232

Chumachenko, D., & Chumachenko, T. (2022). Impact of war on the dynamics of COVID-19 in Ukraine. *BMJ Global Health, 7*(4), e009173. https://doi.org/10.1136/bmjgh-2022-009173

Karmaker, C. L., Ahmed, T., Ahmed, S., Ali, S. M., Moktadir, M. A., & Kabir, G. (2021). Improving supply chain sustainability in the context of COVID-19 pandemic in an emerging economy: Exploring drivers using an integrated model. *Sustainable production and consumption, 26*, 411–427. https://doi.org/10.1016/j.spc.2020.09.019

Kesavan, N., Seenuvasaragavan, S., Burnette, K., Heim, A. B., Patel, R. J., Johnston, J., Cao, X., Sakandar, H. A., Zhuang, Y., Oi, K. K., Zhi, Y., Diao, Y., & Strielkowski, W. (2022). Pandemic-inspired policies. *Science, 377*(6601), 22–24. https://doi.org/10.1126/science.add2046

Low, M. P., & Bu, M. (2022). Examining the impetus for internal CSR practices with digitalization strategy in the service industry during COVID-19 pandemic. *Business Ethics, the Environment & Responsibility, 31*(1), 209–223. https://doi.org/10.1111/beer.12408

Panneer, S., Kantamaneni, K., Akkayasamy, V. S., Susairaj, A. X., Panda, P. K., Acharya, S. S., & Pushparaj, R. R. B. (2022). The great lockdown in the wake of COVID-19 and its implications: Lessons for low and middle-income countries. *International Journal of Environmental Research and Public Health, 19*(1), 610. https://doi.org/10.3390/ijerph19010610

Pereira, P., Zhao, W., Symochko, L., Inacio, M., Bogunovic, I., & Barcelo, D. (2022). The Russian-Ukrainian armed conflict impact will push back the sustainable development goals. *Geography and Sustainability, 3*(3), 277–287. https://doi.org/10.1016/j.geosus.2022.09.003

Ranjbari, M., Esfandabadi, Z. S., Zanetti, M. C., Scagnelli, S. D., Siebers, P. O., Aghbashlo, M., & Tabatabaei, M. (2021). Three pillars of sustainability in the wake of COVID-19: A systematic review and future research agenda for sustainable development. *Journal of Cleaner Production, 297*, 126660. https://doi.org/10.1016/j.jclepro.2021.126660

Rausser, G., Strielkowski, W., & Štreimikienė, D. (2018). Smart meters and household electricity consumption: A case study in Ireland. *Energy & Environment, 29*(1), 131–146. https://doi.org/10.1177/0958305X17741385

Rausser, G., Strielkowski, W., & Korneeva, E. (2021). Sustainable tourism in the digital age: Institutional and economic implications. *Terra Economicus, 19*(4), 141–159. https://doi.org/10.18522/2073-6606-2021-19-4-141-159

Rausser, G., Chebotareva, G., Smutka, L., Strielkowski, W., & Shiryaeva, J. (2022). Future development of renewable energy in Russia: A case of solar power. *Frontiers in Energy Research, 10*, 862201. https://doi.org/10.3389/fenrg.2022.862201

Schumpeter, J. A. (1942). *Capitalism*. Harvard University Press, Cambridge, MA.

Shah, M. I., Foglia, M., Shahzad, U., & Fareed, Z. (2022). Green innovation, resource price and carbon emissions during the COVID-19 times: New findings from wavelet local multiple correlation analysis. *Technological Forecasting and Social Change, 184*, 121957. https://doi.org/10.1016/j.techfore.2022.121957

Simionescu, M., Bordea, E. N., & Pellegrini, A. (2022). How did the COVID-19 pandemic impact the stress vulnerability of employed and non-employed nursing students in Romania? *PLoS One, 17*(3), e0264920. https://doi.org/10.1371/journal.pone.0264920

Strielkowski, W. (2022). New trends in international education: Impact of COVID-19 and digitalization on higher education and student mobility. In B. Akgün & Y. Alpaydın (Eds.), *Education policies in the 21st century* (Maarif global education series). Palgrave Macmillan. https://doi.org/10.1007/978-981-19-1604-5_8

Strielkowski, W., Zenchenko, S., Tarasova, A., & Radyukova, Y. (2022a). Management of smart and sustainable cities in the post-COVID-19 era: Lessons and implications. *Sustainability, 14*(12), 7267. https://doi.org/10.3390/su14127267

Strielkowski, W., Firsova, I., Azarova, S., & Shatskaya, I. (2022b). Novel insights in the leadership in business and economics: A post-coronavirus update. *Economies, 10*(2), 48. https://doi.org/10.3390/economies10020048

Tülüce, N. S., & Yurtkur, A. K. (2015). Term of strategic entrepreneurship and Schumpeter's creative destruction theory. *Procedia-Social and Behavioral Sciences, 207*, 720–728. https://doi.org/10.1016/j.sbspro.2015.10.146

Tvaronavičienė, M., Mazur, N., Mishchuk, H., & Bilan, Y. (2022). Quality of life of the youth: Assessment methodology development and empirical study in human capital management. *Economic Research-Ekonomska Istraživanja, 35*(1), 1088–1105. https://doi.org/10.1080/1331677X.2021.1956361

Xu, L. (2021). Stock return and the COVID-19 pandemic: Evidence from Canada and the US. *Finance Research Letters, 38*, 101872. https://doi.org/10.1016/j.frl.2020.101872

Part I
Leadership as a Social, Economic and Cultural Phenomenon in the Post-COVID Era

Chapter 2
Cooperation as a Driver of Socioeconomic Leadership and Sustainable Development in Local Communities: Main Trends and Perspectives of the Interdisciplinary Research

Olena Krukhmal, Yuriy Petrushenko, and Oksana Duvanova

Abstract Nowadays, one of the priorities of the human development strategy is to promote the sustainable development of local communities. Therefore, it is essential to identify additional factors contributing to it. In this study, we expand the scope of research on alternative mechanisms for ensuring the sustainable development of territorial communities. We hypothesize that cooperation can become a factor in mobilizing leadership at the level of local government, thus contributing to the sustainable development of territorial communities. The development of cooperation processes makes it possible to combine economic, social, and environmental goals effectively. This chapter is devoted to the analysis and theoretical substantiation of the relationship between the concept of cooperation and the concept of sustainable development. To do this, a combination of the Scopus Toolkit and VOSviewer tools was used. The main trends in cross-sector scientific research are identified, and the directions of interdisciplinary research on the relationship between economic cooperation and sustainable development of territorial communities are clustered.

Keywords Sustainable development · Economic cooperation · Territorial community

JEL Classification A12, Q01, P13

O. Krukhmal · Y. Petrushenko · O. Duvanova (✉)
Sumy State University, Sumy, Ukraine
e-mail: o.krukhmal@biem.sumdu.edu.ua; y.petrushenko@biem.sumdu.edu.ua; o.ponomarenko@biem.sumdu.edu.ua

W. Strielkowski (ed.), *Leadership, Entrepreneurship and Sustainable Development Post COVID-19*, Springer Proceedings in Business and Economics, https://doi.org/10.1007/978-3-031-28131-0_2

2.1 Introduction

Today, the paradigm of sustainable development is recognized as the leading ideological concept of the twenty-first century (Brundtland, 1987). Modern society faces a number of challenges.

It is evident that local self-government plays an essential role in strengthening statehood and its defense capability. Therefore, ensuring the sustainable development of local communities is of particular relevance in the current geopolitical situation.

In the works of Bhowmik (2021) and Zaloha et al. (2020), economic cooperation's positive impact on overcoming disparities in regions' development is noted. A number of scholars have studied the effectiveness of public-private partnerships and their social and economic benefits in ensuring sustainable local development (Hrytsenko & Boyarko, 2015; Midor et al., 2021; Ovcharova & Grabowska, 2022; Shipko et al., 2020; Spremberg et al., 2017; Wiebe et al., 2018). Ensuring the financial viability of local budgets is an important component of sustainable development (El Amri et al., 2020; Guliyeva et al., 2021; Liuta et al., 2015).

Another factor that led to negative consequences for society was the challenge of the COVID-19 pandemic (Kuzmenko et al., 2020). As a result of quarantine restrictions, a particular blow fell on the management of human resources and personnel in general. New models of cooperation and leadership can cope with economic and social impacts and motivate employees in organizations (Alkubaisy, 2020).

A number of scientists have studied the role of social cooperation in the formation of leadership and its positive impact on the development of the business environment and on improving the efficiency of management (Abeysekera, 2020; Kaya, 2020; Mercado & Vargas-Hernández, 2019; Pietraszewski, 2020; Ropuszynska-Surma & Weglarz, 2018; Strielkowski et al., 2020; Yarovenko et al., 2021). Gutiérrez et al. (2011) proved in their study that cooperation is the key to the sustainable development of fishing communities. Strong leadership and social cohesion were other significant factors contributing to the effective development of communities.

Corporate social responsibility is defined as a prerequisite for the sustainable development of local communities and industries in general (Bijańska et al., 2018; Khoshnava et al., 2019; Lu et al., 2019; Navickas et al., 2021; Taliouris & Trihas, 2017).

Kamara et al. (2017) invite municipalities to influence local economic development through vocational training, and Karyy and Panas (2016) offer to consider volunteers as a substitute for a leadership resource to improve the efficiency of local government functioning.

Thus, the relationship between leadership and sustainable development is very important because without leadership, there is no development. Following the literature review, we find that leadership is enhanced through cooperation.

The analysis of scientific research revealed the absence of a fundamental theoretical justification for the fact that cooperation as a category in its various manifestations is associated with the sustainable development of territorial communities.

We hypothesize that cooperation is an important factor in sustainable local development.

This chapter is devoted to the analysis and theoretical substantiation of the relationship between the concept of cooperation and the concept of sustainable development. To do this, a combination of the Scopus Toolkit and VOSviewer tools was used.

2.2 Methods and Data

In order to substantiate the role of cooperation as a factor in the sustainable development of territorial communities, we will conduct a bibliometric analysis of studies on this topic.

First, we conduct a decomposition analysis of the substantive aspects of the theory of sustainable development, which includes the traditional determinants of sustainable development: economic, ecological, and social components. Further, to formalize the content-contextual dimension of the relationship between research on cooperation and sustainable development of communities, scientific publications were selected. They were separately related to each component of the concept of sustainable development.

The data of the study were articles indexed by the scientometric database (Scopus, 2022). The following key parameters were selected for analysis:

- The period of analysis: 1990–2020.
- The language of articles: English.
- A combination of keywords: "local community *and* environment *and* sustainable development," "local community *and* social *and* sustainable development," "local community *and* economic *and* sustainable development," and "community *and* cooperation *and* development."

For the purposes of our study, we use VOSviewer, which is a software tool for building and visualizing bibliometric network links (Van Eck & Waltman, 2010; VOSviewer, 2022).

Using VOSviewer, three network connections of the concept of "sustainable development" and "local community" with components of sustainable development were built.

It was established that the concept of "cooperation" is found in the map of relationships for each individual determinant of sustainable development.

2.3 Research Results

Figure 2.1 shows a map of the relationship between the determinant "environment," the concept "sustainable development," and the "local community." The relationship between the concepts of cooperation and international cooperation is revealed.

Figure 2.2 shows a map of the relationship between the determinants of "economic development," the concept "sustainable development," and the "local community." The relationship with the concept of cooperation is revealed.

Figure 2.3 shows a map of relationship of the determinant "social development," the concept "sustainable development," and the "local community." The relationship with the concept of international cooperation is revealed.

Table 2.1 shows the main results of the research. It is indicated the clusters in which the concept of "cooperation" is fixed. The context of the concept is indicated, as well as the total strength of links of the concept.

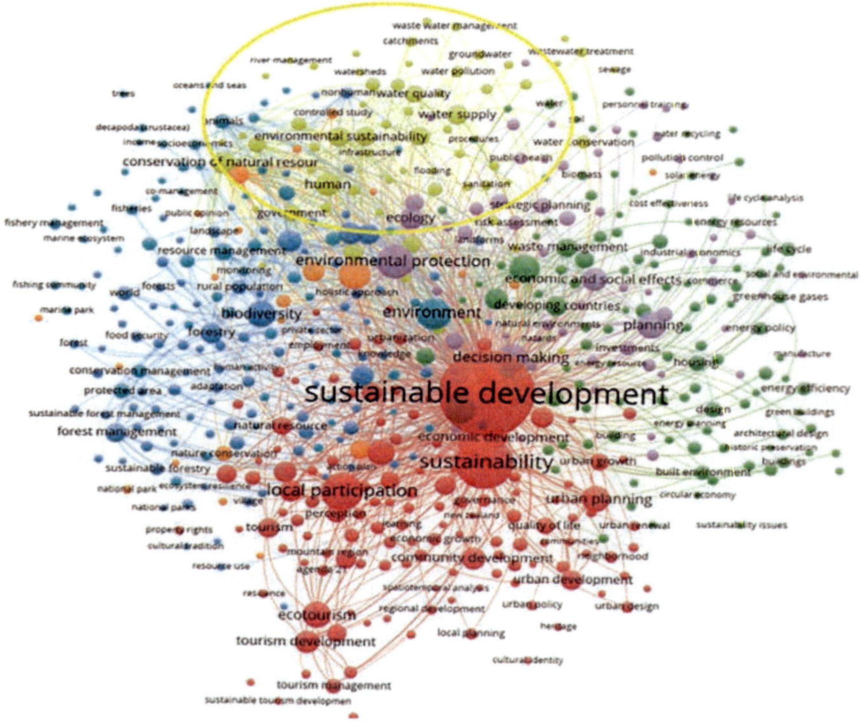

Fig. 2.1 Analysis of publications for the keywords local community *and* environment *and* sustainable development. (Source: Own results)

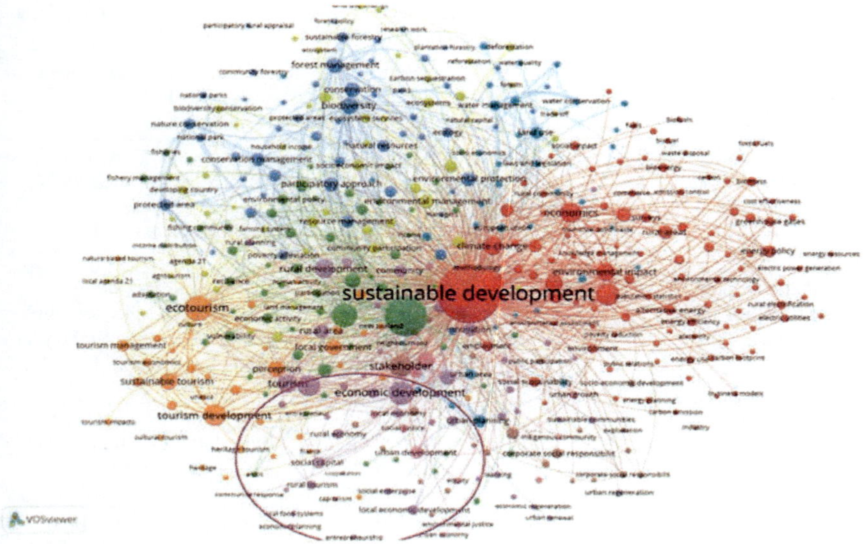

Fig. 2.2 Analysis of publications by keywords local community and economic and sustainable. (Source: Own results)

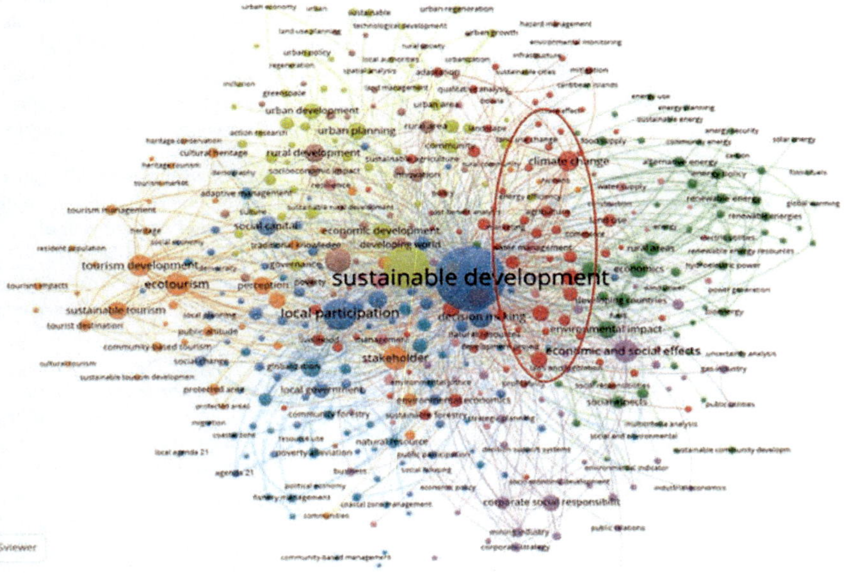

Fig. 2.3 Analysis of publications by keywords local community and social and sustainable development. (Source: Own results)

Table 2.1 Identification of conceptual and substantive relationships of "cooperation" and "determinants of the concept of sustainable development"

Keyword combination	Scopus articles	Keyword co-occurrence	Clusters and their density	Established an interconnection with the concept of cooperation
Local community *and* economic *and* sustainable	1884	442	9 clusters, 16,218 links, and 30,382 total link strength	*Cooperation*: Cluster 5, 27 links, total link strength 24, and average year of publications 2011, 25
Local community *and* social *and* sustainable	1991	460	9 cluster, 16,868 links, and 31,597 total link strength	*International cooperation*: Cluster 1, 26 links, total link strength 31, and average year of publications 2011
Local community *and* environment *and* sustainable	1758	566	7 cluster, 29,204 links, and 54,561 total link strength	*Cooperation*: Cluster 4, 67 links, total link strength 74, and average year of publications 2011, 67
				International cooperation: Cluster 4, 90 links, total link strength 147, and average year of publications 2009, 64
Community *and* cooperation *and* sustainable development	842	293	7 cluster, 9895 links, and 18,404 total link strength	*Transboundary cooperation*: Cluster 1, 59 links, total link strength 93, and average year of publications 2008, 93
				Cooperation: Cluster 2, 188 total link strength 521, and average year of publications 2012, 16 *Cooperative behavior*: Cluster 2, 66 links, total link strength 125, and average year of publications 2014, 07
				Regional cooperation: Cluster 3, 39 links, total link strength 53, and average year of publications 2011, 33
				International cooperation: Cluster 4, 244 links, total link strength 1062, and average year of publications 2008, 74
				Cooperative sector: Cluster 5, 53 links, total link strength 66, and average year of publications 2016, 33 *Development cooperation*: Cluster 5, 23 links, total link strength 26, and average year of publications 2007, 67

Next, in Fig. 2.4, we will identify dominant trends in cross-sectoral research that characterize the relationship between cooperation, sustainable development, and local communities.

Based on the results of clustering, seven clusters of cross-sectoral scientific research were identified, which characterize the general trends of research on issues of cooperation (economic cooperation) in the context of ensuring sustainable local development.

Cluster 1 is devoted to the study of the relationship between cooperation and community development, regional development, and cross-border cooperation, as well as issues of corporate social responsibility in the context of biodiversity conservation, alternative approaches to agricultural development, and the use of biofuels.

Cluster 2 is devoted to the study of the relationship between cooperation and community capacity building, financial planning, the development of mechanisms for community participation in economic development, planning and monitoring of efficient nature management, the formation of cooperative behavior, and the methodology for monitoring control groups.

Cluster 3 is dedicated to the study of the relationship between cooperation and competition, economic efficiency and decision-making mechanisms, regional cooperation, ensuring social responsibility, implementation of the principles of

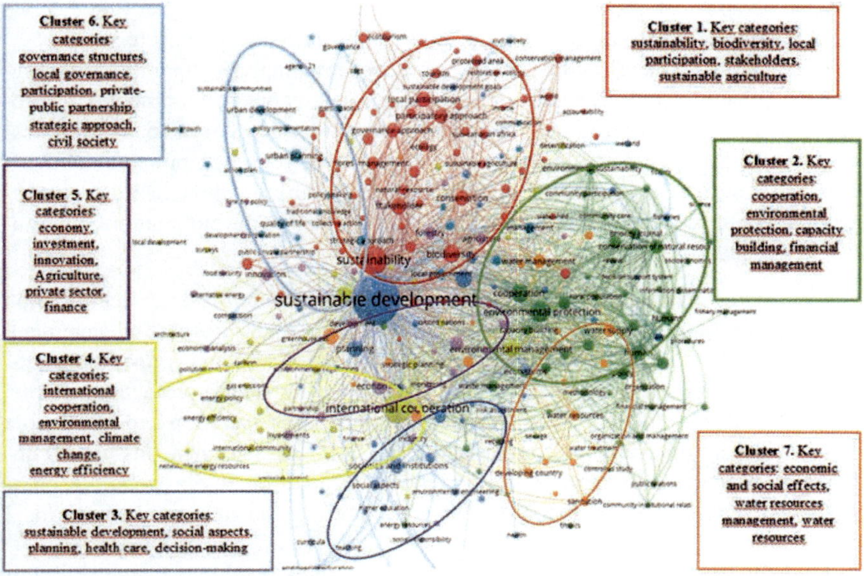

Fig. 2.4 Identification of clusters of cross-sectoral scientific research on "cooperation" and "sustainable local development." (Source: Own results)

sustainable development and rational environmental management, and ensuring the appropriate level of education on these issues and health policy.

Cluster 4 is devoted to the study of the relationship between cooperation and environmental management, energy efficiency and renewable resources in the context of global climate change, and regional and international cooperation.

Cluster 5 is devoted to research on the development of cooperation, the cooperative and private sector in the context of decentralization, the study of financial and economic aspects of community development, the introduction of innovative mechanisms, and the improvement of the investment environment.

Cluster 6 is dedicated to studying of the relationship between cooperation and the development of public society, program documents and action plans to ensure an effective decision-making system, Agenda 21 program documents, governance structures, and public-private partnership.

Cluster 7 is dedicated to the study of the relationship between cooperation and project management, public health, economic and social efficiency, and ensuring the availability of water resources.

It should be noted that clusters 3 and 6 most of all penetrate into other clusters. Since cluster 3 contains the concept of sustainable and local governance, it is interconnected with others. Cluster 6 includes the categories of governance, local self-government, and strategic planning, which is quite logical because the issues of sustainable development of territorial communities are reflected primarily in political documents, such as programs and strategies for the socioeconomic development of regions or UTCs.

In the context of expanding the evolutionary-temporal aspects, in Fig. 2.5, we will consider the main stages in the development of interdisciplinary research.

Within the framework of this block of bibliometric analysis, the main meaningful determinants of research on cooperation and sustainable development of territorial communities are ranked by time for the period from 2008 to 2016. The gradient in the figure changes from blue, the earliest studies, to yellow, the modern work.

In this way, it was possible to identify four stages during which the main emphases in the field of research on cooperation and sustainable development of communities changed:

1. *Until 2008–2010*—the emphasis on environmental warehouse: direct attention was given more importance to state management, environmental management, resource management, biodiversity conservation, international cooperation, strategic planning and social aspects of development, and health protection.
2. *In 2011–2012*—the emphasis shifted to the study of cooperation and sustainable development, the study of community decision-making mechanisms in the context of decentralization processes, and increased attention to the economic, commercial, and investment components of planning the development of local communities.
3. *In 2013–2015*—the mechanisms for achieving the capacity and sustainability of communities were studied, in particular through a participatory approach and

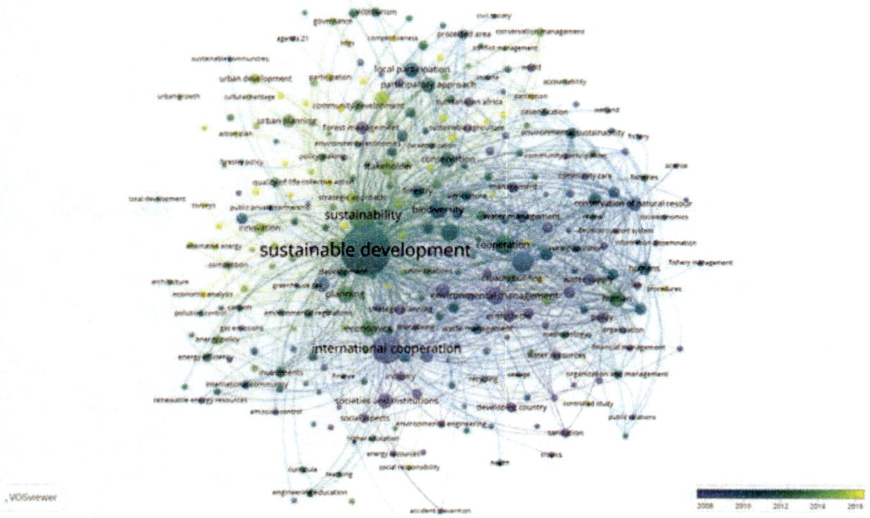

Fig. 2.5 The results of a chronological analysis of the transformation of the content of cross-sectoral research on cooperation and sustainable development of communities according to the Scopus database for the period 1990–2020. (Source: Own results)

cooperative behavior, strategic approaches to community development, increased interest and involvement of stakeholders, and climate change.

4. *Since 2015*—by part—in the field of scientific interests, the question of how cooperation is related to the goals of sustainable development, the level of resilience of communities, collective actions, and socioeconomic conditions, and cooperative management. In the same period, studies based on such methods as questionnaires and surveys became relevant.

So according to the results of the analysis of the evolutionary-temporal block of bibliometric analysis research, it can be noted that the interrelationships of cooperation and the achievement of the sustainable development goals of local communities are an actual subject of research in the modern scientific community.

In Fig. 2.6, we analyze the number of works on the study of cooperation and sustainable development at the level of territorial communities in the international scientific database Scopus for 1990–2020.

An analysis of the dynamics of research between the determinants of sustainable development and cooperation has shown that interest in this topic has been growing in scientific papers since 1991 (in the Scopus database), which is associated with the emergence of the very concept of sustainable development. Its spread was preceded by numerous eco-centric theories (e.g., ecotopia and environmental protection).

As it is known, in 1987, the term "sustainable development" was used in the report of the UN International Commission on Environment and Development (WCED), "Our Common Future," also known as the Brundtland Report.

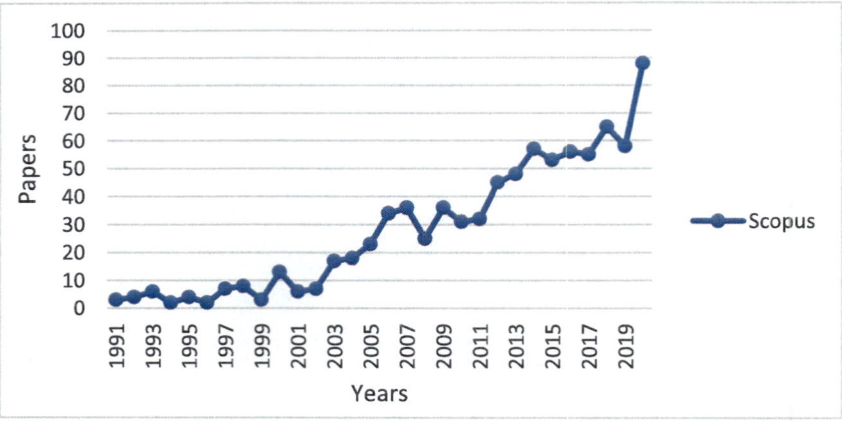

Fig. 2.6 Dynamics of publication activity on issues of cooperation and sustainable local development according to data from the Scopus scientometric database. (Source: Own results)

A significant date is 1992, when the Agenda 21 action plan was adopted at the UN Conference on Environment and Development in Rio de Janeiro, Brazil.

In 1997, the UN General Assembly held a special session to assess the status of Agenda 21 (Harangozó & Zilahy, 2015).

In 2000, another historic event in the concept of sustainable development took place—the UN Millennium Declaration (Resolution 55/2) was adopted at the UN Summit.

The number of works in terms of cooperation at the level of sustainable development of territorial communities began to grow in 2002 when world leaders undertook the commitment through the Convention on Biological Diversity to achieve a significant reduction in the rate of biodiversity loss by 2010.

Rapid growth has also been observed from 2012 to 2015. Note that in 2015, the current 17 Sustainable Development Goals (SDGs) were adopted (UN Resolution 70/1).

2.4 Discussion

The processes of ensuring the sustainability of communities and their ability to cope with today's challenges require the concentration of a significant amount of labor and financial resources, which can be achieved with the help of cooperation.

Recently, society faced the challenges of the COVID-19 pandemic and still has not overcome its consequences. Small producers around the world have lost markets for their products due to the COVID-19 pandemic and quarantine. Each of them today is forced to find a new way to the consumer.

Economic cooperation is considered in world practice as a tool for stimulating business activity and, therefore, the growth of well-being. Territorial units and enterprises are more effectively involved in the processes of economic exchange precisely through cooperation mechanisms.

The term "cooperation" comes from the Latin *cooperante* and means collaboration. For a long time, cooperation has gone through several stages of development, starting from the simplest combination of individual physical efforts to complex modern forms of cooperation between economic entities. In the modern scientific environment, the theory of cooperation is being rethought in the light of general civilizational processes. Cooperation is considered as the core and center of social institutions, thanks to which spatial and temporal transformations are carried out in them, which leads to the emergence of new social subjects.

The tool of economic cooperation, implemented in many territories worldwide through organizational, economic, and financial incentives, has become a real catalyst for forming capable communities. It ensures a high concentration of capital, the development of infrastructure, and favorable conditions for the attraction of investments, which leads to the growth of the gross product.

Thus, cooperation contributes to the activation of the social capital of the local communities; in particular, it stimulates motivation for leadership. The development of agricultural cooperation at the local level contributes to the following:

- Creation of a favorable investment climate in the region
- Creation of new jobs
- Increasing salaries and payments to local budgets
- Formation of an active civic position and leadership in the community

Kustepeli et al. (2020) investigated the impact of cooperation in the form of agricultural cooperatives on social capital formation and livelihood improvement in eight villages in the region of Turkey. They used empirical research methods with questionnaire data collection. The authors proved that membership in a cooperative is an essential factor in the development of agriculture due to its influence on increasing the level of trust and strengthening social capital.

Ma and Abdulai (2017) considered the economic effect of agricultural cooperatives. Thus, membership in agricultural cooperatives positively affects production price, gross income, farm profit, and profitability. Some studies of an ecological orientation are directed to the analysis of supply chains of sources of raw materials and biofuels for impoverished territorial communities.

Heikrlik conducted numerous studies on the development of agricultural territories. It was emphasized that the cooperative movement is a phenomenon that attracts the attention of the national government and international donors to support the functioning of new groups of farmers. Determinants of success were found to be the size of the cooperative, the proportion of women in the cooperative, the balance of nuclear families, the active participation in the cooperative, and the type of product grown (Ahado et al., 2021).

Traditionally, the modern scientific community, when researching issues of sustainability, emphasizes the importance of taking into account the interaction between economic, ecological, and social parameters. Increasingly, however, scholars are also focusing their research on the integration of low-educated and impoverished farmers into supply chains. Hall and Matos (2010) emphasize the need to research entrepreneurial dynamics in disadvantaged communities and provide primary business education to farmers. They propose cooperatives to combine knowledge asymmetry between buyers and suppliers.

Practice shows how important shared leadership is in terms of regional development. Thanks to it, individuals, enterprises, and public authorities mobilize efforts to solve problems of a regional nature. Leadership is based on collective values, trust, and a sense of commitment, often forming the basis for developing cooperative relations (Horlings & Padt, 2013).

The successful practice of neighborhood self-government can help small municipalities solve their citizens' problems and provide minimal services by exchanging best practices between representatives of the local executive power and conducting training. Even with a low level of trust in local authorities, residents demonstrate a willingness to participate actively when familiar with the successful experiences of neighboring communities. And the vast majority agree that citizens can influence changes at the local level. These results demonstrate significant potential for increasing self-determination and citizen participation. Citizens want changes and solutions to their problems and are ready to participate in solving local issues.

It is necessary to emphasize the relationship between the concept of leadership and sustainable development. Sustainable leadership has become a modern trend in the development of this area. This is a newly created area of effective leadership that was created to address issues related to sustainable development. For example, it explores methods for incorporating sustainable leadership into sustainable performance (Iqbal & Ahmad, 2021).

Ratner (2012) in his research pays attention to cooperation, community cooperation, and cooperatives in the global era. Based on an evolutionary perspective, the author demonstrates how the principles of cooperation can make a social system not only more efficient in terms of spending time and resources but also more democratic, thanks to the process of empowering everyone involved.

Identifying additional factors that will contribute to sustainable development is extremely important.

Therefore, the cooperation in modern conditions can become a natural tool for consolidation and rallying the entire society's efforts to revive the country's economy and sustainable development.

Summarizing a composite analysis of the components of the concept of sustainable development and comparative and bibliometric research methods allows for establishing structural and functional relationships between separate economic concepts: sustainable local development and cooperation. Also, it contributes to deeper penetration into the content of the categorical-conceptual apparatus in this research direction.

At the first stage of the research, with the help of a systematic combination of the VOSviewer v.1.6.10 toolkit and the Scopus Citation Overview tool, the relationship between research on the concept of cooperation and the structural and functional environment of the theory of sustainable development was established. The subject levels of the study of cooperation (economic cooperation) in the theory of sustainable development of local communities have been identified as follows:

- Cooperation at the level of business entities
- Cross-border cooperation
- Interregional cooperation
- Cooperation at the international level

At the second level, clustering of scientific patterns of research was carried out according to the meaningful criterion, according to their density interconnection. Seven clusters of cross-sectoral scientific research characterizing the general trends of research on cooperation issues in the context of ensuring sustainable local development are singled out. It was found that the problem of cooperation is most often investigated as follows:

- In combination with issues of development of local communities and regional and cross-border cooperation
- In the context of building community participation in economic development, planning and monitoring of effective nature management, and cooperative behavior
- Ensuring social responsibility, implementing rational nature management, and ensuring an appropriate level of education on these issues, in the context of regional cooperation
- Energy efficiency and international cooperation and development of cooperation, cooperative, and private sector in the context of decentralization
- Development of civil society, program documents, and action plans to ensure an effective decision-making system

Thus, one of the effective mechanisms for activating the sustainable development of local territories is cooperation, in particular the creation of agricultural cooperatives. This is because agricultural cooperatives are not only an economic tool but also a mechanism that positively affects the growth of social capital in newly created territorial communities and stimulates the development of leadership and the formation of thinking oriented toward sustainable development among local residents.

As a result of bibliometric analysis, the set of various content contexts in research of cooperation within the framework of sustainable local development was revealed. It indicates the need to develop an integrated approach to research and the formation of a concept for building cooperative relationships in the territorial community.

2.5 Conclusions

Thence, our research hypothesis was confirmed on the basis of the obtained results. It can be argued that cooperation can become a factor in mobilizing leadership and sustainable local development. As a result of the co-occurrence analysis in terms of the content dimension, three maps of the relationship between the concept of sustainable development and its three main determinants (economic, social, and environmental) were formed. The presence of a network connection with the concept of cooperation in all visualization maps was revealed. Thus, the categorical-conceptual apparatus of the theory of sustainable development was improved.

The bibliometric analysis of scientific publications, carried out using the VOSviewer tools, identified and described the main trends in interdisciplinary research on cooperation and sustainable local development.

The results of clustering according to the content criterion made it possible to identify seven scientific patterns of research on this topic. It has been established that the third and the sixth clusters are most connected. The third one is dedicated to the analysis of relationships between cooperation, decision-making mechanisms, ensuring social responsibility, and the implementation of the principles of sustainable development; and the sixth one shows relationship between cooperation and the development of civil society, public-private partnerships, program documents, and plans of state structures.

Thus, the development of cooperative processes at the local level contributes to the activation of leadership, entrepreneurship, and sustainable development. It can help manage the consequences of the epidemic and strengthen local self-government to face challenges of regional security.

Acknowledgments This research was funded by a grant from the Ministry of Education and Science of Ukraine, "Convergence of economic and educational transformations in the digital society: modeling the impact on regional and national security" (state registration number: 0121U109553) and "Reforming the lifelong learning system in Ukraine to prevent labor emigration: a cooperative model of institutional partnership" (state registration number: 0120U102001).

References

Abeysekera, R. (2020). Co-production related to business counselling in the microfinance sector as a demonstration of social cooperation: An interpersonal relationship approach. *SocioEconomic Challenges, 4*(2), 56–67. https://doi.org/10.21272/sec.4(2).56-66.2020

Ahado, S., Chkhvirkia, L., & Hejkrlik, J. (2021). Is the success of rural cooperatives conditioned by the group characteristics and their value chain? Evidence from new farmer groups in Georgia. *The European Journal of Development Research, 34*, 677–702. https://doi.org/10.1057/s41287-021-00382-4

Alkubaisy, A. (2020). Corporate social responsibility practice in the Gulf cooperation council countries amidst the COVID-19 pandemic. *Business Ethics and Leadership, 4*(4), 99–104. https://doi.org/10.21272/bel.4(4).99-e104.2020

Bhowmik, D. (2021). Gulf cooperation council (GCC) and regional integration in Asia. *Financial Markets, Institutions and Risks, 5*(1), 61–79. https://doi.org/10.21272/fmir.5(1).61-79.2021

Bijańska, J., Kuzior, A., & Wodarski, K. (2018). Social perception of hard coal Mining in Perspective of Region's sustainable development. *Management Systems in Production Engineering, 26*(3), 178–183. https://doi.org/10.1515/mspe-2018-0029

Brundtland, G. H. (1987). *Report of the World Commission on Environment and Development: Our common future* (pp. 1–300). UN. https://sustainabledevelopment.un.org/content/documents/5987our-common-future.pdf

El Amri, A., Boutti, R., & Rodhain, F. (2020). Sustainable finance at the time of institutions: Performativity through the lens of responsible Management in Morocco. *Financial Markets, Institutions and Risks, 4*(2), 52–64. https://doi.org/10.21272/fmir.4(2).52-64.2020

Guliyeva, A., Rzayeva, U., & Jafarova, N. (2021). Analysis of some indicators by means of fuzzy logic on the example of Azerbaijani energy enterprises. *E3S Web of Conferences, 250*, 02001. https://doi.org/10.1051/e3sconf/202125002001

Gutiérrez, N. L., Hilborn, R., & Defeo, O. (2011). Leadership, social capital and incentives promote successful fisheries. *Nature, 470*(7334), 386–389. https://doi.org/10.1038/nature09689

Hall, J., & Matos, S. (2010). Incorporating impoverished communities in sustainable supply chains. *International Journal of Physical Distribution & Logistics Management, 40*(1/2), 124–147. https://doi.org/10.1108/09600031011020368

Harangozó, G., & Zilahy, G. (2015). Cooperation between business and non-governmental organizations to promote sustainable development. *Journal of Cleaner Production, 89*, 18–31. https://doi.org/10.1016/j.jclepro.2014.10.092

Horlings, I., & Padt, F. (2013). Leadership for sustainable regional development in rural areas: Bridging personal and institutional aspects. *Sustainable Development, 21*(6), 413–424. https://doi.org/10.1002/sd.526

Hrytsenko, L., & Boyarko, I. (2015). Reconciliation of investors' interests while public-private partnership projects portfolio forming. *Economic Annals-XXI, 151*(5-6), 45–48.

Iqbal, Q., & Ahmad, N. H. (2021). Sustainable development: The colors of sustainable leadership in learning organization. *Sustainable Development, 29*(1), 108–119. https://doi.org/10.1002/sd.2135

Kamara, R. D., Leonard, S., & Haines, R. (2017). Enhancing the capabilities of municipalities for local economic development through skills training. *SocioEconomic Challenges, 1*(4), 35–44. https://doi.org/10.21272/sec.1(4).35-44.2017

Karyy, O., & Panas, Y. (2016). Preconditions for volunteers' participation in social and economic development of communities. *Marketing and Management of Innovations, 2*, 156–169. https://doi.org/10.21272/mmi.2016.2-14

Kaya, H. D. (2020). Business friendliness, firm performance and owner's optimism. *Financial Markets, Institutions and Risks, 4*(3), 13–23. https://doi.org/10.21272/fmir.4(3).13-23.2020

Khoshnava, S. M., Rostami, R., Zin, R. M., Štreimikienė, D., Yousefpour, A., Strielkowski, W., & Mardani, A. (2019). Aligning the criteria of green economy (GE) and sustainable development goals (SDGs) to implement sustainable development. *Sustainability, 11*(17), 4615. https://doi.org/10.3390/su11174615

Kustepeli, Y., Gulcan, Y., Yercan, M., & Yıldırım, B. (2020). The role of agricultural development cooperatives in establishing social capital. *The Annals of Regional Science*. https://doi.org/10.1007/s00168-019-00965-4

Kuzmenko, O., Vasylieva, T., Vojtovič, S., Chygryn, O., & Snieška, V. (2020). Why do regions differ in vulnerability to COVID-19? Spatial nonlinear modeling of social and economic patterns. *Economics and Sociology, 13*(4), 318–340. https://doi.org/10.14254/2071-789X.2020/13-4/20

Liuta, O., Pihul, N., & Kubakh, T. (2015). Financial capacity of local budget as a basis for sustainable functioning of a territory. *Economic Annals-XXI, 1-2*(1), 78–81.

Lu, J., Ren, L., Qiao, J., Yao, S., Strielkowski, W., & Streimikis, J. (2019). Corporate social responsibility and corruption: Implications for the sustainable energy sector. *Sustainability, 11*(15), 4128. https://doi.org/10.3390/su11154128

Ma, W., & Abdulai, A. (2017). The economic impacts of agricultural cooperatives on smallholder farmers in rural China. *Agribusiness, 33*(4), 537–551. https://doi.org/10.1002/agr.21522

Mercado, M. P. S. R., & Vargas-Hernández, J. G. (2019). Analysis of the determinants of social capital in organizations. *Business Ethics and Leadership, 3*(1), 124–133. https://doi.org/10.21272/bel.3(1).124-133.2019

Midor, K., Kuzior, A., Płaza, G., Molenda, M., & Krawczyk, D. (2021). Reception of the Smart City concept in the opinion of local administration officials – A case study. *Management Systems in Production Engineering, 29*(4), 320–326. https://doi.org/10.2478/mspe-2021-0040

Navickas, V., Petroke, I., Baciuliene, V., & Vasylieva, T. (2021). The impact of the sharing economy as an ecosystem on the tourism sector. *Journal of Tourism and Services, 12*(22), 66–88. https://doi.org/10.29036/jots.v12i22.241

Ovcharova, N., & Grabowska, M. (2022). Implementation of public-private partnership in the healthcare management system. *Health Economics and Management Review, 3*(1), 86–95. https://doi.org/10.21272/hem.2022.1-09

Pietraszewski, D. (2020). The evolution of leadership: Leadership and followership as a solution to the problem of creating and executing successful coordination and cooperation enterprises. *The Leadership Quarterly, 31*(2), 101299. https://doi.org/10.1016/j.leaqua.2019.05.006

Ratner, C. (2012). *Cooperation, community, and co-ops in a global era* (1st ed.). Springer.

Ropuszynska-Surma, E., & Weglarz, M. (2018). A virtual power plant as a cooperation network. *Marketing and Management of Innovations, 4*, 136–149. https://doi.org/10.21272/mmi.2018.4-13

Scopus. (2022). https://www.scopus.com

Shipko, A., Demikhova, N., Pajak, K., & Motrechko, V. (2020). Health management at the regional level: Multivariable performance assessment. *Health Economics and Management Review, 1*(2), 8–15. https://doi.org/10.21272/hem.2020.2-01

Spremberg, E., Tykhenko, V., & Lopa, L. (2017). Public-private partnership in the implementation of National Environmental Projects. *SocioEconomic Challenges, 1*(4), 73–81. https://doi.org/10.21272/sec.1(4).73-81.2017

Strielkowski, W., Volchik, V., Maskaev, A., & Savko, P. (2020). Leadership and effective institutional economics design in the context of education reforms. *Economies, 8*(2), 27. https://doi.org/10.3390/economies8020027

Taliouris, E., & Trihas, N. (2017). Public policy for corporate social responsibility and governance for sustainable tourism development in Greece. *Business Ethics and Leadership, 1*(4), 49–57. https://doi.org/10.21272/bel.1(4).49-57.2017

Van Eck, N. J., & Waltman, L. (2010). Software survey: VOSviewer, a computer program for bibliometric mapping. *Scientometrics, 84*(2), 523–538. https://doi.org/10.1007/s11192-009-0146-3

VOSviewer Vizualizing scientific landscapes. (2022). https://www.vosviewer.com/download

Wiebe, I., Oliinyk, V. M., & Halynska, Y. V. (2018). Innovative instrument of collaborative alliance management in the 'state-region-enterprise' system of withdrawal of the rent income in the extracting industry. *Marketing and Management of Innovations, 2*, 247–261. https://doi.org/10.21272/mmi.2018.2-20

Yarovenko, H., Bilan, Y., Lyeonov, S., & Mentel, G. (2021). Methodology for assessing the risk associated with information and knowledge loss management. *Journal of Business Economics and Management, 22*(2), 369–387. https://doi.org/10.3846/jbem.2021.13925

Zaloha, O., Vakulenko, I., & Jasnikowski, A. (2020). Prerequisites and possibilities of creating a medical engineering cluster as the part of regional territorial marketing. *Health Economics and Management Review, 1*(2), 51–60. https://doi.org/10.21272/hem.2020.2-06

Chapter 3
Economic Development Scenarios for Effective Organization Management in Post-COVID Conditions

Aida Guliyeva, Ulviyya Rzayeva, and Aysel Guliyeva

Abstract With the spread of COVID-19, business leaders are challenged to ensure business stability under difficult conditions and prepare for a future appearing now in a new light. The scenarios of economic situations proposed in this chapter can serve as a means of short-term and medium-term strategic, operational, and financial planning. The scenario represents a possible future development with associated economic consequences, including trends in epidemiology, society, technological development, politics, and the environment. The economic schemes under consideration are hypothetical scenarios designed to form the basis for planning discussions. The potential opportunities for improving the efficiency and quality of the manager's work, which primarily consist in organizing his professional activities, are discussed.

Keywords Coronavirus pandemic · Organization · Development scenario · Stability · Efficiency

JEL Classification H12, M10, Q54

A. Guliyeva (✉) · U. Rzayeva
Faculty of Digital Economy, The Department of Digital Technologies and Applied Informatics, Azerbaijan State University of Economics (UNEC), Baku, Azerbaijan
e-mail: a.guliyeva@unec.edu.az; ulviyya.rzayeva@unec.edu.az

A. Guliyeva
International School of Economy, Department of International Economics, Azerbaijan State University of Economics (UNEC), Baku, Azerbaijan
e-mail: aguliyeva@unec.edu.az

W. Strielkowski (ed.), *Leadership, Entrepreneurship and Sustainable Development Post COVID-19*, Springer Proceedings in Business and Economics, https://doi.org/10.1007/978-3-031-28131-0_3

3.1 Introduction

In modern management science, there are quite a lot of different models of company management. The choice of one or another of them is an independent and very important task for the success of the enterprise (Al-Omari et al., 2020). However, not all decisions concerning the system and methods of management at the enterprise are strategic. A strategy that optimizes the organization's behavior in a recession, a steady decline in the main financial indicators of the organization's activities, and the threat of bankruptcy is a kind of guarantee for the stable development of the organization (Xu, 2019). Of course, one cannot say that the strategy allows any business company to completely avoid crises. It only allows it to reduce the likelihood of their occurrence and, in the event of onset, to ensure that they are overcome as quickly and painlessly as possible. In addition, the strategy makes it possible to shorten the decision-making time on how to overcome the crisis, which is an extremely important factor (Brożyna et al., 2020).

This chapter presents some anti-crisis scenarios during the COVID-19 pandemic and the features of their application. The crisis caused by the COVID-19 pandemic has been a shock to our way of life and worldview (Simionescu et al., 2022). In addition to the requirement to adapt to the new order, many were faced with the impossibility of working under the old rules. In this situation, it was business that reacted most flexibly to the difficult managerial and economic situation under the circumstances of the crisis. The COVID-19 pandemic has significantly increased the importance of organizational activity, since successful business adaptation to new realities is impossible without competent human resource management (Kesavan et al., 2022). This looks into these issues with a scrutiny and discusses some mandatory changes in management during the crisis period to support the income of the enterprise.

3.2 Methodology

There are various approaches to the formation, maintenance, and development of management activities in the organization. In order to do this, at the initial stage, tools of analysis, diagnostics, and scientific research are used. When comparing these tools in management processes, scientific research is preferred. Comparison of the presented models allows us to state the unity of their methodological foundations and hence their compatibility within the framework of a single integrative model.

The materials of using the methods of scientific research on the attributes of management in the management of an organization during a pandemic are presented. The technology of management research in the process of managing an organization has been developed. A feature of the presented material is the complex nature of approaches to the study of the problem under consideration, which creates a creative field for the interaction of various fields of knowledge in the field of managing the development of an organization in a crisis period.

3.3 Literature Review

In his study, Herman (2021) examines the impact of the COVID-19 pandemic on Indonesian public health at a time when a vaccine against it has not yet been developed. The implementation of non-pharmaceutical measures by the government is considered in detail. Public communication was chosen as one of the measures to prevent the spread of the virus, which, according to the results of the article, was not recognized as an effective management method during the pandemic.

Strielkowski et al. (2022) focus on the management of smart cities in the post-COVID-19 era and provide some interesting insights and inspirational ideas for coping with the results of the pandemic.

Comlek and Ordu presented in their work (Comlek & Ordu, 2021) results of a study of 83 patients without symptoms of SARS-CoV-2 for 30 years who were scheduled for appropriate medical treatment. As the authors themselves state, the limitations of this work are the retrospective design and the relatively small sample size. The study found that effective diagnostic/imaging tests aimed at detecting infected patients and healthcare workers, as well as modern telemedicine practices, contribute to the provision of safe and feasible treatments during the SARS-CoV-2 pandemic.

Martin (2010) in practical terms describes factors that affect the ability and readiness of technical medical staff to work during a pandemic. The author paid much attention to the management of a medical enterprise. According to the author, the job of managers should be to anticipate factors that can reduce the propensity and desire of medical staff to work during an obesity pandemic. Preparing to supply staff during favorable times can help keep the workforce healthy when they need it most. Even though this work is more than a decade long, many implications can be used for the current situation with COVID-19 pandemic and are reflected in the current literature (see, e.g., Kruszyńska-Fischbach et al. (2021) or Yeap et al. (2021)).

In his article, Sahin (2020) considers new approaches to neurorehabilitation in a sample of neurological diseases that includes stroke, multiple sclerosis, dementia, Parkinson's disease, and neuromuscular diseases in the context of the COVID-19 pandemic.

The aim of the study by Nair (2021) is an empirical analysis of real data from nine 5-star hotels in Qatar. The approach of four widely used strategies for managing the hotel business during the crisis period was applied. The results show that most of the hypotheses are supported, which served as the basis for incentivizing hotel finance and marketing managers and increasing customer satisfaction and maximizing hotel revenue during COVID-19 or any other pandemic.

In order to manage forced changes more effectively and adjust business continuity plans in the context of COVID-19, the work (Trofimova, 2022) presents practical advice, such as focusing on ongoing changes in the external and internal environment of the enterprise, which are crucial for business performance.

Fainshmidt et al. (2021) recommends adapting management to new ways to improve enterprise performance, including creating virtual project teams, remote meetings, group brainstorming, or using a variety of collaboration tools to support remote collaboration.

The Donthu's study (Donthu & Gustafsson, 2020) provides practical tips for establishing communication in a consistent communication within the enterprise and with all stakeholders. The authors argue that creating a structured communication format for telecommuting will help staff stay focused and productive during times of uncertainty.

As a follow-up to the article above, Boiral's study (Boiral et al., 2021) proposes to enhance employees' communication capabilities for feedback and problem solving, which can help management assess communication effectiveness, acceptance levels, and areas of resistance to change. Also in this article, quite a lot of attention is paid to ensuring continuous improvement in personnel management. Training and development can help employees adapt, embrace new ways of working, and perform their roles more effectively.

Obrenovic in an article (Obrenovic et al., 2020) puts forward very interesting proposals for improving the elasticity of planning. The author argues that as business continuity plans are adjusted to the current situation, it should be ensured that the program is sustainable in the long term with the ability to make timely changes in the event of a new crisis and create a foundation for communication and business continuity through virtual and remote management. A business continuity plan includes understanding how to make decisions in the face of uncertainty and what approach is best, to focus on execution, collaboration, and support of the business in a remote work environment.

3.4 Anti-Crisis Scenarios

This chapter presents the following anti-crisis scenarios during the COVID-19 pandemic and the features of their application:

1. Organizational changes
2. Financial strategies
3. Cost reduction
4. Reduction of assets
5. Profit creation

The characteristics of these scenarios are as follows:

1. Organizational changes. Companies in various fields of activity during the pandemic are quite severely influenced by environmental factors, such as economic, political, social, and technological. This requires management to pay special

attention within organizations to the change management processes of companies when developing measures to adapt them to these factors or neutralize them.

Organizational change management includes two key processes: the planning process and the change implementation process (Fløvik et al., 2019). At the same time, the implementation of these processes should minimize the resistance on the part of employees and the costs for the organization to maximize the effectiveness of change actions.

Resistance to change is a common reaction on the part of the organization's personnel, so it must be motivated in order for the change to be perceived as a necessary process. Resistance can be reduced by informing employees about the positive results that will be achieved by implementing a change program in the company and by involving employees in the process of designing and implementing changes. Effective change management requires the transition of the organization from its current state to the desired future state at minimal cost to the organization, so the change management process that exists in companies on an ongoing basis allows them to be more successful and resilient in crisis situations, as companies with successful change management practices maintain a certain kind of stability in various crisis situations due to the ongoing assessment of the factors of change and the availability of developed schemes for the transition to a new state, which allow streamlining the implementation of changes (Higgs & Rowland, 2010). Regardless of which process is affected by the changes, the system itself, only certain business processes of the company, and strategies, it is obvious that successful management of this process is impossible without the involvement of individual employees of the company in the process of transformation according to the new rules, coordination of the actions of all participants in order to achieve sustainable functioning in the market, and increasing competitiveness. Creating a structured teleworking communication approach that supports employees will help them perform their duties effectively during the COVID-19 pandemic and ensure that employees and customers are aware of the current situation, how it affects them, and new business, processes, and procedures necessary to carry out their work. This will help to minimize or completely eliminate fears within the team and, ultimately, maintain business continuity. Recent research has shown that the ultimate goal of change management is to drive organizational performance and results by engaging employees and inspiring them to adopt a new way of working.

2. Financial strategies. The development of a financial strategy for overcoming the crisis must be in full agreement with the strategic goal of increasing the value of net assets (or bringing it in line with the sales forecast), which ultimately leads to maximizing the market value of the business and maintaining its investment attractiveness (Melecky & Podpiera, 2020). Due to the complexity and time lag associated with attracting external sources of funding, management is initially interested in assessing the rate of growth that can be supported with internal sources of funding. Companies should test their financial plans for sustainability in various scenarios to assess how the crisis could affect financial results and how long it will last. If business plans and budgeting assumptions become out of

date due to the effects of the crisis, they will need to be revised to adapt to the changes. And if the impact of a pandemic threatens to be devastating, the minimum necessary to support operations in terms of staffing, interaction with suppliers, and deployment of facilities and technologies should be determined.

The capital structure optimization strategy is aimed at achieving an acceptable ratio of debt and ownership, as a rule, in the total amount of long-term financing sources (which is achieved by increasing the value of net assets), minimizing the costs of attracted capital, and, ultimately, increasing the market value of the business (Agnihotri, 2014). The problem of attracting investments that can create a powerful impetus for financial and economic development is relevant for most enterprises. Options such as an additional issue of shares, issuance of bonds, the attraction of long-term loans and borrowings, foreign capital, subsidies, grants from the state, and other funds, as well as the use of leasing as a method of updating fixed assets, are considered. The first step toward optimizing the capital structure is to achieve growth in the value of net assets. Only after painstaking and effective work on the "improvement" of property, giving it the status of a resource that promises to make a profit in the future, can one begin to consider options for attracting additional sources of financing.

There may also be problems with short-term capital replenishment to ensure continued operations. Based on the results of the analysis, it is possible to consider the possibility of attracting short-term capital, refinancing debt, attracting additional loans from banks and investors, and seeking government support. In addition, it is necessary to comprehensively analyze operating costs in order to reduce, if possible, all noncritical business items.

3. Cost reduction. Due to the worsening financial and economic situation at enterprises, exacerbated by the impact of negative trends as a result of the pandemic, many companies began to reduce their costs in a panic. The management of companies sees one of the main recipes for survival in times of crisis in reducing costs. Therefore, in the pursuit of savings, even entire cost items began to be "cut off" indiscriminately. Decisions, of course, need to be made in such situations very quickly, practically, and instantly. And decisions regarding costs and expenses are some of the easiest decisions that management will have to make, as they relate to what the company actually owns, cash, as opposed to making, for example, strategic, marketing, innovative decision-making which, due to the growth of uncertainty, has become even more complex. Therefore, it is clear that the easiest way is to "tighten the belts" and "tighten the nuts." However, an ill-considered reduction in spending, "tightening the belts," can lead to negative strategic consequences and tactical losses. In making hasty decisions to reduce costs, the categories of "good" or highly productive costs that bring a company a huge multiplier economic effect are also often overlooked.

   It is necessary to fight for reducing not all, but only unproductive, inefficient, irrational costs. You need to ask yourself the following questions:

   How can you intensify and increase the return and efficiency of the costs that the company is going to reduce?

How will the reduction of this or that item of expenditure affect the enterprise in a year, 2, 3, 5, and 10 years?

What risks are associated with certain costs; how will the reduction in costs affect the likelihood of these risks?

Does the volume of a certain cost item have a "critical weight"; what are the tasks and functions assigned to these costs?

What are the alternative solutions? What can be done to compensate for the reduction in costs?

The strategy is aimed at increasing the value of net assets (equity capital) as a necessary condition for complying with the strategic direction of recovery: "financial restructuring—an increase in the value of net assets while reducing losses and increasing profits—achieving financial indicators that characterize a profitable and growing business." Many enterprises still bear the burden of expenses for the maintenance of social and cultural property that does not belong to them. Therefore, in order to increase the liquidity of assets, it is necessary first of all to transfer these objects to the balance of local authorities and also to audit unused (surplus) intangible and production assets. To improve the structure of current assets, it is necessary to collect overdue receivables (possible claim work, appeal to the arbitration court to declare the debtor bankrupt, sale of debts, assignment of rights, etc.).

Particularly noteworthy is the management of the cost structure in terms of variable and fixed, the strength of the impact of the production leverage, the break-even level, and the marginal security (financial strength zones) (Kulchania, 2016). Optimization of accounting policy is closely related to the improvement of financial, tax, and management accounting, identifying the advantages of calculating limited costs (based on production and variable costs), increasing pricing flexibility, and profit management. To adequately reflect the activities of the enterprise, it is necessary to correct existing accounting errors, increase the responsibility of accountants for the work performed, and expand their functions in analytical (management) accounting.

4. Restructuring of assets. This strategy is the most effective measure to bring the company out of insolvency at the current time. It also creates the prerequisites for the profitable operation of this company in the future (Girgis, 2013).

   A significant variety of the company's assets creates the prerequisites for their change in order to influence the increase in its profitability.

   To make a decision on options for restructuring the company's assets, it is necessary to analyze the effectiveness of their use by the following elements:

- Fixed assets
- Intangible assets
- Construction in progress and uninstalled equipment
- Productive reserves
- Long-term and short-term financial investments
- Accounts receivable

Decisions to restructure a company are usually made with the financial need to rescue the firm and ensure a transition to more efficient operations at the lowest possible cost to those most affected, in particular the workforce and creditors. However, without well-defined conditions (including goals, necessary time frames, and long-term plans), as well as strict regulatory enforcement mechanisms, information transparency, and control, restructuring can have the effect of mere protection without efficiency within the company. Therefore, the restructuring of companies should be provided with clear requirements and methodology if it is planned to achieve the goal.

5. Creation of profit. This strategy includes tightening control over the profitability of products and the cost of operating assets, using the possibilities of pricing policy (arbitrage pricing), and optimizing the assortment policy (Nekhoda et al., 2018). The strategy aims to identify the added value "hidden" within the enterprise in the possible reserves of units (responsibility centers) that can increase their contribution to covering fixed costs and obtaining the desired profit.

   In a crisis period, it is necessary to quickly and in detail analyze all factors of business productivity growth, if possible, excluding all operations that do not bring value to business processes. This refers to both technological and operational processes. In most cases, it is quite difficult to increase the speed of technological processes, as quality may suffer. At the same time, most organizational processes have huge reserves of acceleration. And here it is not necessary to thoughtlessly reduce staff, saving costs. Usually on the contrary, in order to increase productivity, additional costs must be incurred.

3.5 Anti-Crisis Management

Two-thirds of organizations have responded to the COVID-19 pandemic by changing their policies and procedures. Half have begun implementing remote audits or effectively circulating information about the company's coronavirus response plans. Thirty-one percent of them tightened the rules in the field of information security. The anti-crisis management traditionally includes three stages as follows:

1. Pre-crisis stage. The first stage of anti-crisis management is the prevention of possible crises. This process includes the following:

   - Strategic planning. At present, considerable experience has been accumulated in strategic planning and development of strategies for territories at various levels. At the same time, the methodological issues of planning, adaptability of strategies to changes in the external environment, shock resistance, and the ability to withstand emerging risks and threats do not lose their relevance.
   - Strategic planning is one of the functions of strategic management, which is the process of choosing the organization's goals and ways to achieve them. Strategic planning provides the basis for all management decisions. The

functions of the organization, motivation, and control are focused on the development of strategic plans. Without taking advantage of strategic planning, organizations as a whole and individuals will be deprived of a clear way to assess the purpose and direction of the corporate enterprise. The strategic planning process provides the framework for managing the members of an organization.
- Recruitment and training of the crisis management team. For effective work with personnel in a crisis, the key is a system of measures that ensures the effective development of organizational changes. A manager who does not own the tools for managing the process of introducing organizational innovations can only aggravate the crisis situation. Consequently, among the most important issues of personnel working in this situation are methods of overcoming resistance to innovations on the part of personnel. The most common of them can be called the method of adaptive changes, the method of forced organizational changes, the method of managing resistance, and the crisis method.
- Conducting simulations to implement the strategic plan. This stage involves simulation of the process of implementing the chosen plan for the strategic development of the enterprise. There is an opinion that the pandemic and the possibility of remote work have forced companies to radically reconsider the methods of motivation and personnel management. But it is not so. The diversification of the workforce in the HR community was talked about many years ago: the format of work, when someone performs tasks from home or is freelancing, did not appear yesterday. But during the pandemic, there were more such distributed teams at the same time, and the issue of managing staff in the face of blurred corporate boundaries turned out to be not just relevant, but urgent and important. Employers have had to decide in a short time how to support workflows, communication, and motivation within teams in a fully remote work environment.

2. Crisis stage. The second step is to respond to it and implement the crisis management plan that was created in the pre-crisis period. Any action taken at this stage must be completed promptly. For example, during the pandemic, many businesses implemented a work-from-home policy to reduce the risk of transmission of the coronavirus, as required by the government. Obviously, no one knew in advance how long it would take to apply this policy, so for the purpose of crisis management in the second stage, the company must already have an application that supports the efficiency and continuity of remote work. Each level of the organizational structure has a role to play in responding to a crisis by ensuring that best practices are implemented.
3. Post-crisis stage. The final stage comes when the crisis passes or has already passed. However, the process of anti-crisis management is still ongoing. At this stage, it is advisable to evaluate the applied crisis exit strategies, whether they were effective or whether they need to be improved. This will help the company to better prepare for a new crisis in the future. In the post-crisis phase, the

organization returns to its normal business. But the current crisis situation still requires attention. First, PR (public relations) specialists after the crisis should provide more detailed analytical information to the public about what happened. Second, the organization should issue reports on the recovery of work after the crisis and the identification of the causes of the crisis. It is also important to provide full information to the media. It is important to evaluate the effectiveness of the work done to address the crisis and identify what works well and what needs improvement.

The pandemic is undermining the income and liquidity of businesses, as well as their ability to meet tax requirements. Most OECD countries have developed strong fiscal measures to support businesses through the provision of loan guarantees, credits, tax holidays, and assistance with the payment of wages to workers. But in some regions, the situation is quite different: many enterprises have begun their journey to recovery. Wherever possible, as enterprises open up, they research opportunities for development, taking win-win strategic actions that will help them become stronger in anti-crisis competition.

3.6 Conclusions

Today's business leaders are rightly focused on the huge business continuity challenges associated with COVID-19, and above all, they must continue to provide the following:

- Maximum sustainability of the organization
- Implemented successful financial strategies
- Strengthened critical systems to support continued employee engagement

COVID-19 has already become a powerful accelerator of digital transformation projects. The pandemic has drawn attention to the costs of slowing down digital transformation and IT development and has highlighted the opportunities associated with enterprise modernization. Digital transformation is one of the main trends in the IT industry in recent years. It is expected that timely "transformed" enterprises will not only leave behind competitors in existing markets but will also create new, previously inaccessible products and services. However, for a radical transformation of business, technology alone is not enough—the readiness for change on the part of company management plays a decisive role.

Thus, in order to ensure the sustainability of enterprises in the post-crisis world, the following strategic recommendations can be made:

1. It is necessary to optimize organizational structures. A well-built organizational structure makes it possible to optimize the number of personnel and the number of departments, simplify the interaction of departments, evenly distribute the load on staff, avoid duplication of functions, eliminate double and triple subordination, delineate the scope of activities of managers, determine their powers

and responsibilities, and increase labor productivity. The key takeaway is that companies need to take swift, measured action, including optimizing or reducing traditional ex ante controls, while keeping accountability in mind. Mechanisms are needed to account for and analyze costs and actions related to COVID-19. It is important to consider how follow-up forms of control (such as audits) can provide accountability in contexts where rapid responses are required. Optimization of the company's business climate should be complemented by measures aimed at solving long-term goals, such as creating high-quality jobs or improving the living standards of workers. Thus, firms' economic recovery prospects will depend on management's ability to use the crisis as an opportunity for deeper change.

2. A strategic reallocation of resources is needed. Among the many strategic activities that are carried out in the process of planning and implementing the strategy, an important place is occupied by the mobilization and distribution of strategic resources. This work serves as a necessary condition for the implementation of the strategy and at the same time is one of the elements of the strategic choice. The system of plans, programs, and projects, in addition to performing the basic function of management, is also a necessary tool for distributing strategic and tactical resources. In fact, a preliminary indicator of the quality of a plan or program is the willingness of management to allocate resources for its implementation. Plans help to allocate resources to areas that management believes are most effective and lead to the achievement of set goals. At the same time, the plans do not give a complete answer to the question of what specific resources and in what quantity are required.
3. It is necessary to continue to actively introduce digital technologies and automation. As part of government-imposed travel restrictions and social distancing measures, businesses and consumers are actively embracing digital solutions to continue working remotely. Digitalization facilitates the transition to the online environment of medicine, work, and education, allows you to make online purchases, get more data on the spread of the virus, and share information about research. The development of this trend speaks not only of an urgent need but also of the created material base for the widespread use of digital technologies. Accelerated by the pandemic, the digital future is becoming a reality faster than ever, maybe even faster than we can imagine. Digital technologies have played a key role during today's crisis. They have enabled local and regional governments to continue to provide essential public services, enabled entrepreneurs to operate, and connected people as public health measures are implemented during the COVID-19 pandemic.

The COVID-19 pandemic dealt a crushing blow to all components of our usual life, making it fully felt that what was planned is not always realized, and plans can change instantly, whether you like it or not. But at the same time, the coronavirus has given an incredible impetus to the development of organizations: it has provided an opportunity to evaluate employees, discover their new talents and abilities, learn new things, and implement technologies and solutions that will continue to improve work efficiency in the future. And most importantly, the pandemic made it clear that by speaking openly about the existing difficulties and solving them together, you can cope with any problem. We at MMK have seen that the pandemic, despite all the difficulties it brings with it, can also be a source of inspiration and a stimulus for the emergence of new ideas and approaches and the development of individual employees and the entire organization as a whole.

References

Agnihotri, A. (2014). Impact of strategy-capital structure on firms' overall financial performance. *Strategic Change, 23*(1–2), 15–20. https://doi.org/10.1002/jsc.1956

Al-Omari, Z., Alomari, K., & Aljawarneh, N. (2020). The role of empowerment in improving internal process, customer satisfaction, learning and growth. *Management Science Letters, 10*(4), 841–848. https://doi.org/10.5267/j.msl.2019.10.013

Boiral, O., Brotherton, M.-C., Rivaud, L., & Guillaumie, L. (2021). Organizations' management of the COVID-19 pandemic: A scoping review of business articles. *Sustainability, 13*, 3993. https://doi.org/10.3390/su13073993

Brożyna, J., Strielkowski, W., Fomina, A., & Nikitina, N. (2020). Renewable energy and EU 2020 target for energy efficiency in the Czech Republic and Slovakia. *Energies, 13*(4), 965. https://doi.org/10.3390/en13040965

Comlek, S., & Ordu, A. (2021). Interventional pain management during SARS-CoV-2 pandemic. *Signa Vitae., 17*(4), 178–184. https://doi.org/10.22514/sv.2021.008

Donthu, N., & Gustafsson, A. (2020). Effects of COVID-19 on business and research. *Journal of Business Research., 117*, 284–289. https://doi.org/10.1016/j.jbusres.2020.06.008

Fainshmidt, S., Andrews, D. S., Gaur, A., & Schotter, A. (2021). Recalibrating management research for the Post-COVID-19 scientific enterprise. *Journal of Management Studies, 58*(5), 1416–1420. https://doi.org/10.1111/joms.12702

Fløvik, L., Knardahl, S., & Christensen, J. O. (2019). The effect of organizational changes on the psychosocial work environment: Changes in psychological and social working conditions following organizational changes. *Frontiers in Psychology, 10*, 2845. https://doi.org/10.3389/fpsyg.2019.02845

Girgis, J. (2013). Corporate restructuring, the evolution of corporate assets and the public interest. *International Insolvency Review, 22*(1), 29–54. https://doi.org/10.1002/iir.1205

Herman, A. (2021). Indonesian government's public communication management during a pandemic. *Problems and Perspectives in Management, 19*(1), 244–256. https://doi.org/10.21511/ppm.19(1).2021.21

Higgs, M., & Rowland, D. (2010). Emperors with clothes on: The role of self-awareness in developing effective change leadership. *Journal of Change Management, 10*(4), 369–385. https://doi.org/10.1080/14697017.2010.516483

Kesavan, N., Seenuvasaragavan, S., Burnette, K., Heim, A. B., Patel, R. J., Johnston, J., Cao, X., Sakandar, H. A., Zhuang, Y., Oi, K. K., Zhi, Y., Diao, Y., & Strielkowski, W. (2022). Pandemic-inspired policies. *Science, 377*(6601), 22–24. https://doi.org/10.1126/science.add2046

Kruszyńska-Fischbach, A., Sysko-Romańczuk, S., Rafalik, M., Walczak, R., & Kludacz-Alessandri, M. (2021). Organizational E-Readiness for the digital transformation of primary healthcare providers during the COVID-19 pandemic in Poland. *Journal of Clinical Medicine, 11*(1), 133. https://doi.org/10.3390/jcm11010133

Kulchania, M. (2016). Cost structure and payout policy. *Financial Management, 45*(4), 981–1009. https://doi.org/10.1111/fima.12133

Martin, S. D. (2010). Ability and readiness of nurses to work during pandemic influenza. *Journal of Nursing Management, 19*(1), 98–108. https://doi.org/10.1111/j.1365.-834.2010.01190.x

Melecky, M., & Podpiera, A. M. (2020). Financial sector strategies and financial sector outcomes: Do the strategies perform. *Economic Systems, 44*(2), 100757. https://doi.org/10.1016/j.ecosys.2020.100757

Nair, G. (2021). Data-driven pricing strategies for hotels during COVID-19 pandemic. *International Journal of Revenue Management, 12*(1), 34–55. https://doi.org/10.1504/ijrm.2021.10036359

Nekhoda, Y., Redchikova, N., & Tyuleneva, N. (2018). Business models of companies: From profit to sustainable development and value creation. *Upravlenets, 9*(4), 9–19. https://doi.org/10.29141/2218-5003-2018-9-4-2

Obrenovic, B., Du, J., Godinic, D., Tsoy, D., Aamir, M., Khan, S., & Jakhongirov, I. (2020). Sustaining enterprise operations and productivity during the COVID-19 pandemic: "Enterprise Effectiveness and Sustainability Model". *Sustainability, 12*, 5981. https://doi.org/10.3390/su12155981

Sahin, E. (2020). Management of neurorehabilitation during the COVID-19 pandemic. *Journal of Duzce Faculty of Medicine, 22*(S1), 10–13. https://doi.org/10.18678/dtfd.775214

Simionescu, M., Strielkowski, W., & Gavurova, B. (2022). Could quality of governance influence pollution? Evidence from the revised environmental Kuznets Curve in Central and Eastern European countries. *Energy Reports, 8*, 809–819. https://doi.org/10.1016/j.egyr.2021.12.031

Strielkowski, W., Zenchenko, S., Tarasova, A., & Radyukova, Y. (2022). Management of smart and sustainable cities in the post-COVID-19 era: Lessons and implications. *Sustainability, 14*(12), 7267. https://doi.org/10.3390/su14127267

Trofimova, N. N. (2022). Managing organizational change and ensuring business continuity during the COVID-19 pandemic. *International Research Journal, 1–4*(115), 87–89.

Xu, D. (2019). Research on enterprise performance improvement under the innovative business ecosystem. *Malaysian E-Commerce Journal, 3*(1), 26–29. https://doi.org/10.26480/mecj.01.2019.26.29

Yeap, C. F., Suhaimi, N., & Nasir, M. K. M. (2021). Issues, challenges, and suggestions for empowering technical vocational education and training education during the COVID-19 pandemic in Malaysia. *Creative Education, 12*(8), 1818–1839. https://doi.org/10.4236/ce.2021.128138

Chapter 4
Participation of Business in Leading the Development of the Social Sphere: Best World Practices in the Post-Pandemic World

Gulnara Alekseeva, Anna Shmatko, and Vladimir Yamashev

Abstract Our research focuses on the participation of business in taking the leading position in the development of the social sphere. In particular, we pinpoint the best world practices in social sphere in the post-pandemic world. Through an examination of the contributions currently being made in the literature on social innovation, some questions of the future of this field in the wake of the pandemic are raised. This research is centered around the innovations in development and humanitarian efforts in the context of the coronavirus pandemic. In particular, in this chapter, we seek to share a series of emerging lessons on how innovation for development efforts has been successful in the COVID-19 pandemic and to propose ways that such efforts can be made more effective. In doing so, we recognize that the COVID-19 pandemic presents a unique opportunity for the social innovation and impact-driven entrepreneurial fields. Social innovations resonate with the challenges the pandemic presents and provide new opportunities for fostering leadership in business and economics.

Keywords Social sphere · Leadership and development · COVID-19

JEL Classification B55 · D7 · O10

G. Alekseeva (✉)
Audit and Corporate Reporting Department, Financial University under the Government of the Russian Federation, Moscow, Russian Federation
e-mail: GAlekseeva@fa.ru

A. Shmatko
North-West Institute of Management, Saint Petersburg, Russian Federation
e-mail: shmatko-ad@ranepa.ru

V. Yamashev
Volga Region State University of Service, Togliatti, Russian Federation

W. Strielkowski (ed.), *Leadership, Entrepreneurship and Sustainable Development Post COVID-19*, Springer Proceedings in Business and Economics, https://doi.org/10.1007/978-3-031-28131-0_4

4.1 Introduction

The COVID-19 pandemic has initiated a period of profound uncertainty, producing impacts at a scale that has, and will, transformed economies and societies across different settings (Khan et al., 2022). Moreover, the pandemic has exposed the vulnerabilities of the global economy to external shocks and renewed interest in a green agenda for mitigation and adaptation in the private sector to climate disruption (Diop et al., 2021). As a consequence of the evolving and changing nature of the pandemic, decisions on innovative efforts, like those for COVID-19 responses, must take place amid uncertainty, sometimes based on mixed evidence. Existing challenges for how small and large business companies' approach and execute on their innovation ambitions are amplified in the current context (Peters & Buijs, 2022). The following challenges, for instance, are framed as trade-offs and are particularly important in terms of how the world's social infrastructure, business supply chains, and housing conditions are likely to shift post-pandemic. Societal changes are abundant in a post-pandemic world: in a way, the pandemic is reminding us of neglected societal challenges, which include increasing economic inequality, the finite nature of global resources, the burden placed on the environment by Earth as limited capacity to absorb carbon dioxide (CO_2), and maintaining health and safety (Simionescu et al., 2022; Tonne, 2021). As our world moves successfully from a pandemic phase into an endemic (the situation when we will have to co-exist with COVID-19 which would become something like a normal flu), the landscape for healthcare innovation is more fundamentally shifting, with health, well-being, and healthcare being brought more fundamentally into the digital era. Technology will be a driving force for innovations that will redefine healthcare around the patients' experience and navigability as well as other spheres—from education to business (Bird et al., 2021). Integration will drive innumerable innovations, all of them united in their ability to streamline patients' experiences. Innovations, in particular across the healthcare sector, will lead to the reconsideration, reinvention, and reimagining of sustained efforts to safeguard patients' health outside of the exam room (Nanda et al., 2021). Rebooting the sector for social innovation will be an enormous task, one which needs younger talent and effective capacity-building programs now more than ever. Researchers and stakeholders need to map out the strategic approaches for making social innovation central to our economic and societal restarting processes. Alongside various business operations, many technological companies nowadays are committed to pursuing innovation across each of the energy sectors in ways that balance with social goals to address societal and environmental challenges over the long term (Duricin & Vuksanovic-Herceg, 2021). These business operations are global as are pursued using the principle of the economies of scale and other forms of efficiencies. Innovations in better, greener technologies that are energy-efficient paved the way to lower the costs of energy and also to reduce reliance on energy itself (Strielkowski et al., 2022). There are multiple examples of innovations in the way development investments are being made to respond to COVID-19, both to the direct health impacts of the pandemic and also to indirect economic and social impacts (Pu et al., 2021). Following an overview of the overall role of innovation in COVID-19 responses, this chapter reviews innovation efforts that are under way in the international development and humanitarian responses to the pandemic. It explores the ongoing international development and governmental responses to the pandemic, how these efforts are working, and how they may need strengthening in order to meet pressing health, social, and economic challenges, as well as ensuring society's long-term resilience.

Since it began in Wuhan, China, in 2020, COVID-19 pandemic has been at the forefront of recent innovations and advances in healthcare. The pandemic has had profound impacts on young people, particularly on their health, where this shared experience has also highlighted ways that they can become an active agent of change in solving these health needs and challenges with social innovation (Altillo et al., 2021). The aim was to spark young people's ideas for the future and gather social innovations led by young people related to the pandemic. Many international organizations conduct lots of open crowdsourcing calls and an online citizen hackathon to collect ideas from the local youth on the post-pandemic future, as well as innovative solutions for the identified healthcare needs in times of pandemic. These calls receive commendable submissions on current and proposed social innovations addressing a broad array of emerging health issues during the pandemic, with several focused on mental health (Almeida & Wasim, 2022). In addition, these calls see to address critical gaps in our understanding of the dynamic and complex interactions between medium- to long-term societal impacts of the COVID-19 pandemic. These innovations explore the existing opportunities in the areas of self-care, convenience, and gratification, which reflects consumers' behaviors both now and in the future. In the post-pandemic world, however, a recognition that too much emphasis is placed on short-term returns (short-termism) is building momentum for taking a longer-term perspective on public institutional reforms and infrastructure reconstruction (long-termism) (Carvalho & Fernandes, 2022). The pandemic has demonstrated the value of bottom-up innovation and collective wisdom. As long-term challenges mount for governments across the globe, public officials will have to reconsider the way they innovate, gather data, create policies, and serve their populations in order to promote leadership in business and economics.

4.2 The Impact of COVID-19 on Public Sector

In general, there are many sectors including healthcare, airline travel, consumer retail, and asset management that are planning long term, even as those sectors, including healthcare, face the immediate devastation caused by the COVID-19 pandemic (Caiazza et al., 2021). This summarizes key economic responses governments are taking to limit the human and economic impacts of the coronavirus. Governments and decision-makers worldwide approved measures for revenues and spending aimed at combating the COVID-19 outbreak and minimizing its adverse economic impacts. The pandemic and the containment measures had significant economic impacts, reducing the GDP by about 10% in 2020 (Bekkers & Koopman, 2022). During the early stage of the crisis, while many OECD countries were quarantining, severe decreases in mobility and social proximity caused by the crisis itself stranded many sectors. Low-educated workers are concentrating in precarious jobs across the Europe. The second wave of the pandemic has hit temporary workers in particular hard, with the impact of job creation being limited, even temporarily, by the COVID-19 pandemic (Greve et al., 2021). The effect on hours worked of people working temporary contracts was significant and was mostly concentrated in the destruction of jobs - especially among low-educated workers. Male workers were most likely to be employed in sectors that experienced high net job destruction in the early phases of the pandemic, during the second quarter of 2020. The adverse effects on retail employment, which were caused by the crisis itself, led to concerns that large losses in jobs and earnings for women could potentially erode recent gains made to narrow the gender gap (Guliyeva et al., 2021; Jung et al., 2021). The financial impacts throughout the small business sector, and particularly within restaurants, were likely particularly acute for businesses owned by migrants and ethnic minorities.

In economic terms, the impacts of COVID-19 are different in different regions, at least at its early stages. In 2020, the joint OECD-European Committee of the Regions (CoR) survey of 300 representatives from regional and local governments across 24 countries of the EU (CoR-OECD, 2020) indicates that, in the short to medium term, the COVID-19-related socioeconomic crisis is expected to adversely affect the finances of the majority of subnational governments, with the perilous scissors effect, with increased spending and decreased revenues (Canto et al., 2022). In countries with high levels of decentralization, the effect on the spending of subnational governments would be higher, especially in spending areas most affected by COVID-19 (in addition to health and social responsibility, subnational governments are involved in core areas affected by the crisis, including education (24% as the top expenditure), public administration (15%), economic development and transportation (13%), public order and safety (7%), and utilities (waste, water, etc.)) (Martin et al., 2022).

Supply chain disruptions are also having material effects on consumer prices, particularly in the automobile sector. Data also indicate that input shortages are holding back business activity in certain sectors. Entire industries, which shrank sharply during the pandemic, like the hotel and food service sectors, are now struggling to open. Meanwhile, many sectors—including real estate, business services, and construction—have seen a sharp bounce, with economies starting to open up in the third quarter of 2020 and hours worked returning to levels seen in the year before the crisis itself (Su et al., 2022). Some of that can be influenced by the cultural factors (Cabelková & Strielkowski, 2013). The auto industry also made up a third of the overall rise in prices across the economy from the same period the year before. The COVID-19 outbreak has significantly tempered expectations for 2020 and 2021. The prospects of COVID-19's effects have led the US Congress to consider more aid to help farm producers adapt to the disruptions of the domestic and global agriculture markets. For instance, USDA announced round two of the

Coronavirus Food Assistance Program (called CFAP 2), providing an additional $14 billion to farmers facing COVID-19-related market disruptions during the second half of 2020 (Blessley & Mudambi, 2022). Packages in Albania adopted in April 2020 included 7 billion Leks (0.4% of GDP) to cover a one-time transfer of 40,000 Leks to employees of small businesses affected by the COVID-19 pandemic not covered by the first package, employees of large businesses laid off due to the pandemic, and employees in the tourism sector and a sovereign guarantee of 15 billion Leks (0.9% of GDP) to provide working capital loans to all private companies that were tax compliant and solvent before the pandemic (Mano et al., 2021). Yet other countries have also eased requirements to access unemployment benefits and rental housing assistance to individuals whose economic and social situations have deteriorated as a result of the COVID-19 pandemic. In general, it can be seen that mitigating the COVID-19 outbreak requires a broad set of measures that address aspects of the food supply and demand system and their various stakeholders and local and international leaders.

4.3 Pandemic and Public Expenditures

As the COVID-19 pandemic rages, governments and central banks across the globe implemented policies designed to alleviate at least some of the adverse economic effects of the pandemic. Four important federal laws were passed in the United States in March and April of 2020, designed to respond to the public health emergency created by the coronavirus pandemic as it evolved into the global disaster, as well as directly aid households, businesses, and nonfederal governments affected by the economic disruption (López-Santana & Rocco, 2021). The four major federal laws provided grants to state and local governments and to tribal and territorial governments, too—for costs related to the pandemic. Federal aid to state and local governments would help cover increased costs related to the coronavirus pandemic in 2020, when those governments have lower tax revenues. The federal CARES (Coronavirus Aid, Relief, and Economic Security) Act signed into law on March 27, 2020, is the third and largest of several federal emergency aid packages intended to deal with the economic impact resulting from the crisis (McKinnon-Crowley, 2022). The people in the United States (as well as in the other countries) deserve an urgent, strong, and professional response to the growing public health and economic crises caused by the coronavirus outbreak and performed by the true leaders.

Regulatory and supervisory responses in banking sector to address impacts of the coronavirus also included the challenges posed by COVID-19 disruptions for the banking sector and the possible regulatory and supervisory responses that could keep the balance between maintaining financial stability, maintaining bank system soundness, and supporting economic activity. Pandemic responses for debt management that seeks to provide guidance on areas where sovereign debt managers may need to address challenges stemming from the COVID-19 crisis (Karim et al., 2022). Central banks' support for financial markets during coronavirus pandemic provides broad guidance for national authorities regarding potential responses by central banks to disturbances to money, securities, and currency markets likely to arise following financial disruptions, including a COVID-19 shock. Monetary and financial policy responses for emerging markets and emerging economies provide an overview of the appropriate central bank policy responses to the serious economic and financial impacts of the COVID-19 pandemic for emerging markets and developing economies (Berger et al., 2021).

While a comprehensive, inclusive approach to public health management is needed in the post-pandemic era, many authors focus on local and state public health departments' experiences and perspectives during COVID-19. Given variations in the models for local and state health departments' governance, a lack of uniform decision-making, which persisted during the duration of COVID-19, has contributed to variations in public health department responses. Although the health departments were critical for national responses to the pandemic (e.g., developing guidelines, conducting tests, and tracking), there were multiple challenges for the public health sector, resulting from reasons that were both old (e.g., gaps in information technology) and new (e.g., politicization and distrust of public health leaders and guidelines) (Burwell-Naney et al., 2021; Tagoe et al., 2021). Subnational governments, regions, and municipalities are on the front lines of the crisis as governance and recovery, and they are challenged by the asymmetrical impacts of COVID-19 on health, economy, society, and finances across countries, as well as across regions and subnational areas. Beyond health and social responsibilities, subnational governments are involved in the main areas affected by COVID-19, including education (the top spender with 24%), public administration (15%), economic development and transportation (13%), public order and security (7%), and utilities (waste, water, etc.). The territorial impacts of the COVID-19 health crisis, as with all pandemics, have a spatial dimension which needs to be managed (Amdaoud et al., 2021). While SARS has affected the community in ways far greater than its effect on affected individuals' health, the COVID-19 pandemic has affected those who suffer from major health conditions most severely: diabetes, high blood pressure, and obesity, for instance, common across much of the global economy, are all associated with increased risk for COVID-19 pandemic exposures. While many developing countries have seen relatively few cases, these countries have weakened healthcare systems that are likely to worsen the pandemic and its effects on their economies.

Many countries, as well as the European Union Member States, have redirected government funds toward crisis priorities, supporting healthcare, small- and medium-sized businesses, vulnerable populations, and regions that are especially affected by COVID-19. For instance, the US healthcare costs increased rapidly in response to the COVID-19 pandemic, mainly driven by increased federal expenditures, including provider financial assistance to replace lost revenues via the Provider Relief Fund ($122 billion in 2020) and Paycheck Protection Program ($53 billion in 2020), as well as increased federal spending on public health ($114.9 billion), including vaccine development, COVID testing, and healthcare facility readiness (Hartman et al., 2022). As a result of the impacts of the COVID-19 pandemic, health spending increased in 2020 at the fastest pace seen since 2002. For nearly all medical services, particularly hospitals, physicians, and nursing homes, increased federal expenditures to respond to the COVID-19 pandemic significantly exceeded the negative or slower growth of private health insurance, Medicare, and out-of-pocket expenditures, which were associated with lower utilization of medical services in 2020. When US federal spending on public health and other federal programs is removed (which includes COVID-19 pandemic supplemental funds), the expenses on the national health's growth is just 1.9%, which is slower than the 4.3% increase in 2019, due in large part to decreased use of medical goods and services because of the pandemic (Alfonso et al., 2021). In those days, President Biden believed that the federal government should move quickly and aggressively to help protect and support our families, small businesses, first responders, and essential care providers who are helping to meet the pandemic challenge, those who are most vulnerable to the health and economic impacts, and our larger communities—not to rebuke others or to rescue corporations. This appeared to be the right approach which paid off well.

4.4 Uncertainty in the Social Sphere in the Post-COVID-19 Era

The COVID-19 pandemic has sent shockwaves throughout societies and economies across the globe. From its COVID-19 origins at the end of 2019, the COVID-19 pandemic has caused chaos across the globe. On March 11, 2020, the World Health Organization declared COVID-19 to be a pandemic and a global threat. The coronavirus pandemic has exposed that our institutions, domestically and globally, are inadequate for the challenges we face. The effects of the coronavirus pandemic, and resulting economic downturn, are better understood within the context of what has happened to the world's economy in the last several decades.

Subnational governments and regions and municipalities are on the front line in managing the crisis and rebuilding, and they are facing asymmetric impacts on health, economy, society, and finance from COVID-19 across countries, as well as between regions and subnational areas (Gao & Zhang, 2022).

There is now a pressing need to explore how COVID-19, as both health and development crises, has played out the way that it has, to think about opportunities for transformation after the pandemic and to rethink development more generally. The immense global health and development crisis wrapped up in the COVID-19 pandemic has exposed the limits of conventional frames for development, whether in the north or in the south. Drawing on more than a decade of epidemic studies, we argue that the origins, emergence, and impacts of the COVID-19 pandemic demand analyses that tackle both the structural political-economic conditions and much less orderly, messy processes reflecting complexity, uncertainty, contingencies, and context specificity. Second is the way economies work, and the COVID-19 crisis has revealed the limits of a standard model of economic growth and the larger capitalist framework it is embedded within, as well as the need for pluralism and negotiations about viable post-capitalist alternatives. Lessons from earlier outbreaks are limited in their salience to addressing COVID-19's effects on global economies. We acknowledge companies are at varying stages in responding to the COVID-19 crisis, and thus impacts differ across geographies and sectors. The COVID-19 pandemic adaptation has been very successful, because before the crisis, the technology was already highly strategically valuable to companies, so within a very short time, new remote working models were implemented, employees were trained via digital platforms in new digital solutions that they were not used to working on, procedures were adjusted, and collaborative work modes were encouraged through digital means (Krammer, 2022). Prior to the COVID-19 pandemic, some sectors within the media-IT corporations had already been pushing for normalization of certain technologies, in which individual management capabilities were only theoretical. Whereas organizations had merely theorized new forms of work in the past, the COVID-19 crisis forced them into actions, like the actual transition into entirely new ways of working, or even the reinvention of the new forms of work required by new circumstances due to the pandemic. The workplaces of the future must become more digital, less hierarchical, and more agile, while respecting health, safety, and technology conditions required by pandemics evolving like COVID-19. Within the future workplace, technology is one of the factors that will ensure that employees can return safely to their work, considering the likelihood of a future pandemic or the absence of vaccines against COVID-19, and companies should determine how employees will get to the workplace, along with related aspects for health and safety, such as cleanliness or social distancing. COVID-19 is likely to speed up the AI adoption within healthcare, possibly more than in any other industry. Research from past pandemics indicates COVID-19 will durably diminish young generations' trust in political institutions and leaders, as well as their governments' healthcare policies (Sarker et al., 2021). A new canvassing of experts in the fields of technology, communications, and social change found many expected similar impacts from the 2020 COVID-19 pandemic. Beyond recent societal disruptions, COVID-19 is planting seeds of longer-term turmoil. The pandemic represents the largest global economic shock in living memory—it surpasses the global financial crisis both in terms of sheer scale and the uncertainties ahead (Bolleter et al., 2022). At COVID-19's inception, the pandemic has created an important bounce back in confidence across countries worldwide, reversing a long-term slide in confidence in governments and institutions across many countries. The COVID-19 outbreak occurred in a context of the significant HIV outbreak in South Africa, where one out of five people worldwide is living with HIV. The rapid global spread of COVID-19 has rapidly surpassed other recent pandemics in both scale and reach. In addition to its lethal human cost and destruction of millions of lives, economic harm is already substantial and widespread. The combined global GDP losses over 2020 and 2021 due to this rare calamity may total approximately $9 trillion, larger than the economies of Japan and Germany combined (Hossain, 2021). Assuming the pandemic recedes by the second half of 2020, and assuming policy actions taken globally are effective at preventing widespread bankruptcies, prolonged unemployment, and system-wide financial strain, it was possible to predict that some form of recovery would come sooner or later as early as in 2021. In the same time, COVID-19 has brought the "creative destruction" and helped to attract financing into the promising leading fields of research in healthcare and pharmaceuticals (Kesavan et al., 2022). For instance, amid COVID-19, market caps for a number of biotech companies working on developing vaccines are exceeding expectations.

4.5 Conclusions

Our study suggests several important outcomes related to the business landscape after the pandemic, offering fresh perspectives on digital transformation, the future of work, transparency, and resilience. If there were lingering skepticisms about the need for digital transformation for the long-term viability of businesses, the pandemic has put to rest. The coronavirus pandemic has exposed major flaws in the economic models that governments are now following globally. Around the world, many countries are now debating how their economic systems should be changed as a response to hard lessons learned about supply chains, government capacity, and social resilience in the wake of the coronavirus pandemic. Organizations resting on their existing digital laurels may be outpaced by those investing in adjusting their digital capabilities to a pandemic future—a future that looks far different than the pre-pandemic world. Under the COVID-19, the world has, by necessity, moved to lockdown which mean the new normal in those days.

National and local authorities, and philanthropic organizations and other private actors, adopted a broad range of measures to support the creative and cultural sectors through the pandemic. During and after the COVID-19 pandemic, policymakers and employers must broaden their commitments to improving and maintaining workers' health, taking into account factors that will enable workers and enterprising industries to continue to prosper. The International Labor Organization, Organization for Economic Cooperation and Development, and other professional labor organizations have developed concrete recommendations on policies and practices for workers' health and well-being that will promote restoration in a post-pandemic, as well as recommendations on workers' health practices. In addition, integrated, enterprise-based, workplace-based approaches that address interactions between these multidimensional drivers will strengthen organizational and worker resilience to address ongoing changes to work-related and worker safety, health, and well-being in the post-pandemic world. The benefits of such approaches are likely to extend beyond the COVID-19 pandemic, building the resiliency and adaptation needed by enterprises to address future challenges to public health as well as future changes to work. Applying a systems-level model could help to explain the variation the pandemic has had on jobs and workers across countries, within a context of differences in pre-existing social supports, resources available, and existing or emerging policies for public health and labor protection. Partners from different sectors are crucial to workers' post-pandemic health and safety. Policies that are designed to support businesses and workers in times of a pandemic may not adapt well to nontraditional business models and forms of work within cultural and creative sectors. This means that most people working in the creative industries (e.g., artists, solo entrepreneurs, or licensed professionals) are independent workers. Creative jobs are typically precarious forms of employment, and the pandemic has highlighted the ways that these jobs frequently fall through the cracks when it comes to government support. In India and Indonesia, the worsening conditions in labor markets due to the coronavirus pandemic have led to the collaboration between labor organizations and informal worker groups. Businesses and workers are eligible for common policies that seek to alleviate some of the financial strains caused by the pandemic. For instance, in many countries, businesses and employees have benefitted from expanded credit facilities and guarantee schemes, income support, and flexible mechanisms that free them from financial and administrative obligations. During the COVID-19 pandemic, many consumers switched to online services, and businesses and industries responded by remaining competitive in this new market environment. Some behaviors developed in the crisis, including widespread adoption of digital technologies, will endure beyond the pandemic, long after restrictions on activities are lifted. Detecting these signals of change sooner rather than later will be critical for optimizing the customer experience and revising customer value propositions in light of evolving preferences and needs. Digital transformation is needed even more in a pandemic, not less. The pandemic will continue to increase economic and social divisions across the globe. It is far too early to expect the economic effects of the pandemic, as well as the rifts in inequality and healthcare inequalities that it has exposed, will recede by 2025. The lessons learned from COVID-19 (along with its many setbacks) are likely to prove helpful should another pandemic strike the world in the future. As we look forward, our clients around the world are struggling with how they might plan, communicate, and reenter their hundreds of thousands of employees back to their offices, following new guidelines and policies, in order to keep employees' and companies' healthy. These teams are challenged by the logistics of scheduling and implementing new ways of working that balance business continuity with the safety and security of employees—right now and for months to come. Faced with some challenges, and still an uncertain set of risks, corporate leaders are legitimately worried about how their companies will be affected and what they should do next.

All in all, it becomes clear that both small and medium business enterprises and larger companies should become the leaders and trend-setters in the development of the social sphere. The cooperation between business and social domain is therefore desirable and will deepen in the post-COVID era. In order to achieve that, the best world practices in the post-pandemic world should be adapted and shared—some mutual learning programs and sharing experience should be established and maintained using the information and communication technologies as well as other novel digital approaches that were fostered by the pandemic.

References

Alfonso, Y. N., Leider, J. P., Resnick, B., McCullough, J. M., & Bishai, D. (2021). US public health neglected: Flat or declining spending left states ill equipped to respond to COVID-19: Study examines US public health spending. *Health Affairs, 40*(4), 664–671. https://doi.org/10.1377/hlthaff.2020.01084

Almeida, F., & Wasim, J. (2022). The role of data-driven solutions for SMES in responding to COVID-19. *International Journal of Innovation and Technology Management, 20*(1), 2350001. https://doi.org/10.1142/S0219877023500013

Altillo, B. S., Gray, M., Avashia, S. B., Norwood, A., Nelson, E. A., Johnston, C., & Mercer, T. (2021). Global health on the front lines: An innovative medical student elective combining education and service during the COVID-19 pandemic. *BMC Medical Education, 21*(1), 1–12. https://doi.org/10.1186/s12909-021-02616-9

Amdaoud, M., Arcuri, G., & Levratto, N. (2021). Are regions equal in adversity? A spatial analysis of spread and dynamics of COVID-19 in Europe. *The European Journal of Health Economics, 22*(4), 629–642. https://doi.org/10.1007/s10198-021-01280-6

Bekkers, E., & Koopman, R. B. (2022). Simulating the trade effects of the COVID-19 pandemic: Scenario analysis based on quantitative trade modelling. *The World Economy, 45*(2), 445–467. https://doi.org/10.1111/twec.13063

Berger, A. N., Demirgüç-Kunt, A., Moshirian, F., & Saunders, A. (2021). The way forward for banks during the COVID-19 crisis and beyond: Government and central bank responses, threats to the global banking industry. *Journal of Banking & Finance, 133*, 106303. https://doi.org/10.1016/j.jbankfin.2021.106303

Bird, M., McGillion, M., Chambers, E. M., Dix, J., Fajardo, C. J., Gilmour, M., & Carter, N. (2021). A generative co-design framework for healthcare innovation: Development and application of an end-user engagement framework. *Research Involvement and Engagement, 7*(1), 1–12. https://doi.org/10.1186/s40900-021-00252-7

Blessley, M., & Mudambi, S. M. (2022). A trade way and a pandemic: Disruption and resilience in the food bank supply chain. *Industrial Marketing Management, 102*, 58–73. https://doi.org/10.1016/j.indmarman.2022.01.002

Bolleter, J., Edwards, N., Cameron, R., Duckworth, A., Freestone, R., Foster, S., & Hooper, P. (2022). Implications of the COVID-19 pandemic: Canvassing opinion from planning professionals. *Planning Practice and Research, 37*(1), 13–34. https://doi.org/10.1080/02697459.2021.1905991

Burwell-Naney, K., Mortiboy, M., Zitta, J. P., Stevens, E., Patterson, K., Salter, J. C., & Jenkins, R. (2021). Local government approaches to combating COVID-19 inequities: A Durham County Department of Public Health Perspective. *International Journal of Environmental Research and Public Health, 18*(12), 6544. https://doi.org/10.3390/ijerph18126544

Cabelková, I., & Strielkowski, W. (2013). Is the level of taxation a product of culture? A cultural economics approach. *Society and Economy, 35*(4), 513–529. https://doi.org/10.1556/SocEc.2013.0007

Caiazza, R., Phan, P., Lehmann, E., & Etzkowitz, H. (2021). An absorptive capacity-based systems view of COVID-19 in the small business economy. *International Entrepreneurship and Management Journal, 17*(3), 1419–1439. https://doi.org/10.1007/s11365-021-00753-7

Canto, O., Figari, F., Fiorio, C. V., Kuypers, S., Marchal, S., Romaguera-de-la-Cruz, M., & Verbist, G. (2022). Welfare resilience at the onset of the COVID-19 pandemic in a selection of European countries: Impact on public finance and household incomes. *Review of Income and Wealth, 68*(2), 293–322. https://doi.org/10.1111/roiw.12530

Carvalho, F. L., & Fernandes, S. C. (2022). COVID-19 impact on tourism: The system thinking approach applied to the case of Portugal. *Journal of Tourism, Sustainability and Well-being, 10*(2), 75–87.

CoR-OECD (2020). *Results of the joint CoR-OECD survey on "The impact of COVID-19 on EU regional and local governments.* Available at: https://cor.europa.eu/en/news/Pages/econ-cor-oecd-survey-covid-19-results.aspx. Accessed 20 Sep 2022

Diop, S., Asongu, S. A., & Nnanna, J. (2021). COVID-19 economic vulnerability and resilience indexes: Global evidence. *International Social Science Journal, 71*(S1), 37–50. https://doi.org/10.1111/issj.12276

Duricin, D., & Vuksanovic-Herceg, I. (2021). The great reset of Serbia's economy during and after the COVID-19 crisis. *Ekonomika Preduzeća, 69*(3–4), 117–136. https://doi.org/10.5937/EKOPRE2103117D

Gao, J., & Zhang, P. (2022). Mechanisms of the Chinese Government's efforts to fight COVID-19: Integration of top-down interventions and local governance. *Health Security, 20*(4), 348–356. https://doi.org/10.1089/hs.2021.0161

Greve, B., Blomquist, P., Hvinden, B., & van Gerven, M. (2021). Nordic welfare states-still standing or changed by the COVID-19 crisis? *Social Policy and Administration, 55*(2), 295–311. https://doi.org/10.1111/spol.12675

Guliyeva, A., Rzayeva, U., & Jafarova, N. (2021). Analysis of some indicators by means of fuzzy logic on the example of Azerbaijani energy enterprises. *E3S Web of Conferences, 250*, 02001. https://doi.org/10.1051/e3sconf/202125002001

Hartman, M., Martin, A. B., Washington, B., Catlin, A., & National Health Expenditure Accounts Team. (2022). National health care spending in 2020: Growth driven by federal spending in response to the COVID-19 pandemic: National health expenditures study examines US health care spending in 2020. *Health Affairs, 41*(1), 13–25. https://doi.org/10.1377/hlthaff.2021.01763

Hossain, M. S. (2021). Economic scar tissue of COVID-19 puzzle: An analysis, evidence and suggestion on global perspective. *Global Business Review*, 09721509211055960. https://doi.org/10.1177/09721509211055960

Jung, H. S., Jung, Y. S., & Yoon, H. H. (2021). COVID-19: The effects of job insecurity on the job engagement and turnover intent of deluxe hotel employees and the moderating role of generational characteristics. *International Journal of Hospitality Management, 92*, 102703. https://doi.org/10.1016/j.ijhm.2020.102703

Karim, S., Akhtar, M. U., Tashfeen, R., Raza Rabbani, M., Rahman, A. A. A., & AlAbbas, A. (2022). Sustainable banking regulations pre and during coronavirus outbreak: The moderating role of financial stability. *Economic Research-Ekonomska Istraživanja, 35*(1), 3360–3377. https://doi.org/10.1080/1331677X.2021.1993951

Kesavan, N., Seenuvasaragavan, S., Burnette, K., Heim, A. B., Patel, R. J., Johnston, J., Cao, X., Sakandar, H. A., Zhuang, Y., Oi, K. K., Zhi, Y., Diao, Y., & Strielkowski, W. (2022). Pandemic-inspired policies. *Science, 377*(6601), 22–24. https://doi.org/10.1126/science.add2046

Khan, K., Su, C. W., & Zhu, M. N. (2022). Examining the behaviour of energy prices to COVID-19 uncertainty: A quantile on quantile approach. *Energy, 239*, 122430. https://doi.org/10.1016/j.energy.2021.122430

Krammer, S. M. (2022). Navigating the new normal: Which firms have adapted better to the COVID-19 disruption? *Technovation, 110*, 102368. https://doi.org/10.1016/j.technovation.2021.102368

López-Santana, M., & Rocco, P. (2021). Fiscal federalism and economic crises in the United States: Lessons from the COVID-19 pandemic and great recession. *Publius: The Journal of Federalism, 51*(3), 365–395. https://doi.org/10.1093/publius/pjab015

Mano, R., Anamali, A., & Shosha, B. (2021). Covid-19, business failure and bank loan collapse: Albanian case. *Universal Journal of Accounting and Finance, 9*(5), 1058–1066. https://doi.org/10.13189/ujaf.2021.090516

Martin, R., Martinelli, F., & Clifton, J. (2022). Rethinking spatial policy in an era of multiple crises. *Cambridge Journal of Regions, Economy and Society, 15*(1), 3–21. https://doi.org/10.1093/cjres/rsab037

McKinnon-Crowley, S. (2022). A snapshot of financial aid practice during COVID-19. *Community College Journal of Research and Practice, 46*(1–2), 93–100. https://doi.org/10.1080/10668926.2021.1972361

Nanda, A., Xu, Y., & Zhang, F. (2021). How would the COVID-19 pandemic reshape retail real estate and high streets through acceleration of E-commerce and digitalization? *Journal of Urban Management, 10*(2), 110–124. https://doi.org/10.1016/j.jum.2021.04.001

Peters, K., & Buijs, P. (2022). Strategic ambidexterity in green product innovation: Obstacles and implications. *Business Strategy and the Environment, 31*(1), 173–193. https://doi.org/10.1002/bse.2881

Pu, G., Qamruzzaman, M., Mehta, A. M., Naqvi, F. N., & Karim, S. (2021). Innovative finance, technological adaptation and SMEs sustainability: The mediating role of government support during COVID-19 pandemic. *Sustainability, 13*(16), 9218. https://doi.org/10.3390/su13169218

Sarker, S., Jamal, L., Ahmed, S. F., & Irtisam, N. (2021). Robotics and artificial intelligence in healthcare during COVID-19 pandemic: A systematic review. *Robotics and Autonomous Systems, 146*, 103902. https://doi.org/10.1016/j.robot.2021.103902

Simionescu, M., Strielkowski, W., & Gavurova, B. (2022). Could quality of governance influence pollution? Evidence from the revised environmental Kuznets curve in central and eastern European countries. *Energy Reports, 8*, 809–819. https://doi.org/10.1016/j.egyr.2021.12.031

Strielkowski, W., Zenchenko, S., Tarasova, A., & Radyukova, Y. (2022). Management of smart and sustainable cities in the post-COVID-19 era: Lessons and implications. *Sustainability, 14*(12), 7267. https://doi.org/10.3390/su14127267

Su, C. W., Dai, K., Ullah, S., & Andlib, Z. (2022). COVID-19 pandemic and unemployment dynamics in European economies. *Economic Research-Ekonomska Istraživanja, 35*(1), 1752–1764. https://doi.org/10.1080/1331677X.2021.1912627

Tagoe, E. T., Sheikh, N., Morton, A., Nonvignon, J., Sarker, A. R., Williams, L., & Megiddo, I. (2021). COVID-19 vaccination in lower-middle income countries: National stakeholder views on challenges, barriers, and potential solutions. *Frontiers in Public Health, 9*, 709127. https://doi.org/10.3389/fpubh.2021.709127

Tonne, C. (2021). Lessons from the COVID-19 pandemic for accelerating sustainable development. *Environmental Research, 193*, 110482. https://doi.org/10.1016/j.envres.2020.110482

Chapter 5
Public Sector Leadership as a Core Prerequisite for National Security Resistance to COVID-19

Alina Vysochyna, Yaryna Samusevych, and Yaroslav Reshetniak

Abstract The COVID-19 pandemic was a crash test not only for the health-care system but also for the entire system of public administration. In this study, a hypothesis that pre-pandemic leadership and high efficiency of the public administration are one of the core prerequisites for national security resistance to shocks caused by the pandemic is developed. In order to test this hypothesis, the National Security Index based on a systematic combination of the principal components analysis, Fishburn formula, and additive-multiplicative convolution are developed. This integrated indicator takes into account the economic, social, environmental, resource, and energy perspectives of national security. Based on the literature review, a set of indicators for quantitative assessment of leadership and efficiency of the public administration sector has been identified. In this study, the strength and character of the relationship between the level of national security and indicators of leadership and efficiency of the public administration sector in the pre-pandemic and pandemic periods are empirically substantiated. The study was conducted on the basis of data generated from the World Data Bank for 2000–2021 (or last available period). The findings will reveal specific patterns of the level of resistance of countries to the destructive effects of COVID-19, as well as their perceptiveness to regulatory interventions to overcome the pandemic crisis.

Keywords COVID-19 · Public sector · Leadership · Public administration · National security

JEL Classification C23, F52, H51, I18

A. Vysochyna (✉) · Y. Samusevych · Y. Reshetniak
Sumy State University, Sumy, Ukraine
e-mail: a.vysochyna@uabs.sumdu.edu.ua; y.samusevych@uabs.sumdu.edu.ua; ya.reshetniak@ntsa.sumdu.edu.ua

W. Strielkowski (ed.), *Leadership, Entrepreneurship and Sustainable Development Post COVID-19*, Springer Proceedings in Business and Economics, https://doi.org/10.1007/978-3-031-28131-0_5

5.1 Introduction

At the end of 2019—the beginning of 2020—the world faced a serious threat in the form of the COVID-19 pandemic, which led to significant human losses (according to the latest data, during the pandemic, more than 606 million cases of COVID-19 were recorded in the world, of which almost 6.5 million cases were lethal) (Worldometer, 2022). The pandemic revealed the unpreparedness of the health-care systems and economic systems of the countries for threats of such a scale. The socioeconomic crisis caused by the pandemic acted as a trigger for improving the existing mechanisms of state regulation, early response, and prevention of similar shocks in the future (Fabbrini, 2022). At the same time, the large-scale economic consequences, consisting of the loss of a significant amount of world GDP during the acute stage of the spread of the COVID-19 pandemic, forced adaptation to new realities of doing business and a radical transformation of not only business models but also approaches to public administration (Strielkowski et al., 2022).

In such conditions, scientists and representatives of the public administration sector focused their attention on finding the reasons for the variation in the scale of the destructive impact of the pandemic on various components of the state's national security. An equally important direction of theoretical and practical research is the study of prerequisites for faster post-pandemic recovery. Within this context, the architecture and pre-pandemic efficiency of the health-care system and the margin of economic strength formed in the pre-pandemic period are of unquestionable importance.

At the same time, a more in-depth study is needed to study how pre-pandemic leadership and high efficiency of the public administration influenced national security resistance to shocks caused by the pandemic (Guliyeva & Rzayeva, 2018). This direction of scientific research is relevant, because it is political leadership and the effectiveness of the functioning of public institutions that determine the trust of the population and business in state interventions aimed at overcoming the consequences of the pandemic and post-pandemic recovery and the readiness of economic agents and households to comply with the requirements and recommendations of regulators. All this determines the importance and relevance of this study.

5.2 Literature Review

Considering the novelty of the issues of this study, which consists of testing the hypothesis regarding the relevance of the influence of leadership in the public administration sector on the resistance of national security to the destructive impact of the COVID-19 pandemic, it is advisable to present the block of literature review in several meaningful blocks, namely, (1) an overview of the latest researches on the impact of the pandemic on various perspectives of the socioeconomic sphere; (2) analysis of the components of the state's national security determined by scientists, which will be the basis for determining the parameters of its quantitative assessment; and (3) substantiation quantification indicators of leadership in the public administration sector.

Research on the impact of the COVID-19 pandemic is most often focused on identifying specific patterns of the functioning of various business sectors during this period (Alkubaysi, 2020; Ray, 2021; Dutta et al., 2020; Hinrichs & Bundtzen, 2021; Grebosz-Krawczyk & Siuda, 2022; Midor et al., 2021), reasons for regional and national differentiation of the negative consequences of the pandemic (Kuzmenko et al., 2020; Vasylieva et al., 2020b), and the impact of vaccination of the population on the spread of the pandemic (Kuzmenko et al., 2021).

Therefore, the main parameters affecting the national security of the state are identified by scientists as GDP growth rates (Lisin et al., 2018; Bhowmik, 2020; Tiutiunyk et al., 2021; Tiutiunyk et al., 2022), entrepreneurship development, business innovativeness and business network density (Audretsch et al., 2014; Kosheleva et al., 2019; Fila et al., 2020; Melnyk et al., 2022), unemployment rate (Vasilyeva et al., 2021a; Lyeonov et al., 2018), the number of births and deaths of the population (Vasylieva et al., 2020b), migration processes (Sardak et al., 2018; Didenko et al., 2020; Didenko et al., 2021; Zhang et al., 2022), state of the labor market (Smiianov et al., 2020b), current account balance (Zolkover & Renkas, 2020), investment attractiveness (Rekunenko et al., 2022), rates of capital accumulation and the scale of lending (Vasilyeva et al., 2016; Vasylieva et al., 2020a), inflation rate (Djalilov & Hartwell, 2022), environmental sustainability (Keliuotytė-Staniulėnienė & Daunaravičiūtė, 2021; Kuzior et al., 2021a; Mentel et al., 2020; Smiianov et al., 2020a), energy security (Lisin & Strielkowski, 2014; Mentel et al., 2020; Simionescu et al., 2020; Letunovska et al., 2021), development of the healthcare system and its financial support (Samoilikova & Kunev, 2020; Shipko et al., 2020; Serpeninova et al., 2020; Lyeonov et al., 2021b, c), the quality of the education system and its financing (Khanin et al., 2019; Samusevych et al., 2021), and tax potential of the state (Čábelková et al., 2022; Lyulyov et al., 2021).

Among the parameters characterizing leadership in the public administration sector and the effectiveness of the functioning of state institutions, researchers identify publicity, accountability, and transparency (Fomina & Vynnychenko, 2017; Kuzior et al., 2021b; Yarovenko et al., 2021), the quality of state regulation (Kostel et al., 2017), effectiveness of government and institutional environment (Lyeonov et al., 2021a), good governance (Vasilyeva et al., 2021b), and trust in the government (Brychko et al., 2021).

At the same time, the lack of comprehensive studies aimed specifically at determining the relevance of the influence of leadership in the public administration sector on the resistance of national security to the destructive impact of the COVID-19 pandemic determines the relevance of this study.

5.3 Method and Data

The main task of this study is to test the hypothesis regarding the relevance of the influence of leadership in the public administration sector on the resistance of national security to the destructive impact of the COVID-19 pandemic. In order to implement the task, it is advisable to form an integral indicator of the quantitative assessment of the level of national security of the state. The generalization of existing scientific developments proved that among the main components of national security, both theoreticians and practitioners point out economic, ecological, energy, and social components.

Thus, for the formation of an integral indicator of national security, the following blocks of indicators were selected, which characterize the following:

1. Resource and social perspective—agricultural land (% of land area) (R_1); birth rate, crude (per 1000 people) (S_1); death rate, crude (per 1000 people) (S_2); life expectancy at birth, total (years) (S_3); unemployment, total (% of total labor force) (modeled ILO estimate) (S_4); vulnerable employment, total (% of total employment) (modeled ILO estimate) (S_5); school enrollment, secondary (% gross) (S_6); and Gini index (S_7).
2. Environmental perspective—CO_2 emissions (metric tons per capita) (En_1), alternative and nuclear energy (% of total energy use) (En_2), combustible renewables and waste (% of total energy) (En_3), and electric power consumption (kWh per capita) (En_4).
3. Economic perspective—GDP growth (annual %) (Ec_1); foreign direct investment, net inflows (% of GDP) (Ec_2); gross capital formation (% of GDP) (Ec_3); general government final consumption expenditure (% of GDP) (Ec_4); government expenditure on education, total (% of GDP) (Ec_5); military expenditure (% of GDP) (Ec_6); current health expenditure (% of GDP) (Ec_7); current account balance (% of GDP) (Ec_8); high-technology exports (% of manufactured exports) (Ec_9); medium- and high-tech manufacturing value added (% manufacturing value added) (Ec_10); inflation, consumer prices (annual %) (Ec_11); new business density (new registrations per 1000 people ages 15–64) (Ec_12); tax revenue (% of GDP) (Ec_13); and trade (% of GDP) (Ec_14).

All indicators are selected for the 15 studied European countries (Albania, Bulgaria, Croatia, Czech Republic, Estonia, Hungary, Latvia, Lithuania, Moldova, Poland, Romania, Serbia, Slovakia, Slovenia, and Ukraine) for 2000–2021 (or the latest available period) from the World Development Indicators collection of the World Bank Group (World Bank DataBank, 2022). The choice of the geographic structure of the sample is due to several circumstances: the availability of a complete array of statistical data on all indicators of the characteristics of national security, as well as the European integration vector of the development of the Ukrainian economy. At the same time, one of the promising vectors for the further dissemination of the scientific value of this study can be an increase in the number of countries under study.

Considering the fact that the abovementioned indicators are not comparable (they have different units of measurement), an important stage of this research is their normalization. For this purpose, all indicators were divided into stimulators (those whose growth ensures an improvement in the national security of the state) and destimulators (characterized by an inverse relationship with the integral level of the country's national security). Thus, it is proposed to consider as destimulators within the scope of this study such indicators as death rate, crude (per 1000 people), unemployment (% of total labor force) (modeled ILO estimate), Gini index, CO_2 emissions (metric tons per capita), electric power consumption (k Wh per capita), inflation, and consumer prices (annual %). The rest of the indicators are considered stimulants.

At the same time, the stimulus indicator is normalized by dividing the current value of this indicator for a certain country in a given year by the maximum value of this indicator for the entire sample. Instead, the normalization of destimulators is carried out by dividing the minimum value of the corresponding indicator for the whole sample by the current value of this parameter for a certain country in a certain time range. Thus, after carrying out the procedure of bringing to a comparable form, all components of the country's national security belong to the range [0; 1], while a larger value of the indicator indicates its better influence on the integral indicator of the country's national security.

It is also worth noting that the components of the integral indicator of the state's national security do not have an equal influence on the generalized indicator, which makes it necessary to determine their weighting factors. It is proposed to implement the task on the basis of a combination of the method of principal components and the Fishburn formula.

Thus, the first stage of this block of research involves the application of the method of principal components to an array of selected 26 individual indicators. The basis for further ranking of partial indicators is the arithmetic mean of the absolute eigenvalues for the corresponding indicator according to the principal components that explain more than 70% of the total variation of the indicators. Technical application of the method of principal components was carried out using the built-in block in the software product Stata 12/SE. The next step of this research block is the ranking of partial indicators using the built-in ranking function in MS Excel. Those indicators that have a greater impact on national security have a higher ranking value. Determination of the weighting coefficients for each of the partial indicators is carried out by dividing the corresponding rank by the sum of all ranks for all individual indicators. At the same time, it is proposed to combine indicators into an integral indicator using additive convolution and the Fishburn formula:

$$\mathrm{NSI} = \sum_{i=1}^{n} \frac{R_i}{\sum R} q_i \tag{5.1}$$

NSI—National Security Index; R_i—individual rank of the partial indicator of national security of the state; q_i—the actual value of the partial indicator of the state's national security.

It is worth noting that for the 26 partial indicators of the state's national security, the sum of all ranks will be 351 (the denominator in formula 5.1).

The next block of this study is the testing of the basic hypothesis: public sector leadership is a core prerequisite for national security resistance to COVID-19. As indicators characterizing the public sector leadership, it is suggested to use indicators from the collection Worldwide Governance Indicators of the World Bank Group (World Bank DataBank, 2022), including the control of corruption (CC), government effectiveness (GE), political stability and absence of violence/terrorism (PS), regulatory quality (RQ), rule of law (RL), and voice and accountability (VA). It should be noted that the values of all of the determined indicators are in the range [−2.5; 2.5].

To confirm or refute this hypothesis, in the work, two models will be developed, where the dependent variable is the National Security Index and the corresponding Worldwide Governance Indicators are independent variables that characterize public sector leadership. Moreover, the first model characterizes the relationships between variables in the pre-pandemic period (2000–2019), and the second model describes the relationship between them in the pandemic period (2020–2021). Both models will be built using panel data random effects GLS regression in Stata 12/SE software. The hypothesis will be confirmed if the quality of the second model or the significance of the coefficient for the independent variable model is better. Thus, it will confirm the validity of the assumption that the public sector leadership is more important in the pandemic period than in the pre-pandemic period. In fact, the margin of safety formed at the expense of political leadership in the period before the pandemic will allow ensuring resistance to the destructive impact of COVID-19 on the country national security.

5.4 Results and Discussion

Therefore, the first stage of this research on testing the relevance of the influence of leadership in the public administration sector on the resistance of national security to the destructive impact of the COVID-19 pandemic is the formation of an integral indicator of the state's national security. After bringing 26 individual indicators of the state's national security characteristics to a comparable form by normalization according to the minimax method, the method of principal components was used to determine the relevance and significance of the contribution of each of these indicators in the integral one. In particular, five principal components were selected for further research, which cumulatively explain the variation of 71% of the variables, which is a satisfactory indicator of the quality of the conducted analysis.

Averaging absolute eigenvalues for partial indicators in the section of selected principal components, their further ranking from 1 to 26 using the appropriate MS

Table 5.1 Principal components analysis results

Variable	Comp 1	Comp 2	Comp 3	Comp 4	Comp 5	Average	Rank	Weight
R_1	−0.104	−0.334	0.285	−0.088	0.005	0.1632	13	0.0370
S_1	−0.211	0.276	0.241	0.008	0.066	0.1604	12	0.0342
S_2	−0.206	0.306	0.194	−0.221	−0.066	0.1986	23	0.0655
S_3	0.148	0.404	0.047	−0.226	−0.015	0.168	14	0.0399
S_4	0.117	0.011	0.318	0.123	−0.145	0.1428	7	0.0199
S_5	−0.353	0.065	0.076	−0.147	0.087	0.1456	8	0.0228
S_6	0.195	0.248	−0.238	−0.122	−0.104	0.1814	21	0.0598
S_7	0.197	−0.047	0.345	−0.058	−0.044	0.1382	6	0.0171
En_1	−0.320	0.176	0.106	−0.006	0.161	0.1538	10	0.0285
En_2	0.041	0.083	0.160	−0.369	−0.231	0.1768	20	0.0570
En_3	0.036	0.301	−0.204	0.337	0.238	0.2232	25	0.0712
En_4	−0.336	−0.009	0.079	0.101	0.156	0.1362	5	0.0142
Ec_1	−0.109	−0.035	−0.059	0.169	−0.379	0.1502	9	0.0256
Ec_2	0.002	−0.017	0.033	0.135	−0.274	0.0922	1	0.0028
Ec_3	−0.132	0.157	0.115	0.341	−0.401	0.2292	26	0.0741
Ec_4	0.281	−0.214	−0.079	0.021	0.029	0.1248	4	0.0114
Ec_5	0.064	−0.269	0.240	0.248	0.248	0.2138	24	0.0684
Ec_6	0.025	−0.096	−0.428	−0.232	−0.074	0.171	16	0.0456
Ec_7	0.082	−0.187	0.213	−0.074	0.231	0.1574	11	0.0313
Ec_8	0.219	0.057	0.013	−0.275	0.312	0.1752	18	0.0513
Ec_9	0.253	0.233	0.127	0.119	0.149	0.1762	19	0.0541
Ec_10	0.278	0.036	0.183	−0.218	−0.210	0.185	22	0.0627
Ec_11	0.046	0.127	0.074	0.011	0.321	0.1158	3	0.0085
Ec_12	0.179	0.290	−0.014	0.263	0.097	0.1686	15	0.0427
Ec_13	0.197	−0.030	−0.051	0.276	−0.012	0.1132	2	0.0057
Ec_14	0.256	0.104	0.292	0.103	−0.107	0.1724	17	0.0484

Excel function and the determination of weighting factors based on Formula (5.1) allows us to note that in the structure of the integral indicator of national security, the most significant are such indicators as gross capital formation (% of GDP), combustible renewables and waste (% of total energy), and government expenditure on education (% total of GDP). At the same time, such indicators as foreign direct investment, net inflows (% of GDP), tax revenue (% of GDP) and inflation, and consumer prices (annual %) are characterized by the lowest level of significance. More detailed results are presented in Table 5.1.

After determining the weighting coefficients for each of the components of the state's national security, a corresponding integral indicator (*NSI*) was formed, descriptive statistics of which (dependent variable) and determinants of leadership in the field of public administration (independent variables) are presented in Table 5.2.

So, according to the descriptive statistics presented in Table 5.2, it can be noted that the National Security Index fluctuates in countries during 2000–2021 in the

Table 5.2 Descriptive statistics

Variable	Obs	Mean	Std. dev.	Min	Max
NSI	330	0.42	0.056	0.291	0.536
CC	330	0.09	0.606	−1.177	1.688
GE	330	0.355	0.591	−0.87	1.343
PS	330	0.394	0.592	−2.021	1.303
RQ	330	0.587	0.578	−0.856	1.698
RL	330	0.291	0.638	−1.272	1.382
VA	330	0.577	0.461	−0.671	1.214

range [0.291; 0.536]. In fact, none of the 15 European countries recorded the maximum possible value of the integral indicator—1. On average, the level of national security in this geographic sample is 0.42. The countries with the highest level of national security in the sample are the Czech Republic, Estonia, and Hungary. The maximum value of the integral indicator of national security was recorded in Hungary in 2016, and the minimum value was recorded in Moldova in 2000. In Ukraine, the level of national security varies in the range [0.377; 0.469]. The maximum level of the indicator was recorded in 2004 and the minimum in 2009.

In the context of characterizing the variation of factor variables that characterize leadership in the field of public administration, among the selected countries, the most positive shift was recorded in the field of regulatory quality and control of corruption, and the least successful is the state policy in the field of ensuring political stability and the rule of law. For Ukraine, the worst indicator is political stability in 2014 (−2.0208), and the best progress was achieved by the same parameter in 2007 (0.1731). At the same time, most of the values of indicators of political leadership for Ukraine are in the range of negative values, although each of the six selected parameters can potentially fluctuate in the range [−2.5; 2.5].

The next stage of this study is to determine the relevance of the influence of indicators of political leadership on the integral indicator of national security in the pre-pandemic period and pandemic period using panel tools data random effects GLS regression. The simulation results are shown in Table 5.3.

So, based on the modeling results, the following conclusions can be drawn:

- An increase in the pre-pandemic period of control of corruption by 1 unit leads to an improvement in the integral level of national security by 0.029 units, while the independent variable explains only 18.3% of the variation of the dependent variable.
- Instead, the growth of the same indicator in the pandemic period determines the strengthening of the level of national security by 0.053 units, and the coefficient of determination of the model increases to 57%.
- Less significant is the difference in the effect of the government efficiency on the level of national security of the state in the pre-pandemic and pandemic periods, namely, an increase in this indicator of political leadership by 1 unit leads to an increase in the integral level of national security by 0.05 and 0.056 units,

Table 5.3 Panel data regression analysis results (random effects GLS regression)

	Model 1 (pre-pandemic period—2000–2019)			Model 2 (pandemic period—2020–2021)		
	Coefficient	Standard error	*p*-value	Coefficient	Standard error	*p*-value
	Independent variable—control of corruption estimation					
CC	0.029	0.010	0.005	0.053	0.014	0.000
Constant	0.414	0.008	0.000	0.443	0.010	0.000
Overall R^2	0.183			0.570		
	Independent variable—government effectiveness estimation					
GE	0.050	0.009	0.000	0.056	0.014	0.000
Constant	0.399	0.008	0.000	0.432	0.011	0.000
Overall R^2	0.264			0.670		
	Independent variable—political stability and absence of violence/terrorism estimation					
PS	0.004	0.007	0.618	0.030	0.018	0.090
Constant	0.415	0.008	0.000	0.444	0.014	0.000
Overall R^2	0.148			0.427		
	Independent variable—regulatory quality estimation					
RQ	0.034	0.010	0.000	0.053	0.017	0.002
Constant	0.397	0.010	0.000	0.421	0.015	0.000
Overall R^2	0.197			0.583		
	Independent variable—rule of law estimation					
RL	0.055	0.009	0.000	0.056	0.012	0.000
Constant	0.401	0.008	0.000	0.432	0.010	0.000
Overall R^2	0.283			0.716		
	Independent variable—voice and accountability estimation					
VA	−0.005	0.012	0.702	0.066	0.021	0.002
Constant	0.419	0.010	0.000	0.419	0.015	0.000
Overall R^2	0.159			0.645		

respectively, but the quality of the second model is significantly better (the coefficient of determination is 0.264 against 0.67 in the pandemic period).

- In the pre-pandemic period, a statistically significant relationship between the level of political stability and national security for the studied 15 European countries was not confirmed, while in the period of the pandemic, the relevance of this factor increases, namely, an improvement in the state of political stability leads to an increase in the integral level of national security by 0.03 units at a 90% confidence interval.
- Similarly, no statically significant influence of transparency and accountability on the level of national security was established in the pre-pandemic period, while an increase in this indicator by 1 unit during the pandemic period leads to an increase in the integral level of national security by 0.066 units at the 99% confidence interval.

- Improvement in the regulatory quality in the pre-pandemic period provided an increase in the integral level of national security by 0.034 units and in the pandemic period by 0.053 units.
- Finally, the improvement of the state of the rule of law has an almost identical effect on the integral level of national security both in the pre-pandemic period and in the pandemic period—0.055 and 0.056 units, respectively, although the quality of the second model is much better.

5.5 Conclusions

Summarizing the results of the conducted research, it can be noted that in the scientific literature, there is no unified view on the assessment of the components of the state's national security; however, researchers agree that this concept is complex and covers resource, economic, social, and environmental perspectives. The pluralism of scientific positions determined the formation of the author's approach to the assessment of the integral level of national security of the state, which takes into account 26 main indicators and is built on the basis of a complex combination of the method of principal components, the determination of weighting factors according to the Fishburn approach and the compounding of indicators using additive convolution. The practical implementation of this stage of the research made it possible to establish that among the analyzed 15 European countries, the Czech Republic, Estonia, and Hungary have the highest level of national security. In Ukraine, the level of national security fluctuates and is noticeably lower than the sample average. The post-crisis years of 2008 and 2009, as well as 2013–2014, were the most critical for Ukraine, which were characterized by the aggravation of the military-political crisis in our country. The close correlation and correspondence of the results of the integral assessment of the state's national security level to real economic and political processes attest to the satisfactory quality and predicative properties of the approach proposed by the authors to the assessment of the state's integral level of national security.

According to the results of testing the hypothesis regarding the relevance of the influence of leadership in the public administration sector on the resistance of national security to the destructive impact of the COVID-19 pandemic, it was established that five out of six selected factor variables have a significantly higher power of positive influence on the level of national security of the state precisely during the pandemic, and only the influence of the rule of law on the outcome parameter has almost not changed compared to the pre-pandemic period. It is also worth emphasizing that the quality of all models during the pandemic period is significantly higher, which indicates their better prognostic and predicative potential.

Detected using a panel data random effects GLS regression patterns, it allows us to confirm the basic hypothesis that leadership in the public administration sector has a positive effect on the resistance of national security to the destructive impact of the COVID-19 pandemic. Especially important in the context of overcoming the negative consequences of the pandemic for the national security of the state is the improvement of the state of accountability and transparency, the rule of law, government efficiency, and control of corruption.

The obtained results can be useful to both scientists and representatives of state authorities. The results can be applied to adjust national strategies to overcome the negative impact of the COVID-19 pandemic on the national security of the country and to develop roadmaps for post-pandemic recovery, as well as to prevent similar crises in the future.

Acknowledgments This contribution was undertaken as a part of the research projects granted by the Ministry of Education and Science of Ukraine: "Socio-economic recovery after COVID-19: modeling the implications for macroeconomic stability, national security and local community resilience" (registration number 0122U000778) and "De-shadowing and regulatory efficiency of environmental taxation: optimization modeling to ensure national security and rational use of nature" (registration number 0122U000777).

References

Alkubaysi, A. (2020). Corporate social responsibility practice in the Gulf cooperation council countries amid the COVID-19 pandemic. *Business Ethics and Leadership, 4*(4), 99–104. https://doi.org/10.21272/bel.4(4).99-104.2020

Audretsch, D. B., Coad, A., & Segarra, A. (2014). Firm growth and innovation. *Small Business Economics, 43*(4), 743–749. https://doi.org/10.1007/s11187-014-9560-x

Bhowmik, D. (2020). Trends, cycles and seasonal variations of Ukrainian gross domestic product. *Financial Markets, Institutions and Risks, 4*(3), 80–94. https://doi.org/10.21272/fmir.4(3).80-94.2020

Brychko, M., Bilan, Y., Lyeonov, S., & Mentel, G. (2021). Trust crisis in the financial sector and macroeconomic stability: A structural equation modeling approach. *Economic Research-Ekonomska Istrazivanja, 34*(1), 828–855. https://doi.org/10.1080/1331677X.2020.1804970

Čábelková, I., Mareš, D., Strielkowski, W., Zenchenko, S., & Radyukova, Y. (2022). Income tax progressivity and nonreligion in central and Eastern Europe: A case of the Czech Republic. *Religions, 13*(4), 358. https://doi.org/10.3390/rel13040358

Didenko, I., Volik, K., Vasylieva, T., Lyeonov, S., & Antoniuk, N. (2020). Migration, environment, and country security: Analysis of touchpoints. *E3S Web of Conferences, 202*, 03028. https://doi.org/10.1051/e3sconf/202020203028

Didenko, I., Volik, K., Vasylieva, T., Lyeonov, S., & Antoniuk, N. (2021). Environmental migration and country security: Theoretical analysis and empirical research. *E3S Web of Conferences, 234*, 00010. https://doi.org/10.1051/e3sconf/202123400010

Djalilov, K., & Hartwell, C. A. (2022). The spirit is willing, but the institutions are weak: Disclosure of corporate social responsibility and the financial sector in transition. *Eurasian Business Review*. https://doi.org/10.1007/s40821-022-00224-1

Dutta, P., Dutta, U., Hasan, S., Sarkar, S., & Sana, T. (2020). Educate, empower and protect people through timely authentic information: Explore how Bangladeshi newspapers response to COVID-19 pandemic. *SocioEconomic Challenges, 4*(3), 93–103. https://doi.org/10.21272/sec.4(3).93-103.2020

Fabbrini, F. (2022). The legal architecture of the economic responses to COVID-19: EMU beyond the pandemic. *JCMS: Journal of Common Market Studies, 60*(1), 186–203. https://doi.org/10.1111/jcms.13271/

Fila, M., Levicky, M., Mura, L., Maros, M., & Korenkova, M. (2020). Innovations for business management: Motivation and barriers. *Marketing and management of Innovations, 4*, 266–278. https://doi.org/10.21272/mmi.2020.4-22

Fomina, J., & Vynnychenko, N. (2017). Fiscal transparency: Cross-country comparisons. *Business Ethics and Leadership, 1*(2), 39–46. https://doi.org/10.21272/bel.1(2).39-46.2017

Grebosz-Krawczyk, M., & Siuda, D. (2022). The COVID-19 pandemic impact on the TV commercials content in European countries. *Marketing and management of Innovations, 2*, 76–85. https://doi.org/10.21272/mmi.2022.2-07

Guliyeva, A., & Rzayeva, U. (2018). The role of E-government in reduction of information asymmetry in developing countries on the example of Azerbaijan. *Advances in economics and business, 6*(4), 209–217. https://doi.org/10.13189/aeb.2018.060402

Hinrichs, G., & Bundtzen, H. (2021). Impact of COVID-19 on personal insurance sales – Evidence from Germany. *Financial Markets, Institutions and Risks, 5*(1), 80–86. https://doi.org/10.21272/fmir.5(1).80-86.2021

Keliuotytė-Staniulėnienė, G., & Daunaravičiūtė, K. (2021). The global green bond market in the face of the COVID-19 pandemic. *Financial Markets, Institutions and Risks, 5*(1), 50–60. https://doi.org/10.21272/fmir.5(1).50-60.2021

Khanin, I., Shevchenko, G., Bilozubenko, V., & Korneyev, M. (2019). A cognitive model for managing the national innovation system parameters based on international comparisons (the case of the EU countries). *Problems and perspectives in management, 17*(4), 153–162. https://doi.org/10.21511/ppm.17(4).2019.13

Kosheleva, T. N., Tarhanova, N. P., Strielkowski, V., & Ksenofontova, T. Y. (2019). Development of customer motivation system in the field of entrepreneurship: Travel business. *Journal of advanced research in dynamical and control systems, 11*(11 Special Issue), 82–89. https://doi.org/10.5373/JARDCS/V11SP11/20192932

Kostel, M., Leus, D., Cebotarenco, A., & Mokrushina, A. (2017). The sustainable development goals for eastern partnership countries: Impact of institutions. *SocioEconomic Challenges, 1*(3), 79–90. https://doi.org/10.21272/sec.1(3).79-90.2017

Kuzior, A., Ma ń ka - Szulik, M., & Krawczyk, D. (2021a). Management w management electronic uslugami publicznymi w metropolii during pandemii COVID-19. *Polish Journal of Management Studies, 24*(2), 261–275. https://doi.org/10.17512/pjms.2021.24.2.16

Kuzior, A., Lyulyov, O., Pimonenko, T., Kwilinski, A., & Krawczyk, D. (2021b). Post-industrial tourism as a driver of sustainable development. *Sustainability, 13*(15), 8145. https://doi.org/10.3390/su13158145

Kuzmenko, O., Vasylieva, T., Vojtovič, S., Chygryn, O., & Snieška, V. (2020). Why do regions differ in vulnerability to COVID-19? Spatial nonlinear modeling of social and economic patterns. *Economics and Sociology, 13*(4), 318–340. https://doi.org/10.14254/2071-789X.2020/13-4/20

Kuzmenko, O. V., Smiianov, V. A., Rudenko, L. A., Kashcha, M. O., Vasilyeva, T. A., Kolomiiets, S. V., & Antoniuk, N. A. (2021). Impact of vaccination on the COVID-19 pandemic: Bibliometric analysis and cross country forecasting by Fourier series. *News Lekarskie, 74*(10), 2359–2367. https://doi.org/10.36740/wlek202110101

Letunovska, N., Saher, L., Vasylieva, T., & Lieonov, S. (2021). Dependence of public health on energy consumption: A cross-regional analysis. *E3S Web of Conferences, 250*, 04014. https://doi.org/10.1051/e3sconf/202125004014

Lisin, E., & Strielkowski, W. (2014). Modelling new economic approaches for the wholesale energy markets in Russia and the EU. *Transformations in Business & Economics, 13*(2B), 566–580.

Lisin, E., Strielkowski, W., Chernova, V., & Fomina, A. (2018). Assessment of the territorial energy security in the context of energy systems integration. *Energies, 11*(12), 3284. https://doi.org/10.3390/en11123284

Lyeonov, S. V., Vasylieva, T. A., & Lyulyov, O. V. (2018). Macroeconomic stability evaluation in countries of lower-middle income economies. *Naukovyi Visnyk Natsionalnoho Hirnychoho University, 1*, 138–146. https://doi.org/10.29202/nvngu/2018-1/4

Lyeonov, S., Vasilyeva, T., Bilan, Y., & Bagmet, K. (2021a). Convergence of the institutional quality of the social sector: The path to inclusive growth. *International Journal of Trade and Global Markets, 14*(3), 272–291. https://doi.org/10.1504/IJTGM.2021.115712

Lyeonov, S. V., Kuzmenko, O., Koibichuk, V. V., Rubanov, P. M., & Smianov, V. A. (2021b). Behavioral, social, economic and legal dimensions of the public health system of Ukraine: Descriptive, canonical and factor analysis. *News Lekarskie, 74*(12), 3126–3134. https://doi.org/10.36740/wlek202112102

Lyeonov, S., Bilan, S., Yarovenko, H., Ostasz, G., & Kolotilina, O. (2021c). Country's health profile: Social, economic, behavioral and healthcare determinants. *Economics and Sociology, 14*(3), 322–340. https://doi.org/10.14254/2071-789X.2021/14-3/17

Lyulyov, O., Lyeonov, S., Tiutiunyk, I., & Podgórska, J. (2021). The impact of tax gap on macroeconomic stability: Assessment using panel VEC approach. *Journal of International Studies, 14*(1), 139–152. https://doi.org/10.14254/2071-8330.2021/14-1/10

Melnyk, L., Matsenko, O., Kubatko, O., Korneyev, M., & Tulyakov, O. (2022). Additive economy and new horizons of innovative business development. *Problems and perspectives in Management, 20*(2), 175–185. https://doi.org/10.21511/ppm.20(2).2022.15

Mentel, G., Vasilyeva, T., Samusevych, Y., Vysochyna, A., Karbach, R., & Streimikis, J. (2020). The evaluation of economic, environmental and energy security: Composite approach. *International Journal of Global Environmental Issues, 19*(1-3), 177–195. https://doi.org/10.1504/ijgenvi.2020.10037581

Midor, K., Kuzior, A., Płaza, G., Molenda, M., & Krawczyk, D. (2021). Reception of the smart city concept in the opinion of local administration officials-a case study. *Management Systems in Production Engineering, 29*(4), 320–326. https://doi.org/10.2478/mspe-2021-0040

Ray, A. (2021). Containing airborne transmission of COVID-19 and its implications for global economic recovery. *Business Ethics and Leadership, 5*(1), 81–88. https://doi.org/10.21272/bel.5(1).81-88.2021

Rekunenko, I., Zhuravka, F., Nebaba, N., Levkovych, O., & Chorna, S. (2022). Assessment and forecasting of Ukraine's financial security: Choice of alternatives. *Problems and perspectives in Management, 20*(2), 117–134. https://doi.org/10.21511/ppm.20(2).2022.11

Samoilikova, A., & Kunev, R. (2020). The impact of health care financing on the economic growth: EU countries analysis. *Health Economics and Management Review, 1*(2), 24–32. https://doi.org/10.21272/hem.2020.2-03

Samusevych, Y. V., Novikov, V. V., Artyukhov, A. Y., & Vasylieva, T. A. (2021). Convergence trends in the "economy – education – digitalization – national security" chain. *Naukovyi Visnyk Natsionalnoho Hirnychoho University, 6*, 177–183. https://doi.org/10.33271/NVNGU/2021-6/177

Sardak, S., Korneyev, M., Dzhyndzhoian, V., Fedotova, T., & Tryfonova, O. (2018). Current trends in global demographic processes. *Problems and perspectives in Management, 16*(1), 48–57. https://doi.org/10.21511/ppm.16(1).2018.05

Serpeninova, Y., Makarenko, I., Plastun, A., Babko, A., & Gasimova, G. (2020). Mapping of the responsible investments instruments in sdg 3 "good health and well-being" financing: EU and US experience. *Health Economics and Management Review, 1*(1), 106–115. https://doi.org/10.21272/hem.2020.1-10

Shipko, A., Demikhova, N., Pajak, K., & Motrechko, V. (2020). Health management at the regional level: Multivariable performance assessment. *Health Economics and Management Review, 1*(2), 8–15. https://doi.org/10.21272/hem.2020.2-01

Simionescu, M., Strielkowski, W., & Tvaronavičienė, M. (2020). Renewable energy in final energy consumption and income in the EU-28 countries. *Energies, 13*(9), 2280. https://doi.org/10.3390/en13092280

Smiianov, V. A., Vasilyeva, T. A., Chygryn, O. Y., Rubanov, P. M., & Mayboroda, T. M. (2020a). Socio-economic patterns of labor market functioning in the public health: Challenges connected with COVID-19. *News Lekarskie, 73*(10), 2181–2187. https://doi.org/10.36740/wlek202010114

Smiianov, V. A., Lyulyov, O. V., Pimonenko, T. V., Andrushchenko, T. A., Sova, S., & Grechkovskaya, N. V. (2020b). The impact of the pandemic lockdown on air pollution, health and economic growth: System dynamics analysis. *News Lekarskie, 73*(11), 2332–2338. https://doi.org/10.36740/wlek202011102

Strielkowski, W., Zenchenko, S., Tarasova, A., & Radyukova, Y. (2022). Management of smart and sustainable cities in the post-COVID-19 era: Lessons and implications. *Sustainability, 14*(12), 7267. https://doi.org/10.3390/su14127267

Tiutiunyk, I., Drabek, J., Antoniuk, N., Navickas, V., & Rubanov, P. (2021). The impact of digital transformation on macroeconomic stability: Evidence from EU countries. *Journal of International Studies, 14*(3), 220–234. https://doi.org/10.14254/2071-8330.2021/14-3/14

Tiutiunyk, I. V., Zolkover, A. O., Lyeonov, S. V., & Ryabushka, L. B. (2022). The impact of economic shadowing on social development: Challenges for macroeconomic stability. *Naukovyi Visnyk Natsionalnoho Hirnychoho University, 1*, 183–191. https://doi.org/10.33271/nvngu/2022-1/183

Vasilyeva, T., Sysoyeva, L., & Vysochyna, A. (2016). Formalization of factors that are affecting the stability of the Ukrainian banking system. *Risk Governance and Control: Financial Markets and Institutions, 6*(4), 7–11. https://doi.org/10.22495/rcgv6i4art1

Vasilyeva, T., Kuzmenko, O., Kuryłowicz, M., & Letunovska, N. (2021a). Neural network modeling of the economic and social development trajectory transformation due to quarantine restrictions during covid-19. *Economics and Sociology, 14*(2), 313–330. https://doi.org/10.14254/2071-789X.2021/14-2/17

Vasilyeva, T., Ziółko, A., Kuzmenko, O., Kapinos, A., & Humenna, Y. (2021b). Impact of digitalization and the COVID-19 pandemic on the AML scenario: Data mining analysis for good governance. *Economics and Sociology, 14*(4), 326–354. https://doi.org/10.14254/2071-789X.2021/14-4/19

Vasylieva, T., Jurgilewicz, O., Poliakh, S., Tvaronavičienė, M., & Hydzik, P. (2020a). Problems of measuring country's financial security. *Journal of International Studies, 13*(2), 329–346. https://doi.org/10.14254/2071-8330.2020/13-2/22

Vasylieva, T., Kuzmenko, O., Rashid, M. N., Vojtovic, S., Kascha, M., & Lieonov, H. (2020b). Innovations in government management of the healthcare system: Forecasting of covid-19 consequences in social, investment and business development. *Marketing and management of Innovations, 4*, 11–25. https://doi.org/10.21272/mmi.2020.4-01

World Bank DataBank. (2022). https://databank.worldbank.org/home.aspx

Worldometer. (2022). Coronavirus cases. https://www.worldometers.info/coronavirus/#countries

Yarovenko, H., Bilan, Y., Lyeonov, S., & Mentel, G. (2021). Methodology for assessing the risk associated with information and knowledge loss management. *Journal of Business Economics and Management, 22*(2), 369–387. https://doi.org/10.3846/jbem.2021.13925

Zhang, L., Chen, Y., Lyulyov, O., & Pimonenko, T. (2022). Forecasting the effect of migrants' remittances on household expenditure: COVID-19 impact. *Sustainability, 14*(7), 4361. https://doi.org/10.3390/su14074361

Zolkover, A., & Renkas, J. (2020). Assessing the level of macroeconomic stability of EU countries. *SocioEconomic Challenges, 4*(4), 175–182. https://doi.org/10.21272/sec.4(4).175-182.2020

Chapter 6
Social Problems Within the Context of Leadership and Public Policies in the Post-COVID Era

Zuzana Horváthová

Abstract This chapter focuses on the social problems within the context of leadership and conducting public policies for supporting the social sphere after the negative impacts of the COVID-19 pandemic. The coronavirus pandemic has highlighted, yet again, the importance of social inequalities in times of great crises, and this reality has obvious implications for public policy and its main leaders. In addition, it impacted on the problems of leadership by questioning the leading role of the state authorities versus the private business companies. In particular, COVID-19 has exposed, intensified, and reinforced the existing social inequalities as well as created some new, for example, issues on the labor market and with the requalification of employees or the pressing need for digitalization of public administration and the introduction of e-government. It can be noted that COVID-19 restrictions thus impacted the personal social networks and the structure of the larger networks within the society. This chapter observes that social connections generate emergent properties in the community, especially in the sense of social networks. This chapter also provides recommendations to governments and other stakeholders for supporting safe reopening of various activities and supporting social services and other public policies and initiatives throughout the outbreak of COVID-19 and beyond.

Keywords Social problems · Leadership · Public policies · COVID-19

JEL Classification B55 · J18 · H55

Z. Horváthová (✉)
Department of Legal Specialisations and Public Administration, Metropolitan University Prague, Prague, Czech Republic
e-mail: zuzana.horvathova@mup.cz

W. Strielkowski (ed.), *Leadership, Entrepreneurship and Sustainable Development Post COVID-19*, Springer Proceedings in Business and Economics, https://doi.org/10.1007/978-3-031-28131-0_6

6.1 Introduction

The coronavirus disease, proclaimed as a pandemic by the World Health Organization (WHO) in March 2020, attacked our societies at the very heart. Coronavirus became an unprecedented public health emergency, with associated major economic impacts, that is affecting all developed and developing countries. The existence of highly infectious diseases that become pandemics, like COVID-19, poses serious health and economic challenges due to various global social and environmental transformations that have occurred due to economic development (Yadav et al., 2020). The social responses to these challenges can often be spontaneous and self-organized. For instance, neighborhood-based networks emerge due to geographical restrictions (e.g., stay-at-home orders), leading to an increase in neighborhood-based support and volunteerism (Haesler et al., 2021). Indeed, many studies have shown that relationships with neighbors show the largest net increase in relational quality ratings, relative to a number of relation types (e.g., partners, colleagues, friends). Much of this is built through spontaneous personal interactions in the local communities, all of which collectively promote a sense of community experienced by many (Long et al., 2022). COVID-19 anti-pandemic restrictions thus affect personal social networks as well as in terms of the size and complexity of relationships. For example, local area-based networks emerged due to geographic restrictions (e.g., stay-at-home orders, shutdowns, as well as lockdowns), resulting in increases in unneighborly support and local volunteering (Ye et al., 2021). In fact, it appears that relationships with neighbors displayed the largest net gain in ratings of relationship quality compared with a range of relationship types.

Nevertheless, in the case of the global emergencies such as the COVID-19 pandemic (and future pandemics that would inevitably be here), public health theories and practices must recognize the importance of social policies for the direct efficacy of the specially targeted public policies as well as for the longer-term social and economic impacts of the pandemic (Leal Filho et al., 2021). Thence, there should be the measures to ensure public sector continuity, both nationally and at subnational levels, during and following the COVID-19 pandemic. Addressing the COVID-19 pandemic requires an unprecedented level of collaboration among governments and citizens, and in different sectors of society, in order to achieve spatial distancing and other policy measures (Cyr et al., 2021). If not adequately addressed with policies, the societal crises created by the COVID-19 outbreak could also exacerbate inequalities, exclusion, discrimination, and global unemployment over the medium to long term. Policy changes should move beyond an acute response to the COVID-19 pandemic and sustainably address both the pressing social needs and underlying systemic issues that caused them (Leach et al., 2021).

In general terms, the availability of rights-based safety nets to buffer unwanted or unexpected effects from crises like the COVID-19 pandemic must be considered as a policy priority at national and international levels within sustainable development policies. Governments and international bodies should monitor the effects of COVID-19 on the vulnerable groups of citizens carefully and take action to mitigate the pandemic effects on them (Elliott et al., 2021). In addition, governments should ensure public information campaigns focusing on how the citizens who were hit hard by the pandemic may access public services, and they should ensure services are accessible to all vulnerable citizens, including those living in areas with restrictions on movement or quarantine, as well as to those who have contracted COVID-19.

In keeping this agenda, the governments worldwide must prioritize the resilience of the public health system, a component of the national security, at the forefront of all policy implementations in order to respond effectively to the COVID-19 pandemic. In the same time, it can be observed that some social policies in pandemic response public health measures implemented by some countries in order to combat the COVID-19 pandemic have led to numerous collaterals, spillover, socioeconomic, and political damages (Ayuningtyas et al., 2021). To mitigate a multitude of social impacts, some governments (e.g., in the EU countries, namely, Germany) established significant social policy measures, including the social protection packages, aimed at mitigating social and economic impacts of the COVID-19 pandemic. Notably, executive efforts included the creation of a pandemic-wide health inequality task force, which focused on underlying social vulnerabilities.

In the same time, in the case of the COVID-19 crisis, its severe impacts on existing social inequalities are undisputed, regardless of whether one is focused on its economic, social, or public health impacts (Semplici, 2022). Overall, this topical question draws attention to the connections between the COVID-19 crisis, social inequality, and public policies across the globe. While the coronavirus pandemic has focused our attention on the main types of inequalities, it has also created space for a debate on changing policies and the best ways to better tackle issues like poverty, ethno-racial discrimination, health inequalities, and unemployment, among others (De Jesus et al., 2022). Some countries feature their own particularities of public policies and services to protect economic, social, and cultural rights as contributing to recovery from the COVID-19 pandemic.

6.2 Literature Review

Social policies are concerned with how societies around the world address the needs of people in terms of safety, education, employment, health, and well-being. Social policies are designed to enhance human welfare and to address human needs for education, health, housing, and economic security (Cowie & Myers, 2021). Any social policy addresses the ways in which states and societies are responding to the global challenges posed by changes in social, demographic, and economic conditions, as well as by poverty, migration, and globalization (Čajka et al., 2014; Hujo, 2021). Social policy pays particular attention to different configurations of state political actors (such as the state, households, markets, and civil society) involved in providing welfare services under varying circumstances (Guliyeva et al., 2020). Important areas of social policy are well-being and social security, poverty alleviation, welfare, equity, unemployment insurance, living conditions, animal rights, retirement, healthcare, public housing, family policy, social services, child protection, social isolation, education policy, crime and criminal justice, urban development, and labor issues (Fischer et al., 2022). The purpose here is to make the sense of some of the main programs and issues of the social welfare system, particularly involving healthcare and human services. These services and supports include support for children and families, schooling and education, housing and neighborhood rehabilitation, maintaining incomes and alleviating poverty, supporting unemployment and job training, retirement, health, and social services. In addition, they support job placement and training programs for individuals seeking jobs (Alderwick et al., 2021).

Another definition of a welfare state is it is the modern state understanding which assumes duty of providing fair income distribution; protecting groups and classes that require protection; driving the social security practices and employment policies; practicing policies to address society's essential needs, such as education, health, and housing; and taking measures to regulate work lives through taxation and wage policies that are followed by it (Kelly & Lobao, 2021). Consistent with this perspective, many social welfare policies are identified as crucial means to promoting equity. These policies may also have important effects on public health. It is possible to argue that it is not just economic indicators but the changes taking place in population structure and society, which have an impact in shaping the policies of welfare states. These changes are also different depending on social structures and the policies of a country. Other important examples of social policy developments are the Bismarckian welfare states of nineteenth-century Germany, social insurance policies introduced into the USA by the New Deal from 1933 to 1935, and both the Beveridge Report and the National Health Service Act of 1946 in Britain (Edling, 2019).

In the USA, for example, the public sector supports a variety of healthcare and social services at the state and local level, via individual state social services block grants. Government funding of social services by third-party agencies has increased substantially since the 1960s. In the current push to privatize social services, nonprofits as well as for-profit providers are competing for public contracts (Benavides & Nukpezah, 2020). Yet, this is a trend that has a number of critical issues that must be addressed by policymakers, social workers, and others who provide healthcare and human services. There is a growing agreement that policies can address the backlog. This is because the core challenge of social workers today is to contribute significantly to developing policies and programs that will best meet the country's core social problems and challenges. Moving forward with a healthcare policy agenda reflective of the themes outlined here and herein will therefore require social workers and public health professionals to continue framing problems and developing policy solutions that reflect those values and working to cultivate a policy climate that is, in turn, conducive to those solutions (Morley et al., 2020). The social work-based approach to policy might provide a pathway to making policies that better address social determinants of health and that reduce inequalities worldwide. Broadening the reach of social policies that redistribute resources and alleviate poverty could prove especially effective, particularly given the relatively narrow scope of the welfare state in many countries. A social welfare perspective would argue the need for expanding these types of redistributive policies, given their benefits and the relatively small size of the social welfare state in some states (Wienk et al., 2022).

To make up for the government's inadequate responses to growing social problems, charity societies and voluntary organizations took on the task of dealing with urbanization, poverty, and immigration. All issues, such as healthcare and elder's and children's care, struggles against unemployment and poverty, woman's participation in working life, environmental protection, and gender discrimination, became issues that the state fought on behalf of through social policies. Instead of seeing social benefits as a means to assist those who otherwise might have lost social footing, public aid was framed as the support for maintaining a socially disruptive and morally reprehensible way of life (Suresh et al., 2020). Policymakers and stakeholders both in the public and private sectors who advocated greater state implementation of social policies generally did not structure their proposals around typical conceptions of welfare or benefits. Rather, in cases such as Medicare and Medicaid, years ago President Lyndon B. Johnson presented a package called The Great Society, which structured the larger vision around poverty and quality of life. Over the years, America's business community has looked toward the replacement for higher taxes and bigger government when it comes to social welfare promotion (Zelizer, 2015).

Social security, like other social insurance, is an example of universal programs, meaning that any citizens are entitled to participate in the program as a matter of social entitlement (Viola et al., 2018). In other words, participation in the social welfare programs is not based on economic necessity. Another definition of a welfare state is a modern understanding of the nation as having a responsibility to provide a just distribution of income; protect groups and classes that require protection; guide social security practices and employment policies; engage in politicization in order to provide for basic needs in society, such as education, health, and housing; and take measures for the regulation of the working life through taxation and wage policies it follows. There is a question, however, whether the welfare state is viable now and that the COVID-19 pandemic is nearly gone. Strong and popular concept in the European Union (EU) member states, the welfare state in the post-COVID era, faces serious challenges due to the massive spending and financial support of citizens, employees, and entrepreneurs during the pandemic years (Yousefi et al., 2021).

6.3 Social Policy and the State

In the twenty-first century which has been already marked by the threatening global climate change, the COVID-19 global pandemic, as well as the recent war in Ukraine, social policy becomes a complicated matter subject in every state to the local and state governments and supranational political influences. State and local efforts worldwide are addressing social welfare policies as well as the national programs (Adamczyk et al., 2022).

Given the number of social workers involved in health policy practice and research, these individuals may be of significant assistance in every phase of the health policy process (identifying problems, developing policy solutions, and implementing and evaluating policies). The preparation of these workers is a challenge as well. Students pursuing their studies in the area of social policy need to learn to work with instructors who are experts in poverty, labor markets, social demography, family structures, education inequality, political engagement, organizational dynamics, and healthcare, as well as the social welfare policy (Fajardo-Bullón et al., 2021). Graduates need to leave the study programs equipped with the tools needed to engage in policy advocacy to improve other lives. These programs need to provide students with the theories and models needed to organize communities, which can be used to mobilize individuals to support policy changes on the local level. They need to learn how to involve decision-makers to improve conditions for communities through the creation and implementation of policy changes. Federal, state, and local decision-makers must continue to incorporate social welfare and health into all policy strategies, doubling down on efforts like the national prevention strategy. A focus on values and the importance of social environments might indicate an urgency to such policies that emphasize efforts at basic needs protections and health-supportive environments (Testa & Kelly, 2020). Moving forward a social policy agenda reflective of the themes articulated here will therefore require social workers and public health practitioners to continue framing problems and developing policy solutions reflective of those values and working to foster a policy climate that is, in turn, conducive to those solutions (Roberts et al., 2018).

Social policy addresses how states and societies respond to the global challenges of social, demographic, and economic change, as well as poverty, migration, and globalization. Thence, social policy can be seen as the study of policy solutions to problems in education, inequality, poverty, crime, and other issues facing families and children in society (Sharma et al., 2020). Social programming is the process through which decision-makers, legislators, public institutions, planners, and, often, funding agencies seek to address public problems or improve conditions within communities by developing and implementing policies intended to achieve specific outcomes. Sometimes, the impetus for efforts at health and community development comes from social planners and policymakers (Guarnieri & Trojan, 2019). Once communities have expressed their needs, social workers can build awareness of these issues—educating everyone from legislators to nonprofit advocacy groups—and empower community members to push for policy reform. Social work advocacy can range from small actions affecting individuals to big programs designed to benefit whole communities and the larger community. Social policy is concerned with how societies around the world address the needs of individuals in terms of safety, education, employment, health, and well-being (McNeely et al., 2020). Social policies are designed to enhance human well-being and to address human needs for education, health, housing, and economic security. Social welfare policy which guides the government programs to help those who need help is built around the notion that government has the duty of protecting its citizens from harm (Camargo & Vázquez-Maguirre, 2021).

Public policies are designed for the well-being of individuals, addressing a broad set of areas, from education, health, and employment to crime and justice, which becomes very important in the post-COVID era. Important areas of social policies are welfare and well-being, poverty alleviation, welfare, equity, unemployment insurance, living conditions, animal rights, retirement, healthcare, public housing, family policy, social welfare, child protection, social isolation, education policy, crime and criminal justice, urban development, and labor issues. The entitlement programs (also known as the distributional social welfare initiatives) seek to provide benefits to all equally, with assistance based on taxes. According to Marxists, even when social policies do indeed improve working-class lives, the benefits are limited or cut off due to changes in the state and to the general capitalist program (Cepić et al., 2022). It is possible to argue that it is not just the economic indicators, but the changes taking place in the demography and the social structures, which have played their part in shaping welfare state policies. These changes are also different depending on social structures and the policies of a country. Social workers need to be able to identify problems with existing policies, identify ways in which these policies can be improved, and develop specific proposals for accomplishing that goal. The rights advocates have also claimed that mandatory changes made during the industrial revolution are valid even in our globalized and information technology age, which emerged during the latter part of the twentieth century, and that social policies should be developed by supranational organizations following a shift to the welfare state.

6.4 Public Policies Within the Context of Individual's Responsibility

Social welfare policies direct programs offering government services, along with financial and other aid, aimed at meeting particular needs. In the USA, state and local efforts besides the national programs have addressed social welfare policy. The federal government must establish minimum, uniform standards and guidelines for social welfare programs and must assume primary responsibility for funding programs designed to assist with basic needs for individuals and families. Federal, state, and local policymakers should continue to incorporate health into all policy strategies, doubling down on efforts like the national prevention strategy (Pramling Samuelsson et al., 2020). A focus on values and the importance of social environments will indicate an urgent need for these policies to focus efforts on the provision of essential needs and on promoting healthy, supporting environments.

Social work has developed a view of the relationship between policies and health, driven by a core mission to promote human well-being. This view is informed by social work training, which stresses systemic influences on behaviors, health, and well-being (Kaushik & Walsh, 2019). The need for public health, social work, and other fields of inquiry to identify and articulate more clearly the nature of the social structures and environments that contribute to poor health and health inequalities is also important. In line with this, the social work approach would recommend that we should promote policy that motivates and equips providers of health services to evaluate and make active use of social relationships in order to promote better health outcomes.

Each government has a principal role and ultimate responsibility in providing for the social progress and welfare of its people, in planning social development measures within a holistic development plan, in encouraging and coordinating or consolidating all domestic efforts toward that end, and in mandating the needed changes in the social fabric (Nugmanovna, 2021). Although the duties and spheres of responsibility for each social welfare state vary according to each country's social, cultural, economic, and demographic conditions, in general, they include providing for the protection of children, persons with disabilities, families, elderly persons, and women, creating employment, providing educational and vocational training, fighting against poverty and under-income, and improving working conditions. It is possible to create individual provisions on who needs protection in terms of social policies, for children and young people, the disabled, families on lower incomes, and elders. The broadest definition of social insurance includes both programs supported through special taxes and other programs (including tax credits) that provide income support; that help individuals obtain or afford necessities, such as food, shelter, and medical insurance; or that provide services or benefits that enhance economic opportunities, such as education and job training, and childcare (Gerard et al., 2020).

The social insurance system, which we broadly define to include both programs supported by dedicated taxes and other state programs that provide income support, help with essential needs, or services to improve economic opportunity, is a good example to illustrate all of the above. There are various initiatives for improving the social insurance system, including ways that different supports can be better targeted, ways that programs can be made more effective, and ways that the system can better reach the most vulnerable families and individuals both in times of economic hardship and during more normal times. Within context, some researchers define social policy as the body of policies developed by states for ensuring welfare so it can receive services and revenue. Another definition of a welfare state is it is a modern state understanding which assumes a responsibility of providing fair income distribution; protecting groups and classes that require protection; driving the social security practices and employment policies; practicing policies that address basic needs of the community, such as education, health, and housing; and taking measures for the regulation of the working life through taxation and wage policies that follow from the social welfare policies—which direct government programs to help those who are in need—which are built around the notion that government has the duty of protecting its citizens against harm (Roosma & Van Oorschot, 2020). Within this framework, the public authorities need to protect those affected with basic human needs—work, income if employment is unavailable, healthcare, food, education, childcare, and shelter—while protecting and expanding the ability of the government to respond, preserving and increasing funding for essential services, rather than depleting revenues with inappropriate tax cuts (Razavi et al., 2020). Quite often, it becomes necessary to oppose the budget and appropriation provisions that attempt to shift responsibility for government assistance programs from the governments of states to for-profit private corporations. The implementation and effects of reform efforts at the state level also need to be carefully monitored, ensuring benefits are provided where needed and recipients' civil rights are protected.

6.5 Conclusions

Overall, it can be concluded that if left unaddressed by policies, the social crises created by the COVID-19 outbreak could also exacerbate inequalities, exclusion, discrimination, and global unemployment over the medium to long term, thus threatening the concept of leadership. Viewed as diverse, with an emphasis on developed economies, this chapter highlighted both the shared trends and the cross-national variation of the public policy responses to the COVID-19 pandemic, which has had such negative economic consequences. This chapter analyzed the social problems within the context of leadership and public policies in the post-pandemic era highlighting the major social, economic, and financial impacts of COVID-19, along with recommendations for government policies.

In response to an ongoing outbreak of the COVID-19 virus, the public authorities have been quickly identifying and implementing waivers of local and state requirements, as well as making changes in policies and service delivery that will help address the needs of their clients and staff. Those authorities who managed to tackle the threat quickly became the true leaders of the post-pandemic recovery. During the first year of the COVID-19 pandemic, the government had no uniform approach regarding COVID-19 disparities; multiple countries, in different settings, developed or expanded innovative policies to respond on an emerging basis to the social needs and health disparities associated with the pandemic. Our scan of peer-reviewed literature identified several countries that developed more comprehensive and state-driven interventions that emerged early in the COVID-19 pandemic, addressing emerging health determinants of health and social needs. The dramatic disparities in healthcare policy responses to the COVID-19 pandemic have led to renewed health policy investments and interventions addressing social needs and social determinants of health in an effort to increase health equity.

It is crucial to keep the ongoing programs to support the implementation of institutional reforms that will enable successful leading responses to COVID-19. In order to do so, the contingency measures need to be taken to ensure public sector continuity, both nationally and at subnational levels, during and following the COVID-19 pandemic. To lead a successful COVID-19 response, governments will need to be swift, creative, agile, efficient, transparent, and accountable, just as the true leaders should be. In addition, addressing the COVID-19 pandemic requires an unprecedented level of collaboration among governments and citizens, and in all aspects of society, in order to achieve spatial distancing and other policy measures. From healthcare to social services, transportation to IT, the "new normal" is coming, and many governments are going to have to take on responsibility related to monitoring and managing the pandemic. The new challenges, nature of responsibility, and shape and conditions of working in pandemic-dominated realities underscore an urgent need for new expertise that, however, has no prospect of being developed through the traditional, off-line approach. Past experiences with the public responses to disasters, both natural and man-made, including health emergencies such as Ebola, offer good lessons for the stakeholders facing the COVID-19 pandemic. While most governments adopted measures to protect workers from labor market shocks related to the COVID-19 pandemic, there was a great deal of variation in responses and the policy packages implemented across countries. In response to the economic impacts of the pandemic, many governments introduced labor market legislation to protect workers against reduced hours and lost income. Governments invested heavily in employment protection schemes to curb unemployment. In response to the COVID-19 crisis, most OECD countries took proactive measures to expand existing short-time work (STW) schemes, to introduce new ones, or to establish temporary salary subsidies to save jobs and maintain incomes. During the early stage of the crisis, while many OECD countries were locked down, severe cuts to mobility and social proximity caused by the crisis itself stranded many sectors, especially in the case of the low-educated workers who were concentrating in precarious jobs across the EU. The second wave of coronavirus has hit temporary workers in particular hard, with the impact of job creation being limited, even temporarily, by the COVID-19 pandemic. The effect on hours worked of people working temporary contracts was significant and was mostly concentrated in the destruction of jobs—especially among low-educated workers.

During the period when the pandemic was spreading across the globe, work from home (WFH) transformed traditional working practices and transformed into the future of potential employment. Previous research has suggested that telework may simultaneously decrease the conflict between home and work and also increase it, negating the benefits of WFH. Furthermore, confinement, which was enforced during the initial phase of the COVID-19 pandemic, could have exacerbated such a conflict, as home obligations impeded employment obligations. Workplace isolation is another crucial critical characteristic of WFH in the pandemic. Although prior studies highlighted social isolation as a major disadvantage of remote work, it has unavoidably increased in incidence in this time.

The pandemic has increased the trend toward employers playing an expanded role in the financial, physical, and psychological well-being of their employees. Beyond pay, the promotion of healthy workplaces and prevention of workplace accidents and illnesses will likely get increased attention as part of the crisis itself.

All in all, the public sector has been the focus of efforts in combating the COVID-19 virus, from mandating social distancing to constructing hospitals, working with industries to supply needed medical equipment, and providing economic relief for affected individuals. At the same time, the EU member states are taking measures to mitigate the socioeconomic impacts of the virus and to support the recovery. To support the recovery of EU citizens, businesses, and countries from the economic decline caused by the virus, EU leaders agreed to work on Europe's Recovery Plan. EU leaders committed to establishing an EU Recovery Fund aimed at mitigating the effects of the economic recession which is a good example of the leadership action. Subsidies for electrical power and telecom services to businesses that were more affected by the pandemic, which had already received state rent or mortgage support, and to those businesses whose workers were on temporary work stoppages or under temporary short-term employment arrangements can help a lot in this situation.

As the pandemic has slowed down now, many countries already eased their requirements to access unemployment benefits and rental assistance to individuals whose economic and social situations have deteriorated as a result of the COVID-19 pandemic. However, with the new emerging challenges and threats, such as the war in Ukraine, the energy crisis, and the everlasting global climate change, social problems are still present, and there is a need for the public policies to tackle them using the leadership approaches and solutions.

References

Adamczyk, A., Trojanowska-Strzęboszewska, M., Kowalewska, D., & Bartłomiejski, R. (2022). Ukrainian migrants in Poland and the role of an employer as the channel of information during the COVID-19 pandemic. *Sustainability, 14*(9), 5338. https://doi.org/10.3390/su14095338

Alderwick, H., Hutchings, A., Briggs, A., & Mays, N. (2021). The impacts of collaboration between local health care and non-health care organizations and factors shaping how they work: A systematic review of reviews. *BMC Public Health, 21*(1), 1–16. https://doi.org/10.1186/s12889-021-10630-1

Ayuningtyas, D., Haq, H. U., Utami, R. R. M., & Susilia, S. (2021). Requestioning the Indonesia Government's public policy response to the COVID-19 pandemic: Black box analysis for the period of January-July 2020. *Frontiers in Public Health, 9*, 612994. https://doi.org/10.3389/fpubh.2021.612994

Benavides, A. D., & Nukpezah, J. A. (2020). How local governments are caring for the homeless during the COVID-19 pandemic. *The American Review of Public Administration, 50*(6–7), 650–657. https://doi.org/10.1177/0275074020942062

Čajka, P., Jaroszewicz, M., & Strielkowski, W. (2014). Migration incentives and flows between Belarus, Moldova, Ukraine and the European union: A forecasting model. *Economics & Sociology, 7*(4), 11–25. https://doi.org/10.14254/2071-789X.2014/7-4/1

Camargo, B. A., & Vázquez-Maguirre, M. (2021). Humanism, dignity and indigenous justice: The Mayan train megaproject, Mexico. *Journal of Sustainable Tourism, 29*(2–3), 372–391. https://doi.org/10.1080/09669582.2020.1758707

Cepić, D., Doolan, K., & Dolenec, D. (2022). Class analysis as systemic critique: A historical case study. *Communist and Post-Communist Studies, 55*(2), 104–119. https://doi.org/10.1525/j.postcomstud.2022.55.2.104

Cowie, H., & Myers, C. A. (2021). The impact of the COVID-19 pandemic on the mental health and well-being of children and young people. *Children & Society, 35*(1), 62–74. https://doi.org/10.1111/chso.12430

Cyr, J., Bianchi, M., González, L., & Perini, A. (2021). Governing a pandemic: Assessing the role of collaboration on Latin American responses to the COVID-19 crisis. *Journal of Politics in Latin America, 13*(3), 290–327. https://doi.org/10.1177/1866802X211049250

De Jesus, M., Moumni, Z., Sougui, Z. H., Biswas, N., Kubicz, R., & Pourtau, L. (2022). "Living in confinement, stopped in time": Migrant social vulnerability, coping and health during the COVID-19 pandemic lockdown in France. *International Journal of Environmental Research and Public Health, 19*(16), 10084. https://doi.org/10.3390/ijerph191610084

Edling, N. (Ed.). (2019). *The changing meanings of the welfare state: Histories of a key concept in the Nordic countries* (1st ed.). Berghahn Books.

Elliott, S., Satterfield, S. J., Solorzano, G., Bowen, S., Hardison-Moody, A., & Williams, L. (2021). Disenfranchised: How lower income mothers navigated the social safety net during the COVID-19 pandemic. *Socius, 7*, 23780231211031690. https://doi.org/10.1177/23780231211031690

Fajardo-Bullón, F., Pérez-Mayo, J., & Esnaola, I. (2021). The Association of Interpersonal Relationships and Social Services with the self-rated health of Spanish homelessness. *International Journal of Environmental Research and Public Health, 18*(17), 9392. https://doi.org/10.3390/ijerph18179392

Fischer, J., Frisina Doetter, L., & Rothgang, H. (2022). Comparing long-term care systems: A multi-dimensional, actor-centred typology. *Social Policy & Administration, 56*(1), 33–47. https://doi.org/10.1111/spol.12742

Gerard, F., Imbert, C., & Orkin, K. (2020). Social protection response to the COVID-19 crisis: Options for developing countries. *Oxford Review of Economic Policy, 36*(Supplement_1), S281–S296. https://doi.org/10.1093/oxrep/graa026

Guarnieri, P., & Trojan, F. (2019). Decision making on supplier selection based on social, ethical, and environmental criteria: A study in the textile industry. *Resources, Conservation and Recycling, 141*, 347–361. https://doi.org/10.1016/j.resconrec.2018.10.023

Guliyeva, A., Rzayeva, U., & Abdulova, A. (2020). Impact of information technologies on HR effectiveness: A case of Azerbaijan. *International Journal of Advanced Computer Science and Applications, 11*(2), 81–89. https://doi.org/10.14569/IJACSA.2020.0110212

Haesler, S., Schmid, S., Vierneisel, A. S., & Reuter, C. (2021). Stronger together: How neighborhood groups build up a virtual network during the COVID-19 pandemic. *Proceedings of the ACM on Human-Computer Interaction, 5*(CSCW2), 1–31. https://doi.org/10.1145/3476045

Hujo, K. (2021). Social protection and inequality in the global south: Politics, actors and institutions. *Critical Social Policy, 41*(3), 343–363. https://doi.org/10.1177/02610183211009899

Kaushik, V., & Walsh, C. A. (2019). Pragmatism as a research paradigm and its implications for social work research. *Social Sciences, 8*(9), 255. https://doi.org/10.3390/socsci8090255

Kelly, P., & Lobao, L. (2021). Whose need matters?: The local welfare state, poverty, and variation in US Counties' social service provisioning. *Social Currents, 8*(6), 566–590. https://doi.org/10.1177/23294965211047886

Leach, M., MacGregor, H., Scoones, I., & Wilkinson, A. (2021). Post-pandemic transformations: How and why COVID-19 requires us to rethink development. *World Development, 138*, 105233. https://doi.org/10.1016/j.worlddev.2020.105233

Leal Filho, W., Price, E., Wall, T., Shiel, C., Azeiteiro, U. M., Mifsud, M., & LeVasseur, T. J. (2021). COVID-19: The impact of a global crisis on sustainable development teaching. *Environment, Development and Sustainability, 23*(8), 11257–11278. https://doi.org/10.1007/s10668-020-01107-z

Long, E., Patterson, S., Maxwell, K., Blake, C., Pérez, R. B., Lewis, R., & Mitchell, K. R. (2022). COVID-19 pandemic and its impact on social relationships and health. *Journal of Epidemiol Community Health, 76*(2), 128–132. https://doi.org/10.1136/jech-2021-216690

McNeely, C. L., Schintler, L. A., & Stabile, B. (2020). Social determinants and COVID-19 disparities: Differential pandemic effects and dynamics. *World Medical & Health Policy, 12*(3), 206–217. https://doi.org/10.1002/wmh3.370

Morley, C., Le, C., & Briskman, L. (2020). The role of critical social work education in improving ethical practice with refugees and asylum seekers. *Social Work Education, 39*(4), 403–416. https://doi.org/10.1080/02615479.2019.1663812

Nugmanovna, M. A. (2021). Action strategy-the basis of a new stage of national development of Uzbekistan. *Academicia: An International Multidisciplinary Research Journal, 11*(1), 774–785. https://doi.org/10.5958/2249-7137.2021.00241.X

Pramling Samuelsson, I., Wagner, J. T., & Eriksen Ødegaard, E. (2020). The coronavirus pandemic and lessons learned in preschools in Norway, Sweden and the United States: OMEP policy forum. *International Journal of Early Childhood, 52*(2), 129–144. https://doi.org/10.1007/s13158-020-00267-3

Razavi, S., Behrendt, C., Bierbaum, M., Orton, I., & Tessier, L. (2020). Reinvigorating the social contract and strengthening social cohesion: Social protection responses to COVID-19. *International Social Security Review, 73*(3), 55–80. https://doi.org/10.1111/issr.12245

Roberts, C., Geels, F. W., Lockwood, M., Newell, P., Schmitz, H., Turnheim, B., & Jordan, A. (2018). The politics of accelerating low-carbon transitions: Towards a new research agenda. *Energy Research & Social Science, 44*, 304–311. https://doi.org/10.1016/j.erss.2018.06.001

Roosma, F., & Van Oorschot, W. (2020). Public opinion on basic income: Mapping European support for a radical alternative for welfare provision. *Journal of European Social Policy, 30*(2), 190–205. https://doi.org/10.1177/0958928719882827

Semplici, S. (2022). Ethics in emergency times: The case of COVID-19. *Philosophies, 7*(3), 70. https://doi.org/10.3390/philosophies7030070

Sharma, G. D., Talan, G., & Jain, M. (2020). Policy response to the economic challenge from COVID-19 in India: A qualitative enquiry. *Journal of Public Affairs, 20*(4), e2206. https://doi.org/10.1002/pa.2206

Suresh, R., James, J., & RSj, B. (2020). Migrant workers at crossroads-the COVID-19 pandemic and the migrant experience in India. *Social Work in Public Health, 35*(7), 633–643. https://doi.org/10.1080/19371918.2020.1808552

Testa, M. F., & Kelly, D. (2020). The evolution of federal child welfare policy through the family first prevention services act of 2018: Opportunities, barriers, and unintended consequences. *The Annals of the American Academy of Political and Social Science, 692*(1), 68–96. https://doi.org/10.1177/0002716220976528

Viola, E., Biondo, E., & Mosso, C. O. (2018). The role of social worker in promoting immigrants' integration. *Social Work in Public Health, 33*(7–8), 483–496. https://doi.org/10.1080/19371918.2018.1546252

Wienk, M. N., Buttrick, N. R., & Oishi, S. (2022). The social psychology of economic inequality, redistribution, and subjective Well-being. *European Review of Social Psychology, 33*(1), 45–80. https://doi.org/10.1080/10463283.2021.1955458

Yadav, U. N., Rayamajhee, B., Mistry, S. K., Parsekar, S. S., & Mishra, S. K. (2020). A syndemic perspective on the management of non-communicable diseases amid the COVID-19 pandemic in low-and middle-income countries. *Frontiers in Public Health, 8*, 508. https://doi.org/10.3389/fpubh.2020.00508

Ye, Q., Ozbay, K., Zuo, F., & Chen, X. (2021). Impact of social media on travel behaviors during the COVID-19 pandemic: Evidence from New York City. *Transportation Research Record*, 03611981211033857. https://doi.org/10.1177/03611981211033857

Yousefi, H., Karami, J., & Keshishyan, G. (2021). International security in the post-Covid world. *International Journal of Political Science, 11*(4), 1–15.

Zelizer, J. E. (2015). *The fierce urgency of now: Lyndon Johnson, congress, and the battle for the great society* (1st ed.). Penguin.

Chapter 7
Social and Economic Stability of the State in the Post-COVID Era: The Evolution of Theoretical Approaches and Leadership Practices

Vitaly Kaftan and Igor Molodtsov

Abstract This chapter deals with the factors of the social and economic stability of the state and state governance in the post-COVID era. In particular, it focuses on the evolution of theoretical approaches and leadership practices. After the COVID-19 pandemic, the traditional globalization has ended, and the so-called new globalization representing the new era of international economy and international relations has already started. This chapter presents a more detailed examination of economic globalization over the past 20 years and provides the framework for understanding the implications of the trade tensions between some global players. In addition, it describes the implications stemming from the shutdown caused by the COVID-19 pandemic for the whole world's economic order. Specifically, this chapter extends the timeline to explore the rise and decline of economic globalization prior to and following the global financial crisis in 2008, and it examines the specific impacts of the COVID-19 pandemic within the historical process. Furthermore, we argue that in this new era of altered globalization, the leading role of Brazil, Russia, India, China, and South Africa (BRICS countries) will be rising and that these countries will be increasingly more responsible for promoting the Sustainable Development Goals (SDGs) and contributing to the stability of the global economic and social systems as the leaders of the post-COVID era. In addition, we reviewed a number of publications related to COVID-19 control and the effects of COVID-19 in the areas of society, economy, and politics. While efforts are underway to develop drug interventions for the coronavirus pandemic, the science of socio-behavioral behavior may offer valuable insights to managing the global pandemic and its impacts. We discuss topics broadly related to multiple phases of the current pandemic, in order to help decision-makers, leaders, and the public better understand how to manage

V. Kaftan (✉) · I. Molodtsov
Department of Mass Communications and Media Business, Financial University under the Government of the Russian Federation, Moscow, Russian Federation
e-mail: vvkaftan@fa.ru; inmolodtsov@fa.ru

W. Strielkowski (ed.), *Leadership, Entrepreneurship and Sustainable Development Post COVID-19*, Springer Proceedings in Business and Economics, https://doi.org/10.1007/978-3-031-28131-0_7

threats, navigate diverse social and cultural settings, enhance scientific communications, align individual and collective interests, engage in effective leadership, and provide social and emotional support for the citizens.

Keywords Socioeconomic stability · Leadership · Post-COVID-19 era · New globalization

JEL Classification E63 · F60 · H70

7.1 Introduction

It has now become apparent that COVID-19, quite similar to the past disease outbreaks, has repeatedly shown underlying causes and consequences of environmental breakdowns and mirror structural social and political-economic conditions. Given the interdependence of economic and social systems, the COVID-19 crisis has also highlighted the need to enhance international cooperation on evidence-based responses to systematic threats, building upon existing frameworks of emergency preparedness (Zabaniotou, 2020). Many authors noted the consequences of failing to appreciate and address the features of complex global systems, and challenges could be enormous (Hofman et al., 2021; Kesavan et al., 2022; Wickert et al., 2021). Given the interdependence of economic and social systems, the COVID-19 crisis also highlights the need for strengthened international cooperation (building on existing frameworks for emergency preparedness) based on evidence, to tackle systemic threats and help avert systemic collapse (de León et al., 2021).

It is critical to have a framework in place to respond effectively and efficiently to major shocks, like a pandemic. The COVID-19 pandemic is a means to cope with other emergencies, like climate change. A systems-based resilience approach is proposed to prepare socioeconomic systems for future shocks. Such characteristics are highlighted in the economic responses to COVID-19 and other severe shocks, suggesting the potential for a more sustainable, post-capitalist alternative (Butler et al., 2021; Čábelková et al., 2022). Enhanced approaches could be beneficial to the small and medium enterprises (SMEs) and large entrepreneurs practicing a streamlined approach, particularly in critical periods of distress like the COVID-19 pandemic, and within an emerging economy context like the one emphasized in the present study. Entrepreneurship could offer such requirements and serves as a resilience approach to addressing socioeconomic impacts of COVID-19 pandemic on entrepreneurial enterprises all around the world, both in the developed and in the developing countries (Khlystova et al., 2022; Korneeva et al., 2021; Strielkowski et al., 2021). It might represent a means of curbing waste around social-economic losses within complex challenges arising from the COVID-19 pandemic onset within SMEs, which is the sector that is under study. Reckoning has been inclined to generate a demand among researchers and practitioners in lean enterprise research and practice to constantly seek approaches that could deal with the spatial implications of savage ecological challenges, such as the socioeconomic shocks arising from the COVID-19 pandemic (Baral, 2021). Suleimany et al. (2022) suggest public education, the provision of appropriate technologies, and public assessment of the social-economic impacts of COVID-19 pandemic. Entrepreneurship practitioners across different sectors are facing a crucial challenge of dealing with social-economic disruptions and other impacts from COVID-19 pandemic (Gregurec et al., 2021). The safety measures adopted to deal with the recent COVID-19 pandemic had a variety of effects on individuals, depending on the role played by society.

The COVID-19 crisis was the first global pandemic in the twenty-first century, leading to a crisis of considerable loss of life, but also to a large part of the world economy, at high social costs. It showed that we were facing an uncertain future, in which anticipating and resisting large shocks should be a central challenge for research and development practice (Strielkowski et al., 2022). There is a pressing need to explore how COVID-19—as both a health and development crisis—emerges in the ways that COVID-19 has and to think about possibilities for post-pandemic transformations and a more general redesign of development. Development—as conceived of progressive social, economic, and political changes—is being quickly reversed, with COVID-19 threatening collective futures (Habicher et al., 2022).

It is expected that governments that perform better in managing a crisis would do so economically, too, that is, that a good management of a pandemic in healthcare is beneficial for a country's society and economy (Guliyeva & Rzayeva, 2018). Although ideological differences in the months of the pandemic did not significantly influence policy decisions, early responses had greater advantages over rigid delays, that is, earlier, flexible actions for the control of COVID-19 had better results than delayed, rigid actions (Stephan et al., 2022). A good example of adaptation is the South Korean case of managing the Middle East respiratory syndrome (MERS) in 2015, in which, having overcame the pandemic, it implemented new policies and institutional changes anticipating future pandemics, which favored management of a novel coronavirus by early implementation of effective measures (Alnuqaydan et al., 2021). This increased focus may contribute to managing the present situation as well as other potential pandemics and outbreaks.

7.2 COVID-19, Poverty, Hunger, and Political Instability

The COVID-19 pandemic has had a devastating impact on hunger and poverty around the world—particularly for the poorest and most vulnerable populations. The pandemic increased global food insecurity in nearly all countries, by decreasing incomes and disrupting food supply chains. Supply chain disruptions from COVID-19 and increased consumer demand for food dramatically increased food prices around the world—exacerbating the severity of food insecurity among millions of people experiencing hunger in low-income countries, where they are already spending the majority of their incomes on food (Jaacks et al., 2021).

The pandemic erased years of progress toward poverty reduction and caused a sharp rise in economic inequality. Even before the war in Ukraine, recovery in emerging and developing economies following pandemic downturns was weak, returning the levels of inequality between rich and poor countries to levels last seen just a decade before, according to many studies on the related fields (Dreger, 2022). The contrasting economic recoveries meant that the COVID-19 crisis had directly counteracted falling inequality across countries. In contrast, the African countries are experiencing slower-than-average economic growth, so the COVID-19 effect may set back development by a few years. It is not possible yet to quantify the effect on global inequality, but modeling suggests that a mere 1% rise in intra-country inequality will lead 32 million more people living on less than $1.90 per day in 2021 and raise the gap between the income growth of the lowest-income 40% and the highest-income 40% by 4.7%, with no change in intra-country inequality (Wan et al., 2021). A total of 650 million people are thought to have been living in extreme poverty as of 2019, and given probable growth paths, the path to poverty is a stable decline across most countries and across aggregates (Hotez et al., 2021). In fact, some countries with higher levels of poverty may see their 2030 poverty numbers surpass their 2020 numbers. India is an especially interesting case, having large numbers of very vulnerable people who only recently emerged from poverty, coupled with a very large projected drop in economic growth. These include high levels of poverty for a high middle-income country, due in large part to extreme income inequality, meaning a significant portion of the population is far poorer than the average, and mass unemployment (Rasul et al., 2021). Many of the factors associated with high levels of poverty, income inequality, and hunger are intrinsic to the socioeconomic and political status quo. The problem of hunger in the world is partially the result of decades-long neoliberal policies resulting in a high rate of income inequality in many countries. As with several other countries around the world, the effects of the ongoing coronavirus pandemic, and its related closures, have significantly worsened the high levels of poverty and income inequality that were already present in many places (Bundervoet et al., 2022). The rapid destruction caused by the ongoing coronavirus pandemic has increased poverty levels by large numbers since 2020. The socioeconomic impact of the ongoing coronavirus pandemic, and its consequent shutdown, has significantly worsened poverty and income inequality all around the world. For example, poverty and extreme poverty in Latin America reached levels in 2020 that had not been seen for the past 12 years and 20 years, respectively, while the regions' inequality index worsened, alongside the rates of employment and labor force participation, especially among women, as a consequence of the COVID-19 pandemic, and in spite of the extraordinary social protection measures, countries adopted to stem this phenomenon (Bargain & Aminjonov, 2021). The COVID-19 pandemic has emerged amid complex economic, social, and political scenarios, of lower growth, increasing poverty and increasing social tensions.

The global financial crisis in 2008, a slow and uneven recovery, and now an economic shock caused by the pandemic of a novel coronavirus disease, COVID-19, deepened these trends and challenged decision-makers to respond. The coming years will be fraught, as countries in the United States and European Union Member States concentrate on domestic impacts from the COVID-19 pandemic, just as the disease is likely to spread to impoverished, war-torn countries. The devastation from the coronavirus pandemic will also have ramifications for inequality and social mobility over the longer term (Goudeau et al., 2021).

In addition, the implications for the poor children all around the world should also be considered. These children are more likely to experience consequences from COVID-19 outbreaks since the COVID-19 pandemic and its economic and social consequences will harm some children more than others. Poverty puts children at higher risk for exposure to the COVID-19 pandemic. On average, in the OECD countries, one in seven children grows up in poverty which constitutes a serious issue for today's globalized world (Aizer et al., 2022). The severity of problems in health, sanitation, household income, housing, and schools is especially acute for children from developing countries or from the poorest areas of countries with large income disparities. Poor countries are grappling with deeper, more sustained crises, which are increasing global poverty and upending recent trends toward declining inequality. Economic inequality within countries remains especially high in developing regions, home to roughly two-thirds of the world's extreme poor (Miguel & Mobarak, 2022). Extreme poverty has increased rapidly in lower-income countries, delaying progress for 8–9 years, and in middle-income countries, it has been delayed for 5–6 years. For most countries, though, economic harm may be longer-lasting, a real risk to families pushed under the poverty line. As is true of so many other countries across the globe, but particularly for the less wealthy and developing countries, the lasting effects of this challenging year will continue to cause economic difficulties to many ordinary, poor people long into the future.

7.3 COVID-19 and Structural Changes in Various Dimensions

As the coronavirus disaster has settled, it is now interesting to study the governments' responses aimed at addressing the disparate health outcomes, as well as coronavirus socioeconomic, cultural, and ecological impacts on individuals, communities, and industries. The trade-offs between economy and health led to a major debate about the most effective measures to contain the effects of COVID-19 (Alamineh, 2022). Thence, a profound and in-depth analysis needs to be done monitoring the factors that are likely to hamper the economic rebound due to the COVID-19 pandemic. Beyond the health and human tragedies caused by COVID-19, the widespread recognition today is that the pandemic has caused the worst economic crisis since the Second World War (WWII). The rapid global spread of COVID-19 has rapidly surpassed other recent epidemics both in scale and extent. In addition to its lethal human cost and destruction of millions of lives, economic harm is already substantial and widespread (Goldstein et al., 2021). In summary,

COVID-19 has intensified existing inequalities and disparities, as well as created new ones, while it also unmasked the key social needs and strengths. COVID-19 responses have affected different populations differently, frequently amplifying existing structural inequality in income and poverty, socioeconomic inequalities in education and skills, and generational inequality—with specific impacts on children (including vulnerable children), families with children, and youth (Sharma et al., 2021).

Subnational governments, regions, and municipalities are on the front lines in managing the crisis and rebuilding and are facing asymmetric impacts on health, economy, society, and finance from COVID-19 across countries, as well as across regions and local areas.

Among high-income countries, several European states, which are performing well on a variety of governance dimensions, as measured by the World Governance Index (WGI) Annual World Governance Index, are also doing a good job of responding to the COVID-19 pandemic (Alfano & Ercolano, 2022). Given the fact that supposedly developed, rich countries have been struggling, and are struggling, dealing with the effects of coronavirus, one has to wonder what impact this has had on developing countries in Latin America or Africa and on their capacity to respond to this crisis and protect vulnerable people and poor people. In the case of Black Death, exposure (along with the uncertainties surrounding the epidemiology of the virus and its impacts on public health and economic activities) has led to considerable variation in the methods adopted by various governments in dealing with COVID-19 infections and in the strictness of their preventive societal restrictions (Glatter & Finkelman, 2021). Social choices are further complicated by the fact that collective responses to novel pandemics, like that of the Black Death, are greatly affected by prevailing political systems and differing governing objectives. The governments' choice of policies for controlling the Black Death was substantially affected by political pressures, whose character changed as the Spanish flu pandemic continued. As the COVID-19 pandemic has progressed, some leaders may have ordered restrictive measures which made public health sense during the height of the crisis and then extended those measures hoping to quell dissent as the disease has receded. Looking forward, governments will need to decide whether they favor more collaborative approaches to managing the COVID-19 pandemic, not just in terms of global public health but as a policy and security challenge (Nichols et al., 2022). It is expected that governments that perform better at managing this crisis will do so in economic terms, too, namely, good healthcare management for COVID-19 is beneficial to a country's society and economy. In that respect, the pace and nature of the post-COVID-19 economic recovery will be hindered by the government's actions in controlling the pandemic, for example, multiple disruptions of supply chains as a consequence of lockdowns and reduced demand as a consequence of reduced activities in order to avoid infection. Such austerity measures, combined with other economic effects of the COVID-19 pandemic—such as a loss of tourism to regions heavily dependent on foreign visitors—could lead to economic disruptions lasting far beyond the immediate crisis, creating the possibility of long-lasting labor unrest and social instability. While COVID-19 has already placed short-term strains on healthcare and social spending, as well as various categories of income, more severe impacts are expected over the medium term. The pandemic's public health and economic impacts are likely to put pressure on the relationship between governments and citizens, particularly in cases of weakened healthcare services; maintaining public order may prove difficult as security forces are stretched thin, and populations grow increasingly disenchanted with government responses to the COVID-19 pandemic. The novel coronavirus pandemic will place tremendous strain on government institutions in severely affected countries—particularly in healthcare systems, but also on a range of other critical public functions, from education and food supply chains to law enforcement and border controls. There is an urgent need to explore how COVID-19—as both a health and development crisis—has played out the way that COVID-19 has and to think about opportunities for post-pandemic transformations, as well as a more general redesign of development. Some leaders might also view the COVID-19 pandemic as a cover for undertaking destabilizing foreign adventures, either to distract from domestic resentment or because they perceive that there will be little resistance amid a global health crisis (Gibson-Fall, 2021). Beyond that evocative experience of escaping risky isolation, they are looking deeper at policies for social policy measures and ideological maneuvering and making a case for proper social protection provisions that will enable migrants to escape from poverty and guarantee their health and security, in spite of billions of dollars that have been pledged by the governments to deal with the economic and social consequences of COVID-19.

7.4 The Role of BRICS in the Post-COVID-19 Era

With regard to all of the above, it appears relevant to pinpoint the increasingly important and leading role of Brazil, Russia, India, China, and South Africa (BRICS) countries. Due to the recent post-COVID-19 changes in the global economic and social system, BRICS will be increasingly more responsible for promoting the sustainable development tools and principles aimed at implementing the Sustainable Development Goals (SDGs) and therefore contributing to the stability of the new global economic and social system (Nair et al., 2021; Skivko et al., 2021).

Within this context, the role of digital economy as a major instrument to modernize and transform industries, promote inclusive economic growth, support global trade and smooth commercial operations, and thus contribute to BRICS domestic economies to meet the sustainable development goals also needs to be mentioned. We acknowledge that suitable instruments, which focus on creating an enabling investment environment, domestically, within BRICS, and internationally, can contribute to international trade, sustainable development, and inclusive growth (Awosusi et al., 2022). The BRICS countries also acknowledge the positive role trade and investments may play in contributing to sustainable development, domestic and regional industrialization, and transition to sustainable patterns of consumption and production.

BRICS economies have a significant role to play at the G20, shaping global economic policies and contributing to financial stability (Larionova & Shelepov, 2021). Having been victims of the global financial crisis, and suffering from the effects of high and volatile capital flows, as well as the so-called tourist dollar, the BRICS countries are being driven by the shared goal of reforming the international financial and monetary systems, with the firm wish of building a fairer, more balanced international order, reflecting the dynamics of the contemporary global economy, which serves everyone's interests fairly. Brazil, India, and South Africa all supported BRICS's common statements about the need for reforming the existing system for the last 20 years. Valuing the continued role of the G20 as a leading forum for international economic cooperation and for the coordination of actions for meeting current global challenges, we pledge to continue coordination and cooperation efforts in the G20 on issues of common interest among all the BRICS countries and to work toward the advancement in the G20 of the interests and priorities of emerging and developing economies, including updating the G20 Action Agenda (Cooper, 2021).

Going beyond the economy into politics, one need to carefully examine the factors that have shaped BRICS's formation, in order to discuss the potential impact that BRICS could have on international institutions and global governance. During the BRICS summit, the representative of South Africa reaffirmed their interest in working on a mechanism in the BRICS for local currency settlements and stressed that volatility in the developing countries' currency markets is not actually due to any dynamics within South Africa but rather because of dynamics within the world economy. The comments expressed not only South Africa's frustration with the lack of self-government to gain economic security on world markets but a shared frustration of developing countries both within BRICS groupings and outside. Therefore, BRICS countries need to implement various working mechanisms for ensuring its leading position within the post-COVID-19 era.

7.5 Conclusions

To sum our results and findings up, it can be stated that the economic impacts of the COVID-19 crisis vary between regions, depending on a region's exposure to the tradable sectors and the global value chains. In terms of trade and investment, the COVID-19 pandemic has hit at a time when manufacturing and investments are heavily transnational, with decentralized manufacturing processes spread throughout global value chains. The recovery in many countries has been constrained by a renewed COVID-19 pandemic, uneven vaccination, and a partial rollback in state economic support measures. Despite accelerating, global GDP levels are projected to fall well below pre-pandemic forecasts in 2021, and the GDP per capita in many emerging and developing economies is projected to remain below its pre-COVID-19 peak for a sustained period. Global growth is projected to accelerate to 5.6% in 2021, driven in large part by strength in large economies like the United States and China. If this is indeed the case, we will need to study the global economic order transition and the degree to which economic globalization has been decelerated due to the COVID-19 pandemic in greater detail, once economic data becomes available. In addition, it also becomes apparent that localization and regionalization fill a vacuum in the retreating globalization, providing us with a better picture, or at least plausible hypothesis, about the pandemic's potential longer-term effects on the global economic order. Therefore, we find that the political (number of treaties and membership of international organizations) dimension of globalization has had the largest effect among all the other subdimensions of globalization. This is a very surprising finding given the calls for international cooperation and coordination from a range of international organizations.

The territorial implications of COVID-19 mean that the COVID-19 crisis has left governments across the globe operating in an environment of acute uncertainty, faced with hard trade-offs given the health, economic, and societal challenges that COVID-19 poses. The COVID-19 pandemic has the potential to cause serious disruptions in vulnerable states, spark widespread disruptions, and seriously challenge international crisis management systems. The disastrous economic effects of the disease may occur regardless of whether a country in question has experienced major disease outbreaks, though the dangers will be amplified in those that have. The health, economic, political, and social crises now facing the world will leave health scars in international relations as well as the ongoing process of globalization itself. If health, with its attendant consequences for the economy, politics, and society, were of similar magnitude as in the crisis of 2008, then we expect changes to be similar to those observed for all variables and dimensions. The effect of external shocks to foreign direct investments (FDI) would depend on the reconfiguration of multinational production, as well as on its medium-term effects on international financial markets. Such austerity measures, combined with the other economic effects of the COVID-19 pandemic—such as a loss of tourism to regions heavily dependent on foreign visitors—could result in an economic shock lasting far beyond the immediate crisis, creating potential for long-lasting labor unrest and social instability. There is a consensus on the adverse economic effects that health will have on the world economy, in the short term as well as in the medium term. Pessimists see little hope for effective international cooperation within the present framework (underlying anarchy) of global governance, on the one hand, and they argue that COVID-19 would cause substantial, sustained structural damage to the global economic order, on the other hand. It also needs to be stressed that COVID-19 pandemic can also lead to destabilization because the global health crisis tends to push many important issues and disputes between countries on the back burner. Countries with higher levels of state efficiency, as well as policies and conditions that tend to promote or support globalization (e.g., trade policies, policy connectivity, and participation in international policy cooperation), are also less likely to implement restrictions on travel. These findings indicate that countries with higher levels of government effectiveness and greater globalized social, interpersonal, and cultural flows are less likely to enact policies implementing travel restrictions during a pandemic crisis and thus might postpone doing so. This is likely because it is relatively easier to execute domestic development programs in countries with greater globalization, because legally binding international travel and trade agreements and regulations, as well as the possibility of large economic losses, will also hamper implementation of international travel restriction policies compared with domestic plans for development and post-COVID recovery that would ensure building the leading position in the world's economy and politics.

Acknowledgments This chapter was prepared based on the results of research carried out at the expense of budgetary funds under the state assignment of the Financial University VTK-GZ-PI-40-22.

References

Aizer, A., Hoynes, H., & Lleras-Muney, A. (2022). Children and the US social safety net: Balancing disincentives for adults and benefits for children. *Journal of Economic Perspectives, 36*(2), 149–174. https://doi.org/10.1257/jep.36.2.149

Alamineh, G. A. (2022). The nexus between coronavirus and tourism: Tourism as peace sensitive industry. *Cogent Arts & Humanities, 9*(1), 2014110. https://doi.org/10.1080/23311983.2021.2014110

Alfano, V., & Ercolano, S. (2022). Stay at home! Governance quality and effectiveness of lockdown. *Social Indicators Research, 159*(1), 101–123. https://doi.org/10.1007/s11205-021-02742-3

Alnuqaydan, A. M., Almutary, A. G., Sukamaran, A., Yang, B. T. W., Lee, X. T., Lim, W. X., & Chellappan, D. K. (2021). Middle East Respiratory Syndrome (MERS) virus-pathophysiological axis and the current treatment strategies. *American Association of Pharmaceutical Scientists, 22*(5), 1–23. https://doi.org/10.1208/s12249-021-02062-2

Awosusi, A. A., Adebayo, T. S., Altuntaş, M., Agyekum, E. B., Zawbaa, H. M., & Kamel, S. (2022). The dynamic impact of biomass and natural resources on ecological footprint in BRICS economies: A quantile regression evidence. *Energy Reports, 8*, 1979–1994. https://doi.org/10.1016/j.egyr.2022.01.022

Baral, P. (2021). Health systems and services during COVID-19: Lessons and evidence from previous crises: A rapid scoping review to inform the United Nations research roadmap for the COVID-19 recovery. *International Journal of Health Services, 51*(4), 474–493. https://doi.org/10.1177/0020731421997088

Bargain, O., & Aminjonov, U. (2021). Poverty and COVID-19 in Africa and Latin America. *World Development, 142*, 105422. https://doi.org/10.1016/j.worlddev.2021.105422

Bundervoet, T., Dávalos, M. E., & Garcia, N. (2022). The short-term impacts of COVID-19 on households in developing countries: An overview based on a harmonized dataset of high-frequency surveys. *World Development, 153*, 105844. https://doi.org/10.1016/j.worlddev.2022.105844

Butler, J. R., Davila, F., Alders, R., Bourke, R. M., Crimp, S., McCarthy, J., & Walker, D. (2021). A rapid assessment framework for food system shocks: Lessons learned from COVID-19 in the Indo-Pacific region. *Environmental Science & Policy, 117*, 34–45. https://doi.org/10.1016/j.envsci.2020.12.011

Čábelková, I., Smutka, L., & Strielkowski, W. (2022). Public support for sustainable development and environmental policy: A case of the Czech Republic. *Sustainable Development, 30*(1), 110–126. https://doi.org/10.1002/sd.2232

Cooper, A. F. (2021). China, India and the pattern of G20/BRICS engagement: Differentiated ambivalence between 'rising' power status and solidarity with the Global South. *Third World Quarterly, 42*(9), 1945–1962. https://doi.org/10.1080/01436597.2020.1829464

de León, E. A., Shriwise, A., Tomson, G., Morton, S., Lemos, D. S., Menne, B., & Dooris, M. (2021). Beyond building back better: Imagining a future for human and planetary health. *The Lancet Planetary Health, 5*(11), e827–e839. https://doi.org/10.1016/S2542-5196(21)00262-X

Dreger, C. (2022). Economic impact of the Corona pandemic: Costs and the recovery after the crisis. *Asia and the Global Economy, 2*(1), 100030. https://doi.org/10.1016/j.aglobe.2022.100030

Gibson-Fall, F. (2021). Military responses to COVID-19, emerging trends in global civil-military engagements. *Review of International Studies, 47*(2), 155–170. https://doi.org/10.1017/S0260210521000048

Glatter, K. A., & Finkelman, P. (2021). History of the plague: An ancient pandemic for the age of COVID-19. *The American Journal of Medicine, 134*(2), 176–181. https://doi.org/10.1016/j.amjmed.2020.08.019

Goldstein, I., Koijen, R. S., & Mueller, H. M. (2021). COVID-19 and its impact on financial markets and the real economy. *The Review of Financial Studies, 34*(11), 5135–5148. https://doi.org/10.1093/rfs/hhab085

Goudeau, S., Sanrey, C., Stanczak, A., Manstead, A., & Darnon, C. (2021). Why lockdown and distance learning during the COVID-19 pandemic are likely to increase the social class achievement gap. *Nature Human Behaviour, 5*(10), 1273–1281. https://doi.org/10.1038/s41562-021-01212-7

Gregurec, I., Tomičić Furjan, M., & Tomičić-Pupek, K. (2021). The impact of COVID-19 on sustainable business models in SMEs. *Sustainability, 13*(3), 1098. https://doi.org/10.3390/su13031098

Guliyeva, A., & Rzayeva, U. (2018). The role of E-government in reduction of information asymmetry in developing countries on the example of Azerbaijan. *Advances in Economics and Business, 6*(4), 209–217. https://doi.org/10.13189/aeb.2018.060402

Habicher, D., Windegger, F., Heiko, A., & Pechlaner, H. (2022). Beyond the COVID-19 crisis: A research note on post-pandemic scenarios for South Tyrol 2030+. *Technological Forecasting and Social Change, 180*, 121749. https://doi.org/10.1016/j.techfore.2022.121749

Hofman, J. M., Watts, D. J., Athey, S., Garip, F., Griffiths, T. L., Kleinberg, J., & Yarkoni, T. (2021). Integrating explanation and prediction in computational social science. *Nature, 595*(7866), 181–188. https://doi.org/10.1038/s41586-021-03659-0

Hotez, P. J., Fenwick, A., & Molyneux, D. (2021). The new COVID-19 poor and the neglected tropical diseases resurgence. *Infectious Diseases of Poverty, 10*(1), 1–3. https://doi.org/10.1186/s40249-020-00784-2

Jaacks, L. M., Veluguri, D., Serupally, R., Roy, A., Prabhakaran, P., & Ramanjaneyulu, G. V. (2021). Impact of the COVID-19 pandemic on agricultural production, livelihoods, and food security in India: Baseline results of a phone survey. *Food Security, 13*(5), 1323–1339. https://doi.org/10.1007/s12571-021-01164-w

Kesavan, N., Seenuvasaragavan, S., Burnette, K., Heim, A. B., Patel, R. J., Johnston, J., Cao, X., Sakandar, H. A., Zhuang, Y., Oi, K. K., Zhi, Y., Diao, Y., & Strielkowski, W. (2022). Pandemic-inspired policies. *Science, 377*(6601), 22–24. https://doi.org/10.1126/science.add2046

Khlystova, O., Kalyuzhnova, Y., & Belitski, M. (2022). The impact of the COVID-19 pandemic on the creative industries: A literature review and future research agenda. *Journal of Business Research, 139*, 1192–1210. https://doi.org/10.1016/j.jbusres.2021.09.062

Korneeva, E., Skornichenko, N., & Oruch, T. (2021). Small business and its place in promoting sustainable development. *E3S Web of Conferences, 250*, 06007. https://doi.org/10.1051/e3sconf/202125006007

Larionova, M., & Shelepov, A. (2021). BRICS, G20 and global economic governance reform. *International Political Science Review, 43*(4), 512–530. https://doi.org/10.1177/01925121211035122

Miguel, E., & Mobarak, A. M. (2022). The economics of the COVID-19 pandemic in poor countries. *Annual Review of Economics, 14*, 253–285. https://doi.org/10.1146/annurev-economics-051520-025412

Nair, R., Viswanathan, P. K., & Bastian, B. L. (2021). Reprioritising sustainable development Goals in the Post-COVID-19 Global context: Will a mandatory corporate social responsibility regime help? *Administrative Sciences, 11*(4), 150. https://doi.org/10.3390/admsci11040150

Nichols, C. M., Ewing, E. T., Gaston, K. H., Marinari, M., Lessoff, A., & Huyssen, D. (2022). What came next?: Reflections on the aftermath (s) of the 1918–19 flu pandemic in the age of COVID. *The Journal of the Gilded Age and Progressive Era, 21*(2), 111–149. https://doi.org/10.1017/S1537781421000682

Rasul, G., Nepal, A. K., Hussain, A., Maharjan, A., Joshi, S., Lama, A., Gurung, P., Ahmad, F., Mishra, A., & Sharma, E. (2021). Socio-economic implications of COVID-19 pandemic in South Asia: Emerging risks and growing challenges. *Frontiers in Sociology, 6*, 629693. https://doi.org/10.3389/fsoc.2021.629693

Sharma, S., Walton, M., & Manning, S. (2021). Social determinants of health influencing the New Zealand COVID-19 response and recovery: A scoping review and causal loop diagram. *System, 9*(3), 52. https://doi.org/10.3390/systems9030052

Skivko, M., Korneeva, E., & Korableva, O. (2021). From sustainable habits to sustainable energy use: Social sustainability in the context of digitization. *E3S Web of Conferences, 250*, 06002. https://doi.org/10.1051/e3sconf/202125006002

Stephan, U., Zbierowski, P., Pérez-Luño, A., Wach, D., Wiklund, J., Alba Cabañas, M., & Zahid, M. M. (2022). Act or wait-and-see? Adversity, agility, and entrepreneur wellbeing across countries during the COVID-19 pandemic. *Entrepreneurship Theory and Practice*, 10422587221104820. https://doi.org/10.1177/10422587221104820

Strielkowski, W., Guliyeva, A., Rzayeva, U., Korneeva, E., & Sherstobitova, A. (2021). Mathematical modeling of intellectual capital and business efficiency of small and medium enterprises. *Mathematics, 9*(18), 2305. https://doi.org/10.3390/math9182305

Strielkowski, W., Zenchenko, S., Tarasova, A., & Radyukova, Y. (2022). Management of smart and sustainable cities in the post-COVID-19 era: Lessons and implications. *Sustainability, 14*(12), 7267. https://doi.org/10.3390/su14127267

Suleimany, M., Mokhtarzadeh, S., & Sharifi, A. (2022). Community resilience to pandemics: An assessment framework developed based on the review of COVID-19 literature. *International Journal of Disaster Risk Reduction, 80*, 103248. https://doi.org/10.1016/j.ijdrr.2022.103248

Wan, G., Wang, C., & Zhang, X. (2021). The poverty-growth-inequality triangle: Asia 1960s to 2010s. *Social Indicators Research, 153*(3), 795–822. https://doi.org/10.1007/s11205-020-02521-6

Wickert, C., Post, C., Doh, J. P., Prescott, J. E., & Prencipe, A. (2021). Management research that makes a difference: Broadening the meaning of impact. *Journal of Management Studies, 58*(2), 297–320. https://doi.org/10.1111/joms.12666

Zabaniotou, A. (2020). A systemic approach to resilience and ecological sustainability during the COVID-19 pandemic: Human, societal, and ecological health as a system-wide emergent property in the anthropocene. *Global Transitions, 2*, 116–126. https://doi.org/10.1016/j.glt.2020.06.002

Chapter 8
Institutional Problems of Leadership Development in the Developing Economies: A Case of Azerbaijan

Afet Abbasova, Ali Agha Ismayilzada, and Kamala Jabbarova

Abstract The current stage of economic development is characterized by the continuous growth of the role of human resources and their leadership qualities in the production process. The ever-increasing level of the vocational, general, educational, and cultural level of employees increases their motivation (interest) in the results of production, in its economic and social development. Until the 1960s, the employee was considered as an "economic person" (Adam Smith) and then after that at the same time as a "social person" (Frederick Herzberg), that is, who loves his work, is ready to do it on his own, demonstrating high consciousness and devotion, interested in the results of the activities of his enterprise, and aware of the social responsibility of this activity. These are the qualities that define the essence of leadership. They are especially pronounced in critical periods of the development of society. During the years of the pandemic, when the production activity of people was limited, the potential of these qualities has increased significantly. In the aftermath of the COVID-19 pandemic, to accelerate economic growth, this energy must be allowed to fully manifest. The experience of foreign countries and our study shows that all employees have leadership qualities; they are organically inherent in every person. For their manifestation, it is necessary to use appropriate methods. In our opinion, procedural factors are the most suitable for these purposes. In countries with developing market economies, the main emphasis in stimulating the personnel of enterprises is placed on the use of content factors. These factors are based on rigid, administrative methods of management that limit the possibility of realizing the potential abilities of the staff. While procedural factors imply an impact on the cognitive prerequisites of employees, the behavior of employees acts as the implementation of actions aimed at achieving their expectations. The main reason for this approach to stimulating leadership qualities in developing countries is the lack of relevant regulations in them. In our opinion, in order to create objective conditions

---

A. Abbasova (✉) · A. A. Ismayilzada · K. Jabbarova
Azerbaijan State University of Economics (UNEC), Baku, Azerbaijan
e-mail: afetabbasova@unec.edu.az; Aliaga_Ismayilzada@unec.edu.az; kamala_jabbarova@unec.edu.az

W. Strielkowski (ed.), *Leadership, Entrepreneurship and Sustainable Development Post COVID-19*, Springer Proceedings in Business and Economics, https://doi.org/10.1007/978-3-031-28131-0_8

for stimulating the leadership qualities of personnel, it is necessary to adopt relevant regulations. Based on our results, it becomes clear that the leadership motivation can provide a significant impetus to the effective business development in the emerging economies.

Keywords Leadership · Stimulation · Personnel · Developing economies

JEL Classification F63 · M12 · O10

8.1 Introduction

The world around us is in the continuous development and change as the globalization is going on and the technology and production are evolving at an enormous speed. The most rapid development is observed in the economy, as the basis as well as the essence of the life of society. It is becoming more and more creative, which is reflected in the widespread introduction of the achievements of the Fourth Industrial Revolution, primarily knowledge and technology, digitalization of production processes, their automation and robotization, and the growing speed of technological processes (Gruzina et al., 2021; Guliyeva et al., 2020; Shvab, 2018; Wang et al., 2020).

There is also a transformation in the approach to determining the role of the human factor in the production process. The human factor in the organization is increasingly viewed not as some kind of amorphous community but as a complex set of relationships between employees, with a high level of professional and general education and specific sociopsychological characteristics (Nardo et al., 2020). For instance, back in the 1970s and 1980s, in such economically developed countries as Japan, Germany, and the United States, managers faced the problem of a sharp decrease in the role of material incentives for workers as a means of motivating them to work more productively (Makhlouf, 2016).

It became more and more clear that the further increase in the efficiency of personnel and the introduction of innovations will depend on the general production culture and the psychological climate in the team (Ravino, 2014). Relevant institutions should predetermine the development of a high production culture, create a favorable atmosphere for the emergence of new leaders who are able to generate new ideas, and lead the team, without the use of administrative and other levers of influence (Ahlstrom et al., 2020). Motivating and shaping up the new employees, as well as making them true leaders, both at their respective workplaces and beyond, is a complex and cumbersome process that needs to be handled with great care and accuracy (Henderson, 2021). This is not to mention that there are a number of serious obstacles to the development of leadership qualities, which are analyzed in this study.

This chapter focuses on the institutional problems of leadership development in the developing economies using a case study of Azerbaijan. We study the experience of that country in order to convey lessons and best practices for the other countries both in the region and all around the world.

8.2 Literature Review

In the context of the intensive development of the creative economy, the problem of leadership has acquired important theoretical and practical significance. The deepening of the development of human resources at the end of the twentieth century was accompanied by the development of the leadership qualities of workers, highlighting them to the fore (Akdere & Egan, 2020; Shuck et al., 2019).

What are leadership qualities and leadership in the modern economy? What role does leadership play in the modern economy? How do different motivational institutions affect the development of leadership qualities? Different authors define the essence of leadership in different ways. Stephen Covey believes that leadership is the ability to convey to people the idea of their dignity and potential so clearly that they can see these qualities in themselves (Covey, 2006). For example, Gibbins-Klein, founder of REAL Thought Leaders, believes that leadership is the ability to direct others without force in a direction or decision that leaves them feeling strong and successful; Cash Hanson, CEO of Snuggwugg, sees leadership as a quality that provides vision and motivation for a team, bringing people together to achieve a common goal. According to Garfield, president of the Garfield Group, leadership consists of three things: listening, inspiring, and empowering (Business News Daily, 2015).

Despite the lack of a single definition of leadership, most leaders of the world's leading companies recognize the decisive role in the leadership of organizations of leaders who identify promising problems and mobilize the energy of their colleagues to achieve production goals (Brock & Von Wangenheim, 2019; Leithwood et al., 2020).

In general terms, modern production is characterized by a continuous increase in the cost of factors of production (Guliyeva & Rzayeva, 2017; Kwade et al., 2018). The greatest dynamism in this regard is the human factor, the value of which (due to the growth of knowledge and experience) is growing at a faster pace. Obviously, in this aspect, employees who demonstrate leadership qualities will be of the greatest importance for the organization. Today, the brand of the company, first of all, is determined by the name of their leader. The change in the place of work of such leaders in the company is accompanied by a change in the entire team of managerial employees and radical changes in the work of the entire company, if the appropriate continuity is not ensured. Recently, the use of the term "technocratic leader" has become very popular, which is closely linked with the development of an innovative, creative economy (Blok, 2018).

The above concepts of leadership allow us to give the following more thorough definition of it: leadership – a set of knowledge, experience, and strong-willed qualities of an employee, allowing to unite employees, as well as to convince them to direct their actions to achieve production goals.

It goes without saying that the best leaders develop their technical and career skills over time, becoming invaluable experts in their fields and adept at leading their teams. Rather than constantly reinventing the wheel, the best leaders learn how to do new, unexpected things using tools they have already got, encouraging others to do so as well (Zeike et al., 2019). Great leaders have long-term visions for the future, avoiding getting bogged down in the here-and-now. Truly great leaders acknowledge this, and they promote risk-taking and innovation in their organizations. While they may get tactical when needed, they keep a strategic perspective that is essential for steering their company toward the best possible future. Instead, effective leaders are accountable in their decisions, retain an optimistic attitude, and focus on mapping out new courses of action. They also help others deal with the changes in the organization, and they deal with problems early, to prevent problems from festering and becoming more serious (Hayes et al., 2020). Moreover, good leaders possess self-awareness, gain trust, focus on relationship-building, demonstrate action-taking tendencies, demonstrate humility, enable others, remain authentic, show up consistently and consistently, are role models, and are completely present. As a leader, it is essential that one leads authentically, whether to establish trust with the team or create model behaviors for others within the organization (Qiu et al., 2019). If one wants to lead and thrive in today's fast-paced economy, she or he needs to cultivate these leadership skills within herself/himself and at every level of the organization. Leadership development in this age of New Economy is the critical task of most organizations. Change management is not a new concept to leaders looking to differentiate themselves from competitors and positioning their organizations for the future (Rogers & Euchner, 2022). In response, organizations and their leaders need to adopt new methods of continuous adaptation. Organizational leaders are reviewing, prioritizing, and rolling out initiatives to deal with these challenges, but relying only on centralized change management efforts for projects is insufficient for managing the experiences of individuals as they transition. As enablers to change, leaders, including designers who practice leading, are encouraged to design and embed new methodologies that can mobilize and empower diverse stakeholders to own their collective success, instead of providing short-term fixes that simply boost the organizations' bottom line (El-Adaway & Jennings, 2022). If successful, this leadership approach would allow larger organizations to adopt new value systems and sustain a shift toward a New Economy. For instance, by acknowledging that the new power structures will redefine what it means to drive change, leaders tomorrow will no longer see themselves as providers of standardized offerings, nor executors of conventional approaches (Bag & Pretorius, 2022). In order to achieve all of these, leaders in different fields, including design, must redefine the conceptions of what leadership is. Having an understanding means leaders need to be able to mold their organizations cultures to meet changing needs from stakeholders. Equally critical is for organizations to have a culture, structure, and governance processes that nurture leaders with emerging competencies.

Generally, in order to develop needed leadership competencies, organizations can provide more varied, developmental assignments for individuals, promote individuals to senior roles at earlier and later stages in their careers, provide opportunities for leaders with less traditional backgrounds to manage businesses and initiatives, and celebrate the capacity of their workers and leaders at all levels, early and later in their careers, to redefine, challenge, and grow the businesses in which they are engaged (Fitzsimmons & Callan, 2020). The competencies required for being a successful leader are evolving, too. The new zeitgeist will demand executives who have instincts for dealing with shifting external forces, an ability to sniff out fresh economic opportunities, and skills for leading and managing in another age. Future leaders will be adept at operating with agility; they will be IT-savvy, digitally literate, and capable of building business models based on data analytics. Leaders of the twenty-first century will have to possess futuristic outlook in order to guide their path toward strategic execution. Organizations know that they also must develop leaders with the capabilities needed for the demands of the fast-evolving, tech-driven business landscape – capabilities like leading through uncertainty, managing increasing complexity, being well-educated in modern technology, managing changing demographics of customers and employees, and managing cross-national and cultural differences (Benitez et al., 2022). Organizations need to continually identify leaders with such traits or must train their existing leadership teams to gain such traits. Having leaders who are culturally competent within an organization is important as it helps develop a sense of curiosity about differences within a workplace among employees. Cultural intelligence requires leaders to play a crucial role in leading diverse values in bringing them in line with business. Because change is a constant, leaders who are culturally smart are in a position to leverage their cultural intelligence to lead organizations – and the people they lead – to discover innovative strategies and solutions for cross-cultural challenges (Nguyen et al., 2019). More organizations today are reconsidering what they are all about and the importance of leadership for them. Modern leaders now face a daunting challenge in leading their organizations through the complicated and constantly shifting landscape of global economy. Our research shows that, although organizations are expecting new leadership capabilities, they are still predominantly pushing conventional models and thinking – when they should be developing skills and measuring leadership in ways that help leaders navigate greater ambiguity effectively, assume responsibility for fast-moving changes, and connect with stakeholders both externally and internally. To meet key organizational goals, leaders must be open-minded, must be willing to collaborate, must be capable of bringing different perspectives into alignment, and must possess the charm to lead teams of diverse motives and needs.

In the conditions of the development of a creative economy, where deep knowledge is required, the need to make nonstandard decisions, the risk for their consequences, and the presence of leaders become a necessary condition for the successful development of any business. In 2011, the American Company Boston Consulting Group conducted a sociological study of the main promising areas in the field of human resource management – Creating People Advantage. The main respondents of this study were executives and top managers of large companies. The main purpose of the study was to determine the main areas of work with personnel for top managers of the world's leading companies for the next 20 years. The survey was conducted among representatives of companies from 39 countries. This survey analyzed 416 responses from respondents from Russia, as one of the major representatives of countries with emerging market economies. All the companies that accepted in the survey worked in the areas of high technology and communications, healthcare, etc. Many of them were transnational in terms of the geography of their activities. It was revealed that the most important topics in the field of people management are strategic issues aimed at future development.

In the course of our research, the following most important top three topics in human resource management in the world have been identified:

(a) Talent management.
(b) Improved leadership development.
(c) Transforming HR into a strategic partnership.

For Russia, as a representative of countries with developing economies, the following most important topics in human resource management were identified:

(a) Improving performance management.
(b) Compensation and remuneration of employees.
(c) Talent management (Nikolaeva, 2011).

In our opinion, in countries with a developing market economy, a specific situation has developed when the regulation of property relations is still far from complete; state property still retains significant positions in the national economy, thereby allowing the bureaucracy to concentrate significant financial and material resources in its hands. As a result, market competition is limited; private companies, if they are not affiliated with companies in the public sector or with companies from the representatives of this sector, are doomed to bankruptcy. As a result, many workers develop a sense of alienation, a desire for greater benefits than their contribution to the bottom line. Labor productivity is slowly growing, the quality of manufactured products is growing, and the development of innovations is at a low level. The production culture requires further development; manual, monotonous labor is widely used. In these countries, a typical, average worker corresponds to the X employee model of McGregor's theory. These are employees who have little interest in the results of their labor activity; they lack ambition and the desire to take responsibility and show initiative and independence. They do not consider the goals of production as their own. These employees prefer to work in teams, their expectations are many times greater than what they receive, and in accordance with the current distribution system, they see it as far from social justice (McGregor, 1957). In this model, the system of motivation of employees and their leadership qualities should combine such elements of substantive theories as material motivation, improvement of working conditions, and elements of procedural theories, such as participation in the management of the company and the distribution of profits and the formation of partnerships in the team. All this requires deep institutional changes.

The issues of the impact of Japanese quality circles on labor productivity, stimulation of innovation, leadership qualities among employees, and the motivation system used are discussed in detail in the work of Munchus (1983). This system made a very strong impression on American managers, who consider it quite applicable to American conditions.

The issues of development of the institute of material motivation for leadership qualities of middle and top managers in the United States are considered by Gerchikova. In many large companies, the captains (leaders) of American business have an annual salary exceeding 200 thousand dollars (Gerchikova, 2012). As a result of such an effective system of organization of production and bonus systems, the United States for a long time continues to be the world leader both in terms of productivity and wages. She believes that thanks to this institution, the United States manages to maintain its leadership position for such a long time both in terms of productivity and wages.

The issues of employee's participation in the system of making important management decisions are considered in the work of Jecchinis (1979). Based on the results of European studies, the idea was put forward on the need to create institutional mechanisms in the existing system of relations between workers and administration in Canada as well.

The impact of staff's participation in company management on their productivity, the efficiency of the entire company, and the development of leadership qualities among employees are also considered in the work of Galeazzo et al. (2021). Based on a study of the experience of 330 firms from 15 countries of the world, they revealed the positive impact of involving employees in management on the development of their responsibility and the efficiency of the entire production (Galeazzo et al., 2021).

8.3 Materials and Methods

Increasing the motivation of workers, labor initiative, and leadership qualities of workers in developing economies, including Azerbaijan, is one of the key institutional problems for the development of the economies of these countries. Important factors hindering the development of the above processes is the low level of staff motivation and the uneven distribution of income in the country, both at the macroeconomic and microeconomic levels (between employees within the company). This fact, of course, has a negative impact on the motivation of employees in the results of labor activity and the development of their professional and qualification characteristics and leadership qualities. The highly unequal distribution of income in developing countries can be seen in Azerbaijan (Gulaliyev et al., 2018). This process is most accurately reflected in the dynamics of the Gini coefficient.

Table 8.1 Gini coefficient in Azerbaijan (2009–2016)

	$\sum\sum\lvert yi - yj\rvert$	n^2	$\overline{y}$	$G = \frac{\sum_{i=1}^{n}\sum_{j=1}^{n} \# yi - yj\#}{2 * \overline{y} * n^2}$
2005	178.22	300	5	0.5722
2009	186.572	400	5	0.4654
2010	2234.867	400	5	0.5587
2012	2163.084	361	5.26	0.5692
2014	2460.618	361	5.26	0.6475
2015	1989.712	484	4.55	0.4522
2016	3417.817	529	4.35	0.7430

Source: Own results

Table 8.1 shows that the Gini coefficient for 2005–2009 increased from 0.5722 to 0.7430, that is, by more than 29%. Thus, the oil boom contributed to an increase in the income of the population, while at the same time increasing the unevenness of its distribution. In the economically developed countries of the world, with the general high well-being of the population, the Gini coefficient is not high, as is its growth rate. Thus, the low level of economic development is one of the causes of income inequality.

Another widely used method for measuring income inequality is the Theil index (***T1***; ***T0***). Both indices are based on the idea of calculating the entropy in income inequality with the variables such as the income of an individual (or population group), the average income, and the size of the population (or individuals in the group).

The Theil index can vary in the interval (0 – lnN), which theoretically can increase from an equal distribution of income ("0") to absolute inequality (lnN). Table 8.2 shows that the Theil index for Azerbaijan for the last 7 years is in the range (0–8086), which indicates a significant inequality in the distribution of income in the country.

The level of motivation of employees is also insufficient. Thus, the average salary in Azerbaijan is no more than 800 manats per month (less than $500). Suffice it to say that in the developed countries of the world the minimum wage is $1500–2000 per month.

The results of studies conducted by such well-known consulting companies as Aon Hewitt, Gallup International, Korn Ferry Hay Group, and Towers Watson confirm our hypothesis about the importance of the involvement of company employees in production and economic activities, which in modern conditions is a key factor in the successful development of an organization and its financial well-being.

According to these studies, in companies with a high level of employee's involvement, as a rule:

- The level of profitability exceeded 22–43%.
- The level of customer satisfaction was also higher by 10%.
- Labor productivity was higher on average by 18–21%.

Table 8.2 The Theil index in Azerbaijan (2009–2015)

	Population (thousand people)	Population income (million manat)	Average income (manat)	Theil index
2009	8897.0	22601.1	2540.306	0.8086
2010	8997.6	25607.0	2845.981	0.6588
2011	9111.1	30524.6	3350.265	0.8829
2012	9235.1	34769.5	3764.929	0.7985
2013	9356.5	37562.0	4014.535	0.8543
2014	9477.1	39472.2	4165.008	0.7730
2015	9593.0	41744.8	4351.59	0.5692

Source: Own results

- Staff stability was higher by 35–50%.
- -Product quality was higher by 40–60%.
- The number of accidents caused by violation of safety regulations decreased by 48% (Artamonov, 2020).

An important role in the development of leadership is called upon to play the use of perfect forms of stimulating employees. Research suggests that motivation must be continuous and consistent. Motivation of employees should be based on specific results, timely, personal, concrete, and public. This encourages employees not to rest on their laurels but to move on. This is achieved through the participation of employees in the distribution of profits of the enterprise. In this case, everyone gets the opportunity to see the result of his work and his contribution to the achievement of production goals; the specific personal income received from him creates a positive psychological climate in the team. According to the coefficient of Losada, a Chilean psychologist, the success and efficiency of teamwork are closely dependent on the ratio of positive and negative information in it. According to his research, a positive effect in a team is achieved if the number of positive events (thanks, praises, rewards) is greater than the number of negative ones, and their ratio should be in the range from 2.9 to 11. The conclusion of his research is that the set small daily praises are better than one big bonus at the end of the year (Shper, 2013).

In a creative economy, the role and importance of motivated leaders who determine goals and objectives, strategy, and tactics of development, who have a sense of foresight, increase significantly. The survival of firms and companies in the face of growing competition depends on the creativity of progressive, competent leaders. In the context of the growing availability of material factors of production and the management of a firm, a company is increasingly shifting toward the management of human resources, especially the formation and development of leadership qualities in them. In developing countries, as well as in Azerbaijan, leadership has not become a priority goal of management. One of the most important reasons for this, in our opinion, are the lack of objective and institutional prerequisites for the development of leadership and the lack of demand from the owners of enterprises. Business in the republic mainly relies on outdated technical and technological systems and old administrative methods of management. It is no coincidence that, for example, handicraft production is the most conservative production in its management system. The experience of domestic and foreign firms and companies shows that leaders are able to accelerate the process of technical renewal of production, especially in the context of growing integration and globalization of the modern economy. Singapore, the United Arab Emirates, Turkey, Malaysia, Qatar, and the Republic of Korea are clear evidence of what leaders can do with fundamental institutional reforms in a historically short time.

The way out of this situation, in our opinion, is seen in the widespread use in developing countries of the American system of employee motivation. The specified system is focused on the development of individual, professional qualification and leadership qualities of employees, and stimulation of competition and rivalry within the team and between teams of workers. Motivation in this system is focused on specific individual results of the staff. As a result, there is a significant differentiation of employees in terms of wages; employees of managerial personnel have the opportunity to receive large bonuses for specific achievements of the departments they lead. Bonuses are also paid for all proposals aimed at improving the efficiency of production and management. Individualism dominates in decision-making and personal responsibility in their implementation.

An important direction in improving the stimulation of leadership qualities of employees, increasing their motivation in putting forward initiatives for the development of production, is the development of the institution of social partnership at enterprises. In a market economy, the normal process is the constant clash of opposing interests of workers and employers. Therefore, achieving the goals of effective management requires the identification of trade-off solutions for these interests. Social partnership is, today, the most effective form of interaction between the interests of the owners of enterprises and their employees. Within the framework of this partnership, issues of remuneration of employees, various social guarantees, and their active participation in making the most important management decisions and in the distribution of profits are considered.

The need to create the Institute of Social Partnership is indicated in the documents of various international organizations. Thus, in the European Social Charter, signed by all countries of the European Council, including the Republic of Azerbaijan, the following main principles are indicated:

- Parity and equality of interests of the parties
- The need to ensure decent working conditions and pay
- Development of the system of social security and social protection

workers

- Development of employees' awareness and their participation in decision-making.
- The need for employees to participate in the governing bodies of the enterprise and distribution of his income (CVCE, 1996).

The development of the social partnership in Azerbaijan is at the initial stage of its development. Periodically, every two years, at the level of the republic, a General

Collective Agreement is concluded between the Cabinet of Ministers; the Confederation of Trade Unions of Azerbaijan, which represents the interests of workers; and the National Confederation of the Organization of Entrepreneurs (Employers) of the Republic of Azerbaijan. The last such agreement was concluded in March 2020 (Vesti, 2020), which reflects the most important issues as follows:

- In the sphere of economic policy.
- In the field of improving wages, incomes, and living standards of the population.
- In the development of the labor market and guarantees of employment of the population.
- In the field of strengthening the social protection of the population.

Industry agreements and agreements at the level of enterprises in their content are guided by this General Agreement and are drawn up. A significant part of the clauses of these agreements are declarative in nature; they do not provide a mechanism that would allow personnel to control the provisions specified in it. All this does not contribute to the development of leadership in labor collectives, the promotion of advanced ideas, the orientation of workers to achieve production goals, and the formation of social responsibility of owners and employees.

In our opinion, the creation of a system of institutional development of leadership in the Republic of Azerbaijan can be achieved through the major institutional changes adapted by the relevant stakeholders and policy makers. In the case of Azerbaijan, it can be achieved as a result of the adoption of the specially targeted law "On Social Partnership," which has to necessarily include such provisions as the participation of employees in the distribution of profits and enterprise management, as well as taking into account the best practices of foreign countries for their implementation at domestic conditions.

8.4 Conclusions

In summary, our results stemming from the literature review and our own analysis demonstrate that numerous studies have proved that in the conditions of the growing creativity of the modern economy and the employees themselves, it is necessary to develop the leadership qualities of employees at all levels of management. This task can be achieved only by creating two important institutions: one institution aimed at ensuring the participation of employees in the company's management system and the other institution of participation of employees in the distribution of profits of the enterprise which would motivate them to spend more of their times and efforts for boosting their respective companies' productions and profits. Creating an atmosphere of social partnership in the company will therefore motivate employees to develop such leadership qualities as promoting innovation, social responsibility, and the interest and deeper insight into the results of the company's activities. In parallel, this might also solve the problem of increasing material incentives for employees and reducing the uneven distribution of income within the company.

In our view, the proper and balanced development of the institution of social partnership will simultaneously contribute to the corporate culture. This will ensure the full transparency of all internal information, boost the active communication between all employees, increase the active participation of employees in the management of the enterprise, and ensure the direct participation in the distribution of the enterprise's profits. Thence, the development of leadership in a company might necessarily involve the creation of the institutions defined and described above, as well as the active involvement of personnel in management at the three levels: information and communication, management, and distribution.

References

Ahlstrom, D., Arregle, J. L., Hitt, M. A., Qian, G., Ma, X., & Faems, D. (2020). Managing technological, sociopolitical, and institutional change in the new normal. *Journal of Management Studies, 57*(3), 411–437. https://doi.org/10.1111/joms.12569

Akdere, M., & Egan, T. (2020). Transformational leadership and human resource development: Linking employee learning, job satisfaction, and organizational performance. *Human Resource Development Quarterly, 31*(4), 393–421. https://doi.org/10.1002/hrdq.21404

Artamonov, A. S. (2020). Employee engagement as a leading success factor for a modern Bank. *Administrative Consulting, 3*, 89–98. https://doi.org/10.22394/1726-1139-2020-3-89-98

Bag, S., & Pretorius, J. H. C. (2022). Relationships between industry 4.0, sustainable manufacturing and circular economy: Proposal of a research framework. *International Journal of Organizational Analysis, 30*(4), 864–898. https://doi.org/10.1108/IJOA-04-2020-2120

Benitez, J., Arenas, A., Castillo, A., & Esteves, J. (2022). Impact of digital leadership capability on innovation performance: The role of platform digitization capability. *Information & Management, 59*(2), 103590. https://doi.org/10.1016/j.im.2022.103590

Blok, V. (2018). Technocratic management versus ethical leadership redefining responsible professionalism in the Agri-food sector in the Anthropocene. *Journal of Agricultural and Environmental Ethics, 31*(5), 583–591. https://doi.org/10.1007/s10806-018-9747-2

Brock, J. K. U., & Von Wangenheim, F. (2019). Demystifying AI: What digital transformation leaders can teach you about realistic artificial intelligence. *California Management Review, 61*(4), 110–134. https://doi.org/10.1177/1536504219865226

Business News Daily. (2015). 30 ways to define leadership.. Available at: www.businessnewsdaily.com

Covey, S. (2006). *Seven habits of highly effective People* (1st ed.). Simon & Schuster.

CVCE. (1996, May 3). *European social charter*, Strasbourg. Available at: https://www.cvce.eu/en/obj/revised_european_social_charter_strasbourg_3_may_1996-en-bcb7bcd5-60c2-41d7-aa90-ba6fa8a59f69.html

El-Adaway, I. H., & Jennings, M. (2022). Professional and organizational leadership role in ethics management: Avoiding reliance on ethical codification and nurturing ethical culture. *Science and Engineering Ethics, 28*(4), 1–30. https://doi.org/10.1007/s11948-022-00385-2

Fitzsimmons, T. W., & Callan, V. J. (2020). The diversity gap in leadership: What are we missing in current theorizing? *The Leadership Quarterly, 31*(4), 101347. https://doi.org/10.1016/j.leaqua.2019.101347

Galeazzo, A., Furlan, A., & Vinelli, A. (2021). The role of employees' participation and managers' authority on continuous improvement and performance. *International Journal of Operations & Production Management, 41*(13), 34–64. https://doi.org/10.1108/ijopm-07-2020-0482

Gerchikova, I. N. (2012). *Management: A textbook for universities* (1st ed.). Unity-Dana Publishing House.

Gruzina, Y., Firsova, I., & Strielkowski, W. (2021). Dynamics of human capital development in economic development cycles. *Economies, 9*(2), 67. https://doi.org/10.3390/economies9020067

Gulaliyev, M., Aga, A., Azizov, A., Kazimov, F., & Mir-Babayev, R. (2018). Assessing the degree of inequality in the distribution of national income and its macroeconomic consequences in Azerbaijan. *Amazonia İnvestiga, 7*(17), 85–108.

Guliyeva, A., & Rzayeva, U. (2017). Analysis of the de-dollarization problem in developing countries on the example of Azerbaijan in the conditions of geopolitical asymmetry. *Technology Audit and Production Reserves, 6*(5(38)), 57–63. https://doi.org/10.15587/2312-8372.2017.116825

Guliyeva, A., Rzayeva, U., & Abdulova, A. (2020). Impact of information technologies on HR effectiveness: A case of Azerbaijan. *International Journal of Advanced Computer Science and Applications, 11*(2), 81–89. https://doi.org/10.14569/IJACSA.2020.0110212

Hayes, T. L., Oltman, K. A., Kaylor, L. E., & Belgudri, A. (2020). How leaders can become more committed to diversity management. *Consulting Psychology Journal: Practice and Research, 72*(4), 247–262. https://doi.org/10.1037/cpb0000171

Henderson, R. (2021). Innovation in the 21st century: Architectural change, purpose, and the challenges of our time. *Management Science, 67*(9), 5479–5488. https://doi.org/10.1287/mnsc.2020.3746

Jecchinis, C. A. (1979). Employee's participation in management: International experiences and the prospects for Canada. *Relations Industrielles/Industrial Relations, 34*(3), 490–515. https://doi.org/10.7202/028988ar

Kwade, A., Haselrieder, W., Leithoff, R., Modlinger, A., Dietrich, F., & Droeder, K. (2018). Current status and challenges for automotive battery production technologies. *Nature Energy, 3*(4), 290–300. https://doi.org/10.1038/s41560-018-0130-3

Leithwood, K., Harris, A., & Hopkins, D. (2020). Seven strong claims about successful school leadership revisited. *School Leadership & Management, 40*(1), 5–22. https://doi.org/10.1080/13632434.2019.1596077

Makhlouf, H. H. (2016). The rewards and challenges of export-led strategies. *Journal of Economics and Political Economy, 3*(3), 440–445.

McGregor, D. (1957). Human Side of Enterprise. *Management Review, 11*, 41–49.

Munchus, G. (1983). Employer-employee based quality circles in Japan: Human resource policy implications for American firms. *The Academy of Management Review, 8*(2), 255–261. https://doi.org/10.5465/amr.1983.4284735

Nardo, M., Forino, D., & Murino, T. (2020). The evolution of man-machine interaction: The role of human in industry 4.0 paradigm. *Production & Manufacturing research, 8*(1), 20–34. https://doi.org/10.1080/21693277.2020.1737592

Nguyen, T., White, S., Hall, K., & Bell, R. (2019). Emotional intelligence and managerial communication. *American Journal of Management, 19*(2), 54–63. https://doi.org/10.33423/ajm.v19i2.2068

Nikolaeva, Z. (2011). *The future of HR: Why companies will hunt for talents*. Available at: https://republic.ru/business/budushchee_hr_pochemu_kompanii_obyavyat_okhotu_za_talantami682751.xhtml

Qiu, S., Alizadeh, A., Dooley, L. M., & Zhang, R. (2019). The effects of authentic leadership on trust in leaders, organizational citizenship behavior, and service quality in the Chinese hospitality industry. *Journal of Hospitality and Tourism Management, 40*, 77–87. https://doi.org/10.1016/j.jhtm.2019.06.004

Ravino, A. V. (2014). *Fundamentals of leadership* (1st ed.). BGTU.

Rogers, D., & Euchner, J. (2022). Digital transformation: An interview with David Rogers Jim Euchner talks with David Rogers about the ways digital technology is changing strategy and what leaders can do to manage the pace and complexity of the digital world. *Research-Technology Management, 65*(5), 11–17. https://doi.org/10.1080/08956308.2022.2095759

Shper, V. L. (2013). Losada coefficient. *Methods of Quality Management, 7*, 56–57.

Shuck, B., Alagaraja, M., Immekus, J., Cumberland, D., & Honeycutt-Elliott, M. (2019). Does compassion matter in leadership? A two-stage sequential equal status mixed method exploratory study of compassionate leader behavior and connections to performance in human resource development. *Human Resource Development Quarterly, 30*(4), 537–564. https://doi.org/10.1002/hrdq.21369

Shvab, K. (2018). *Fourth industrial revolution* (1st ed.). Eksmo.

Vesti. (2020). *Law on social partnership to be adopted in Azerbaijan*. Available at: https://vesti.az/socium/v-azerbajdzhane-budet-prinyat-zakon-o-socialnom-partnerstve-399825

Wang, L., Luo, G. L., Sari, A., & Shao, X. F. (2020). What nurtures fourth industrial revolution? An investigation of economic and social determinants of technological innovation in advanced economies. *Technological Forecasting and Social Change, 161*, 120305. https://doi.org/10.1016/j.techfore.2020.120305

Zeike, S., Bradbury, K., Lindert, L., & Pfaff, H. (2019). Digital leadership skills and associations with psychological well-being. *International Journal of Environmental Research and Public Health, 16*(14), 2628. https://doi.org/10.3390/ijerph16142628

Part II
Novel Leadership Trends in Business and Economics

Chapter 9
Entrepreneurial Ecosystem and Digitalization: Relationship and Synergy of Development

Muslum Mursalov, Hanna Yarovenko, and Tetyana Vasilyeva

Abstract This research aims at determining the nature of the relationship between the entrepreneurial ecosystem, formed by business resources, infrastructure, and innovation, and the digital development of economic and social spheres in countries. The calculations are based on statistics from 127 countries and made for 14 indicators characterizing the country's business ecosystem and 5 indicators of IT development and cybersecurity. The canonical analysis revealed a strong, statistically substantive relationship between the two groups of factors. It is established that changes in the business ecosystem development significantly affect the Digital Development Level. Correlation-regression analysis allowed us to select the most significant and influential indicators, such as Startup Skills, Risk Acceptance, Networking, Human Capital, Internationalization, and Process Innovation. The Principal Component Analysis removed multicollinearity in the data set to perform cluster analysis. As a result, three clusters of countries were identified with the highest, medium, and low levels of the business ecosystem in terms of its influence on the country's digitalization. The obtained results allowed us to build a Predictive Decision Tree with high overall accuracy in general and for each classification group. The model predicts changes in segments for countries due to the interaction

M. Mursalov (✉)
Azerbaijan State University of Economics, Baku, Azerbaijan
e-mail: muslum_mursalov@unec.edu.az

H. Yarovenko
Sumy State University, Sumy, Ukraine

University Carlos III of Madrid, Madrid, Spain
e-mail: h.yarovenko@biem.sumdu.edu.ua

T. Vasilyeva
Sumy State University, Sumy, Ukraine

Silesian University of Technology, Gliwice, Poland

The London Academy of Science and Business, London, UK
e-mail: t.vasylieva@biem.sumdu.edu.ua

W. Strielkowski (ed.), *Leadership, Entrepreneurship and Sustainable Development Post COVID-19*, Springer Proceedings in Business and Economics, https://doi.org/10.1007/978-3-031-28131-0_9

synergy of the business ecosystem environment and the level of countries' digitalization.

Keywords Entrepreneurial · Ecosystem · Digitalization · Synergy · Decision · Tree

JEL Classification C38 · L26 · M15

9.1 Introduction

The development of the business environment is a crucial aspect of the functioning of any country's economy, which forms the foundation for its further sustainable development. The lack of favorable conditions for the existence of enterprises and entrepreneurs also negatively affects the social environment in the country. As a result, the level and quality of population life decrease, leading to increased crime, poverty, and population emigration to other countries. That is why one of the main tasks of the government policy of different countries should be creating appropriate conditions for forming an ecosystem that would allow businesses to function effectively under other conditions.

It is especially relevant in the conditions developed during the global COVID-19 pandemic. Thus, according to the analysis of statistical data at the end of 2020, there was a decrease in the average density of new enterprises opened per 1,000 people of working age compared to 2019. This trend was typical for the countries of Europe and Central Asia, Latin America and the Caribbean, East Asia and the Pacific, the Middle East, and North Africa. The value of this indicator has not changed significantly for the countries of sub-Saharan Africa. The average density of new enterprises was the highest and amounted to 5.1 per 1,000 people for OECD (Organization for Economic Cooperation and Development) high-income countries. For the countries of South Asia, this indicator also increased in 2020, but its value is the smallest among all countries in the world and is less than 1.0 (Meunier et al., 2022). As a result, the global pandemic affected the business environment, and this impact was negative for most countries, except for countries with high economic development.

Over the last decade, one of the priority areas of any country's development is the computerization of its various spheres, which contributes to their transformation into digital and cybernetic space and has a significant impact on its image (Petroye et al., 2020). If we analyze the IT sector share in the European countries' GDP, then in the period from 2010 to 2019, it increased for 68% of countries. Such a positive trend is observed for Bulgaria (+38%), the Czech Republic (+7%), Germany (+12%), Estonia (+25%), Greece (+8%), France (+9%), Latvia (+ 54%), Lithuania

(+46%), Hungary (+9%), Malta (+6%), Austria (+17%), Poland (+14%), Romania (+20%), Slovenia (+6%), Sweden (+3%), Norway (+12%), and Great Britain (+14%) (Eurostat, 2022). What happened during the COVID-19 pandemic? This crisis positively affected the level of business digitalization, as it contributed to the transfer of most companies to Internet platforms and the use of mobile technologies and the organization of remote work for employees. In turn, this made it possible to ensure that the work of most entrepreneurs was carried out in the usual mode adopted before the start of the COVID-19 pandemic. Meunier et al. (2022) proved that the higher the acceptance of new information technologies, the newer businesses are opened. The possibilities of digital development contribute to the effective organization of the entrepreneurial ecosystem, increasing its sustainability (Kuzior et al., 2022). Therefore, in the conditions of complex challenges, such as pandemics, military conflicts, global warming, etc., it is important to predict the potential state of the country's development through the synergy of the interaction of the business ecosystem environment and the level of country digitalization. It will allow the country's government and special government organizations to better understand the potential threats and effects of crises and develop long-term countermeasures promptly.

9.2 Literature Review

During and after the end of global crises, the creation of appropriate conditions is a priority for the formation of a "healthy" business ecosystem. Kaya (2020) proves that the ability to start a new business quickly affects the growth of its income and the owners' optimistic attitude about its future success. Creating the potential for the professional development of human capital, acquiring new knowledge, and improving skills contribute to providing businesses with highly qualified personnel (Gruzina et al., 2021; Guliyeva et al., 2020; Mercado & Vargas-Hernández, 2019). On the other hand, using ethical standards for conducting and organizing business also affects human capital, which ensures the formation of a favorable or unfavorable business environment (Čábelková et al., 2015; Hanić & Jevtić, 2020; Strielkowski et al., 2021). These conditions should be considered when developing a personnel management strategy as the main factor in improving business efficiency (Hossein Khan, 2018; Meresa, 2019).

An essential condition for business evaluation during pandemics or military conflicts is the development of appropriate legal measures for entrepreneurs and introduction of more effective tax instruments to reduce the risks of shadowing the economy (Čábelková et al., 2022; Levchenko et al., 2018). The influence of the state due to a high level of corruption can become a destabilizing factor in the development of business, so it is also crucial to have mechanisms to counter corruption in the country (Kobushko et al., 2021). The creation of opportunities to reduce business risks and increase the level of communication between entrepreneurs forms a favorable image on the international market which is also one of the conditions for creating the country's business ecosystem (Tiutiunyk et al., 2022).

Lu et al. (2019) or Bilan et al. (2020) determined that sustainable business models are those that develop an innovative component, or clean production, or are built on corporate social responsibility and environmental management. In our opinion, the creative and technological potential of these three models serves as a driver of the entrepreneurial environment development in such promising directions as digital healthcare (Kotenko & Bohnhardt, 2021), knowledge economy (Nureev et al., 2020; Polyakov et al., 2019), additive manufacturing (Melnyk et al., 2022), innovation ecosystem of universities (Gontareva et al., 2022), and cognitive and artificial intelligence technologies (Kuzior & Kwilinski, 2022).

It should be noted that information technology is a factor in increasing the efficiency of the national economy in the same way that the national economy is the driving force for developing the IT sphere. It is confirmed by Tiutiunyk et al. (2021), who proved the existence of a close relationship between the level of economic digitalization and the country's macroeconomic stability. Also, knowledge management technologies significantly impact companies' financial results, which is one of the positive arguments for business automation (Brimah et al., 2020). It should also consider the impact of digitalization on improving the quality of human capital, which was proven by Antonyuk et al. (2021). Beyi (2018) also highlights the existence of a triune relationship between a person, his digital personality, and the digital economy in his research. All this ultimately shows the broad possibilities of applying information and computer technologies at all levels of the national economy, which can potentially increase the business environment's efficiency in global crises.

How can the connections between the entrepreneurial ecosystem and the country's digitalization be measured? In conducting economic research, scientists use a wide range of mathematical methods and models. We can single out such specific approaches as the entropy information method (Ardil & Bilgen, 2017), building multiplicative models (Levchenko et al., 2019), structural-functional modeling (Shipko et al., 2020), structural equation modeling (Brychko et al., 2021), blockchain technologies (Kuzior & Sira, 2022), and construction of phase portraits (Lyeonov et al., 2020). Statistical methods, among which correlation-regression analysis stands out (Didenko et al., 2020), are the most common and effective for researching economic problems. However, their use is reasonable in the initial stages of working with data. The methods of intellectual analysis and artificial intelligence have ampler opportunities for solving various economic problems. For example, Njegovanović (2018) used the capabilities of artificial intelligence to assess the development of the country's financial sector. Kuzmenko et al. (2020) applied intelligence and bifurcation analysis to investigate the risk of money laundering through financial institutions. In our opinion, the methods of intellectual analysis will be the most effective for this chapter.

Thus, scientists explored the links between individual elements of the business environment and information technology. This research aims the presence and nature of relationships with various components of the business ecosystem and perspective areas for developing the digital and cybernetic environment. The study will be retrospective for the possible application of its results in the context of global crises and overcoming them.

9.3 Indicators and Research Methodology

9.3.1 Indicators

The Global Entrepreneurship and Development Institute (2014) uses the Global Entrepreneurship Index to measure the entrepreneurial ecosystem, which is determined based on 14 indicators characterizing the development of the economic and social sphere, considering available infrastructure, business resources, technologies, risks, etc. This integral index compares the effectiveness of countries' business environments and determines the level of its influence in the internal and external international arenas. Each indicator component is more informative for conducting research than the index itself since its quantitative assessment in the range from 0 to 1 allows one to assess its impact on the whole business ecosystem. Therefore, individual indicators of the entrepreneurial ecosystem were selected for the study.

Thus, Opportunity Perception (OR) characterizes the level of opportunities for the country's population to engage in entrepreneurial activity and the compliance of the institutional environment with these opportunities. Startup Skills (SS) shows the degree of people's skills to start a new business, considering their education, competencies, experience, etc. Risk Acceptance (RA) assesses the willingness of the population to accept risk in the process of doing business, as well as environmental conditions that can increase or decrease entrepreneurial risk. Networking (NW) allows you to reveal the level of communication between entrepreneurs and the distribution of their networks in a specific geographical area. Cultural Support (CS) characterizes the country's influence on entrepreneurship, especially in terms of a corruption component. Opportunity Startup (OS) allows you to assess the business environment for its opportunities for starting a new business. The technological sector's development level and the speed of its company integration are estimated using the Technology Absorption (TA) indicator. Human Capital (HC) shows the quality of personnel in the entrepreneurial sphere, manifested in their education, professional training, and mobility in the labor market. The ability to create new products and compete with them in the market can be assessed using Competition Indicator (C). Product Innovation (PDI) characterizes the country's ability to create new products and technologies. The use of new technologies and high-quality human capital in the business environment is measured with Process Innovation (PI). High Growth (HG) allows you to assess the potential for business development in the country, as well as the intentions of the business sector to ensure its growth. Internationalization (IN) reflects the capabilities of the country's economy and its entrepreneurial sector regarding its entry into international markets. Risk Capital (RC) assesses the availability of various types of capital.

In contrast to the indicators characterizing the country's entrepreneurial ecosystem, a group of indexes was chosen that reflect the level of its digital development in various aspects. The data source was the information presented by the e-Governance Academy Foundation (2018). The selected Global Cybersecurity Index (GCI) allows assessing the level of cybersecurity for member countries of the International Telecommunication Union, taking into account their technical, legal, organizational measures, capacity building, and cooperation with other countries in the field of cyber protection. The National Cyber Security Index (NCSI) determines the level of a country's national potential to counter cyber threats and manage cyber incidents. The ICT Development Index (ICT DI) characterizes countries' degree of progress and potential in developing information and communication technologies. The Networked Readiness Index (NRI) measures the degree of the country's readiness for applying the latest information and communication technologies, considering human capital, the level of development, and the management of the country's technological component. Digital Development Level (DDL) characterizes the degree of the country's cybernation, considering the development and implementation of new information and communication technologies.

The authors formed an input data array from 14 indicators of the business ecosystem and 5 indicators of digitalization for 127 world countries for 2018. This period is used to test the proposed methodology because the data set contains the least amount of missing information. In the future, the forecast model can be adapted to the data of the crisis or post-crisis period. Because the dimensions of the selected indicators are different, the indexes characterizing the level of computer development of the countries were normalized for the comparability of the data. They are stimulator indicators by nature, and natural normalization (1) was used for this purpose:

$$x_{ij}^{*} = \frac{x_{ij} - x_j^{\min}}{x_j^{\max} - x_j^{\min}}, \tag{9.1}$$

where x_{ij}^{*} is the normalized value of the j-th indicator characterizing the level of digital development for the i-th country (i = 1÷m; j = 1÷n), x_{ij} is the current value of the j-th digital development indicator for the i-th country, and $x_{ij}^{\max}$ and $x_{ij}^{\min}$ are the maximum and minimum values of the j-th digital development indicator. The normalization process result is indicators' value from 0 to 1, with 1 corresponding to a high level of the country's digital development and 0 to a low one. After normalization, the data is ready to study the relationship between the state of the ecosystem and the level of digitization and build a predictive model.

9.3.2 Research Methodology

The authors have developed a research methodology that involves implementing the following stages. In the *first step*, a canonical analysis of two groups of selected factors is performed to study linear dependencies and identify relationships between two sets of variables. It allows for assessing one set's influence on another and justifying its statistical significance. The idea of conducting a canonical analysis involves the construction and implementation of the following Eq. (9.2):

$$a_1y_1 + a_2y_2 + \ldots + a_{14}y_{14} = b_1x_1 + b_2x_2 + \ldots + b_5x_5, \tag{9.2}$$

where $y_1, y_2, \ldots, y_{14}$ is a set of variables that reflect the indicators characterizing the country's entrepreneurial ecosystem, $x_1, x_2, \ldots, x_5$ is a set of variables that reflect indicators characterizing the country's digital development, $a_1, a_2, \ldots, a_{14}$ and $b_1, b_2, \ldots, b_5$ are weighting factors calculated based on the maximum correlation of both sets, and « = » means the existence of a stochastic relationship between linear combinations of both group variables.

This analysis will reveal which indicators are influencing variables and which are dependent on them.

A canonical analysis is conducted in the *second stage*, considering the previous step's results. One group of indicators will include only influencing factors and the other only one dependent variable. At this step, the most influencing factor is identified for further calculations.

In the *third stage*, a regression analysis is conducted to identify relevant independent variables. The evaluation and selection of statistically significant variables are carried out using statistical characteristics, which will determine the most pertinent factors that characterize the entrepreneurial ecosystem and the level of digital development in the country.

Suppose there is an interfactor correlation. In that case, applying the Principal Component Analysis in the *fourth stage* will be necessary, reducing the input data array size and eliminating the interfactor correlation. If there is no interfactor correlation, this step is not mandatory for further calculations.

In the *fifth stage*, a cluster analysis is conducted, for which the selected variables of the business environment and digital development are considered. Clustering is performed using the K-means method. To interpret and check consistency in clusters, we will use the Silhouette method, which identifies how similar an object is to its own group compared to others. Silhouette scores range from −1 to +1. A high value indicates that the thing matches well with other matters in its cluster and poorly with values in neighboring groups.

In the *sixth stage*, a Predictive Decision Tree is built, which will allow us to make forecasts of countries' clusters based on the synergy of indicators characterizing the entrepreneurial ecosystem and the level of digitization of various life spheres in the country. For efficient model building, it is necessary to use oversampling procedures to balance the training data set if necessary. Also, the model should be checked for accuracy.

9.4 Empirical Results

At the *first stage* of the proposed methodology, a canonical analysis was conducted between the variables characterizing the countries' entrepreneurial ecosystem (left set) and the level of the country's digitalization in different directions (right set). The analysis was performed using the STATISTICA analytical package. Its results are presented in Table 9.1.

The redundancy value for the left set, which corresponds to the country's entrepreneurial ecosystem indicators, equals 52.9232% (Table 9.1). It means that the right set, including the digital development variables, explains the variability of the entrepreneurial ecosystem factors by 52.9232%. In turn, 67.9654% of the country's business ecosystem indicators define the variability of the country's digitalization, that is, it depends on the business environment for 67.9654%. At the same time, the canonical roots extract 76.7744% of the variance from the left set and 100.0000% from the right set. The obtained canonical correlation (R = 0.9167) indicates a robust correlation between the selected factors. As a result, an increase in the influence of the business ecosystem causes the growth of countries' digital development. However, an increase in the level of digitalization positively affects the development of the country's business activity. The significance of the correlation coefficient is confirmed by the high value of the Pearson test (chi-square (14) = 323.0800), the significance level of which does not exceed 0.05 (p = 0.0000). Thus, the countries' digital development mainly depends on the entrepreneurial ecosystem. Therefore, the digital development indicators will act as dependent variables and the entrepreneurial ecosystem indicators as independent variables.

It turned out that three canonical roots are statistically significant. It is necessary to have a sample that will exceed 40–60 times the amount of initial data to obtain reliable estimates of canonical factor loadings for three pairs of canonical variables (Halafyan, 2007). Hence, it has been decided that we will consider only the first significant root. The following canonical weight values were obtained for a set of digitalization indicators: GCI = −0.0033, ICT DI = 0.1636, NRI = −0.0409, NCI = 0.1022, and DDL = 0.7960. Thus, Digital Development Level makes the biggest contribution to the value of the first canonical variable. Since this variable is the most important, it is advisable to leave only this variable for further research.

A canonical analysis was carried out in the *second stage* for the entrepreneurial ecosystem indicators (the left set) and individual digital development indicators (the right set). Table 9.2 presents its results.

The set of the entrepreneurial ecosystem variables most affects the Digital Development Level, which is explained by these factors by 83.6413% (Table 9.2).

Table 9.1 The canonical analysis results of the relationship between the indicators of the entrepreneurial ecosystem and the country's digitalization

	Left set	Right set
Variables	EEI[a]	GCI, ICT DI, NRI, NCI, and DDI
Variance extracted	76.7744%	100.0000%
Total redundancy	52.9232%	67.9654%
Statistical characteristics	Canonical R = 0.9167; chi-square (14) = 323.0800; p = 0.0000	

Source: Calculated by the authors
[a]EEI-14 Indicators which characterize entrepreneurial ecosystem

Table 9.2 The canonical analysis results of the relationship between the entrepreneurial ecosystem indicators and individual indicators of the country's digitalization

	Left set	Right set
Variables	EEI	GCI
Variance extracted	55.6746%	100.000%
Total redundancy	30.0458%	53.9668%
Statistical characteristics	Canonical R = 0.7346; chi-square (14) = 91.5450; p = 0.0000	
Variables	EEI	ICT DI
Variance extracted	55.8907%	100.0000%
Total redundancy	45.0255%	80.5598%
Statistical characteristics	Canonical R = 0.8976; chi-square (14) = 193.2600; p = 0.0000	
Variables	EEI	NRI
Variance extracted	55.1093%	100.0000%
Total redundancy	33.4302%	60.6616%
Statistical characteristics	Canonical R = 0.7789; chi-square (14) = 110.0900; p = 0.0000	
Variables	EEI	NCI
Variance extracted	49.8365%	100.000%
Total redundancy	30.3990%	60.9975%
Statistical characteristics	Canonical R = 0.7810; chi-square (14) = 111.1000; p = 0.0000	
Variables	EEI	DDI
Variance extracted	58.9665%	100.000%
Total redundancy	49.3203%	83.6413%
Statistical characteristics	Canonical R = 0.9146; chi-square (14) = 213.6300; p = 0.0000	

Source: Calculated by the authors

The linear relationship between the left and right parts is strong (0.9146), and its statistical significance is confirmed by a high Pearson test value and a significance level of less than 0.5. The obtained data confirm the preliminary conclusion of the first stage. As a result, we have left the entrepreneurial ecosystem indicators as independent variables and only Digital Development Level as a dependent variable.

Implementation of the following stages was carried out using the Python programming language. In the *third stage*, a regression analysis was performed to

```
                                   OLS Regression Results
==============================================================================================
Dep. Variable:      Digital Development Level   R-squared:                       0.826
Model:                                    OLS   Adj. R-squared:                  0.818
Method:                         Least Squares   F-statistic:                     95.15
Date:                        Thu, 23 Jun 2022   Prob (F-statistic):           3.16e-43
Time:                                01:30:39   Log-Likelihood:                 130.86
No. Observations:                         127   AIC:                            -247.7
Df Residuals:                             120   BIC:                            -227.8
Df Model:                                   6
Covariance Type:                    nonrobust
==============================================================================================
                              coef    std err          t      P>|t|      [0.025      0.975]
----------------------------------------------------------------------------------------------
const                       0.3224      0.018     17.624      0.000       0.286       0.359
2. Startup Skills           0.1873      0.033      5.732      0.000       0.123       0.252
3. Risk Acceptance          0.1056      0.044      2.413      0.017       0.019       0.192
4. Networking               0.1496      0.047      3.179      0.002       0.056       0.243
8.Human Capital             0.2355      0.043      5.498      0.000       0.151       0.320
11. Process Innovation      0.1095      0.044      2.473      0.015       0.022       0.197
13. Internationalization    0.1078      0.042      2.539      0.012       0.024       0.192
==============================================================================
Omnibus:                       27.744   Durbin-Watson:                   1.802
Prob(Omnibus):                  0.000   Jarque-Bera (JB):               61.225
Skew:                          -0.878   Prob(JB):                     5.07e-14
Kurtosis:                       5.913   Cond. No.                         9.69
==============================================================================
```

Fig. 9.1 Results of regression analysis. (Source: Calculated by the authors)

select only relevant independent variables. Factors that are not statistically significant were excluded during the evaluation process. The obtained results are presented in Fig. 9.1.

Figure 9.1 shows that Startup Skills, Risk Acceptance, Networking, Human Capital, Process Innovation, and Internationalization are statistically significant and relevant for the country's digital development. The influence of Startup Skills is because most startups are implemented in the IT field today. Many projects launched by the population in this direction positively affect the country's technological development trend. Also, the more the population is ready to take risks, the more opportunities are created for business informatization and cybernation.

Only well-educated personnel with professional skills on the use of IT technologies can contribute to increasing their automation level to solve daily tasks in business activities. A high level of innovation has a positive effect on the technological development. The excellent communication between entrepreneurs and the internationalization of the economy increases the possibilities of digitalization in various spheres of the country's life.

Although the selected variables are statistically significant and relevant, some pairs have an above-average correlation. The calculated value of the correlation matrix determinant is equal to 0.0498. Its value approaches zero. That is, there is multicollinearity between the explanatory variables. Accordingly, the calculated value of the chi-square criterion is 369.5291, which exceeds the tabular value

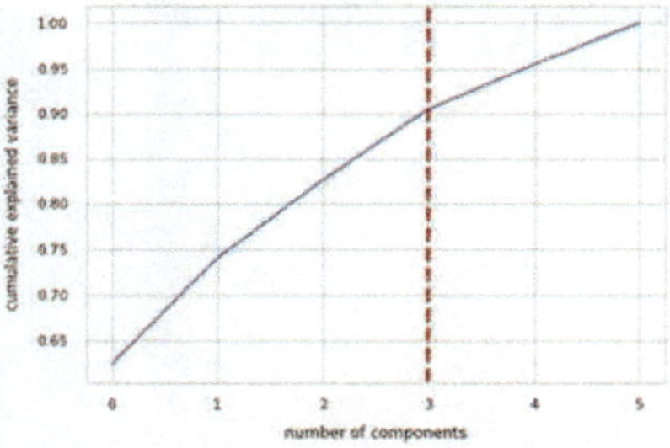

	Cumulative Variance Ratio	Explained Variance Ratio
0	0.623447	0.623447
1	0.739884	0.116437
2	0.825565	0.085681
3	0.904371	0.078806
4	0.954172	0.049801
5	1.000000	0.045828

Fig. 9.2 The results of applying the Principal Component Analysis. (Source: Calculated by the authors)

(24.9958) determined for 0.5 m(m-1) degrees of freedom and significance level $\alpha = 0.05$. It indicates the presence of multicollinearity in the array of variables. The Principal Component Analysis was applied in the *fourth stage* to eliminate the multicollinearity and reduce the dimensionality of the data. The results are presented in Fig. 9.2.

Figure 9.2 demonstrates that the first component explains 62.3447% of the total variation; the second, 11.6437%; the third, 8.5681%; and the fourth, 7.8806%. The other components explain less than 5% of the total variation each. Therefore, it has been decided that we should choose the first four components for further calculations, which together explain 90.4371%.

K-means cluster analysis was performed at the *fifth stage*, for which four components obtained as a result of the canonical analysis were taken. Silhouette scores (Fig. 9.3) were determined, and Silhouette analysis (Fig. 9.4) was performed for the groups with the highest scores to justify the number of clusters.

The obtained Silhouette scores (Fig. 9.3) show that the objects corresponding to clusters 2, 3, 13, and 14 are the most similar. Visualization of the selected clusters in the process of Silhouette analysis (Fig. 9.4) demonstrates that the data divided into 2, 13, and 14 groups have objects that are exceptions. Their values are to the left of zero, which indicates that they do not fall into these clusters. Accordingly, the three-cluster model demonstrates a good classification of the input data. It has been decided to visualize countries according to the three-cluster model and distinguish countries with the highest, medium, and low levels of business ecosystem development, considering its impact on the country's digitalization level. The corresponding results are presented in Fig. 9.5.

Cluster 2 (Fig. 9.5) includes 28 countries: the USA, Canada, Australia, Great Britain, Germany, the Netherlands, Austria, France, Switzerland, Japan, etc. These are economically developed countries characterized by the highest level of evolution of the entrepreneurial ecosystem and the country's digitalization. Cluster 1 annexes 35 countries: Spain, Italy, Portugal, Lithuania, Poland, Romania, Ukraine, China, Turkey, Saudi Arabia, etc. They are actively developing or developed countries, characterized by an average level of influence of the entrepreneurial

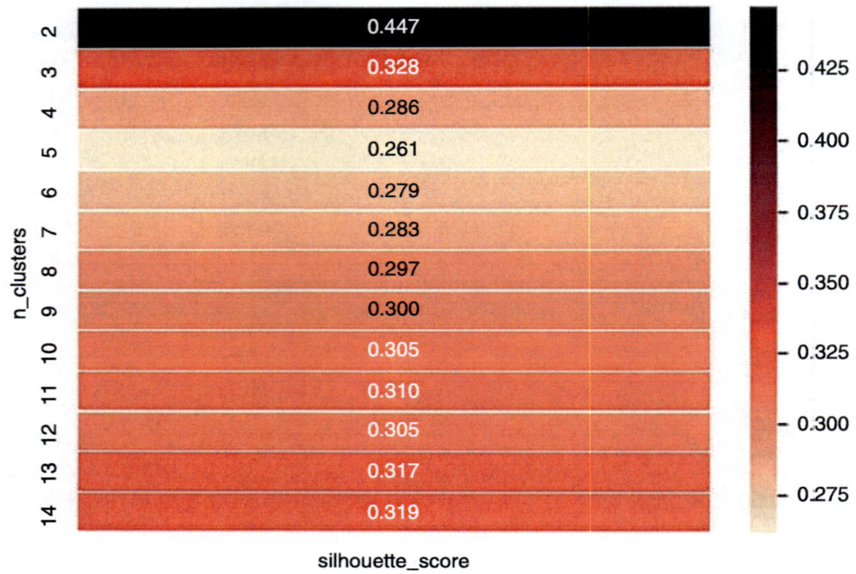

Fig. 9.3 Silhouette scores for 2–14 clusters. (Source: Calculated by the authors)

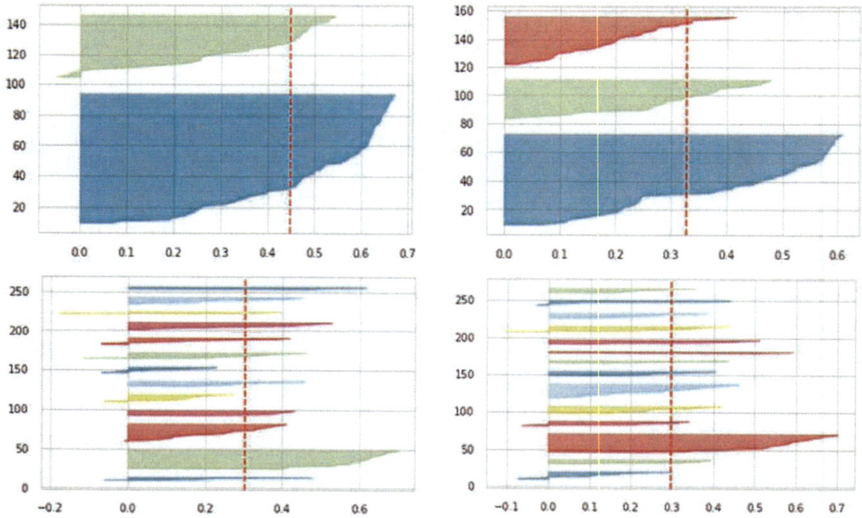

Fig. 9.4 Silhouette analysis for 2, 3, 13, and 14 clusters. (Source: Calculated by the authors)

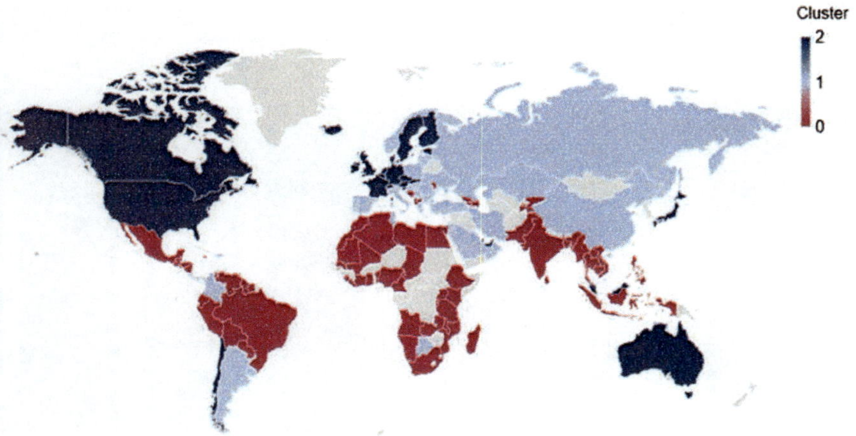

Fig. 9.5 Map of the world countries, distributed according to the cluster analysis results. (Source: Calculated by the authors)

ecosystem on the country's digitalization. A zero group comprises 64 countries, Mexico, Egypt, Thailand, Armenia, Brazil, Benin, Uganda, Chad, Angola, Pakistan, etc., most of which are the least developed or developing countries. They can be classified as countries with a low level of entrepreneurial ecosystem development, under which the growth of their digitization occurs relatively slowly.

At the *sixth stage*, a Predictive Decision Tree was built, which considers the synergy of mutual influences of individual factors of the business ecosystem and the level of the country's digitization. Since the sizes of the identified clusters are different (Fig. 9.6a), the Synthetic Minority Oversampling Technique was applied to balance the data array. The results are presented in Fig. 9.6b.

When building the Decision Tree, it was essential to use a more efficient function to measure the partition's quality. For this purpose, the Gini and Entropy criteria were chosen. As a result, the graphs of the maximum tree depth were plotted for balanced and unbalanced data, which compare the two measures and the corresponding value of their accuracy (Fig. 9.7).

Figure 9.7 demonstrates that the Synthetic Minority Oversampling Technique did not allow us to improve the model's accuracy. It would be better to use unbalanced data to build a Decision Tree. Fig. 9.7a also presents that applying the Gini criterion for most Decision Tree configurations shows a higher partitioning accuracy than Entropy. The maximum value for both measures is 0.9615. At the same time, this value is acquired at the peak depth of the model, which is equal to 4. Further division does not make sense since the accuracy of the Decision Tree will be at the same level, but it will complicate the interpretation of the model. Thus, the Gini criterion and limit of its construction to a depth of 4 have been applied to build a Decision Tree. The model was built based on the synergy of the relevant business

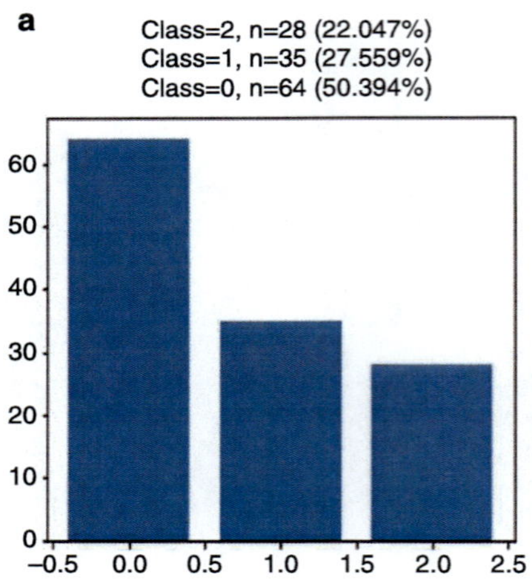

Fig. 9.6 Data before (**a**) and after (**b**) Synthetic Minority Oversampling Technique. (Source: Calculated by the authors)

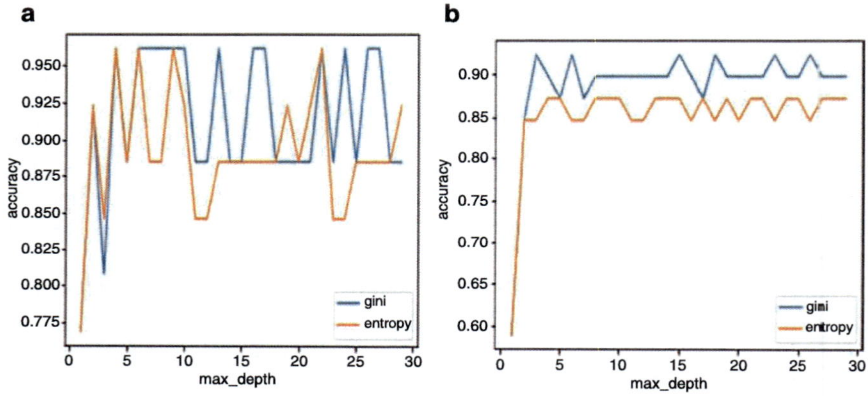

Fig. 9.7 Comparison of the accuracy of splitting Decision Tree branches for unbalanced data (**a**) and balanced data (**b**) with the Gini and Entropy test. (Source: Calculated by the authors)

ecosystem variables and the Digital Development Level, the interaction of which affects the country's development cluster. The result is presented in Fig. 9.8.

The model begins the classification with the Digital Development Level. If its value is less than or equal to 0.723, then the transition occurs to the left node, which corresponds to the Networking indicator, in the opposite case, and to the right one, Process Innovation. If Networking is less than or equal to 0.326, then observations are classified as those corresponding to a low level of development of the business ecosystem and digitalization (cluster 0). The data belong to the null cluster for

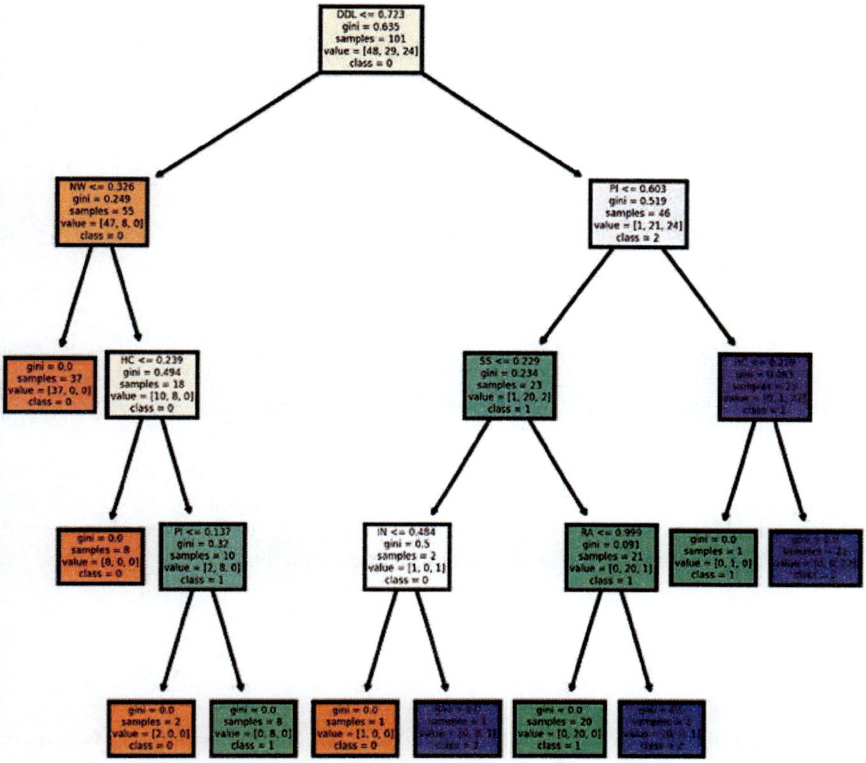

Fig. 9.8 Predictive Decision Tree model. (Source: Calculated by the authors)

Human Capital values that are less than or equal to 0.239. Suppose the value of Process Innovation is less than or equal to 0.603. In that case, the country is identified as one that is characterized by a high level of the business ecosystem and digitalization development (cluster 2). If its value is less than 0.137, then the country will correspond to cluster 1 with an average level of development. If the Startup Skills value is less than or equal to 0.229, then the country is identified by cluster 1. If Human Capital is less than 0.289, the observation belongs to cluster 2. Risk Acceptance less than 0.999 corresponds to cluster 1. Countries for which internationalization is less than or equal to 0.484 belong to the zero group.

Quality estimates of the constructed predictive model were also obtained (Fig. 9.9).

The model's overall quality for all classes is high and equal to 0.9615. The Decision Tree will make a correct prediction with a probability of 96.15%. Accuracy for positive ranks ranges from 0.8 to 1.0, indicating a high likelihood of the model making many accurate optimistic predictions and a lower number of incorrect positive classifications. The sensitivity parameter for all classes is between 0.83 and

```
Classification Report:
              precision    recall  f1-score   support

           0       1.00      1.00      1.00        16
           1       1.00      0.83      0.91         6
           2       0.80      1.00      0.89         4

    accuracy                           0.96        26
   macro avg       0.93      0.94      0.93        26
weighted avg       0.97      0.96      0.96        26

Accuracy: 0.9615384615384616
```

Fig. 9.9 Quality estimates of the Predictive Decision Tree. (Source: Calculated by the authors)

1.00, which confirms the model's high ability to identify positive ranks correctly. Since we did not obtain significantly different values of precision and recall, the F1-score has high values that approach 1, indicating a good combination of the model's precision and recall. Thus, the constructed model is very high quality.

9.5 Conclusions

World crises, like the COVID-19 pandemic and war conflicts, significantly impact countries' economic development. That is why it is necessary to have tools for analyzing the situation and predicting possible consequences in the future. This chapter aims to identify the connections between the indexes that determine the country's entrepreneurial ecosystem and the various directions of its digital development and construct a predictive model based on the synergy of the selected factors. The canonical analysis of the two groups of indicators showed that the entrepreneurial ecosystem has a more significant influence on developing the country's digitization, informatization, and cybersecurity. The better the conditions for business, the more directions for forming the cybernetic environment in the country. The canonical weight investigation, as well as conducting an additional canonical analysis for the group of the entrepreneurial ecosystem indicators and individual indexes of digitization, informatization, and cybersecurity, determined that the Digital Development Level is the most dependent on the entrepreneurial environment.

Regression analysis allowed us to identify relevant factors of the most significant influence on the Digital Development Level (Startup Skills, Risk Acceptance, Networking, Human Capital, Internationalization, and Process Innovation). Thus, digital development is provided mainly by highly qualified personnel, opportunities for opening and running a new business, enterprises' entry into international markets, the innovative component of the country's technological development, risks, and communication. Since multicollinearity exists between the identified set of factors, applying the Principal Component Analysis contributed to its elimination from the input data array for clustering. Conducting Silhouette analysis and Silhouette scores helped to prove the need to divide the data into three clusters of countries with the highest, medium, and low levels of the business ecosystem and digitalization development. At the same time, the cluster with the highest level included countries with a high level of economic growth, the middle-level group had some economic developed and developing countries, and the least developed and developing countries entered the low-level cluster.

The cluster analysis results made it possible to build a Predictive Decision Tree. At the same time, it turned out that the Synthetic Minority Oversampling Technique gives worse estimates regarding the model's accuracy. Therefore, unbalanced data were used to build the model. It is also determined that a more effective partition coefficient is the Gini coefficient at a tree depth of four. As a result, the obtained model will predict the change of countries' segments due to the interaction synergy of the business ecosystem environment and the country's digitalization, both in the normal development conditions and future pandemic requirements. This chapter also proves the high accuracy of the obtained model. This model can be used to improve the state strategy of the country's digital growth in the conditions of existence and counteraction of the global crises' risks, develop a policy on the informatization and cybernation of specific business areas, form priorities for the support of IT startups, and implement information technologies in the business sphere.

Acknowledgments The work was performed within the framework of state budget research: No. 0121 U109559, No. 0121 U109553, and No. 0120 U102001.

References

Antonyuk, N., Plikus, I., & Jammal, M. (2021). Human capital quality assurance under the conditions of digital business transformation and COVID-19 impact. *Health Economics and Management Review, 2*(3), 39–47. https://doi.org/10.21272/hem.2021.3-04

Ardil, C., & Bilgen, S. (2017). Online performance tracking. *SocioEconomic Challenges, 1*(3), 58–72. https://doi.org/10.21272/sec.1(3).58-72.2017

Beyi, W. A. (2018). The trilogy of a digital communication between the real man, his digital individual and the market of the digital economy. *SocioEconomic Challenges, 2*(2), 66–74. https://doi.org/10.21272/sec.2(2).66-74.2018

Bilan, Y., Pimonenko, T., & Starchenko, L. (2020). Sustainable business models for innovation and success: Bibliometric analysis. *E3S Web of Conferences, 159*, 04037. https://doi.org/10.1051/e3sconf/202015904037

Brimah, B. A., Olanipekun, W. D., Bamidele, A. G., & Ibrahim, M. (2020). Knowledge management and its effects on financial performance: Evidence from Dangote flour Mills, Ilorin. *Financial Markets, Institutions and Risks, 4*(2), 34–42. https://doi.org/10.21272/fmir.4(2).34-42.2020

Brychko, M., Bilan, Y., Lyeonov, S., & Mentel, G. (2021). Trust crisis in the financial sector and macroeconomic stability: A structural equation modelling approach. *Economic Research-Ekonomska Istrazivanja, 34*(1), 828–855. https://doi.org/10.1080/1331677X.2020.1804970

Čábelková, I., Abrhám, J., & Strielkowski, W. (2015). Factors influencing job satisfaction in post-transition economies: The case of The Czech Republic. *International Journal of Occupational Safety and Ergonomics, 21*(4), 448–456. https://doi.org/10.1080/10803548.2015.1073007

Čábelková, I., Mareš, D., Strielkowski, W., Zenchenko, S., & Radyukova, Y. (2022). Income tax progressivity and nonreligion in central and Eastern Europe: A case of The Czech Republic. *Religions, 13*(4), 358. https://doi.org/10.3390/rel13040358

Didenko, I., Paucz-Olszewska, J., Lyeonov, S., Ostrowska-Dankiewicz, A., & Ciekanowski, Z. (2020). Social safety and behavioural aspects of populations financial inclusion: A multicountry analysis. *Journal of International Studies, 13*(2), 347–359. https://doi.org/10.14254/2071-8330.2020/13-2/23

E-Governance Academy Foundation. (2018). National Cyber Security Index.. Available at: https://ncsi.ega.ee/ncsi-index/

Eurostat. (2022). *Percentage of the ICT sector in GDP*. Available at: https://ec.europa.eu/eurostat/databrowser/view/isoc_bde15ag/settings_1/table?lang=en

Gontareva, I., Litvinov, O., Hrebennyk, N., Nebaba, N., Litvinova, V., & Chimshir, A. (2022). Improvement of the innovative ecosystem at universities. *Eastern-European Journal of Enterprise Technologies, 1*(13–115), 59–68. https://doi.org/10.15587/1729-4061.2022.251799

Gruzina, Y., Firsova, I., & Strielkowski, W. (2021). Dynamics of human capital development in economic development cycles. *Economies, 9*(2), 67. https://doi.org/10.3390/economies9020067

Guliyeva, A., Rzayeva, U., & Abdulova, A. (2020). Impact of information technologies on HR effectiveness: A case of Azerbaijan. *International Journal of Advanced Computer Science and Applications, 11*(2), 81–89. https://doi.org/10.14569/IJACSA.2020.0110212

Halafyan, A. A. (2007). *Statistical data analysis*. Binom-Press Publishing.

Hanić, A., & Jevtić, D. (2020). Human resource management between economy and ethics – Research of Serbia and Bosnia and Hercegovina. *Business Ethics and Leadership, 4*(3), 127–136. https://doi.org/10.21272/bel.4(3).127-136.2020

Hossein Khan, M. Y. (2018). Strategic human resource practices and its impact on performance towards achieving organizational goals. *Business Ethics and Leadership, 2*(2), 66–73. https://doi.org/10.21272/bel.2(2).66-73.2018

Kaya, H. D. (2020). Business friendliness, firm performance and owner's optimism. Financial Markets. *Institutions and Risks, 4*(3), 13–23. https://doi.org/10.21272/fmir.4(3).13-23.2020

Kobushko, I., Tiutiunyk, I., Kobushko, I., Starinskyi, M., & Zavalna, Z. (2021). The triadic approach to cash management: Communication, advocacy, and legal aspects. *Estudios De Economia Aplicada, 39*(7). https://doi.org/10.25115/eea.v39i7.5071

Kotenko, N., & Bohnhardt, V. (2021). Digital health projects financing: Challenges and opportunities. *Health Economics and Management Review, 2*(1), 100–107. https://doi.org/10.21272/hem.2021.1-10

Kuzior, A., Kettler, K., & Rąb, Ł. (2022). Digitalization of work and human resources processes as a way to create a sustainable and ethical organization. *Energies, 15*(1). https://doi.org/10.3390/en15010172

Kuzior, A., & Kwilinski, A. (2022). Cognitive technologies and artificial intelligence in social perception. *Management Systems in Production Engineering, 30*(2), 109–115. https://doi.org/10.2478/mspe-2022-0014

Kuzior, A., & Sira, M. (2022). A bibliometric analysis of blockchain technology research using VOSviewer. *Sustainability* (Switzerland), *14*(13). https://doi.org/10.3390/su14138206.

Kuzmenko, O., Šuleř, P., Lyeonov, S., Judrupa, I., & Boiko, A. (2020). Data mining and bifurcation analysis of the risk of money laundering with the involvement of financial institutions. *Journal of International Studies, 13*(3), 332–339. https://doi.org/10.14254/2071-8330.2020/13-3/22

Levchenko, V., Boyko, A., Savchenko, T., Bozhenko, V., Humenna, Y., & Pilin, R. (2019). State regulation of the economic security by applying the innovative approach to its assessment. *Marketing and Management of Innovations, 4*, 364–372. https://doi.org/10.21272/mmi.2019.4-28

Levchenko, V., Kobzieva, T., Boiko, A., & Shlapko, T. (2018). Innovations in assessing the efficiency of the instruments for the national economy de-shadowing: The state management aspect. *Marketing and Management of Innovations, 4*, 361–371. https://doi.org/10.21272/mmi.2018.4-31

Lu, J., Ren, L., Yao, S., Qiao, J., Strielkowski, W., & Streimikis, J. (2019). Comparative review of corporate social responsibility of energy utilities and sustainable energy development trends in the Baltic states. *Energies, 12*(18), 3417. https://doi.org/10.3390/en12183417

Lyeonov, S., Żurakowska-Sawa, J., Kuzmenko, O., & Koibichuk, V. (2020). Gravitational and intellectual data analysis to assess the money laundering risk of financial institutions. *Journal of International Studies, 13*(4), 259–272. https://doi.org/10.14254/2071-8330.2020/13-4/18

Melnyk, L., Matsenko, O., Kubatko, O., Korneyev, M., & Tulyakov, O. (2022). Additive economy and new horizons of innovative business development. *Problems and Perspectives in Management, 20*(2), 175–185. https://doi.org/10.21511/ppm.20(2).2022.15

Mercado, M. P. S. R., & Vargas-Hernández, J. G. (2019). Analysis of the determinants of social capital in organizations. *Business Ethics and Leadership, 3*(1), 124–133. https://doi.org/10.21272/bel.3(1).124-133.2019

Meresa, M. (2019). The effect of strategic management practices on the institutional performance; the case of Dedebit credit and saving institution in eastern Tigray. *SocioEconomic Challenges, 3*(3), 80–97. https://doi.org/10.21272/sec.3(3).80-97.2019

Meunier, F., Coste, C., & Maia, R. (2022). How did the COVID-19 pandemic influence the pace of new business formation?. Available at: https://blogs.worldbank.org/developmenttalk/how-did-covid-19-pandemic-influence-pace-new-business-formation

Njegovanović, A. (2018). Artificial intelligence: Financial trading and neurology of decision. *Financial Markets, Institutions and Risks, 2*(2), 58–68. https://doi.org/10.21272/fmir.2(2).58-68.2018

Nureev, R., Volchik, V., & Strielkowski, W. (2020). Neoliberal reforms in higher education and the import of institutions. *Social Sciences, 9*(5), 79. https://doi.org/10.3390/socsci9050079

Petroye, O., Lyulyov, O., Lytvynchuk, I., Paida, Y., & Pakhomov, V. (2020). Effects of information security and innovations on country's image: Governance aspect. *International Journal of Safety and Security Engineering, 10*(4), 459–466. https://doi.org/10.18280/ijsse.100404

Polyakov, M., Bilozubenko, V., Korneyev, M., & Shevchenko, G. (2019). Selection of parameters for multifactor model in the knowledge economy marketing (country level). *Innovative Marketing, 15*(1), 89–99. https://doi.org/10.21511/im.15(1).2019.08

Shipko, A., Demikhova, N., Pajak, K., & Motrechko, V. (2020). Health management at the regional level: Multivariable performance assessment. *Health Economics and Management Review, 1*(2), 8–15. https://doi.org/10.21272/hem.2020.2-01

Strielkowski, W., Guliyeva, A., Rzayeva, U., Korneeva, E., & Sherstobitova, A. (2021). Mathematical modeling of intellectual capital and business efficiency of small and medium enterprises. *Mathematics, 9*(18), 2305. https://doi.org/10.3390/math9182305

The Global Entrepreneurship and Development Institute. (2014). *The Global Entrepreneurship Index*.. Available at: http://thegedi.org/global-entrepreneurship-and-development-index/

Tiutiunyk, I., Drabek, J., Antoniuk, N., Navickas, V., & Rubanov, P. (2021). The impact of digital transformation on macroeconomic stability: Evidence from EU countries. *Journal of International Studies, 14*(3), 220–234. https://doi.org/10.14254/2071-8330.2021/14-3/14

Tiutiunyk, I. V., Zolkover, A. O., Lyeonov, S. V., & Ryabushka, L. B. (2022). The impact of economic shadowing on social development: Challenges for macroeconomic stability. *Naukovyi Visnyk Natsionalnoho Hirnychoho Universytetu, 1*, 183–191. https://doi.org/10.33271/nvngu/2022-1/183

Chapter 10
Greenwashing Detection and Impact on Responsible Business and Investment: Case of Ukrainian Companies (Agriculture Leaders)

Inna Makrenko, Serhiy Makarenko, and Pavlo Rubanov

Abstract Ukrainian agricultural companies are playing a leading role in preventing the global food crisis, having significant production potential and promoting responsible business practices. Ukraine's application for membership in the European Union (EU) will open new opportunities for responsible investment by European investors in these companies and will build new, more sustainable food value chains during the Russian-Ukrainian war. The prerequisite for making responsible investment decisions is transparent and high-quality reporting of Ukrainian agricultural companies, which allows to assess the social, environmental, and managerial aspects of these companies' activity. This study is aimed at recording the facts of unfair disclosure of information by these companies, building a methodology for identifying signs of greenwashing in the reporting of these companies in accordance with EU sustainability reporting requirements. These requirements cover the EU Taxonomy and NFRD (Non-Financial Reporting Directive). Content analysis and PLS –SEM (partial least squares structural equation modeling) method were used for greenwashing detecting in the sample of agriculture companies' sustainability reports. Study results will create a reliable basis for investors' decision-making on responsible investments in Ukrainian companies – leaders of the agricultural sector.

Keywords Agriculture · Leadership · Responsible investment · Greenwashing · Sustainability reporting

I. Makrenko (✉)
School of Business, Prague City University, Prague, Czech Republic

Sumy State University, Sumy, Ukraine
e-mail: i.makarenko@uabs.sumdu.edu.ua

S. Makarenko · P. Rubanov
Sumy State University, Sumy, Ukraine
e-mail: s.makarenko@asp.sumdu.edu.ua; p.rubanov@biem.sumdu.edu.ua

W. Strielkowski (ed.), *Leadership, Entrepreneurship and Sustainable Development Post COVID-19*, Springer Proceedings in Business and Economics, https://doi.org/10.1007/978-3-031-28131-0_10

JEL Classification Q01 · O13 · O44

10.1 Introduction

The war in Ukraine has had a negative impact on global energy, food security, investment allocation, and the sustainability of agri-food value chains. At the same time, progress in achieving the relevant Sustainable Development Goals (SDGs) – SDGs 7, 2, and 12 and other related goals not only in Ukraine but also at the global level – is significantly slowed down.

The agricultural sector, strategically important for Ukraine's image and "green" transition, was significantly affected by the war. According to FAO (Food and Agriculture Organization) estimates, during the blockade of Ukraine seaports, only direct losses of agricultural companies' assets amount to USD 6.5 billion. Currently around 5% of agricultural land is damaged (NCRUCW, 2022). The importance of the agricultural sector in ensuring national food security (its share is 10% of the GDP of Ukraine, pre-war growth was 5–6% every year), overcoming the global food crisis (the production volume of the sector is more than 6% of global calorie consumption), and activation of Ukrainian exports (41% of total Ukrainian exports) is a high priority (NCRUCW, 2022).

In such conditions, unlocking and increasing the potential of the agricultural sector as a driver of Ukraine's post-war recovery on a responsible basis first of all requires investment support. Within the framework of Program 8 of the Post-war Recovery Plan of Ukraine (Ukraine Recovery Conference, 2022), the development of sectors of the economy with added value is underlined. In this direction, promoting the transition of the agro-food sector to "green" growth occupies a prominent place and requires investment resources at the level of USD 37.4 billion.

Ukraine's acquisition of the status of a candidate for EU and joining the Green Deal intensify the importance of investments in the bioenergy direction of the agricultural sector. This has been assisted by other initiatives linked, for example, to migration (Čajka et al., 2014). According to the Bioenergy Association of Ukraine (UABIO), a boom in the construction of biogas plants by agro-food companies was observed in 2019. Capacity capable of producing 31.3 MW of electricity was put into operation. And on the basis of the processing of organic waste from the agricultural sector into biogas, Ukraine is able to completely abandon the import of energy resources (Latifundist, 2022).

Having significant investment attractiveness from both the food and energy sides, companies of the national agricultural sector must demonstrate high standards of information disclosure and transparency for investors to make them informed decisions. In light of the harmonization of Ukrainian legislation with European compliance of agrarian companies regarding the disclosure to investors of social and environmental aspects of activity, ESG (environmental, social, and governance) criteria is in the plane of the Non-Financial Reporting Directive (NFRD, Directive 2014/95/EU), EU Taxonomy (Regulation (EU), 2020/ 852), and countering greenwashing.

The well-known recent scandals with corporate accusations of greenwashing at DWS and its owner Deutsche Bank, Volkswagen (Financial Times, 2022), emphasize the need for more balanced investment screening of information disclosed by companies. Currently, the concept of greenwashing is most specific to companies engaged in resources and agriculture productions.

Greenwashing has a short history in academic circles and is considered mainly at the corporate and product level (Akturan, 2018; Liu & Wang, 2022; More, 2019; Pimonenko et al., 2020). Currently, determining the investment attractiveness of sectors, in particular the agricultural sector, which are decisive for the country's image and green brand, is a poorly studied issue (Us et al., 2022). Some studies focus on the transparency of agricultural companies and ensuring their investment attractiveness, but greenwashing is not considered in this context (Makarenko et al., 2022; Sukhonos et al., 2018).

The purpose of the study is to detect greenwashing and model its impact on the investment attractiveness of Ukrainian companies – the leaders of the agricultural sector.

Identification of greenwashing and its impact on investment attractiveness was carried out for Ukrainian companies – leaders in terms of land bank volume based on two-stage PLS-SEM modeling. Taking into account the consistent modeling of the relationships of explicit and latent variables, detecting of greenwashing and describing its impact on the investment attractiveness of agriculture companies were provided. Investment attractiveness is determined at the operational and financial levels. The modeling results indicate the need for agricultural companies to focus on building the potential for long-term financial investments, the formation of a holistic regulatory environment regarding its sustainability activities, and the regularity of sustainability disclosure.

Based on the identified factors that determine the impact of greenwashing on the investment attractiveness of leading agricultural companies of Ukraine, recommendations have been developed to minimize its signs and increase such attractiveness for investors in the context of the post-war recovery on the sustainable basis, overcoming the food and energy crisis and the European "green" transition. These recommendations are primarily aimed at increasing the volume of "green" investments in the agricultural sector, which is strategic for Ukraine, and increasing the level of trust of "green" investors in the national "green" brand and socially responsible business.

10.2 Literature Review

The investment attractiveness of companies and their sustainability activity and responsible investments are the subject of various scientific discussions (Formankova et al., 2018; Guliyeva et al., 2021; Kozmenko & Vasyl'yeva, 2008; Leonov et al., 2012; Leonov et al., 2014; Lu et al., 2019; Nguedie, 2018; Singh, 2019b; Strielkowski et al., 2021; Us et al., 2020; Vasylieva et al., 2020).

Agricultural companies play a significant role in the investment process accompanying the transition to a "green" economy and the implementation of responsible investment through corporate social responsibility (CSR), investing in communities, soil conservation, biodiversity, bioenergy, organic production, and promoting food security (Abrham et al., 2015; Kiss, 2020; Smith et al., 2017; Smith, 2018; Sokolov et al., 2018; Khoshnava et al., 2019; Singh, 2019a; or Lyulyov et al., 2019).

Today, the source of information investment decision-making regarding agricultural companies is not only financial reporting but also sustainability reporting as a whole. This reporting is intended to comprehensive assessment of the investment attractiveness of these companies taking into account ESG criteria (b; Mentel et al., 2020; Samusevych, Maroušek, et al., 2021a) and CSR (Kaya, 2020; Keliuotytė-Staniulėnienė & Daunaravičiūtė, 2021; Kuzior et al. 2022a, b; Polyakov et al., 2021; Taliento & Netti, 2020; Zhuravka et al., 2021). The challenge for making balanced investment decisions is the sign of greenwashing in these companies' reporting.

Greenwashing is a fairly new and ambiguous concept in academia, especially in the context of investment activity of agricultural companies. The key achievements of scientists in this field can be divided into two groups: (1) those investigating the essential signs of greenwashing, its essence, basic concepts (de Freitas Netto et al., 2020; Montero-Navarro et al., 2021), taxonomy (Yang et al., 2020), terminology (Us et al., 2022), and classification (Vollero, 2022) and (2) those investigating the effects of greenwashing on certain economic variables.

The main conclusion, based on the analyzed systematic reviews of greenwashing, is that its appearance harms the interests of not only the stakeholders (investors) of companies directly but also society in a broad sense and contradicts the sustainability concept and responsible business conduct. Paradoxically, it is the sustainability reporting, as a tool for achieving comprehensive transparency of the company's activities, that is used during greenwashing. Our study uses a broad understanding of greenwashing as any deviation in the disclosure of information regarding the environmental aspects of the company's activities (More, 2019) and other ESG criteria and SDGs.

In the second group of studies, approaches to identifying greenwashing are mostly described by scientists at the corporate level or at the level of company products (Akturan, 2018; Lyeonov et al., 2021; More, 2019; Olasiuk & Bhardwaj, 2019; Pimonenko et al., 2020; Starchenko et al., 2021).

Only the study by Wang et al. (2020) describes the spillover effects between greenwashing and the green brand of companies in the industry. And in the studies of Us et al. (2022) and Renko et al. (2017), greenwashing is associated with the country's "green" brand.

At the same time, the impact of greenwashing on the "green" brand and companies and on the intentions and expectations of consumers is actively investigated (Akturan, 2018; Liu & Wang, 2022; More, 2019; Pimonenko et al., 2020).

The experience of agriculture companies' greenwashing in academic circles is considered fragmentary, while their impact on sustainability value chains, SDGs, and sustainable development initiatives is significant. In particular, the study of Liu and Wang (2022) is one of the few focused on the greenwashing of agricultural companies at the industry level. However, the authors primarily model the impact of greenwashing on the green brand of agricultural companies, in particular its competitiveness. The study of Montero-Navarro et al. (2021) appeals only to the systematization of scientific achievements on greenwashing in the agricultural sector based on a meta-review.

As for the description of the greenwashing research methodology, it should be noted that it is determined primarily by the goal and object of the study. Thus, descriptive methods of bibliometric analysis (Pimonenko et al., 2020; Us et al., 2022) or Preferred Reporting Items for Systematic Reviews and Meta-Analyses (de Freitas Netto et al., 2020) were used for descriptive studies on the essence of greenwashing and the formation of a scientifically based landscape in the study of this phenomenon and its classification features.

The predominant method in detecting greenwashing is the company's consumer survey method, which was used by Akturan (2018) (500 respondents), Wang et al. (2020) (377 respondents), and Liu and Wang (2022).

The methodology of the partial least squares structural equation modeling (PLS-SEM), which is considered the most appropriate in the analysis of economic phenomena described with the help of explicit and latent variables, was used in the study of Pimonenko et al. (2020). However, the scope of this study covered only the three largest manufacturing companies of Ukraine PJSC "ArcelorMittal Kryviy Rih," Metinvest Group, and PJSC "Dneprospetsstal" in 2014–2017.

In our research, using this method, the focus is primarily on identifying greenwashing in the 100 largest agricultural companies of Ukraine – leaders in terms of land bank volume and modeling its impact on the investment attractiveness of these companies for "green" investment, which gives a representative picture of the image of one of the most important strategic sectors' economies of Ukraine in the context of its post-war recovery and "green" European transition.

10.3 Data and Methodology

The 100 largest companies of the Ukrainian agricultural sector, ranked by the volume of the land bank (Latifundist, 2020) in 2020, were selected for the analysis. On the basis of the content analysis of the sites and sustainability reporting, as well as the financial reporting of these companies for the same year, a set of explicit variables was formed to identify greenwashing and its impact on the investment attractiveness of 97 companies. Three companies out of a hundred did not have active sites for conducting a full-fledged research.

To conduct this research, this chapter proposes to use the method of multidimensional data analysis of the second generation – modeling of structural equations based on partial least squares structural equation modeling (PLS-SEM). It is defined as a "causal-prognostic" statistical method that explains numerous complex dependencies between variance-based variables (Chin et al., 2020; Jöreskog & Wold, 1982).

According to Hair et al. (2021), to form a path model within the framework of PLS-SEM, it is necessary to take into account a measurement theory (confirming the reliability and validity of the model), as well as structural theory (which determines relationships in the model). As model constructs or latent variables within the scope of this study, it is proposed to use indicators of greenwashing and investment attractiveness, the relationship between which is defined as a structural model (inner model). Explicit or manifest variables that determine the selected latent of these indicators are as follows (Table 10.1); as a result, two measurement models (outer models) can be defined (Fig. 10.1).

RStudio software was used for computational analysis, which allows obtaining the necessary results based on the use of the R programming language.

As a result, PLS-SEM can be presented in the form of a simplified path model, reflecting the dependencies between greenwashing and investment attractiveness as a structural model and the normative relationships between greenwashing and investment attractiveness with explicit variables as measurement models 1 and 2.

The modeling process includes two stages as follows:

1) Identification of the primary impact of greenwashing on the investment attractiveness of agricultural companies, which is determined at the operational level (based on the volume of the land bank and the presence of investments in biogas plants) for the entire sample of companies.
2) Identification of the complex impact of greenwashing on the investment attractiveness of agricultural companies, which is determined at the operational and financial level (based on the volume of gross profit of companies, the volume of their long-term financial investments, and the volume of the land bank) for a sample of companies, which reveals both financial indicators and sustainability indicators.

10.4 Results

In accordance with the above conditions, we present the results of the analysis in Table 10.2, which contain data on the indicators of convergent validity and multicollinearity of the model.

Table 10.1 Explicit variables in the PLS-SEM model

Variables	Designation	Values for identifying greenwashing	Scale
Measurement model №1 for greenwashing			
Availability of sustainability disclosure on the company's resources	x1	The primary sign, the lack of publicity regarding activities in the field of sustainability or its insufficient presentation, indicates possible intentions to hide true information	0–2
Detailed and accessible sustainability policy	x2	A sign of giving importance to the process of disclosing sustainability information within the formalized regulatory environment of the company's activities in this field	0–2
Regularity of sustainability disclosure	x3	The regular submission of sustainability reports indicates an established process of transparent and systematic communication with investors, which makes sporadic responses to their requests impossible	0–2
Compliance with national and European requirements for the disclosure of social and environmental aspects	x4	Compliance with the minimum disclosure requirements in the management report indicates not only compliance with the basic national legislation on the sustainability disclosure but also the company's compliance with the key requirements of the EU NFRD and EU taxonomy, on which these requirements are based on	0–2
Independent verification of sustainability reporting by auditors	x5	Independent confirmation of the completeness, accuracy, and reliability of information disclosure by auditors is a guarantee of the absence of distortions and greenwashing in it for investors in accordance with generally recognized standards of auditing services	0–2
The type of opinion expressed by the auditors	x6	The presence of a qualified opinion or refusal of the auditor to express an opinion on financial reporting may be indirect evidence of dishonest disclosure of information by companies	0–2
Comprehensive disclosure of information according to ESG criteria	x7	Comprehensive coverage of information of both a positive and negative nature by companies according to all ESG criteria gives a complete picture of its activities according to the key guidelines of the NFRD	0–2
Comprehensive disclosure of information according to SDG	x8	Comprehensive disclosure of information of both a positive and negative nature by companies under the SDGs provides a complete picture of its activities within the limits of targets and indicators in the most relevant areas of sustainability for the company	0–2
Measurement model №2 for investment attractiveness			
Ranking of the land bank	x9	The increase in the volume of the land bank by the company indicates its expansive investment strategy and the increase in the volume of operational activities and characterizes the investment attractiveness at the operational level	1 – The biggest

(continued)

Table 10.1 (continued)

Variables	Designation	Values for identifying greenwashing	Scale
Bioenergy	x10	The company's investment activity in the field of construction of biogas plants testifies to its "green" investment strategy and "green" transition and characterizes the investment attractiveness of taking into account the sustainability values	0–1

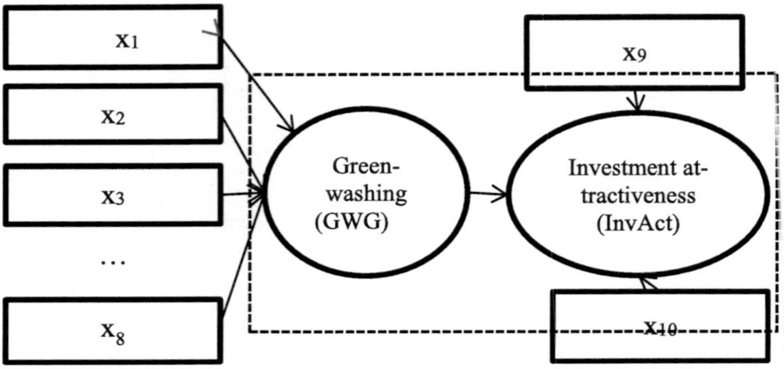

Fig. 10.1 PLS-SEM path model measuring the dependence of greenwashing and investment attractiveness

Table 10.2 Convergent validity and multicollinearity test results for the entire sample of companies

Coefficients	GWG	InvAct
Path Coef.	−0.568	
Alpha	0.871	−0.647
rhoC	0.736	0.008
AVE	0.290	0.621
rhoA	1.000	1.000

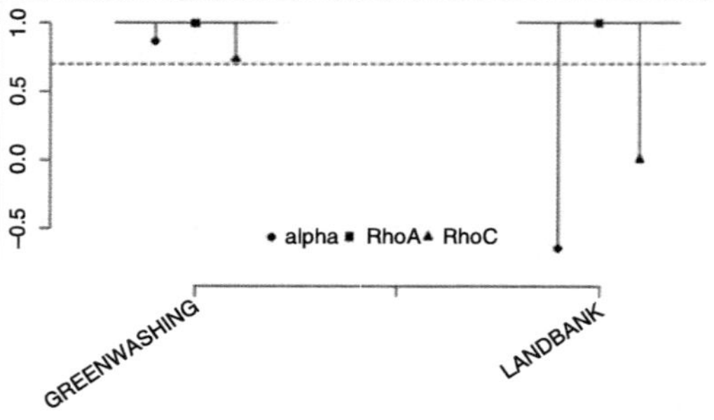

VIF								
	Greenwashing							
	x_1	x_2	x_3	x_4	x_5	x_6	x_7	x_8
	8.199	6.463	3.277	1.937	6.059	6.939	2.207	1.594
	Investment attractiveness							
	x_9	x_{10}						
	1.063	6.063						

The model shows low convergent validity and multicollinearity, which requires the exclusion of individual indicators: x_1, x_5, x_6, and x_{10}

Table 10.3 Results of PLS-SEM analysis at the first stage of the study

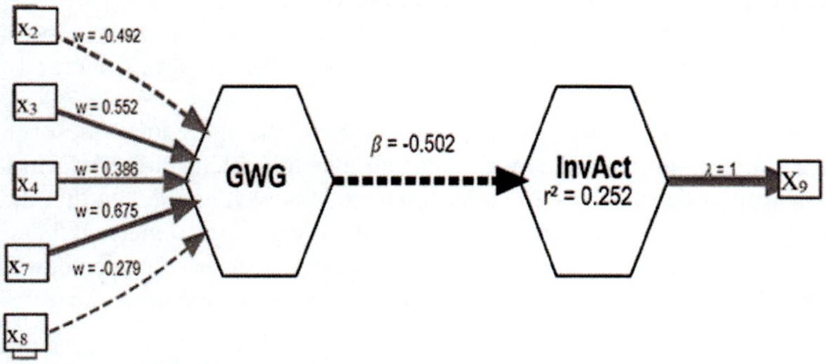

	Original estimate weight	Original estimate loadings	Bootstrap mean	Bootstrap SD	T stat.	2.5% CI	97.5% CI
x2 –> GWG[a]	−0.492	0.307	−0.463	0.227	−2.165[a]	−0.920	−0.033
x3 –> GWG[b]	0.552	0.766	0.464	0.340	1.625	−0.249	1.057
x4 –> GWG[b]	0.386	0.712	0.378	0.239	1.613	−0.060	0.879
x7 –> GWG[a]	0.675	0.710	0.670	0.213	3.165[a]	0.214	1.037
x8 –> GWG	−0.279	0.095	−0.277	0.195	−1.429	−0.627	0.153
x9 –> InvAct	1.000	1.000	1.000	0.000	NA	1.000	1.000

[a]Statistically significant relationships
[b]Admissible relationships

The exclusion of factors x_1, x_5, x_6, and x_{10} for the entire sample of companies is due to a number of factors. The availability of sustainability information in reports or on the company's web resources, unfortunately, is not characteristic of all the studied companies (the share of companies providing such information is 53.0%). At the same time, audit verification of sustainability information is generally carried out by a small number of companies and correlates with both the type of audit service provider and the type of audit opinion expressed (the share of verified reports on sustainable development in the total population is 25.8%). Taking into account the current earnings of companies in the field of bioenergy, they still belong only to the leaders of the agricultural sector (8.2% of the studied companies) and do not constitute a pattern for the entire population.

As a result of repeated construction of the model after the elimination of multicollinear factors at the first stage of the study, the following path model was obtained (Table 10.3), where w are the weighting coefficients of the multiple regression describing the relationships for the formative measurement model. λ is equal to one, because the investment attractiveness construct itself and the x_9 element are equivalent. The coefficient β is a path indicator and characterizes the level of convergent validity, and r^2 is the coefficient of determination.

The obtained data indicate the average quality of the built model; for measurement model No. 1, the influence of the indicators was statistically significant for x_2, x_3, x_4, and x_7.

Taking into account the fact that not all companies of the agricultural sector in Ukraine have a high level of transparency and simultaneous disclosure of information in financial reporting and sustainability reporting, as well as the fact that the investment attractiveness of companies can be described with operational and financial indicators of the activity of agricultural companies, the total sample was adjusted. Two additional variables were added to the structural and measurement model:

x_{11} – long-term financial investments, which illustrate the company's own potential for investment, in particular, in the "green" transition.
x_{12} – gross profit (primary indicator of financial efficiency of companies and ability to generate positive financial results).

At the second stage, 48 companies were included in the sample of companies. Regarding them, we will carry out similar steps of PLS-SEM analysis. As in the first stage, the reliability and validity of formative measurement models are determined (Table 10.4).

Table 10.4 Convergent validity and multicollinearity test results for 48 companies disclosing both financial indicators and sustainability reporting indicators

Coefficients	GWG	InvAct
Path coefficient	0.745	
Alpha	0.837	−0.316
rhoC	0.694	0.251
AVE	0.281	0.387
rhoA	1.000	1.000

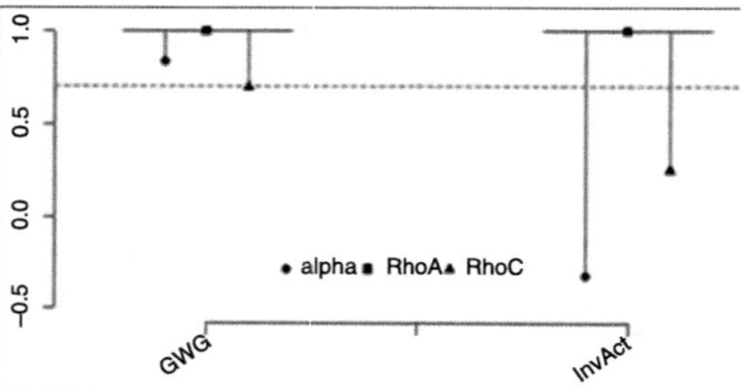

VIF	Greenwashing							
	x_1	x_2	x_3	x_4	x_5	x_6	x_7	x_8
	7.368	5.065	3.241	1.967	3.992	4.330	3.138	1.589
	Investment attractiveness							
	x_9	x_{10}	x_{11}	x_{12}				
	1.307	6.133	1.096	1.213				

The model shows low convergent validity and multicollinearity, which requires the exclusion of individual indicators: x_1 and x_{10}

Table 10.5 Results of PLS-SEM analysis at the second stage of the study

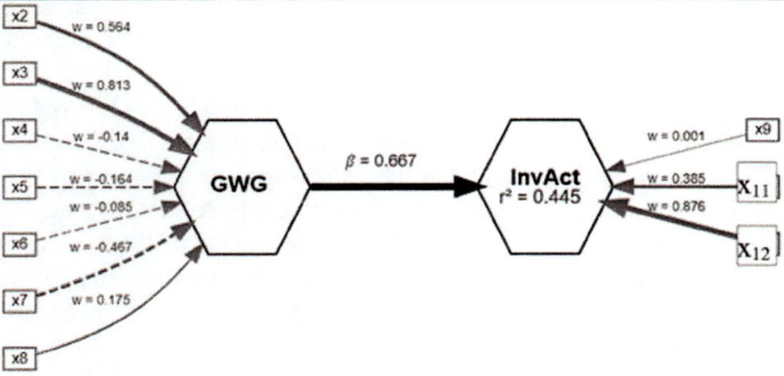

0

	Original estimate weight	Original estimate loadings	Bootstrap mean	Bootstrap SD	T stat.	2.5% CI	97.5% CI
x2 –> GWG[b]	0.564	0.806	0.283	0.588	0.959	−1.041	1.196
x3 –> GWG[a]	0.813	0.837	0.354	0.415	1.958[a]	0.507	1.081
x4 –> GWG	−0.140	0.184	0.073	0.476	−0.295	−0.827	0.967
x5 –> GWG	−0.164	0.065	−0.033	0.754	−0.218	−1.328	1.432
x6 –> GWG	−0.085	0.004	−0.151	0.566	−0.151	−1.240	0.891
x7 –> GWG	−0.467	0.395	−0.190	0.440	−1.060	−0.893	0.856
x8 –> GWG	0.175	0.498	0.182	0.355	0.493	−0.484	0.825
x9 –> InvAct	0.001	−0.447	0.014	0.433	0.003	−0.804	0.854
x10 –> InvAct	0.385	0.495	0.299	0.503	0.766	−0.675	01.105
x11 –> InvAct[b]	0.876	0.924	0.521	0.614	1.427	−1.012	1.194

[a]Statistically significant relationships
[b]Admissible relationships

The analysis conducted proved the model significance and – given the opportunity to move to next study stage – stage of assessment path mode and coefficient for measurement and structural models (Table 10.5).

The obtained data testify the high quality of the built model; for measurement model № 1, the influence of the indicators turned out to be statistically significant for x_2 and x_3 and for measurement model №2 – x_{11}.

Thus, during the initial assessment of the impact of greenwashing on the investment attractiveness of agricultural companies at the first stage of the study, the most statistically significant factors determining the investment attractiveness of companies at the operational level are x_2, x_3, x_4, and x_7. A detailed and accessible sustainability policy, regularity of disclosure of sustainability information, compliance with national and European requirements for disclosure of social and environmental aspects, and comprehensive disclosure of information according to ESG criteria are the most significant factors affecting the level of greenwashing of Ukrainian agricultural companies.

However, the average quality of the built model, in our opinion, is due to the low level of sustainability disclosure by most of companies. A high level of such disclosure is inherent only to individual leaders of the agricultural sector; companies with the largest land bank (Kernel, Astarta, MHP, Nibulon), which demonstrate a high level of compliance with European norms in the field of ESG and CSR, are actively developing the biogas sector and have affordable, regular, verified sustainability reporting and holistic policy in this area. No signs of greenwashing were identified in the reports of such companies.

At the second stage of the study of the impact of greenwashing on the investment attractiveness of agricultural companies, which is determined at the operational and financial levels, the most important role is played by a detailed and accessible sustainability policy and the regularity of sustainability disclosure in terms of direct identification of the level of greenwashing of companies, as well as long-term financial investments of agricultural companies in terms of investment attractiveness.

The obtained results show that in companies that have published a holistic sustainability policy, a significant role is assigned to the process of sustainability disclosure within the formalized regulatory environment of the company as a whole, which minimizes the level of greenwashing thanks to clear rules and procedures.

The regular submission of sustainability reports indicates an established process of transparent and systematic communication with stakeholders, which makes it impossible to provide sporadic answers to their requests and to distort information on the most significant requests of investors.

Agricultural companies making their own long-term financial investments is an important prerequisite for their green transition and an additional signal for investors regarding the feasibility of directing funds to the projects of such companies, in addition to the minimum level of greenwashing in the reporting of such companies. Moreover, the model of the second stage has a higher level of quality, which is due to the comprehensive disclosure by the companies included in the model of both financial indicators and sustainability reporting indicators. In this case, signs of the impact of greenwashing on investment attractiveness are minimal.

10.5 Conclusions

Taking into account the significant investment potential and strategic importance in the formation of food and energy security of Ukraine, agricultural companies can act as drivers within the framework of the Post-War Recovery Plan by attracting investment support and intensifying efforts in the direction of the European "green" transition, harmonizing Ukrainian legislation with European legislation regarding the disclosure of ESG criteria to investors and SDGs.

Facilitating the transition of the agro-food sector to "green" growth requires minimizing the level of greenwashing, significantly increasing the level of transparency of Ukrainian agricultural companies and their attractiveness for investors.

Identification of greenwashing and its impact on investment attractiveness was carried out for Ukrainian companies – leaders in terms of land bank volume based on two-stage PLS-SEM modeling. At the first stage, the conducted modeling made it possible to determine the primary impact of greenwashing on the investment attractiveness of agricultural companies at the operational level. Model is based on the availability of sustainability policies, regularity of sustainability disclosure, compliance with national and European requirements for such disclosure, and consideration of ESG criteria as a main factor influencing greenwashing.

At the second stage of identifying the complex impact of greenwashing on the investment attractiveness of agrarian companies, which is determined at the operational and financial levels, the modeling results indicate the need for agrarian companies to focus on building the potential for long-term financial investments, forming a holistic regulatory environment for their sustainability activities and the regularity of sustainability disclosure.

Taking into account the specified factors, recommendations have been developed to minimize signs of greenwashing and increase the investment attractiveness of companies in the agricultural sector of Ukraine. They are aimed at increasing the volume of attracting "green" investments in the agricultural sector, which is strategic for Ukraine, and increasing the level of trust of "green" investors in the national "green" brand and socially responsible business.

First of all, the specified recommendations relate to the development of a regulatory environment in order to intensify the adoption by agricultural companies of their own sustainability policies as strategic guidelines for their activities in this area. Such policies should be detailed in terms of procedures, directions, and deadlines and should be formed taking into account the most significant requests of key stakeholders, national and European sustainability legislation (Non-Financial Reporting Directive (NFRD, Directive, 2014/95/EU), and EU Taxonomy (Regulation (EU), 2020/852)). An important element of such a policy should be the communication component, ensuring the transparency of the activities of agriculture companies in the field of sustainability, codifying all types of such activities, and establishing monitoring and reporting processes.

The next recommendation concerns ensuring periodicity of reporting in accordance with the needs of stakeholders, complex and purposeful disclosure of basic information according to sustainability dimensions, and ESG criteria and SDGs, relevant for companies and their stakeholders. Developing of information disclosure calendars, its regular updating and active communication with investors, communities, and state authorities as main stakeholders concerning the most material requests will contribute to minimizing the level of greenwashing.

Finally, the search for companies' own sources to make active progress in the field of "green" transition and the activation of the movement in this direction, supported by regular disclosure of information in the financial dimension and regarding the sustainability dimension within the framework of a strategically defined policy, is a prerequisite for increasing the investment attractiveness of agricultural companies of Ukraine at the stage of post-war recovery.

Acknowledgments The research was funded by a grant from the Ministry of Education and Science of Ukraine, 0121 U100468, "Green investing: cointegration model of transmission ESG effects in the chain "green brand of Ukraine – social responsibility of business.""

References

Abrham, J., Strielkowski, W., Vošta, M., & Šlajs, J. (2015). Factors that influence the competitiveness of Czech rural SMEs. *Agricultural Economics, 61*(10), 450–460. https://doi.org/10.17221/63/2015-AGRICECON

Akturan, U. (2018). How does greenwashing affect green branding equity and purchase intention? An empirical research. *Marketing Intelligence & Planning, 36*(7), 809–824. https://doi.org/10.1108/MIP-12-2017-0339

Čajka, P., Jaroszewicz, M., & Strielkowski, W. (2014). Migration incentives and flows between Belarus, Moldova, Ukraine and the European Union: A forecasting model. *Economics & Sociology, 7*(4), 11–25. https://doi.org/10.14254/2071-789X.2014/7-4/1

Chin, W. W., Cheah, J.-H., Liu, Y., Ting, H., Lim, X.-J., & Cham, T. H. (2020). Demystifying the role of causal-predictive modeling using partial least squares structural equation modeling in information systems research. *Industrial Management & Data Systems, 120*(12), 2161–2209. https://doi.org/10.1108/imds-10-2019-0529

de Freitas Netto, S. V., Sobral, M. F. F., Ribeiro, A. R. B., et al. (2020). Concepts and forms of greenwashing: A systematic review. *Environmental Sciences Europe, 32*(19). https://doi.org/10.1186/s12302-020-0300-3

Directive 2014/95/EU of the European Parliament and of the Council of 22 October 2014 amending Directive 2013/34/EU as regards disclosure of non-financial and diversity information by certain large undertakings and groups. Available at: https://eur-lex.europa.eu/legal-content/EN/TXT/?uri=CELEX%3A32014L0095.

Financial Times. (2022). How ESG investing came to a reckoning. Available at: https://www.ft.com/content/5ec1dfcf-eea3-42af-aea2-19d739ef8a55.

Formankova, S., Trenz, O., Faldik, O., Kolomaznik, J., & Vanek, P. (2018). The future of investing – Sustainable and responsible investing. *Marketing and Management of Innovations, 2*, 94–102. https://doi.org/10.21272/mmi.2018.2-08

Guliyeva, A., Rzayeva, U., & Jafarova, N. (2021). Analysis of some indicators by means of fuzzy logic on the example of Azerbaijani energy enterprises. *E3S Web of Conferences, 250*, 02001. https://doi.org/10.1051/e3sconf/202125002001

Hair, J. F., Hult, G. T. M., Ringle, C. M., Sarstedt, M., Danks, N. P., & Ray, S. (2021). An introduction to structural equation modeling. In *Partial least squares structural equation modeling (PLS-SEM) using R* (1st ed.). classroom companion: Business, Springer.

Jöreskog, K. G., & Wold, H. (1982). The ML and PLS techniques for modeling with latent variables: Historical and comparative aspects. In H. Wold & K. G. Jöreskog (Eds.), *Systems under indirect observation, part I* (1st ed.). North-Holland.

Kaya, H. D. (2020). Business friendliness, firm performance and owner's optimism. *Financial Markets, Institutions and Risks, 4*(3), 13–23. https://doi.org/10.21272/fmir.4(3).13-23.2020

Keliuotytė-Staniulėnienė, G., & Daunaravičiūtė, K. (2021). The global green bond market in the face of the COVID-19 pandemic. *Financial Markets, Institutions and Risks, 5*(1), 50–60. https://doi.org/10.21272/fmir.5(1).50-60.2021

Khoshnava, S. M., Rostami, R., Zin, R. M., Štreimikienė, D., Yousefpour, A., Strielkowski, W., & Mardani, A. (2019). Aligning the criteria of green economy (GE) and sustainable development goals (SDGs) to implement sustainable development. *Sustainability, 11*(17), 4615. https://doi.org/10.3390/su11174615

Kiss, L. B. (2020). Examination of agricultural income inequality in the European Union. *Business Ethics and Leadership, 4*(3), 36–45. https://doi.org/10.21272/bel.4(3).36-45.2020

Kozmenko, S., & Vasyl'yeva, T. (2008). Specialized innovative investment banks in Ukraine. *Banks and Bank Systems, 3*(1), 48–56. https://doi.org/10.21511/bbs.3

Kuzior, A., Arefieva, O., Poberezhna, Z., & Ihumentsev, O. (2022a). The mechanism of forming the strategic potential of an enterprise in a circular economy. *Sustainability (Switzerland), 14*(6). https://doi.org/10.3390/su14063258

Kuzior, A., Postrzednik-Lotko, K. A., & Postrzednik, S. (2022b). Limiting of carbon dioxide emissions through rational management of pro-ecological activities in the context of CSR assumptions. *Energies, 15*(5), 18–25. https://doi.org/10.3390/en15051825

Latifundist. (2020). *Top 100 latifundist in Ukraine*. Available at: https://latifundist.com/rating/top100.

Latifundist. (2022). *Green energy.* TOP 5 agricultural companies producing biogas. Available at: https://latifundist.com/rating/energiya-zelenyh-top-5-agrarnyh-kompanij-po-moshchnosti-biogazovyh-ustanovok/.

Leonov, S., Frolov, S., & Plastun, V. (2014). Potential of institutional investors and stock market development as an alternative to households' savings allocation in banks. *Economic Annals-XXI, 146*, 11–12. 65–68.

Leonov, S. V., Vasylieva, T. A., & Tsyganyuk, D. L. (2012). Formalization of functional limitations in functioning of co-investment funds basing on comparative analysis of financial markets within FM CEEC. *Actual Problems of Economics, 134*(8), 75–85.

Liu, H., & Wang, X. (2022). Promoting competitiveness of green brand of agricultural products based on agricultural industry clusters. *Wireless Communications and Mobile Computing*, 7824638. https://doi.org/10.1155/2022/7824638

Lu, J., Ren, L., Yao, S., Qiao, J., Strielkowski, W., & Streimikis, J. (2019). Comparative review of corporate social responsibility of energy utilities and sustainable energy development trends in the Baltic states. *Energies, 12*(18), 3417. https://doi.org/10.3390/en12183417

Lyeonov, S., Pimonenko, T., Chygryn, O., Reznik, O., & Gaynulina, R. (2021). Green brand as a marketing instrument: Principle, features and parameters. *International Journal of Global Energy Issues, 43*(2–3), 147–165. https://doi.org/10.1504/ijgei.2021.115141

Lyulyov, O., Pimonenko, T., Stoyanets, N., & Letunovska, N. (2019). Sustainable development of agricultural sector: Democratic profile impact among developing countries. *Research in World Economy, 10*(4), 97–105. https://doi.org/10.5430/rwe.v10n4p97

Makarenko, I., Plastun, A., Mazancova, J., Juhaszova, Z., & Brin, P. (2022). Transparency of agriculture companies: Rationale of responsible investment for better decision making under sustainability. *Agricultural and Resource Economics: International Scientific E-Journal, 8*(2), 50–66. https://doi.org/10.51599/are.2022.08.02.03

Mentel, G., Vasilyeva, T., Samusevych, Y., Vysochyna, A., Karbach, R., & Streimikis, J. (2020). The evaluation of economic, environmental and energy security: Composite approach. *International Journal of Global Environmental Issues, 19*(1–3), 177–195. https://doi.org/10.1504/ijgenvi.2020.10037581

Montero-Navarro, A., González-Torres, T., Rodríguez-Sánchez, J.-L., & Gallego-Losada, R. (2021). A bibliometric analysis of greenwashing research: A closer look at agriculture, food industry and food retail. *British Food Journal, 23*(13), 547–560. https://doi.org/10.1108/BFJ-06-2021-0708

More, P. V. (2019). The impact of greenwashing on green Brand Trust from an Indian perspective. *Asian Journal of Innovation and Policy, 8*(1), 162–179. https://doi.org/10.7545/AJIP.2019.8.1.162000

NCRUCW. (2022). *The National Council for the recovery of Ukraine from the consequences of the war. Draft Ukraine recovery plan*. Materials of the "new agrarian policy" working group. Available at: https://www.globsec.org/news/ukraines-recovery-plan-2022-the-very-first-draft/

Nguedie, Y. H. N. (2018). Corruption, investment and economic growth in developing countries: A panel smooth transition regression approach. *SocioEconomic Challenges, 2*(1), 63–68. https://doi.org/10.21272/sec.2(1).63-68.2018

Olasiuk, H., & Bhardwaj, U. (2019). An exploration of issues affecting consumer purchase decisions towards eco-friendly brands. *Marketing and Management of Innovations, 2*, 173–184. https://doi.org/10.21272/mmi.2019.2-15

Pimonenko, T., Bilan, Y., Horák, J., Starchenko, L., & Gajda, W. (2020). Green brand of companies and greenwashing under sustainable development goals. *Sustainability, 12*, 1679. https://doi.org/10.3390/su12041679

Polyakov, M., Khanin, I., Bilozubenko, V., Korneyev, M., & Shevchenko, G. (2021). Factors of uneven progress of the European Union countries towards a circular economy. *Problems and Perspectives in Management, 19*(3), 332–344. https://doi.org/10.21511/ppm.19(3).2021.27

Regulation (EU) 2020/852 of the European Parliament and of the Council of 18 June 2020 on the establishment of a framework to facilitate sustainable investment, and amending Regulation (EU) 2019/2088. Available at: https://eur-lex.europa.eu/legal-content/EN/TXT/?uri=CELEX:32020R0852.

Renko, S., Petljak, K., & Stulec, I. (2017). The basic postulates of the green image of a country: The case of Croatia. In S. Renko & A. Pestek (Eds.), *Green economy in the Western Balkans* (1st ed.). Bingley.

Samusevych, Y., Maroušek, J., Kuzmenko, O., Streimikis, J., & Vysochyna, A. (2021a). Environmental taxes in ensuring national security: A structural optimization model. *Journal of International Studies, 14*(2), 292–312. https://doi.org/10.14254/2071-8330.2021/14-2/19

Samusevych, Y., Vysochyna, A., Vasylieva, T., Lyeonov, S., & Pokhylko, S. (2021b). Environmental, energy and economic security: Assessment and interaction. *E3S Web of Conferences, 234*, 00012. https://doi.org/10.1051/e3sconf/202123400012

Singh, S. N. (2019a). Climate change and agriculture in Ethiopia: A case study of Mettu Woreda. *SocioEconomic Challenges, 3*(3), 61–79. https://doi.org/10.21272/sec.3

Singh, S. N. (2019b). Private investment and business opportunities in Ethiopia: A case study of Mettu town in Ethiopia. *Business Ethics and Leadership, 3*(4), 91–104. https://doi.org/10.21272/bel.3

Smith, T. W., Kim, J., & Son, J. (2017). Public attitudes toward climate change and other environmental issues across countries. *International Journal of Sociology, 47*(1), 62–80. https://doi.org/10.1080/00207659.2017.1264837

Smith, M. (2018). A real options approach to evaluating agricultural investments under uncertainty: When to get in and out of sugarcane production. *SocioEconomic Challenges, 2*(1), 21–34. https://doi.org/10.21272/sec

Sokolov, M., Mykhailov, A., & Khandurin, D. (2018). Distribution of investment resources: Where is agriculture in the Ukraine's economy? *Financial Markets, Institutions and Risks, 2*(3), 38–42. https://doi.org/10.21272/fmir.2

Starchenko, L., Lyeonov, S., Vasylieva, T., Pimonenko, T., & Lyulyov, O. (2021). Environmental management and green brand for sustainable entrepreneurship. *E3S Web of Conferences, 234*, 00015. https://doi.org/10.1051/e3sconf/202123400015

Strielkowski, W., Tarkhanova, E., Baburina, N., & Streimikis, J. (2021). Corporate social responsibility and the renewable energy development in the Baltic States. *Sustainability, 13*(17), 9860. https://doi.org/10.3390/su13179860

Sukhonos, V., Makarenko, I., Serpeninova, Y., & Qasimova, G. (2018). Classification and prioritization of stakeholders' information requests according to sustainable development goals: Case of cross-sector partnership in Ukrainian food production industry. *Problems and Perspectives in Management, 16*(4), 126–140. https://doi.org/10.21511/ppm.16

Taliento, M., & Netti, A. (2020). Corporate social/environmental responsibility and value creation: Reflections on a modern business management paradigm. *Business Ethics and Leadership, 4*(4), 123–131. https://doi.org/10.21272/bel.4

Ukraine Recovery Conference. (2022). *Recovery plan.* Available at: https://www.urc2022.com/urc2022-recoveryplan.

Us, Y., Pimonenko, T., Lyulyov, O., & Ziabina, E. (2022). Country's green brand: The main gaps in the scientific treatises. *VADYBA. Journal of Management, 38*(1), 9–15. https://doi.org/10.38104/vadyba.2022.1.01

Us, Y., Pimonenko, T., Tambovceva, T., & Segers, J.-P. (2020). Green transformations in the healthcare system: The covid-19 impact. *Health Economics and Management Review, 1*(1), 48–59. https://doi.org/10.21272/hem.2020.1-04

Vasylieva, T., Kuzmenko, O., Rashid, M. N., Vojtovic, S., Kascha, M., & Lieonov, H. (2020). Innovations in government management of the healthcare system: Forecasting of covid-19 consequences in social, investment and business development. *Marketing and Management of Innovations, 4*, 11–25. https://doi.org/10.21272/mmi.2020.4-01

Vollero, A. (2022). *The state of art: A systematic literature review. Greenwashing* (1st ed.). Emerald Publishing Limited.

Wang, H., Ma, B., & Bai, R. (2020). The spillover effect of greenwashing behaviours: An experimental approach. *Marketing Intelligence & Planning, 38*(3), 283–295. https://doi.org/10.1108/MIP-01-2019-0006

Yang, Z., Thi Thu, H.-N., Hoang, N.-N., Thi Thuy, N.-N., & Thi, T.-C. (2020). Greenwashing behaviours: Causes, taxonomy and consequences based on a systematic literature review. *Journal of Business Economics and Management, 21*(5), 1486–1507. https://doi.org/10.3846/jbem.2020.13225

Zhuravka, F., Botvinov, R., Parshyna, M., Makarenko, T., & Nebaba, N. (2021). Ukraine's integration into the world arms market. *Innovative Marketing, 17*(4), 146–158. https://doi.org/10.21511/im.17

Chapter 11
Development of the Leading Sustainable and Viable e-Government Concept in the Post-COVID Era

Tatiana Goloshchapova, Natalia Skornichenko, and Aksana Turgaeva

Abstract This chapter focuses on the pandemic-inspired leading sustainable development of the concept of viable and effective e-government. With all its negative externalities and lost lives, the coronavirus pandemic provided a strong boost for the development for the online services and tools that constitute the basis of the strong and effective e-government. This chapter discusses the impact the COVID-19 crisis may have on the future of science, technology, and the innovation policy. We argue that COVID-19 crisis and its impacts on socioeconomic activities would inevitably lead to the long-lasting changes in science, technology, and the approach to digitalization and introducing artificial intelligence. Furthermore, this chapter also contributes to broader debates about open-source solution adoption, since open-source software could have an important role to play in public sectors and the development of digital public services. Furthermore, this chapter also evaluates the role played by electronic government services in shaping interactions between citizens and the public administration. Our results might be of a special interest for the policymakers and relevant stakeholders focused on digitalization of administration and creation of the efficient e-government.

Keywords e-Government · Sustainable development · Leadership · COVID-19

JEL Classification G18 · H12 · Q01

T. Goloshchapova (✉) · N. Skornichenko
Volga Region State University of Service, Togliatti, Russian Federation

A. Turgaeva
Financial University Under the Government of the Russian Federation, Moscow, Russian Federation

W. Strielkowski (ed.), *Leadership, Entrepreneurship and Sustainable Development Post COVID-19*, Springer Proceedings in Business and Economics, https://doi.org/10.1007/978-3-031-28131-0_11

11.1 Introduction

The urgent measures to ensure public sector continuity, both nationally and subnationally, at government levels, during and after the COVID-19 pandemic has been and is remaining a priority for the governments worldwide. A holistic recovery from the pandemic requires capacity for coordination at different levels of government and also providers, insurers, industry, and the academia (Bali et al., 2022). Addressing the COVID-19 pandemic requires an unprecedented level of collaboration among governments and citizens, as well as in different sectors of society, in order to achieve spatial distancing and other policy measures (Arif et al., 2022).

With regard to the above, the Digital Agenda for Europe, one of seven flagship initiatives in the Europe 2020 strategy, which seeks to identify key roles to be played by the use of information and communication technologies (ICTs) if Europe is to achieve its ambitions, can be an excellent example and a subject of research on this topic (Kersan-Skabic, 2021). To ensure that there would be a sustainable future for Europe, the newly launched Action Plan on the Digital Agenda sets out the main areas where Europe is lacking the capacity to provide e-services within its borders, including the provision of health services online. The Digital Agenda will require sustained levels of commitment on both European and member state levels (including at regional levels) (Aristovnik et al., 2021).

The guidance provided by European Investment Fund (EIF) is effective on a governance and strategy level and is designed to enhance coordination within the EU and improve how EU countries develop public services with cross-border connectivity (Casiano Flores et al., 2022). Developed by the European Commission, the Agenda provides shared principles and guidelines for developing interoperable, integrated digital public services throughout the EU and beyond. The hypothesis is that the effects on citizens of the coordinated strategy actions of EU-wide efforts to improve digital public services interoperability can be assessed. Introduction digitalization holds tremendous promise to improve public services delivery across the EU (Campmas et al., 2022). The new regulation seeks to establish a new European digital identity, making it easier and safer for citizens to interact online with businesses and public services throughout the EU and ensuring that citizens have greater control over the data that they share and how that data is used. The Digital Services Act seeks to introduce measures for online platforms and online intermediary services, to make the online environment safer and more transparent (Sule et al., 2021). This initiative is expected to have implications for regulation of content on the Internet, as well as potentially having implications for other EU legislations, such as the Digital Services Act. The EU Commission is similarly ambitious, hoping that several EU Commission initiatives can successfully establish the gold standard of digital regulation globally, just as it did for Europe's landmark General Data Protection Regulation (GDPR) (Schünemann & Windwehr, 2021). Given the complexity and rapid pace of developments in the digital domain, it is likely that many European Commission proposals will need to be subjected to extensive review, debate, and detailed negotiations with the European Parliament and Member States before they can receive ratification (Burni et al., 2021). Budgets, and funding allocations among Europe's priorities – social, economic, or environmental – are guided by conclusions from the partnership agreements that are agreed between the Commission and the national authorities of Member States. The EU supports economic development projects within Member States and also economic integration projects at EU level that traverse Europe's borders, internal or external. This has also played a critical role in supporting EU enlargement, through loans used to fund improvements to infrastructure, research, and industrial production, in order to assist countries preparing for possible European Union membership (Jiroudkova et al., 2015; Nugent & Rhinard, 2016). The European Structural Funds and Investment Funds, including the European Regional Development and the European Social Fund, were created in 1975 to help alleviate economic and social differences among the regions of the EU. In addition to these efforts at public sector policy, from the point of view of industrial policy and geopolitics, these policies are discussed at a strategic level at the European Commission and at the Member State capitals (Camagni et al., 2020). In the light of all this, it appears to us that efforts should be strengthened to enhance the quality of electronic public services throughout the EU, to support the common digital agenda of the EU, and to enhance benefits to European citizens. Wider adoption and better exploitation of digital technologies will allow Europe to tackle key challenges in Europe and provide Europeans with better quality of life, through better healthcare, safer, and more efficient transportation solutions, cleaner environments, new media opportunities, and easier access to public services and cultural content, for instance (Lytras & Şerban, 2020). The Green Growth Europe needs to harness digital technologies to create a sustainable, low-carbon, and resource-efficient economy and society. It might be the good time for us to embrace the changes brought by digitization and globalization, with clear vision and responsible leadership (Matos et al., 2022). The question is what we want and how we are going to get there. There is an attempt to create a Europe in which digital technologies, innovation, and AI enable European people to have better jobs, better healthcare, and better public services. A strong, non-fragmented, digitalized Europe, taking leadership to build digital inclusion, green growth, innovation, trust, agile, and mission-based policymaking, promoting prosperity, and creating benefits for European societies, would be the true leader of the open economies (Korneeva et al., 2021).

Therefore, Europe needs to address the suboptimal character of current R & D (research and development) efforts by leveraging more private investment, better coordination and pooling of resources, easier and faster access for digital SMEs to the European Union research funds, joint R & D infrastructures and innovation clusters, and the development of standards and open platforms for new applications and services. Weaknesses in standards setting, government procurement, and coordination among public authorities are hampering Europeans from working together as they should. All of that represents an ambitious project that would foster the e-government and introduce some leading elements into the process of digitalization of administration. And it appears that the recent COVID-19 pandemic acted as a trigger for its development and implementation.

11.2 e-Government and the COVID-19 Health Crisis

People worldwide deserve a prompt, vigorous, professional response to the growing public health and economic crises caused by the coronavirus outbreak and its adverse effects. In order to provide this response, governments all around the world are developing and implementing a remedy to the COVID-19 pandemic (Edwards & Ott, 2021). The coronavirus pandemic is impacting lives across the world, but people in developing countries are likely to suffer most in this global crisis. COVID-19 represents an unprecedented public health emergency, with associated major economic impacts, which is affecting all developed and developing countries. Having existed for only two years now, COVID-19 has caused tremendous human suffering, brought economies to a halt, and challenged governments worldwide in ways that are rarely seen in history (Yu et al., 2022). The UN stated that, in the course of the COVID-19 outbreak, governments across the globe began using digital platforms, such as web portals, social media, and mobile apps, to communicate virus-related information to the public. The World Health Organization (WHO) launched a new information platform called Epidemics Infonetwork (EPI-WIN) following the declaration of the COVID-19 pandemic, which is a public health emergency of international concern. In particular, leveraging e-government information services is crucial and essential to combating a deadly disease caused by the new virus (Utunen et al., 2021).

Policymakers could have relied on the recommendations from users to facilitate faster dissemination and uptake of COVID-19-related informational e-government services. This implies that users' decisions about whether or not to adopt informational e-government services related to COVID-19 will be determined by the utility of the provided information (Strielkowski et al., 2022). The appropriateness of the information and the informational ease of use of the informational e-government services supported findings from other researchers, who found that both were important factors in determining informational utility (Guliyeva & Rzayeva, 2018; Mensah et al., 2021).

Therefore, the present study attempts to investigate the leading factors that influence adoption of information about COVID-19 shared via informational e-government systems. The prompt delivery of the appropriate COVID-19 information, as well as making sure the information is delivered so users can digest it and make use of it with fewer efforts, would have an effect positively on their perceptions about the utility of the COVID-19 information shared through informational e-government (Yuan et al., 2021). During a COVID-19 outbreak, public health decision-makers can utilize e-government and social media platforms as effective tools for improving public participation in protective behaviors (Kesavan et al., 2022). Although social media has been recognized for long as a platform for spreading misinformation about health, as well as spreading panic in times of a health crisis, some studies found that social media was helpful for engaging the public in protective behaviors during the COVID-19 outbreak (Alqahtani et al., 2021). It was considered important to validate the frame of reference within the governments' setting, since electronic governments and social media were used extensively by the public health authorities to communicate about the health status of people during the COVID-19 outbreak. It is clear that various platforms' usage in local health agencies was useful, when comparing such behaviors within a COVID-19-like crisis. Moreover, it is critical to understand the implications of information distribution via Twitter and Facebook regarding the COVID-19 pandemic, as well as digital transformations in the structure of the governing bodies during and after this incident (Mellado et al., 2021).

In the near future, governments need to be able to perceive and respond quickly to changes in the COVID-19-like viruses and similar disasters. From healthcare to social services and transportation to IT, we are going to be living in a new world, and many parts of government are going to have to take on pandemic-related surveillance and management responsibilities (Marome & Shaw, 2021). Many parts of government must continue to operate with strong community contacts – thus the COVID-19 virus. Auditors are themselves constrained by their own limitations on physical access as well as by their need to avoid impeding government's rapid responses to the COVID-19 pandemic. It is crucial that we examine the role of electronic governance in managing the COVID-19 pandemic in order to foster sustained socioeconomic development, exploring the recent experience and insights drawn from the pandemic experience, which represents the path toward successful regional integration. The common practices are now adapting and responding to the global demand for goods and services in order to fight the COVID-19 pandemic (Ullah et al., 2021). Past experiences with the involvement in public responses to disasters, both natural and man-made, including health emergencies such as Ebola, offer good lessons for dealing with the COVID-19 pandemic. In public health, this can offer enormous opportunities for engaging diverse countries to collaborate and act collectively in global efforts to address the COVID-19 pandemic, as well as other communicable or chronic diseases, as well as outbreaks that pose potential threats to both health information management (HIM) and laboratory information management (LIM) systems (Lima de Miranda & Snower, 2022). This crisis offers numerous examples of governments exercising flexibility with workforce management: reallocating workers to areas with high demand, like healthcare and labor, relaxing hiring regulations, and using the full power of digital technologies to accommodate a virtual workplace. To deal with the COVID-19 pandemic, many countries had to adopt stricter social distancing measures, which were maintained through almost a year and were only partially loosened at the end of November and in December 2021, pending improvements in COVID statistics. They have also eased requirements to access unemployment benefits and rental assistance to individuals whose economic and social situations have deteriorated as a result of the COVID-19 pandemic. In some countries, federal governments agreed with the subjects of the federations (provinces, republics, states) to share public health costs incurred by them in treating COVID-19 and fighting against the pandemic using digital tools and techniques (Wu et al., 2020).

11.3 War Against COVID-19 as the Driver of Innovations

It can be observed over the past two years how the COVID-19 pandemic risked long-term damage to innovation systems at a time when science and innovation are needed the most to address the climate emergency, achieve the sustainable development goals, and accelerate the digital transition (Tonne, 2021). The collective experiences of the COVID-19 pandemic have also presented challenges, which hold lessons for the mobilization of science, technology, and innovation for its main goals. In the longer run, COVID-19 today reminds us that we need to cultivate the socially beneficial applications of digital technologies, with an emphasis on improving access and usage in countries that are lacking. In the short term, governments should continue their support of scientific and innovation activities aimed at developing solutions for the COVID-19 pandemic and mitigating its adverse effects, while paying attention to COVID-19's unequal distributional effects (Secundo et al., 2021). Improving the use and effectiveness of digital technologies is another value-added field for an economic response to COVID-19, and the week included an announcement of a program funded by the World Bank targeting developing countries all around the world. Deploying technology-enabled solutions to enhance government service delivery and improve access to online services by businesses and individuals is a critical component of these new funds, which address changing needs in response to COVID-19 (Sapire et al., 2022). Mobilizing business innovation activity on business innovation as it relates to the COVID-19 pandemic in addition to the research conducted by health sectors – usually in conjunction with government research organizations – industries have moved to respond to the challenges of COVID-19 (Hall, 2022). In the middle of the global pandemic, digital technologies have captured our imaginations because of their potential to support us in our battle against COVID-19. We need to explore the innovative potential of three emerging digital technologies (namely, IoT (Internet of Things), AI (artificial intelligence), and distributed ledgers) to address the challenges associated with the pandemic. Drawing on previous studies and applications of these three emerging technologies, with the spotlight being the unprecedented global pandemic, some researchers propose ways that academics and practitioners can efficiently harness their potential (Guggenberger et al., 2021). Thus, even during the unprecedented global pandemic, we have seen a number of companies, and researchers begin combining IoT, AI, and DLT (distributed ledger technology) in order to drive innovation. The mobilization of digital technologies in response to the COVID-19 pandemic opened up opportunities to foster knowledge-sharing, to gather relevant data for enabling science-based policy decisions, and significantly to lower the carbon emission, even temporarily. The concept of tipping points in technological adoption or in the disruptive nature of digital technologies is not new, but the results from this survey indicate that the COVID-19 crisis is a tipping point of historical proportions – and as economic and human conditions change, even greater changes are needed (Hoang et al., 2021). Factors shaping the future of scientific research include the uneven impact of the COVID-19 crisis on research and development (R & D) in different sectors, accelerating adoption of digital tools and technologies, and changes to openness, inclusivity, and the flexibility of R & D ecosystems (Corazza et al., 2021). The uneven effects of the pandemic across and within sectors could increase the disparity in future innovation outcomes between firms that have prospered during the pandemic (e.g., large digital technology firms) and those most affected, who are more cash-strapped and less capable of taking advantage of digital tools. Firms' positive responses to the crisis, in terms of adopting technologies (and introducing new products) in the short run, can contrast with impacts on fewer longer-run innovation activities, such as R & D. Although major platforms optimized their innovation spending and investments during the COVID-19 crisis, there is no reason to think capital flows would not have contributed to the Internet companies' proactive technology innovation and strategies during the post-pandemic period (Wang et al., 2020). More importantly, when we looked into the impact of the crisis on a number of measures about companies, including staffing levels, respondents said funding for digital initiatives has increased more than any other, more than costs, number of people working on digital or other tech roles, and number of customers. In a survey of the British companies, Riom and Sivropoulos-Valero (2020) found that over 60% of firms had adopted new digital technologies (such as remote work technologies or cloud computing) or new managerial practices since the pandemic began and almost 40% had invested in new digital capabilities (such as e-commerce or advanced analytics). Despite the fact that the full effects of the COVID-19 crisis on the market and the innovative and technological potential of the world's largest platform companies have yet to be assessed, the evidence analysis suggests that the research hypothesis that, in the context of the specific organization of platform innovation, the key driver of their further development is not so much their own technological activities but the development of innovation ecosystems and, more broadly, the system of engagement with stakeholders has been partially confirmed (Faulks et al., 2021). In sum, these examples illustrate that, while organizations seek to react to the imperatives in the COVID-19 crisis, they are either pursuing innovations in order to mitigate threats stemming from physical interdependence within their main technologies or buffering against risks associated with such interdependence. Some specific industries, such as hospitals, sports leagues, airlines, meatpacking plants, and ride-sharing companies, are affected by the physical interdependence in their core technologies and provide a good example of their product and process innovations, as they strive to contain the COVID-19 crisis. Researchers and policymakers alike should join their forces to help those industries and to increase their comparative advantage over the others.

11.4 Citizens' Satisfaction with Their Governments During the Pandemic

Research suggests that the COVID-19 crisis has led to stronger support for, or satisfaction with, governments and their policies – largely mediated through levels of trust and partisanship (Brouard et al., 2022).

Views on how each of the countries in the world has handled the pandemic, and on the impact COVID-19 has had on national solidarity, should be best examined within the context of longer-term trend data. Here, the role of the social media and the Big Data becomes crucial. Some studies investigate the roles of governments and public health agencies during the COVID-19 pandemic in many countries, as well as the public's views on policies implemented in each country, by analyzing Twitter data using both quantitative and qualitative methods. Their aim is to analyze the influence factors on citizen's dissatisfaction with public services during the pandemic in order to assist public departments to identify problems within service delivery processes and potential countermeasures (Kada et al., 2022). More studies and assessments examine how various levels of government worked together to address the COVID-19 pandemic, how citizens and stakeholders were involved in the decision-making processes related to crisis responses, and what measures might have helped protect democratic norms (Hai et al., 2021; Northington et al., 2021). Several evaluations note the utility of releasing health and infection data in a unified national dashboard in order to convey current pandemic status in various areas and show a scientific basis for policies responding to the crisis (Shadbolt et al., 2022). These early evaluations suggest that a number of governments have reached similar conclusions, thereby allowing us to identify important insights that could both inform current policy responses to the COVID-19 pandemic and improve resilience in the future. The Federal Member countries (Canada, Belgium, Canada, Switzerland, the United States) placed more importance than non-Federal Member countries on the assessment of cooperation across levels of government, for instance, regarding responsibility allocations across national and subnational governments at various stages of the crisis (Chattopadhyay et al., 2022). A government's policy regarding the COVID-19 pandemic vaccine, including its vaccine procurement, distribution plans, and priority setting methods, may represent further potential public perception drivers related to the governments' COVID-19 pandemic response in 2021. Evidence-based policies and leadership for emergency responses are among the most discussed topics in counties, and this research found that they were associated with increased public mistrust in government expertise and dissatisfaction with its pandemic approach and measures implemented. While the literature on rallying around the flag effect suggests there are compelling reasons for studying policy trust in the pandemic context, it does not give us firm expectations of what actions or strategies from government may produce greater trust (Belchior & Teixeira, 2021). The findings from the Swedish survey illuminate a process by which a short-term rally-around-the-flag effect, found in both Denmark and Sweden in earlier studies, was weakened in the early months of the pandemic. The findings indicate that the

COVID-19 crisis has caused significant changes in attitudes; however, they obscure the individual-level heterogeneity in each of the countries in the sample. Therefore, one is capable of comparing the levels of confidence that a single individual has in Sweden's ability to lead the country through the pandemic in safety across two different points in time (Nielsen & Lindvall, 2021). Perceptions that the government responded too slowly in times of crisis are the strongest factors explaining confidence in government, even controlling for effects of crisis (number of deaths in country) and several other factors. Even in the case of New Zealand, which was least exposed to coronavirus, in terms of deaths, citizens did respond to policies related to COVID-19 and adjust to the governments' recommendations (Blair et al., 2022).

11.5 Conclusions

In summary, we can note that there are several leading public technology trends nowadays with potential to accelerate digital innovation and optimize or transform government services in the post-COVID era. This list of strategic technology trends is directly related to government management and policy challenges government leaders should be grappling with nonprofits and high-tech companies that focus on data governance and Web 2.0.

As the digital economy grows, with new leadership business models, technologies, products, and services, regulatory agencies worldwide may benefit from cooperative approaches, such as joint rulemaking, self-regulation, and international coordination. As emerging technologies fuel new business and service models, governments need to create, adapt, and enforce regulations quickly. As new business models and services such as ride-sharing services and initial coin offerings arise, governmental agencies are challenged with creating or changing regulations, enforcing them, and communicating these regulations to the public at previously unimagined speeds. The issue of speed has gained renewed urgency because of how quickly contemporary innovations are scaled. Digital products, services, and industries have the potential to grow very big and very quickly. Developments like new data technologies, as well as the need to upgrade traditional methods of transportation, communications, and manufacturing, have made it necessary, or at least preferable, to the status quo.

Since modernization involves the social transition from an agrarian to an industrial society, the perspective of technology is important; however, new technologies alone will not transform society. These modernization theories also consider the domestic dynamics, while still taking into account the social and cultural structures, as well as adaptations of the new technologies. Technology makes possible more innovative societies and wider social changes. Today, the open governments' mentioned above methods for transparency, public engagement, and accountability are based in some or other ways on the innovative use of technology.

More than the first few reforms, open government reforms are highly reliant on technology innovations which was very well illustrated by the COVID-19 pandemic. Open governments have reasonable opportunities, such as geographical adaptation, reduced dependence on the central authority, and the participation of new political actors, for breaking with the historical strains and barriers to government governance reform. Moreover, we have already seen how core ideas about accessing government information and citizens' participation have much older antecedents in government management, suggesting perhaps that open government reform has more basic roots in the systems of government. These different ideas for reform at different times provide fresh inspiration to public managers. What cannot be forgotten about the new technologies is the dynamic tension between the liberating effects of these technologies, which have the potential to democratize information access and flows, and the new emerging models of corporate ownership and income generation, which require the control of those very technologies. When they are allied to the interests of war and domestic surveillance, these new technologies are largely insulated from civil regulation and supervision. Relevant stakeholders and authorities are keen on adopting a composable government in order to move beyond existing, siloed approaches to managing services, systems, and data, which limits governments' ability to adapt to the fast-evolving needs of an emerging digital society. Government CIOs can leverage principles and practices from hyper automation to design highly automated, highly connected, end-to-end government processes and services that require little or no human intervention. Hyperconnected public services are a government-wide application of a variety of technologies, tools, or platforms that are used to automate as many of its business and IT processes as possible. GovTech is a whole-of-government approach to public sector modernization. The Office of Electronic Government and Information Technology (E-Gov), headed by the federal governments' chief information officer, develops and provides guidance on using Internet-based technologies to simplify interactions between citizens and businesses with the government, saving taxpayer dollars, and streamlining citizen engagement. More than a half of the technology companies provide products and services to government to offer packaged enterprise capabilities supporting composable applications within a few years from now. In addition, the majority of the new IT investments made by government agencies will be in service in the next several years. Unfortunately, many governments have mostly missed this transition because they have mismanaged their tech investments, where IT projects have far too often cost hundreds of millions more than needed, taken years longer to implement, and delivered technologies that were outdated when completed. At a time of expanding government services, some strategies rely heavily on funding resources to roll out new programs with new staffing levels and technologies. This stratification in technologies led to new emphasis on providing better access for everyone. Media and technology have evolved together, from the first printing presses to today's publications and radio to television and movies. Recent scholarship on new media has offered mixed accounts about its potential for bringing men and women into an equal footing within the realms of technology and public discourse. The activity has focused on accountability tools, such as charters and standards for services, with mechanisms for ensuring compliance and monitoring, as well as the use of technology to promote greater government transparency. Modernization in developing countries' healthcare sectors acknowledges that transitioning from traditional to modern is more than just the advance of technology and adoption of Western practices. The implementation of modern healthcare requires the restructuring of policy agendas, which, in turn, requires increased financing by public health-related funds and resources. In this second stage, contemporary societies face a host of new challenges, whose solutions often appear to lie outside of the purview of traditional nation-states. This very well illustrates how the enterprises need to establish a plan for the administration of human resources (HR) at a governmental level, including creating shared service centers to manage human resources. Thence, governments should balance their duties to protect citizens with the promotion of innovation in emerging technologies and businesses, resisting the impulse for excessive regulation. Those bringing technologies into underdeveloped countries – whether nongovernmental organizations, businesses, or governments – often have an agenda. A functionalist, by contrast, may focus on the ways that technology has created new ways of sharing information about a successful agricultural program or on the economic benefits of opening up a new market for the new technology or innovation.

References

Alqahtani, A. S., Alrasheed, M. M., & Alqunaibet, A. M. (2021). Public response, anxiety and behaviour during the first wave of COVID-19 pandemic in Saudi Arabia. *International Journal of Environmental Research and Public Health, 18*(9), 428. https://doi.org/10.3390/ijerph18094628

Arif, M., Bin, M. F., & Choo Ta, G. (2022). COVID-19 pandemic management: A review of the digitalisation leap in Malaysia. *Sustainability, 14*(11), 6805. https://doi.org/10.3390/su14116805

Aristovnik, A., Kovač, P., Murko, E., Ravšelj, D., Umek, L., Bohatá, M., & Tomaževič, N. (2021). The use of ICT by local general administrative authorities during Covid-19 for a sustainable future: Comparing five European countries. *Sustainability, 13*(21), 11765. https://doi.org/10.3390/su132111765

Bali, A. S., He, A. J., & Ramesh, M. (2022). Health policy and COVID-19: Path dependency and trajectory. *Policy and Society, 41*(1), 83–95. https://doi.org/10.1093/polsoc/puab014

Belchior, A. M., & Teixeira, C. P. (2021). Determinants of political trust during the early months of the COVID-19 pandemic: Putting policy performance into evidence. *Political Studies Review*, 14789299211056193. https://doi.org/10.1177/14789299211056193

Blair, A., de Pasquale, M., Gabeff, V., Rufi, M., & Flahault, A. (2022). The end of the elimination strategy: Decisive factors towards sustainable management of COVID-19 in New Zealand. *Epidemiologia, 3*(1), 135–147. https://doi.org/10.3390/epidemiologia3010011

Brouard, S., Foucault, M., Michel, E., Becher, M., Vasilopoulos, P., Bono, P. H., & Sormani, N. (2022). 'Citizens' attitudes under Covid19', a cross-country panel survey of public opinion in 11 advanced democracies. *Scientific Data, 9*(1), 1–10. https://doi.org/10.1038/s41597-022-01249-x

Burni, A., Erforth, B., & Keijzer, N. (2021). Global Europe? The new EU external action instrument and the European Parliament. *Global Affairs, 7*(4), 471–485. https://doi.org/10.1080/23340460.2021.1993081

Camagni, R., Capello, R., Cerisola, S., & Fratesi, U. (2020). Fighting gravity: Institutional changes and regional disparities in the EU. *Economic Geography, 96*(2), 108–136. https://doi.org/10.1080/00130095.2020.1717943

Campmas, A., Iacob, N., & Simonelli, F. (2022). How can interoperability stimulate the use of digital public services? An analysis of national interoperability frameworks and e-government in the European Union. *Data & Policy, 4*, E19. https://doi.org/10.1017/dap.2022.11

Casiano Flores, C., Rodriguez Müller, A.P., Virkar, S., Temple, L., Steen, T., & Crompvoets, J. (2022). Towards a co-creation approach in the European interoperability framework. Transforming Government. *People, Process and Policy, 16*(4), 519–539. https://doi.org/10.1108/TG-03-2022-0033

Chattopadhyay, R., Knüpling, F., & Chebenova, D. (2022). Impact of COVID-19 on the comparative practice of federalism: Some preliminary observations. *Cadernos Gestão Pública e Cidadania, 27*(87), 1–16. https://doi.org/10.12660/cgpc.v27n87.85110

Corazza, M. V., Moretti, L., Forestieri, G., & Galiano, G. (2021). Chronicles from the new normal: Urban planning, mobility and land-use management in the face of the COVID-19 crisis. *Transportation Research Interdisciplinary Perspectives, 12*, 100503. https://doi.org/10.1016/j.trip.2021.100503

Edwards, F. L., & Ott, J. S. (2021). Governments' responses to the COVID-19 pandemic. *International Journal of Public Administration, 44*(11–12), 879–884. https://doi.org/10.1080/01900692.2021.1936964

Faulks, B., Song, Y., Waiganjo, M., Obrenovic, B., & Godinic, D. (2021). Impact of empowering leadership, innovative work, and organizational learning readiness on sustainable economic performance: An empirical study of companies in Russia during the COVID-19 pandemic. *Sustainability, 13*(22), 12465. https://doi.org/10.3390/su132212465

Guggenberger, T., Lockl, J., Röglinger, M., Schlatt, V., Sedlmeir, J., Stoetzer, J. C., & Völter, F. (2021). Emerging digital technologies to combat future crises: Learnings from COVID-19 to be prepared for the future. *International Journal of Innovation and Technology Management, 18*(04), 2140002. https://doi.org/10.1142/S0219877021400022

Guliyeva, A., & Rzayeva, U. (2018). The role of e-government in reduction of information asymmetry in developing countries on the example of Azerbaijan. *Advances in economics and business, 6*(4), 209–217. https://doi.org/10.13189/aeb.2018.060402\

Hai, T. N., Van, Q. N., & Thi Tuyet, M. N. (2021). Digital transformation: Opportunities and challenges for leaders in the emerging countries in response to COVID-19 pandemic. *Emerging Science Journal, 5*, 21–36. https://doi.org/10.28991/esj-2021-SPER-03

Hall, J. (2022). Public Administration for a new Season: Reflecting on neglected rural concerns. *Public Administration Review, 82*(4), 613–618. https://doi.org/10.1111/puar.13536

Hoang, A. T., Nižetić, S., Olcer, A. I., Ong, H. C., Chen, W. H., Chong, C. T., & Nguyen, X. P. (2021). Impacts of COVID-19 pandemic on the global energy system and the shift progress to renewable energy: Opportunities, challenges, and policy implications. *Energy Policy, 154*, 112322. https://doi.org/10.1016/j.enpol.2021.112322

Jiroudkova, A., Anna, L., Strielkowski, W., & Slosarcík, I. (2015). EU accession, transition and further integration for the countries of central and Eastern Europe. *Economics and Sociology, 8*(2), 11–25. https://doi.org/10.14254/2071-789X.2015/8-2/1

Kada, A., Chouikh, A., Mellouli, S., Prashad, A. J., Straus, S. E., & Fahim, C. (2022). An exploration of Canadian government officials' COVID-19 messages and the public's reaction using social media data. *PLoS One, 17*(9), e0273153. https://doi.org/10.1371/journal.pone.0273153

Kersan-Skabic, I. (2021). Digital trade enablers and barriers in the European Union. *Montenegrin Journal of Economics, 17*(4), 99–109. https://doi.org/10.14254/1800-5845/2021.17-4.9

Kesavan, N., Seenuvasaragavan, S., Burnette, K., Heim, A. B., Patel, R. J., Johnston, J., Cao, X., Sakandar, H. A., Zhuang, Y., Oi, K. K., Zhi, Y., Diao, Y., & Strielkowski, W. (2022). Pandemic-inspired policies. *Science, 377*(6601), 22–24. https://doi.org/10.1126/science.add2046

Korneeva, E., Olinder, N., & Strielkowski, W. (2021). Consumer attitudes to the smart home technologies and the internet of things (IoT). *Energies, 14*(23), 7913. https://doi.org/10.3390/en14237913

Lima de Miranda, K., & Snower, D. J. (2022). The societal responses to COVID-19: Evidence from the G7 countries. *Proceedings of the National Academy of Sciences, 119*(25), e2117155119. https://doi.org/10.1073/pnas.2117155119

Lytras, M. D., & Şerban, A. C. (2020). E-government insights to smart cities research: European union (EU) study and the role of regulations. *IEEE Access, 8*, 65313–65326. https://doi.org/10.1109/ACCESS.2020.2982737

Marome, W., & Shaw, R. (2021). COVID-19 response in Thailand and its implications on future preparedness. *International Journal of Environmental Research and Public Health, 18*(3), 1089. https://doi.org/10.3390/ijerph18031089

Matos, S., Viardot, E., Sovacool, B. K., Geels, F. W., & Xiong, Y. (2022). Innovation and climate change: A review and introduction to the special issue. *Technovation, 102612*, 102612. https://doi.org/10.1016/j.technovation.2022.102612

Mellado, C., Hallin, D., Cárcamo, L., Alfaro, R., Jackson, D., Humanes, M. L., & Ramos, A. (2021). Sourcing pandemic news: A cross-national computational analysis of mainstream media coverage of COVID-19 on Facebook, twitter, and Instagram. *Digital Journalism, 9*(9), 1261–1285. https://doi.org/10.1080/21670811.2021.1942114

Mensah, I. K., Adams, S., Adjei, J. K., & Mwakapesa, D. S. (2021). Drivers of egovernment adoption amidst COVID-19 pandemic: The information adoption model (IAM) approach. *Information Development*, 02666669211010872. https://doi.org/10.1177/02666669211010872

Nielsen, J. H., & Lindvall, J. (2021). Trust in government in Sweden and Denmark during the COVID-19 epidemic. *West European Politics, 44*(5–6), 1180–1204. https://doi.org/10.1080/01402382.2021.1909964

Northington, W. M., Gillison, S. T., Beatty, S. E., & Vivek, S. (2021). I don't want to be a rule enforcer during the COVID-19 pandemic: Frontline employees' plight. *Journal of Retailing and Consumer Services, 63*, 102723. https://doi.org/10.1016/j.jretconser.2021.102723

Nugent, N., & Rhinard, M. (2016). Is the European Commission really in decline? *JCMS: Journal of Common Market Studies, 54*(5), 1199–1215. https://doi.org/10.1111/jcms.12358

Riom, C., & Sivropoulos-Valero, A. V. (2020). Covid has forced many firms to innovate, with possible lasting impacts. *LSE Business Review.* Available at: https://blogs.lse.ac.uk/businessreview/2020/10/07/covid-has-forced-many-firms-to-innovate-with-possible-lasting-impacts

Sapire, R., Ostrowski, J., Maier, M., Samari, G., Bencomo, C., & McGovern, T. (2022). COVID-19 and gender-based violence service provision in the United States. *PLoS One, 17*(2), e0263970. https://doi.org/10.1371/journal.pone.0263970

Schünemann, W. J., & Windwehr, J. (2021). Towards a 'gold standard for the world'? The European general data protection regulation between supranational and national norm entrepreneurship. *Journal of European Integration, 43*(7), 859–874. https://doi.org/10.1080/07036337.2020.1846032

Secundo, G., Shams, S. R., & Nucci, F. (2021). Digital technologies and collective intelligence for healthcare ecosystem: Optimizing internet of things adoption for pandemic management. *Journal of Business Research, 131*, 563–572. https://doi.org/10.1016/j.jbusres.2021.01.034

Shadbolt, N., Brett, A., Chen, M., Marion, G., McKendrick, I. J., Panovska-Griffiths, J., & Swallow, B. (2022). The challenges of data in future pandemics. *Epidemics, 100612*, 100612. https://doi.org/10.1016/j.epidem.2022.100612

Strielkowski, W., Zenchenko, S., Tarasova, A., & Radyukova, Y. (2022). Management of smart and sustainable cities in the post-COVID-19 era: Lessons and implications. *Sustainability, 14*(12), 7267. https://doi.org/10.3390/su14127267

Sule, M. J., Zennaro, M., & Thomas, G. (2021). Cybersecurity through the lens of digital identity and data protection: Issues and trends. *Technology in Society, 67*, 101734. https://doi.org/10.1016/j.techsoc.2021.101734

Tonne, C. (2021). Lessons from the COVID-19 pandemic for accelerating sustainable development. *Environmental Research, 193*, 110482. https://doi.org/10.1016/j.envres.2020.110482

Ullah, A., Pinglu, C., Ullah, S., Abbas, H. S. M., & Khan, S. (2021). The role of e-governance in combating COVID-19 and promoting sustainable development: A comparative study of China and Pakistan. *Chinese Political Science Review, 6*(1), 86–118. https://doi.org/10.1007/s41111-020-00167-w

Utunen, H., Van Kerkhove, M. D., Tokar, A., O'Connell, G., Gamhewage, G. M., & Fall, I. S. (2021). One year of pandemic learning response: Benefits of massive online delivery of the World Health Organization's technical guidance. *JMIR Public Health and Surveillance, 7*(4), e28945. https://doi.org/10.2196/28945

Wang, Y., Hong, A., Li, X., & Gao, J. (2020). Marketing innovations during a global crisis: A study of China firms' response to COVID-19. *Journal of Business Research, 116*, 214–220. https://doi.org/10.1016/j.jbusres.2020.05.029

Wu, J., Wang, J., Nicholas, S., Maitland, E., & Fan, Q. (2020). Application of big data technology for COVID-19 prevention and control in China: Lessons and recommendations. *Journal of Medical Internet Research, 22*(10), e21980. https://doi.org/10.2196/21980

Yu, Z., Razzaq, A., Rehman, A., Shah, A., Jameel, K., & Mor, R. (2022). Disruption in global supply chain and socio-economic shocks: A lesson from COVID-19 for sustainable production and consumption. *Operations Management Research, 15*, 233–248. https://doi.org/10.1007/s12063-021-00179-y

Yuan, Y. P., Tan, G. W. H., Ooi, K. B., & Lim, W. L. (2021). Can COVID-19 pandemic influence experience response in mobile learning? *Telematics and Informatics, 64*, 101676. https://doi.org/10.1016/j.tele.2021.101676

Chapter 12
Leading Information and Communication Model for Social and Economic Sustainability of the Post-Pandemic Small and Medium Business

Elena Korneeva

Abstract This chapter focuses on the leading information and communication model for social and economic sustainability of the post-pandemic small and medium enterprises (SMEs). This chapter suggests the key outcomes about the business landscape after the pandemic, offering fresh perspectives on digital transformation, the future of work, transparency, and resilience. In addition, it discusses how lessons on COVID-19 pandemic digitalization from the sustainable economic development of SMEs can inform resilience to COVID-19 as well as strengthen leadership in the post-COVID era. Furthermore, it suggests additional areas for research that could enhance a successful digital transformation of SMEs in the aftermath of COVID-19. This chapter presents a framework for supporting the post-COVID-19 digital transformation to sustain SMEs' development and suggests practical and research implications for improving the successful digital transformation of post-COVID-19 small and medium business.

Keywords Information and communication · Leadership · Small and medium enterprises · Sustainability · Post-COVID era

JEL Classification D80 · L20 · L15

E. Korneeva (✉)
Department of Mass Communications and Media Business, Financial University under the Government of the Russian Federation, Moscow, Russian Federation
e-mail: ENKorneeva@fa.ru

W. Strielkowski (ed.), *Leadership, Entrepreneurship and Sustainable Development Post COVID-19*, Springer Proceedings in Business and Economics, https://doi.org/10.1007/978-3-031-28131-0_12

12.1 Introduction

As the dust from the coronavirus pandemic has settled and new challenges have started to emerge, numerous commentators, activists, and researchers are discussing whether and how increased consciousness created by COVID-19 could be used to enhance ecological consciousness (Chen & Zhao, 2022) and to reframe economic models (Sibley et al., 2020). A new canvassing of experts on technology, communications, and social change found many expected similar impacts to arise from the 2020 COVID-19 pandemic. The majority of businesses, both micro-, small and medium, and large enterprises had to lay off employees because of the risks presented by COVID-19 and the ensuing containment measures. Evidence from business surveys of the current impact of the COVID-19 pandemic crisis on SMEs indicates significant disruptions and concerns among small businesses. Based on multiple surveys across several countries, it has been found that from 25% to 36% of small businesses may shut down for good due to the disruptions during the first 4 months of the pandemic (Beglaryan & Shakhmuradyan, 2020; Waldkirch, 2021). A majority of micro-enterprises and small businesses, especially those in service sectors, predicted they would need to close down in 1–3 months, if the pandemic continues and the existing restrictions are maintained (Fairlie, 2020). It can be therefore observed that micro-, small-, and medium-sized businesses are feeling squeezed by COVID-19 and related restrictions on businesses. According to other cross-country surveys of small and medium enterprises, two-thirds of micro-enterprises reported the ongoing COVID-19 pandemic has significantly affected their business operations, with a fifth reporting a risk of closing down for good in 3 months (Endris & Kassegn, 2022). Pak et al. (2020) further mentioned in their research that as some countries, for example, China, recover from the pandemic, small businesses are struggling to open up again, while medium-sized enterprises are likely to suffer the worst impact, since they lack alternative arrangements such as electronic trading and IT infrastructure. According to the study conducted on 86,000 small businesses in the United States, more than half of businesses increased their online interactions with customers in order to adjust to the ongoing COVID-19 pandemic (Klein & Todesco, 2021). During post-COVID, the so-called high-performing small businesses invested over double as much in digital tools compared with the so-called low-performing small businesses (Belitski et al., 2022). Although self-employed workers were hit initially more heavily than larger firms in the United States and Europe after the COVID shutdown, there is cause for optimism, as adoption of digital tools is achievable for millions of SMEs who are still lacking skills, technology, and resources. Entrepreneurship in a post-COVID world will be further integrated into the digital economy. Trends in digital skills appear to interact with post-COVID and its societal, political, economic, ecological, and demographic strains, all combining to speed up reconfigurations of the productive and service systems (Kalenkoski & Pabilonia, 2022). The commitment to enterprise flexibility, artificial intelligence, data and analytics, and other emerging technologies has increased in the wake of COVID-19. In addition, COVID-19 has challenges and is likely to bring new digitalization technologies into the supply chain context, many of which impact supply chain resilience. The present research will examine COVID-19 impacts for two types of firms, analyzing how IT infrastructure helps firms sustain economic activities in these situations. A similar analysis was conducted in many recent research papers that classified industries into three categories according to their exposure risks from the ongoing COVID-19 pandemic (immediate risks, near-term risks, and longer-term risks) and examined the small businesses exposure to each of these categories (Gebhardt et al., 2022).

Thence, we argue in this contribution that innovation practices by SMEs under any ecological circumstances, such as COVID-19 pandemic, may contribute positively to firm productivity. The strategy of practices has shown that competitiveness in one firm usually requires a lengthy learning process, which, locally, could be achieved by the joint efforts of all universities, research institutions, investors, and entrepreneurs, creating and developing some strong networks that aim at strengthening certain emerging technologies and creating some entrepreneurial incubators that could support local innovation processes. In the globalized era – a society based on knowledge, respectively – SMEs reveal exclusive competitive advantages in the strength of flexibility and rapid adaptation to new production and global market requirements, exposure to new industrial know-how, the ability to identify and apply the value of new external information, the ability to rapidly adopt modern technologies for new business models, the fast decision-making, and lack of bureaucracy in the process of implementing new products and innovative processes.

12.2 COVID-19 and Its Implications for SMEs

Generally, it has to be acknowledged, based on the analysis of many studies and the data that was available at the time of writing this contribution, that COVID-19 restrictions and shutdowns were more severe for small and medium enterprises compared to larger and global firms (Shafi et al., 2020). From our research, we can see that micro-, small-, and medium-sized enterprises worldwide are feeling squeezed by COVID-19 and related restrictions on businesses. A majority of micro-enterprises, especially those in service sectors, anticipated having to shut down in a very short span of time if the pandemic continued and the existing restrictions were maintained. Evidence of the COVID-19 crisis impacts on SMEs derived from surveys of businesses indicates significant disruptions and concerns for small businesses (Lu et al., 2021). It was found by many researchers that although the small business sector as a whole has been adversely affected by the COVID-19 pandemic, there is a large variability in the size of impacts. This provides an initial assessment of the impact of COVID-19 and subsequent economic recession on small businesses across the world. Many research papers and reports seek to evaluate the effects of COVID-19 on small businesses, as well as provide policy recommendations that will assist micro-enterprises to mitigate their losses and survive in this crisis (Kesavan et al., 2022).

However, more detailed case studies are needed to understand how the entrepreneurs are managing COVID-19 crises and coping with its adverse effects, as well as the governments' policies, across countries, because the vulnerabilities and challenges faced by the SMEs are diverse. In particular, they are required to present estimates on the impact of the national crisis on small firms' financial outcomes. In order to do that, changes in the small businesses cash balance, revenues, and expenses from March 2020 (when the World Health Organization declared COVID-19 a global pandemic event) using a deidentified sample of millions of small firms across various countries need to be studied and assessed.

Another issue is female businesses versus male ones. Female-owned businesses are really concentrated in the consumption-oriented sectors (services, hospitality, retail), which are hit the hardest by demand disruption. For instance, across the OECD, 60–70% of small- and medium-sized enterprises operate in these sectors, making small- and medium-sized enterprises most affected by the COVID-19 crisis (Dey et al., 2022). It is apparent that SMEs are substantially affected by the COVID-19 crisis because they are overrepresented in sectors that are most affected, such as retail, hospitality, food services, leisure services, and construction. On the other hand, SMEs were most affected by COVID-19 relative to larger enterprises, as they mostly do business in sectors overrepresented in SMEs. Agricultural SMEs are also sadly especially hard hit, particularly as the COVID-19 pandemic has had severe impacts on the agriculture sector (Joshi et al., 2021). Especially in the time of COVID-19 crisis, many SMEs stopped functioning because of rising rents and dwindling cash reserves, making them economically vulnerable. Some surveys and palls show that two-thirds of micro-enterprises reported the ongoing COVID-19 pandemic has significantly affected their business operations, with a about one-fifth reporting a risk of closing down for good in 3 months. Some businesses had to improvise and to digitalize by increasing their online interactions with customers in order to adjust to the crisis (Ratten, 2021).

All around the world, businesses are responding to COVID-19-induced economic disruptions in various ways. COVID-19 has impacted many businesses worldwide through full or partial shutdowns and social distancing measures, all essential measures taken to curb the spread of the virus around the globe. The coronavirus disease outbreak has seriously affected the world's economy, so the government authorities had to intervene using the funding support schemes. For example, Scotland set to make grants worth not less than 3000 pounds to small businesses across sectors facing the worst economic impacts from the ongoing COVID-19 pandemic (Mitha, 2020). Another new support to small businesses affected by the COVID-19 was set to assist over 150,000 small businesses, strengthening the Economic Injury Disaster Loan (EIDL) program, which provides long-term, low-cost loans. Stakeholders need to make sure no small business has to begin repaying those loans until 2 years after receiving funding, so small businesses can weather the pandemic without worrying about making payments (Belitski et al., 2022). At a sectoral level, higher proportions of businesses in manufacturing and services reported reduced capacity to pay off outstanding debt as COVID-19 spread, than did those in agriculture. Notably, a relatively higher proportion of small-to-medium-sized businesses in services, in particular, reported declines in access to credit and financial liquidity relative to larger businesses. In short, it has been found that micro-enterprises experienced greater business downturns than mid-market and larger firms – an unsurprising finding given that the majority of the nation's micro-enterprises shut down operations because they were unable to adopt preventative health measures, such as providing employees with housing and sanitary equipment for customers. During the COVID-19 pandemic, as already stated before, women-owned small- and medium-sized businesses faced added stress, since women were frequently asked to concurrently hold traditional caregiving roles while also running businesses. Surely, the post-COVID era presents lots of challenges for SMEs that need to reestablish their position in the world's economy.

12.3 COVID-19 and Capital and Supply Chains

On top of the problems induced on SMEs, COVID-19 pandemic has also impacted capital and supply chains influencing the distribution and availability of products. Unlike past studies, we investigated challenges faced by companies as a result of COVID-19 outbreaks within a supply chain context, as well as investigated the relationships between them. Furthermore, multiple recent studies on COVID-19 indicate that improving supply chain resiliency is crucial for mitigating supply chain-related challenges (Chowdhury et al., 2021; Raj et al., 2022). Supply chain resilience can be enhanced through increasing inventories of key materials, works-in-progress, and finished products; adding production and/or warehouse facilities to enhance production surge capabilities; and increasing and assuring surge capabilities for suppliers of key materials or works-in-progress, in order to mitigate possible disruptions from suppliers. Additional regional capacity would also provide consumers with greater options for purchasing local-produced products by helping to ensure that foods are accessible to consumers and reducing the climate impacts on the food supply chains (Burgos & Ivanov, 2021). Many relevant stakeholders and international organizations invest in solving problems related to the limited capacity of the food industry's various processing, distribution, storage, and aggregation sectors, including the high costs of equipment, the lack of competition, and limited coordination in supply chains and value chains. As a result, many countries have the ability to close pressing gaps in global trade and supply chain infrastructure, working with trade partners to advance a new generation of diverse, sustainable, and competitive supply chains, with an intricate portfolio of off-shore, near-shore, and restored production – an effort that will support jobs, commerce, and the national security (Bahn et al., 2021).

Furthermore, COVID-19 pandemic has led to restrictions on workers' movements, changes in consumer demand, shutdowns of food manufacturing plants, restrictive food trade policies, and financial strains in food supply chains. The pandemic had significant impacts on the food trade, causing interruptions to the food supply chains because of the restrictions in exports. The coronavirus outbreak caused significant increases in food prices related to the restrictions imposed on shipments, which were followed by panic purchasing, and also caused supply chain disruptions (Ben Hassen & El Bilali, 2022). The logistics processes at the fish markets have been also affected by the pandemic. Both export and import prices of fish and shellfish experienced a shock due to reduced demand and supply chain disruptions caused by the COVID-19 pandemic (White et al., 2021). A supply bottleneck at processing plants (caused by the redesign of packing lines) and the COVID-19 shutdowns reduced demand for cattle slaughtered, and prices fell considerably in March 2020, forcing livestock producers either to accept lower prices or to hold back sales of their cattle (Marchant-Forde & Boyle, 2020). In June 2020, with orders coming back from stay-at-home households across the United States, restaurant demand increased, and supply chains began to revert to normal, measured in meat PPI, falling more than 25%, returning to close to the pre-COVID-19 levels. The shock to demand and problems with supply chains contributed to increased volatility in imports, exports, producer, and consumer prices (Hamulczuk & Skrzypczyk, 2021). Despite the demand from consumers for foods, food supply chains remain robust, as many actors within supply chains – including farmers, producers, distributors, and retailers – have worked diligently to renew shelves (Chkanikova & Sroufe, 2021).

The implications also include the importance of COVID-19 on the supply chain. Understanding the impact of a pandemic on the basis of restrictive policies is also essential, as levels of food supply are high and projections of key commodity production are good. The supply mismatch, according to Razdan and Kumar (2020), the supply-side capacity constraints and the price-quantity fluctuations are a significant disruption due to the COVID-19 pandemic. The COVID-19 pandemic outbreak has caused record-low demand for some commodities, which has had significant impacts on manufacturing-sector companies and their upstream suppliers (Sengupta et al., 2021). Adoption of the suboptimal substitutes at the advent of the coronavirus and lockdown restrictions has impeded continuity of operations for suppliers alike. A decades-long focus on streamlining the supply chain – to reduce costs, lower inventories, and increase utilization of assets – has removed buffers and flexibility to withstand disruptions and COVID-19 demonstrates how many companies are unaware of their supply chains' vulnerabilities to global disruptions. Even as companies seek to ramp up production and regain value chains, they should reserve logistics capabilities ahead of time to minimize their exposure to potential price increases. While markets will ultimately correct, it may be a slow process, and the impacts for producers and consumers could be costly. An interesting case was the eggs that are used in culinary and decorative purposes, and demand for them was driven by consumer panic purchasing stemming from the stay-at-home orders implemented in response to the COVID-19 pandemic. For instance, eggs that were destined for recently closed restaurants and food services customers were diverted into supermarket supply channels (Bulgari et al., 2021). The heightened demand by customers for certain goods –healthcare products and equipment, groceries, and home products which frequently shifted geographically by moving from one hotspot to another and a sharp drop-off in demand for other goods, essentially all of non-healthcare items – exposed the inability of supply chains to rapidly shift production and logistics to respond.

12.4 Lessons for the Digitalization of SMEs

In spite of its terrible effects and thousands of lives, the COVID-19 pandemic also brought the digitization lessons for sustainable development of small and medium enterprises. Identifying paths for digital transformation of business models in small- and medium-sized enterprises in a time of pandemic became a must for the survival of many businesses (Bai et al., 2021). This is why during the COVID-19 pandemic, leveraging digital and AI-driven technologies has been a crucial decision for maintaining SMEs' resilience.

With many rapidly deploying digital technologies, like artificial intelligence (AI), many companies embraced them into their production or service operations in order to pursue sustainable development (Guliyeva et al., 2020). COVID-19 has challenges and is likely to bring about new digital technologies in a supply chain context, many of which impact sustainability of supply chains. The main victims of the COVID-19 outbreak are micro-enterprises (MSEs), particularly in developing countries, mostly due to the limited adoption of digital technologies. In some countries, where two-thirds of the population works in micro-, small-, and medium-sized enterprises (MSMEs), local and international support are needed significantly to increase resilience and capacity to survive this interrupted business climate. Thence, the coercive pressures and digitization on micro-enterprises international buyers, foreign investors, professional associations, and transnational institutions have imposed coercive pressures on MSMEs to protect the environment in developing countries (Gupta et al., 2020).

Although research on the role of AI technologies on the survival and development of SMEs in the context of the COVID-19 pandemic has increased, a conceptual framework is lacking for understanding sustainable development in AI-enabled SMEs and for driving future research in a post-pandemic context (Denicolai et al., 2021). This contribution also elaborates the framework for supporting post-COVID-19 digital transformations to sustainably develop SMEs and suggests practical and research implications for strengthening the success of post-COVID-19 digital transformations in these enterprises. Thence, we further suggest additional areas for research that will improve successful digital transformation of MSEs after the novel coronavirus COVID-19 (Quayson et al., 2020). This perspective's contribution further discusses how lessons from the digitalization of the COVID-19 pandemic on the sustainability development of MSMEs could impact the resilience of these small enterprises after COVID-19 embedding the technology for social good perspective. Different MSME owners from various countries as well as international experts from the field related to the field of business need to meet and to share the main tips that MSMEs owners need to maintain and grow their businesses for during and after the new coronavirus. Following digital leadership and strategy training, small entrepreneurs can take various actions to promote digitalization levels for local SME businesses in pilot regions, while developing better knowledge about digital trends after the pandemic, the dynamics of digitization and transformation, next-generation business models, and innovation strategies, as well as designing short-, medium-, and long-term digitalization programs (Rodrigues & Franco, 2021). Governments and relevant stakeholders should seek to determine how ICTs are emerging as a strategy for entrepreneurship and investigate the effects of factors driving the uptake of ICTs on perceived benefits for micro-, small-, and medium-sized enterprises (Gebrihet & Pillay, 2021). During COVID-19, ICT's role as a business strategy was investigated. However, as the MSMEs are diverse; the intensity of pandemic impacts among them varied significantly depending on their characteristics.

Overall, in a digitally enabled economy of post-pandemic times, computational capacity is limited, and it is not always easily accessible or cost-effective for SMEs to account for, although the use of digital services and implementation of AI could be strategically developed and offer multiple sources of new revenues. The impact may be even greater for underrepresented segments. In the United Kingdom, for instance, some studies concluded that women-led SMEs were more likely to reject digital solutions for improving performance. This is, of course, a trend that needs to be reversed with the digitalization should be made available to all SMEs and micro-enterprises regardless of their size, gender of their owners, and other limiting factors.

12.5 Conclusions

All in all, it becomes clear based on our findings and the vast body of literature analyzed and studied that post-pandemic companies now need to rethink their business strategies, incorporating crisis scenarios, and creating their resilient business continuity plans for maintaining their leading positions.

Once the COVID-19 pandemic is under control, companies would want to revisit and update business continuity plans (BCPs). When the COVID-19 outbreak is over, it will become clear which companies have the resiliency and flexibility to reinvent their business strategies for success going forward. However, it is likely that the impacts of the COVID-19 crisis on businesses and their employees will be felt for a long time. The economic downturn caused by the pandemic does not necessarily mean an end to whole industries or businesses. For some, the COVID-19 crisis has put extreme strain on businesses and operational models, even breaking them. The task of reigniting the world economy after an economic downturn caused by COVID-19 is, of course, different. More generally, the pace and depth of the COVID-19 crisis has shown that the central principle underlying the global economy a prioritizing short-term economic growth and efficiency at the expense of longer-term sustainability can come at enormous social costs. As large stimulus packages begin to be unveiled across the globe, governments, businesses, and society at large all have both a responsibility and self-interest in looking beyond short-term measures to bolster livelihoods and employment, but to take a step back and consider the policy and economic drivers that led to the current crisis. From local restaurants to big corporations such as Google, Amazon, or Unilever, companies successfully weathering the COVID-19 pandemic and ensuing downturn often have switched to business models conducive to survival in both short-term and long-term resilience and growth. Technology companies, both small and large global players, are supporting this pivot, providing safe, remote working options that are helping businesses across the globe sustain employees and operations throughout COVID-19.

Our results demonstrate that many businesses were simply not prepared to adjust their work models to meet new regulations put in place to prevent the spread of COVID-19. In addition, it can be seen that COVID-19 has forced many businesses to make radical changes in order to stay in business, actions that they never previously considered. Throughout the COVID-19 pandemic, many businesses were either not prepared for, or could not afford, the security that would ensure their employees safety, thus risking their leading position on the market.

Leading global consultancies released their surveys of businesses plans for rebuilding resilience in their organizations during the post-COVID-19 era. To be successful and to become leaders in business and to remain profitable over the coming months and years, organizations need to create a business plan in the wake of COVID-19. In the absence of a COVID-19 vaccine or treatment, any upturn in business activity can easily be followed by another round of react, reset, reset, and then again, so the imperative is to rapidly incorporate lessons learned and incorporate lasting changes in the business and operational models. Organizations that keep up-to-date business continuity plans or enterprise resilience plans in place will be in a position to mitigate the effects of a crisis in order to sustain ongoing operations and to resume business activities with greater ease once the crisis has subsided. In the post-viral age, businesses may have to undertake significant changes affecting the whole of the business. To address the crisis and its consequences, companies will not only have to rapidly develop digital solutions but will have to adjust their organizations to the new operational models and to provide those solutions at scale, both to customers and employees. In the longer run, companies will have to evaluate how resilient their businesses, leadership teams, and initiatives are to a crisis. Organizations in all industries will need to revisit their plans for preparing for a crisis, so that they can concentrate on continuing to achieve business success and resilience. While no one can blame the companies' chief managers and principals for failing to fully anticipate the extent of the COVID-19 impacts to businesses, many of them are getting strategic now about planning for the future unknown scenarios that could occur. Some are calling the COVID-19 pandemic a "black swan event," as most businesses, even many experienced ones, failed to see it coming and were unprepared for its effects. The global pandemic has dramatically challenged how companies define, evaluate, manage, and track risks, upending long-held beliefs about safety, IT resilience, crisis management, and business continuity planning. All of the evidence stemming from the academic research papers, business reports, and governmental studies and papers indicate that the COVID-19 pandemic contributed to speeding up business automation plans, as companies are now recognizing the urgency of embracing automation to mitigate risks and make strategic investments. This new focus might help small and medium enterprises to effectively face the new disasters and eventual pandemics that might occur in the short-run and in the long-run future for achieving and maintaining the social and economic sustainability of the post-pandemic era.

Acknowledgments This chapter was prepared based on the results of research carried out at the expense of budgetary funds under the state assignment of the Financial University VTK-GZ-PI-40-22.

References

Bahn, R. A., Yehya, A. A. K., & Zurayk, R. (2021). Digitalization for sustainable agri-food systems: Potential, status, and risks for the MENA region. *Sustainability, 13*(6), 3223. https://doi.org/10.3390/su13063223

Bai, C., Quayson, M., & Sarkis, J. (2021). COVID-19 pandemic digitization lessons for sustainable development of micro-and small-enterprises. *Sustainable Production and Consumption, 27*, 1989–2001. https://doi.org/10.1016/j.spc.2021.04.035

Beglaryan, M., & Shakhmuradyan, G. (2020). The impact of COVID-19 on small and medium-sized enterprises in Armenia: Evidence from a labor force survey. *Small Business International Review, 4*(2), e298. https://doi.org/10.26784/sbir.v4i2.298

Belitski, M., Guenther, C., Kritikos, A. S., & Thurik, R. (2022). Economic effects of the COVID-19 pandemic on entrepreneurship and small businesses. *Small Business Economics, 58*(2), 593–609. https://doi.org/10.1007/s11187-021-00544-y

Ben Hassen, T., & El Bilali, H. (2022). Impacts of the Russia-Ukraine War on global food security: Towards more sustainable and resilient food systems? *Food, 11*(15), 2301. https://doi.org/10.3390/foods11152301

Bulgari, R., Petrini, A., Cocetta, G., Nicoletto, C., Ertani, A., Sambo, P., & Nicola, S. (2021). The impact of COVID-19 on horticulture: Critical issues and opportunities derived from an unexpected occurrence. *Horticulturae, 7*(6), 124. https://doi.org/10.3390/horticulturae7060124

Burgos, D., & Ivanov, D. (2021). Food retail supply chain resilience and the COVID-19 pandemic: A digital twin-based impact analysis and improvement directions. *Transportation Research Part E: Logistics and Transportation Review, 152*, 102412. https://doi.org/10.1016/j.tre.2021.102412

Chen, S., & Zhao, Y. (2022). Ecological civilization: A blindspot in global media coverage of China's environmental governance. *Environmental Communication, 16*(2), 195–208. https://doi.org/10.1080/17524032.2021.1981419

Chkanikova, O., & Sroufe, R. (2021). Third-party sustainability certifications in food retailing: Certification design from a sustainable supply chain management perspective. *Journal of Cleaner Production, 282*, 124344. https://doi.org/10.1016/j.jclepro.2020.124344

Chowdhury, P., Paul, S. K., Kaisar, S., & Moktadir, M. A. (2021). COVID-19 pandemic related supply chain studies: A systematic review. *Transportation Research Part E: Logistics and Transportation Review, 148*, 102271. https://doi.org/10.1016/j.tre.2021.102271

Denicolai, S., Zucchella, A., & Magnani, G. (2021). Internationalization, digitalization, and sustainability: Are SMEs ready? A survey on synergies and substituting effects among growth paths. *Technological Forecasting and Social Change, 166*, 120650. https://doi.org/10.1016/j.techfore.2021.120650

Dey, P. K., Malesios, C., Chowdhury, S., Saha, K., Budhwar, P., & De, D. (2022). Adoption of circular economy practices in small and medium-sized enterprises: Evidence from Europe. *International Journal of Production Economics, 248*, 108496. https://doi.org/10.1016/j.ijpe.2022.108496

Endris, E., & Kassegn, A. (2022). The role of micro, small and medium enterprises (MSMEs) to the sustainable development of sub-Saharan Africa and its challenges: A systematic review of evidence from Ethiopia. *Journal of Innovation and Entrepreneurship, 11*(1), 1–18. https://doi.org/10.1186/s13731-022-00221-8

Fairlie, R. (2020). The impact of COVID-19 on small business owners: Evidence from the first three months after widespread social-distancing restrictions. *Journal of Economics & Management Strategy, 29*(4), 727–740. https://doi.org/10.1111/jems.12400

Gebhardt, M., Spieske, A., Kopyto, M., & Birkel, H. (2022). Increasing global supply chains' resilience after the COVID-19 pandemic: Empirical results from a Delphi study. *Journal of Business Research, 150*, 59–72. https://doi.org/10.1016/j.jbusres.2022.06.008

Gebrihet, H. G., & Pillay, P. (2021). Emerging challenges and prospects of digital transformation and stakeholders integration in urban land administration in Ethiopia. *Global Journal of Emerging Market Economies, 13*(3), 341–356. https://doi.org/10.1177/09749101211034097

Guliyeva, A., Rzayeva, U., & Abdulova, A. (2020). Impact of information technologies on HR effectiveness: A case of Azerbaijan. *International Journal of Advanced Computer Science and Applications, 11*(2), 81–89. https://doi.org/10.14569/IJACSA.2020.0110212

Gupta, K. P. (2020). Investigating the adoption of MOOCs in a developing country: Application of technology-user-environment framework and self-determination theory. *Interactive Technology and Smart Education, 17*(4), 355–375. https://doi.org/10.1108/ITSE-06-2019-0033

Hamulczuk, M., & Skrzypczyk, M. (2021). COVID-19, spatial market integration and producer prices: A case study of EU agri-food markets. *Studies in Agricultural Economics, 123*(2), 53–61. https://doi.org/10.7896/j.2137

Joshi, T., Mainali, R. P., Marasini, S., Acharya, K. P., & Adhikari, S. (2021). Nepal at the edge of sword with two edges: The COVID-19 pandemics and sustainable development goals. *Journal of Agriculture and Food Research, 4*, 100138. https://doi.org/10.1016/j.jafr.2021.100138

Kalenkoski, C. M., & Pabilonia, S. W. (2022). Impacts of COVID-19 on the self-employed. *Small Business Economics, 58*(2), 741–768. https://doi.org/10.1007/s11187-021-00522-4

Kesavan, N., Seenuvasaragavan, S., Burnette, K., Heim, A. B., Patel, R. J., Johnston, J., Cao, X., Sakandar, H. A., Zhuang, Y., Oi, K. K., Zhi, Y., Diao, Y., & Strielkowski, W. (2022). Pandemic-inspired policies. *Science, 377*(6601), 22–24. https://doi.org/10.1126/science.add2046

Klein, V. B., & Todesco, J. L. (2021). COVID-19 crisis and SMEs responses: The role of digital transformation. *Knowledge and Process Management, 28*(2), 117–133. https://doi.org/10.1002/kpm.1660

Lu, L., Peng, J., Wu, J., & Lu, Y. (2021). Perceived impact of the COVID-19 crisis on SMEs in different industry sectors: Evidence from Sichuan, China. *International Journal of Disaster Risk Reduction, 55*, 102085. https://doi.org/10.1016/j.ijdrr.2021.102085

Marchant-Forde, J. N., & Boyle, L. A. (2020). COVID-19 effects on livestock production: A one welfare issue. *Frontiers in Veterinary Science, 7*, 585787. https://doi.org/10.3389/fvets.2020.585787

Mitha, S. (2020). UK COVID-19 diary: Policy and impacts. *National Tax Journal, 73*(3), 847–878. https://doi.org/10.17310/ntj.2020.3.10

Pak, A., Adegboye, O. A., Adekunle, A. I., Rahman, K. M., McBryde, E. S., & Eisen, D. P. (2020). Economic consequences of the COVID-19 outbreak: The need for epidemic preparedness. *Frontiers in Public Health, 8*, 241. https://doi.org/10.3389/fpubh.2020.00241

Quayson, M., Bai, C., & Osei, V. (2020). Digital inclusion for resilient post-COVID-19 supply chains: Smallholder farmer perspectives. *IEEE Engineering Management Review, 48*(3), 104–110. https://doi.org/10.1109/EMR.2020.3006259

Raj, A., Mukherjee, A. A., de Sousa Jabbour, A. B. L., & Srivastava, S. K. (2022). Supply chain management during and post-COVID-19 pandemic: Mitigation strategies and practical lessons learned. *Journal of Business Research, 142*, 1125–1139. https://doi.org/10.1016/j.jbusres.2022.01.037

Ratten, V. (2021). COVID-19 and entrepreneurship: Future research directions. *Strategic Change, 30*(2), 91–98. https://doi.org/10.1002/jsc.2392

Razdan, H., & Kumar, A. (2020). *Ramping up the supply chain post COVID-19. KPMG*. Available at: https://home.kpmg/in/en/home/insights/2020/05/ramping-upthe-supply-chain-post-covid-19.html. Accessed 20 Aug 2022

Rodrigues, M., & Franco, M. (2021). Digital entrepreneurship in local government: Case study in municipality of Fundão, Portugal. *Sustainable Cities and Society, 73*, 103115. https://doi.org/10.1016/j.scs.2021.103115

Sengupta, T., Narayanamurthy, G., & Moser, R. (2021). Disruptive technologies for achieving supply chain resilience in COVID-19 era: An implementation case study of satellite imagery and blockchain technologies in fish supply chain. *Information System Frontiers*. https://doi.org/10.1007/s10796-021-10228-3

Shafi, M., Liu, J., & Ren, W. (2020). Impact of COVID-19 pandemic on micro, small, and medium-sized enterprises operating in Pakistan. *Research in Globalization, 2*, 100018. https://doi.org/10.1016/j.resglo.2020.100018

Sibley, C. G., Greaves, L. M., Satherley, N., Wilson, M. S., Overall, N. C., Lee, C. H., & Barlow, F. K. (2020). Effects of the COVID-19 pandemic and nationwide lockdown on trust, attitudes toward government, and well-being. *American Psychologist, 75*(5), 618. https://doi.org/10.1037/amp0000662

Waldkirch, A. (2021). Firms around the world during the COVID-19 pandemic. *Journal of Economic Integration, 36*(1), 3–19. https://doi.org/10.11130/jei.2021.36.1.3

White, E. R., Froehlich, H. E., Gephart, J. A., Cottrell, R. S., Branch, T. A., Agrawal Bejarano, R., & Baum, J. K. (2021). Early effects of COVID-19 on US fisheries and seafood consumption. *Fish and Fisheries, 22*(1), 232–239. https://doi.org/10.1111/faf.12525

Chapter 13
Shadow Economy, Transparency, and Leadership in Business: Short-Run Dynamics and Long-Run Equilibrium

Muslum Mursalov, Serhii Lyeonov, and Inna Tiutiunyk

Abstract One of the most influential inhibitors of business sustainability is the shadow economy. Economic entities direct their efforts to distort their financial, accounting, or tax reporting in the direction of reducing the declared amount of income, hiding the results of their activities. This leads to a decrease in the level of their financial and investment potential, the level of trust in the business environment, transparency, and openness of the business. The findings confirmed that scientists and practitioners have not developed effective tools to increase business transparency. This provokes a more detailed study of the factors of confidence building in business and the development of tools for their effective use. This study explores the short- and long-run relationship between transparency and leadership in business and the level of shadow economy through time series analysis. The information base of the study is the World Bank and the International Monetary Fund data for EU countries and Ukraine for 2003–2020. The results confirm the hypothesis of the negative impact of shadow economy on the dynamics of indicators of transparency and trust in business and prove their dependence in the long run.

Keywords Shadow economy · Transparency · Leadership · Trust · Illicit financial flows

JEL Classification D21 · D81 · O17

M. Mursalov · I. Tiutiunyk (✉)
Azerbaijan State University of Economics, Baku, Azerbaijan
e-mail: muslum_mursalov@unec.edu.az; i.tiutiunyk@biem.sumdu.edu.ua

S. Lyeonov
Sumy State University, Sumy, Ukraine

Silesian University of Technology, Gliwice, Poland

The London Academy of Science and Business, UK, London
e-mail: s.lieonov@uabs.sumdu.edu.ua

W. Strielkowski (ed.), *Leadership, Entrepreneurship and Sustainable Development Post COVID-19*, Springer Proceedings in Business and Economics, https://doi.org/10.1007/978-3-031-28131-0_13

13.1 Introduction

In modern conditions of business operation, which are characterized by the influence of a number of macro- and micro-factors, in particular, the variability of supply and demand, the aggressiveness of the competitive environment, and the low level of investment attractiveness, without effective corporate management and a quality policy for the formation of sustainable competitive advantages, companies are not able to effectively and rationally determine their strategic goals and implement measures to achieve them (Kumar et al., 2020).

Business management is closely related to all elements of the corporate management system, and its quality depends on the quantitative (the level of the financial potential of the company, which is usually understood as the set of financial resources of the business, which are formed at the determined or actual values of resource, labor, investment, and production potentials) and qualitative (leadership, level of trust in business) indicators of the company's functioning (Sosnovska & Zhytar, 2018; Gruzina et al., 2021).

Today, the business of most developed countries is already working in new dimensions of digital society and economic development (Koudelková et al., 2015; Bilan et al., 2022; Fila et al., 2020; Kibaroğlu, 2020; Kuzior & Kwilinski, 2022; Kuzior et al., 2022a; Vasilyeva et al., 2021b), the manifestations of which are the introduction of new types of innovations in the field of corporate management, such as the automation of production processes, the publication of corporate reporting, and information about the company's directions and areas of activity. Under these conditions, the implementation of the policy of openness and transparency, which is ensured with the help of electronic management, become a new element in the system of marketing and strategic management of companies (Strielkowski et al., 2022). The electronic management that includes the ubiquitous digitalization which has been further strengthened by the COVID-19 pandemic through the digital surge it has brought into such areas as economic and social life, business, and education, as well as many others, is becoming the new normal for modern firms in the conditions of interconnectedness, globalization, and the era of information technologies. This, in turn, is considered as an effective tool for strengthening leadership positions in the market, accelerating the flow of funds, strengthening the effectiveness of financial monitoring operations and control over the transparency of financial transactions, increasing the company's level of attractiveness for consumers, international partners, and investors.

13.2 Literature Review

The consequences of the company's loss of competitive positions significantly increase the responsibility of managers for their decisions and financial results of their activities. These managers, especially the ones in the international and multinational corporations, are often no longer the owners of their companies but have to report to the shareholders who are, in turn, interested in the return of their investments at all possible costs. Erroneous decisions of the company's management affect the indicators of its economic and financial development, and in some cases can lead to a decrease in sales volumes, the level of trust of consumers and investors, profitability, and competitiveness of the company (Koudelková & Svobodová, 2014; Lu et al., 2019; Vasylieva et al., 2020, 2021a; Samusevych et al., 2021; Rekunenko et al., 2022; Moskalenko et al., 2022; Djalilov & Hartwell, 2021). In these conditions, the requirements for the quality and transparency of information, on the basis of which management decisions are made, including regarding the volumes of real and potential financial revenues, are significantly increased (Meresa, 2019; Melnyk et al., 2022; Zenchenko et al., 2022).

The complexity and versatility of the category of leadership and the diversification of approaches to its interpretation actualize the need for a more thorough study of the economic content of this concept, its dependence on the quality of corporate financial management, and the indicators of the company's economic and financial development. The raising scientific interest in this definition of leadership is due to the peculiarities of business functioning in conditions of increased competition on the market. With the ongoing globalization as well as the negative effects of the coronavirus pandemic on the supply and retail chains (especially due to the "zero COVID" policy adapted by the People's Republic of China in the recent years following the announcement of the COVID-19 pandemic in March 2020), this is becoming an increasingly important concept that is widely discussed, studied, and followed both by the general public and the relevant stakeholders and policy makers all around the world.

The trend analysis of the number of searches for the concept of "leadership" (Fig. 13.1) we conducted for this chapter allows us to draw a conclusion about the interest in these issues for a long time.

At the same time, despite the important role of leadership in the system of corporate governance, many scientists lack unambiguous interpretations regarding the content and components of this concept (Chebotareva et al., 2020).

The results of the systematization of scientific approaches to the interpretation of leadership allow us to distinguish several of its concepts:

- Resource (quantitative) concept, according to which leadership is interpreted as the ability of companies to maintain the highest volume of financial resources in the company (Bilan et al., 2020; Eddassi, 2020; Kotenko & Bohnhardt, 2021; Kozmenko et al., 2014; Samoilikova & Kunev, 2020)
- Target (qualitative) concept, which interprets leadership as the ability to most effectively achieve the set goals (Starchenko et al., 2021; Shkarlet et al., 2019)
- A functional concept, which under leadership means the result of a constructive influence on the market, as a result of which the company receives support among consumers and other market participants (Kaya, 2020)

At the same time, the majority of scientific schools tend to consider leadership from the position of a functional concept, basing its assessment on a qualitative

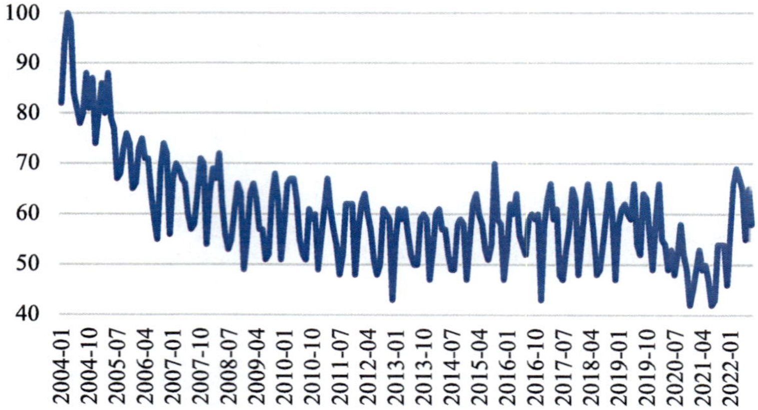

Fig. 13.1 The dynamics of the number of requests for the term "leadership" in Google. (Source: Own results)

rather than a quantitative method of evaluating the results of the company's activities (Polyakov et al., 2019; Serpeninova et al., 2020; Vasilyeva et al., 2021a). Truly, there is a popular statement that a leader is and should be different from a boss with all the skills and approaches to managing the company and treating the employees involved in the process.

Most recently, a significant number of scientists and practitioners increasingly emphasize the importance of achieving and maintaining a company's leadership positions as a key element of ensuring its stable functioning.

However, today, the company's ability to form and obtain sustainable competitive advantages depends on the influence of a number of both exogenous and endogenous factors, the timely identification and consideration of the multiplex effects of which serves as an important component of corporate governance policy (Haseeb et al., 2019; Strielkowski et al., 2021).

For the purpose of a more detailed analysis of the sensitivity of leadership to the variability of the company's operating environment, we will analyze the relationship of leadership with other economic categories using a bibliometric analysis of the dynamics of changes in the interest of the scientific community (VOSviewer tool) to these issues.

At the first stage, we will conduct a dynamic analysis of changes in the number of publications on leadership issues in publications indexed in the Web of Science database. The results presented in Fig. 13.2 testify to their constant growth since 1970. According to the data of the WoS database, the greatest increase in publication activity on leadership issues has occurred since 2005. The precondition for these processes was the aggravation of crisis phenomena in the economy, the increase in the level of bankruptcy of companies, the global economic crisis, etc. The peculiarity of these processes was their spread not only on the territory of a separate country but also in the world as a whole. This led to increased attention to

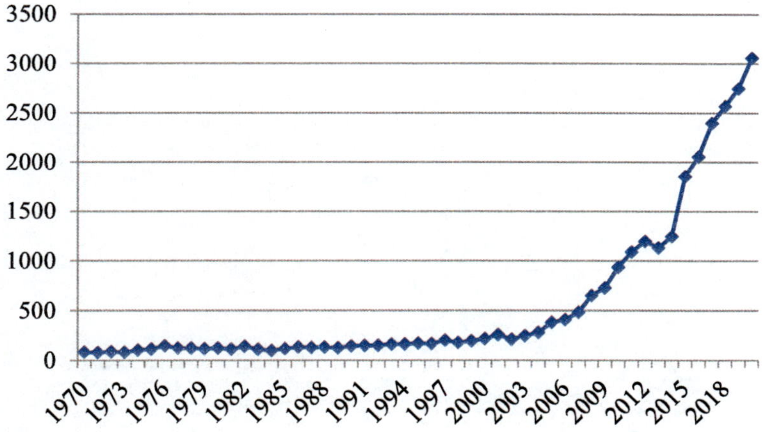

Fig. 13.2 The number of publications indexed in the WoS database on leadership issues. (Source: Own results)

Table 13.1 Top ten journals indexed in the Web of Science database, in which published materials are devoted to leadership

Source	Number of publications	Quarter
Leadership Quarterly	619	Q1
Leadership Organization Development Journal	401	Q1
Leadership	367	Q1
Journal of Business Ethics	354	Q1
Educational Management Administration Leadership	339	Q1
Frontiers in Psychology	327	Q1
Sustainability	243	Q1
Educational Administration Quarterly	209	Q1
Journal of Nursing Management	200	Q1
International Journal of Leadership in Education	196	Q2

Sources: Own results

leadership issues, an increase in the number of studies on the antecedents, and the consequences of a company's loss of leadership positions.

The analysis of scientific publications in which these articles are published allows us to conclude that the majority of publications were made in highly rated journals of the first quarter (Table 13.1). We used the Web of Science, the reference and citation data database from the academic journals and books run by Clarivate Analytics. The choice of WoS has been predetermined by its coverage and its well-established position on the academic market. Looking at Table 13.1, one can see that the largest number of articles (619) was published in *Leadership Quarterly* journal.

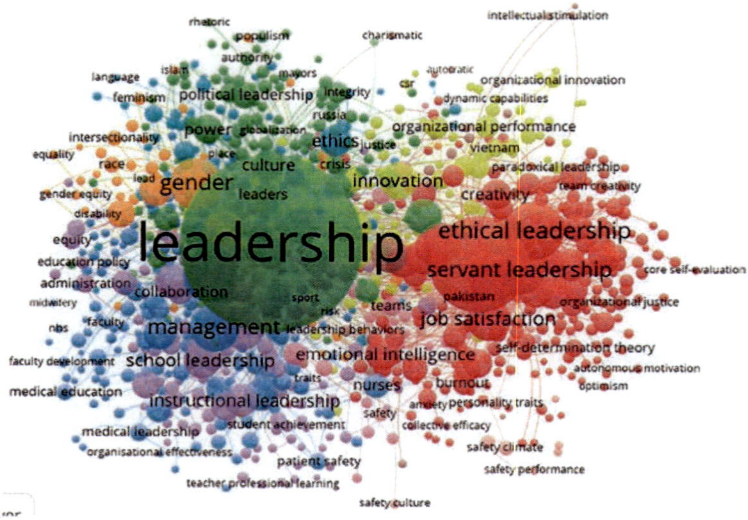

Fig. 13.3 Visualization of the relationship of leadership with other economic categories. (Source: Own results)

The bibliometric analysis of the relationship between the concept of leadership and other categories conducted using the VOSviewer toolkit proved its connection with various economic categories (Fig. 13.3).

The results of the bibliometric analysis and literature review prove the presence of a close connection between the competitive positions of the company and the level of transparency of its activities, the publication of its reports, etc. A significant part of scientists substantiates the connection of leadership with the transparency of the company's activities, the tendency of its owners to carry out unofficial (shadow) activities (Mostenska, 2015; Djalilov & Hartwell, 2022; Lyulyov et al., 2018; Petroye et al., 2020; Vasilyeva et al., 2016; Voronkova et al., 2019). Thus, as evidenced by the results of the structural analysis of the shadow sector of the economy, managers and owners of companies are often the subjects of manipulation by participants in shadow operations. Every year, an increasing number of economic entities try to distort their financial, accounting, or tax reporting in order to reduce the declared amount of income and hide part of the results of their activities. In the final case, this leads to a decrease in the level of their financial potential and trust on the part of consumers, partners, investors, etc. (Zolkover & Georgiev, 2020; Navickas et al., 2021; Lyulyov et al., 2021b; Brychko et al., 2021).

The analysis of the economic content of the concepts of "leadership" and "shadow economy" indicates the existence of a connection between them, in particular, the low level of consideration by potential counterparties of business reputations when making decisions regarding further cooperation (Fig. 13.4). Thus, H. Beckev in his work "Crime and Punishment: an Economic Approach" first proposed the theory of rational behavior of participants in the shadow economy sector

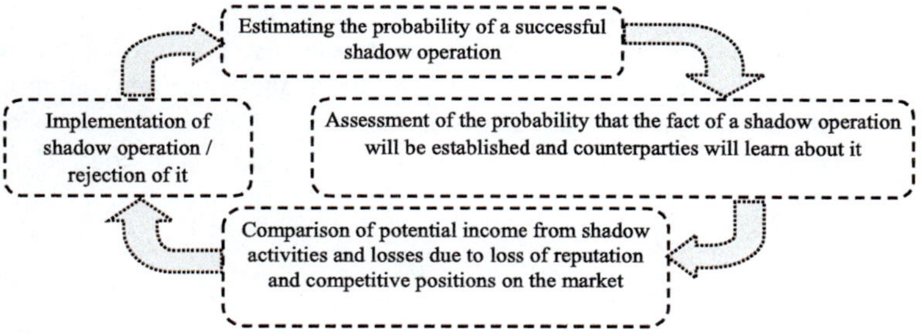

Fig. 13.4 Modeling the relationship between leadership and the shadow activity of economic entities. (Source: Own results)

according to which the main driver of the criminal's actions is the ratio of losses (damages from reputational losses) and income from shadow activity. Even the minimal probability of establishing the facts of the company's implementation of shadow operations will affect its activity in the shadow sector of the economy (Hanić & Jevtić, 2020; Kuzior et al., 2022a, b; Kwilinski et al., 2022; Mercado & Vargas-Hernández, 2019).

Thus, based on the results of the analysis of publication activity on leadership issues, it is possible to conclude about a significant level of relevance of these issues and the need for a more thorough investigation of its connection with the results of the company's activities.

13.3 Method and Data

13.3.1 Data

The annual data of the World Bank (WB) and the International Monetary Fund (IMF) serve as the information base of the present study. The period of the study is 2003–2020. The EU countries and Ukraine, which have been considered as a candidate country for the EU membership (according to the decision of the EU summit of June 23, 2022), were chosen as the object of the study. Formalization, data preprocessing, and econometric analysis were performed using the Stata 14 software package.

13.3.2 Materials and Methods

Empirical confirmation of the short- and long-run relationship between transparency, leadership, and the level of shadow economy will be carried out using time series analysis and methods of system dynamics.

At the first stage, the order of integration of transparency, leadership, and the level of shadow economy data series will be determined using unit root tests (Hadri, Levin et al., (LLC), and Im (IPS)). If the hypothesis about the integration of the first-order data series is confirmed, their cointegration will be evaluated using the Pedroni's and Kao panel tests, which can be formalized as follows (Lyulyov et al., 2021a):

$$L_{it} = a_i + \delta_t + \beta Y_{it} + \varepsilon_{it} \tag{13.1}$$

where L and Y are the natural logarithm of the leadership indicator and factor indicators (shadow economy, transparency), respectively; a is country; δ is time-fixed effects; ε is the standard error; $i = 1 \ldots N$ is the number of countries under analysis; and $t = 1 \ldots t$ is a period.

The presence of a long-term relationship between data series will be assessed using the VEC model:

$$\Delta L_{it} = \theta_{1j} + \sum_{k=1}^{m} \theta_{11ik} \Delta L_{it-k} + \sum_{k=1}^{m} \theta_{12ik} \Delta Y_{it-k} + \tau_{1i} \text{LTR}_{it-1} + u_{1it} \tag{13.2}$$

$$\Delta SE_{it} = \theta_{2j} + \sum_{k=1}^{m} \theta_{21ik} \Delta TG_{it-k} + \sum_{k=1}^{m} \theta_{22ik} \Delta L_{it-k} + \tau_{2i} \text{LTR}_{it-1} + u_{2it} \tag{13.3}$$

$$\Delta T_{it} = \theta_{3j} + \sum_{k=1}^{m} \theta_{31ik} \Delta L_{it-k} + \sum_{k=1}^{m} \theta_{32ik} \Delta Y_{it-k} + \tau_{3i} \text{LTR}_{it-1} + u_{3it} \tag{13.4}$$

where Δ are the first differences of leadership; k is the optimal length of the lags (determined according to the Bayesian Schwartz criterion); θ is the regression parameter that is estimated; LTR is a parameter that analyzes the presence of a long-term effect; and τ is a parameter characterizing the deviation of indicators from their long-term equilibrium values.

Testing the reliability of the hypothesis about the presence of a short-term relationship between indicators will be carried out using the Granger causality test.

13.4 Discussion

The initial stage of the research is to check the analyzed data series for stationarity (Table 13.2). According to the obtained results, the absolute values of McKinnon's τ-statistic for leadership, shadow economy, and transparency are smaller than the absolute critical values at 1%, 5%, and 10% levels of statistical significance. Since the probability that the analyzed data series has a single root and is nonstationary exceeds 50% (p-value>10%), this does not allow us to reject the null hypothesis of the presence of single roots. Thus, we can conclude that the data series for leadership, shadow economy, and transparency are nonstationary. Using the estimation of

Table 13.2 Panel unit root results

Variables	Test statistics		Individual intercept	
			Level	First difference
L	LLC	Statistic	1.3384	−3.3836
		p-value	0.8932	0.0000
	IPS	Statistic	2.68208*	−7.4507
		p-value	0.9736	0.0000
	Hadri	Statistic	0.4719	16.2783
		p-value	0.3070*	0.0000*
SE	LLC	Statistic	−16.6219	−14.4348
		p-value	0.0000*	0.0000
	IPS	Statistic	0.0706	−4.2185
		p-value	0.5164	0.0000
	Hadri	Statistic	0.9379	13.8468
		p-value	0.1644	0.0000*
T	LLC	Statistic	−14.3122	−12.4290
		p-value	0.0000*	0.0000
	IPS	Statistic	0.0553	−3.3048*
		p-value	0.4045	0.0000
	Hadri	Statistic	0.7348	10.8477
		p-value	0.1288	0.0000

Source: Own calculations and results
*$p < 0.01$

the parameters of the first differences, we will transform nonstationary data series into stationary ones (Table 13.2).

The nonstationarity of the first-order data series allows us to hypothesize the presence of a cointegrating relationship between them. The results of testing the reliability of this hypothesis are presented in Table 13.3.

The results of testing the hypothesis about the absence of cointegration between data series with the help of a test indicate the existence of cointegration between indicators of leadership, shadowing of the economy, and transparency. Thus, at the level of statistical significance of 1%, 5%, and 10%, four models do not confirm the null hypothesis of no cointegration: Panel PP-Statistic (−1.9655), Panel ADF-statistic (−1.53918), Panel rho-statistic (−1.41007), and Panel PP-statistic (−2.17727).

The results of testing data series for cointegration based on the Kao residual cointegration test (Table 13.4) allow rejecting the null hypothesis of no cointegration of data at the 5% significance level. This allows us to make a conclusion about the long-term nature of the relationship between leadership, shadow economy, and transparency.

According to the results of evaluating the parameters of the VECM model (Table 13.5), the values of the LTR coefficient for equations (2–4) are statistically significant at the 10% and 5% levels of statistical significance; the value of the LTR indicator testifies to the relatively slow return of the system to long-term

Table 13.3 Results of Pedroni's cointegration test

Test	Statistic	Prob.	Test	Statistic	Prob.
Within dimension (panel statistics)			*Between dimension*		
Panel v-statistic	−1.89509	0.95052	Group rho-statistic	−1.93649	1.69061
Panel rho-statistic	−1.21761	0.10355	Group PP-statistic	−1.41007	−0.00224
Panel PP-statistic	−1.9655	(−0.02147)*	Group ADF-statistic	−2.17727	−1.05662
Panel ADF-statistic	−1.53918	(−0.05602)*			
(Weighted statistic)					
Panel v-statistic	−1.93649	0.95296			
Panel rho-statistic	−1.41007	(−0.07252)**			
Panel PP-statistic	−2.17727	(−0.01249)***			
Panel ADF-statistic	−1.62604	−0.04675			

Source: Author's calculations
*$p < 0.01$; **$p < 0.05$; ***$p < 0.001$

Table 13.4 Results of testing data series for integration based on the Kao residual cointegration test

Trend assumption: no deterministic trend		
	t-statistic	Prob.
ADF	−2.2167	(0.0137)*
Residual variance	0.002278	
HAC variance	0.001531	

Source: Author's calculations
*$p > 0.05$

equilibrium due to the action of exogenous shocks ($L = -0.2367$). In addition, the value of the coefficient of determination (R-squared) for all analyzed indicators exceeds 0.8, which confirms the validity of the hypothesis.

The results of testing the presence of short-term relationships between leadership, shadowing, and transparency based on the Granger test are shown in Table 13.6.

The data in Table 13.6 prove the presence of a one-way Granger causality between the analyzed data series at the level of statistical significance of 10%. Thus, an increase in the level of shadowing of the economy will lead to a deterioration of the company's leadership positions. At the same time, there is a direct relationship between leadership and transparency – a decrease in the company's level of transparency will lead to a decrease in its leadership positions.

13.5 Conclusions

Overall, our study is dedicated to confirming the causal relationships between leadership, business transparency, and the shadow economy. Data series for 2003–2020 were analyzed using the example of the EU and Ukrainian cases. The results of the

Table 13.5 Results of estimating the long-term relationship between indicators using the panel vector error correction estimate

Error correction	Δ(*L*)	Δ(*SE*)	Δ(*T*)
LTR_{t-1}	−0.2367	−0.01276	−0.01174
	[−4.3746]	[−1.06475]	[−1.07635]
Δ(*L*(−1))	−0.2748	0.016253	0.−157362
	[−3.7153]	[0.08765]	[0.06746]
Δ(*Y*(−1))	1.08764	0.001254	0.002235
	[1.03746]	[0.01176]	[0.01423]
C	0.11736	0.014435	0.01653
	[2.73647]	[1.53645]	[1.87564]
R-squared	0.98676	0.87654	0.88731
Adj. R-squared	0.72653	0.55342	0.47655

Source: Author's calculations
t-statistics are provided in square brackets

Table 13.6 Results of evaluating the causal relationship between data series based on the Granger test

	Short run			Long run	Conclusion
Dependent variables	Δ(*L*)	Δ(*SE*)	Δ(*T*)	LTR_{t_1}	
Δ(*L*)	–	−4.52398 (0.0513)*	–	−0.3442 [−4.7162]	$L > SE$
Δ(*SE*)	−0.245637 (0.0615)*	–	–	−0.01436 [−1.8764]	$SE \neq > L$
Δ(*L*)	–		3.62736 (0.0362)*	−0.6564 [−4.17634]	$L > T$
Δ(*T*)	0.176380 (0.0678)*		–	−0.01547 [−1.8367]	$T \neq > L$

Source: Author's calculations
$*p < 0.001$

cointegration evaluation of the data series confirmed the existence of a long-term relationship between the analyzed indicators.

According to the assessment of the short-term relationship between the indicators, the presence of a direct cause-effect relationship between leadership and business transparency and an inverse relationship between the level of the shadow economy was proven. Thus, the results of our econometric analysis confirmed the hypothesis that business competitiveness and the sustainable competitive advantages depend on the effectiveness of the transparency and publicity policy in the company in question.

Therefore, the main results obtained in the course of this study allow us to conclude that the shadow economy affects not only the quantitative indicators of the company's activity but also the qualitative component of its development. The implementation of a policy aimed at strengthening its competitive advantages in the market should include not only measures to improve the quality of products, enter new markets, and increase the volume of production but also to support the business reputation, positioning the company as one that conducts its activities openly and transparently. These results can be of a special value both for the business entrepreneurs and the decision-makers aiming to regulate the economic development and tackle the issues and threats presented by the shadow economy for their countries' development and growth.

Acknowledgments The research was funded by a grant from Ministry of Education and Science of Ukraine, 0120U10047, 0121U109553, 0121U109559, and 0120U102001.

References

Bilan, Y., Pimonenko, T., & Starchenko, L. (2020). Sustainable business models for innovation and success: Bibliometric analysis. *E3S Web of Conferences, 159*, 04037. https://doi.org/10.1051/e3sconf/202015904037

Bilan, S., Šuleř, P., Skrynnyk, O., Krajňáková, E., & Vasilyeva, T. (2022). Systematic bibliometric review of artificial intelligence technology in organizational management, development, change and culture. *Business: Theory and Practice, 23*(1), 1–13. https://doi.org/10.3846/btp.2022.13204

Brychko, M., Savchenko, T., Vasylieva, T., & Piotrowski, P. (2021). Illegal activities of financial intermediaries: A burden of trust crisis. *Journal of International Studies, 14*(1), 172–189. https://doi.org/10.14254/2071-8330.2021/14-1/12

Chebotareva, G., Strielkowski, W., & Streimikiene, D. (2020). Risk assessment in renewable energy projects: A case of Russia. *Journal of Cleaner Production, 269*, 122110. https://doi.org/10.1016/j.jclepro.2020.122110

Djalilov, K., & Hartwell, C. (2021). Do social and environmental capabilities improve bank stability? Evidence from transition countries. *Post-Communist Economies, 34*(5), 624–646. https://doi.org/10.1080/14631377.2021.1965359

Djalilov, K., & Hartwell, C. A. (2022). The spirit is willing, but the institutions are weak: Disclosure of corporate social responsibility and the financial sector in transition. *Eurasian Business Review*. https://doi.org/10.1007/s40821-022-00224-1

Eddassi, H. (2020). Fiscal regime and tax policy in resource-rich countries in the process of globalization: Literature review. *SocioEconomic Challenges, 4*(2), 67–77. https://doi.org/10.21272/sec.4(2).67-77.2020

Fila, M., Levicky, M., Mura, L., Maros, M., & Korenkova, M. (2020). Innovations for business management: Motivation and barriers. *Marketing and Management of Innovations, 4*, 266–278. https://doi.org/10.21272/mmi.2020.4-22

Gruzina, Y., Firsova, I., & Strielkowski, W. (2021). Dynamics of human capital development in economic development cycles. *Economies, 9*(2), 67. https://doi.org/10.3390/economies9020067

Hanić, A., & Jevtić, D. (2020). Human resource management between economy and ethics – Research of Serbia and Bosnia and Hercegovina. *Business Ethics and Leadership, 4*(3), 127–136. https://doi.org/10.21272/bel.4(3).127-136.2020

Haseeb, M., Hussain, H. I., Kot, S., Androniceanu, A., & Jermsittiparsert, K. (2019). Role of social and technological challenges in achieving a sustainable competitive advantage and sustainable business performance. *Sustainability, 11*(14), 3811. https://doi.org/10.3390/su11143811

Kaya, H. D. (2020). Business friendliness, firm performance and owner's optimism. *Financial Markets, Institutions and Risks, 4*(3), 13–23. https://doi.org/10.21272/fmir.4(3).13-23.2020

Kibaroğlu, O. (2020). Self sovereign digital identity on the blockchain: A discourse analysis. *Financial Markets, Institutions and Risks, 4*(2), 65–79. https://doi.org/10.21272/fmir.4(2).65-79.2020

Kotenko, N., & Bohnhardt, V. (2021). Digital health projects financing: Challenges and opportunities. *Health Economics and Management Review, 2*(1), 100–107. https://doi.org/10.21272/hem.2021.1-10

Koudelková, P., & Svobodová, P. (2014). Knowledge creation & sharing as essential determinants of SMEs innovation. *International Economics Letters, 3*(1), 12–20. https://doi.org/10.24984/iel.2014.3.1.3

Koudelková, P., Strielkowski, W., & Hejlová, D. (2015). Corruption and system change in the Czech Republic: Firm-level evidence. *Danube: Law, Economics and Social Issues Review, 6*(1), 25–46. https://doi.org/10.1515/danb-2015-0002

Kozmenko, S. M., Korneyev, M. V., & Makedon, V. V. (2014). Financialisation of economy and its influence on the indicators of countries' socioeconomic development. *Actual Problems of Economics, 161*(11), 290–298.

Kumar, V., Singh, D., Purkayastha, A., Popli, M., & Gaur, A. (2020). Springboard internationalization by emerging market firms: Speed of first cross-border acquisition. *Journal of International Business Studies, 51*(2), 172–193. https://doi.org/10.1057/s41267-019-00266-0

Kuzior, A., & Kwilinski, A. (2022). Cognitive technologies and artificial intelligence in social perception. *Management Systems in Production Engineering, 30*(2), 109–115. https://doi.org/10.2478/mspe-2022-0014

Kuzior, A., Kettler, K., & Rąb, Ł. (2022a). Digitalization of work and human resources processes as a way to create a sustainable and ethical organization. *Energies, 15*(1), 172. https://doi.org/10.3390/en15010172

Kuzior, A., Kettler, K., & Rabą, Ł. (2022b). Great Resignation—Ethical, cultural, relational, and personal dimensions of generation Y and Z employees' engagement. *Sustainability, 14*(11), 6764. https://doi.org/10.3390/su14116764

Kwilinski, A., Lyulyov, O., Pimonenko, T., Dzwigol, H., Abazov, R., & Pudryk, D. (2022). International migration drivers: Economic, environmental, social, and political effects. *Sustainability (Switzerland), 14*(11), 6413. https://doi.org/10.3390/su14116413

Lu, J., Ren, L., Qiao, J., Yao, S., Strielkowski, W., & Streimikis, J. (2019). Corporate social responsibility and corruption: Implications for the sustainable energy sector. *Sustainability, 11*(15), 4128. https://doi.org/10.3390/su11154128

Lyulyov, O., Chygryn, O., & Pimonenko, T. (2018). National brand as a marketing determinant of macroeconomic stability. *Marketing and Management of Innovations, 3*, 142–152. https://doi.org/10.21272/mmi.2018.3-12

Lyulyov, O., Lyeonov, S., Tiutiunyk, I., & Podgórska, J. (2021a). The impact of tax gap on macroeconomic stability: Assessment using panel VEC approach. *Journal of International Studies, 14*(1), 139–152. https://doi.org/10.14254/2071-8330.2021/14-1/10

Lyulyov, O., Paliienko, M., Prasol, L., Vasylieva, T., Kubatko, O., & Kubatko, V. (2021b). Determinants of shadow economy in transition countries: Economic and environmental aspects. *International Journal of Global Energy Issues, 43*(2-3), 166–182.

Melnyk, L., Matsenko, O., Kubatko, O., Korneyev, M., & Tulyakov, O. (2022). Additive economy and new horizons of innovative business development. *Problems and Perspectives in Management, 20*(2), 175–185. https://doi.org/10.21511/ppm.20(2).2022.15

Mercado, M. P. S. R., & Vargas-Hernández, J. G. (2019). Analysis of the determinants of social capital in organizations. *Business Ethics and Leadership, 3*(1), 124–133. https://doi.org/10.21272/bel.3(1).124-133.2019

Meresa, M. (2019). The Effect of Strategic Management Practices on the institutional Performance; the case of Dedebit credit and saving institution in Eastern Tigray. *SocioEconomic Challenges, 3*(3), 80–97. https://doi.org/10.21272/sec.3(3).80-97.2019

Moskalenko, B., Lyulyov, O., & Pimonenko, T. (2022). The investment attractiveness of countries: Coupling between core dimensions. *Forum Scientiae Oeconomia, 10*(2), 153–172. https://doi.org/10.23762/FSO_VOL10_NO2_8

Mostenska, T. (2015). Informal market as a result of shadow economy. *Czech Journal of Social Sciences, Business and Economics, 4*(2), 23–29. https://doi.org/10.24984/cjssbe.2015.4.2.3

Navickas, V., Petroke, I., Baciuliene, V., & Vasylieva, T. (2021). The impact of the sharing economy as an ecosystem on the tourism sector. *Journal of Tourism and Services, 12*(22), 66–88. https://doi.org/10.29036/jots.v12i22.241

Petroye, O., Lyulyov, O., Lytvynchuk, I., Paida, Y., & Pakhomov, V. (2020). Effects of information security and innovations on Country's image: Governance aspect. *International Journal of Safety and Security Engineering, 10*(4), 459–466. https://doi.org/10.18280/ijsse.100404

Polyakov, M., Bilozubenko, V., Korneyev, M., & Shevchenko, G. (2019). Selection of parameters for multifactor model in the knowledge economy marketing (country level). *Innovative Marketing, 15*(1), 89–99. https://doi.org/10.21511/im.15(1).2019.08

Rekunenko, I., Zhuravka, F., Nebaba, N., Levkovych, O., & Chorna, S. (2022). Assessment and forecasting of Ukraine's financial security: Choice of alternatives. *Problems and Perspectives in Management, 20*(2), 117–134. https://doi.org/10.21511/ppm.20(2).2022.11

Samoilikova, A., & Kunev, R. (2020). The impact of health care financing on the economic growth: EU countries analysis. *Health Economics and Management Review, 1*(2), 24–32. https://doi.org/10.21272/hem.2020.2-03

Samusevych, Y. V., Novikov, V. V., Artyukhov, A. Y., & Vasylieva, T. A. (2021). Convergence trends in the "economy – education – digitalization – national security" chain. *Naukovyi Visnyk Natsionalnoho Hirnychoho Universytetu, 6*, 177–183. https://doi.org/10.33271/NVNGU/2021-6/177

Serpeninova, Y., Makarenko, I., Plastun, A., Babko, A., & Gasimova, G. (2020). Mapping of the responsible investments instruments in SDG 3 'good health and well-being' financing: EU and US experience. *Health Economics and Management Review, 1*(1), 106–115. https://doi.org/10.21272/hem.2020.1-10

Shkarlet, S., Kholiavko, N., Dubyna, M., & Zhuk, O. (2019). Innovation, educational, research components of the evaluation of information economy development (as exemplified by Eastern Partnership countries). *Marketing and Management of Innovations, 1*, 70–83. https://doi.org/10.21272/mmi.2019.1-06

Sosnovska, O., & Zhytar, M. (2018). Financial architecture as the base of the financial safety of the enterprise. *Baltic Journal of Economic Studies, 4*(4), 334–340. https://doi.org/10.30525/2256-0742/2018-4-4-334-340

Starchenko, L. V., Samusevych, Y., & Demchuk, K. (2021). Social and eco-friendly enterpreneurship: The keys to sustainability. *Business Ethics and Leadership, 5*(1), 118–126. https://doi.org/10.21272/bel.5(1).118-126.2021

Strielkowski, W., Guliyeva, A., Rzayeva, U., Korneeva, E., & Sherstobitova, A. (2021). Mathematical modeling of intellectual capital and business efficiency of small and medium enterprises. *Mathematics, 9*(18), 2305. https://doi.org/10.3390/math9182305

Strielkowski, W., Zenchenko, S., Tarasova, A., & Radyukova, Y. (2022). Management of smart and sustainable cities in the post-COVID-19 era: Lessons and implications. *Sustainability, 14*(12), 7267. https://doi.org/10.3390/su14127267

Vasilyeva, T., Sysoyeva, L., & Vysochyna, A. (2016). Formalization of factors that are affecting stability of Ukraine banking system. *Risk Governance and Control: Financial Markets and Institutions, 6*(4), 7–11. https://doi.org/10.22495/rcgv6i4art1

Vasilyeva, T., Kuzmenko, O., Kuryłowicz, M., & Letunovska, N. (2021a). Neural network modeling of the economic and social development trajectory transformation due to quarantine restrictions during covid-19. *Economics and Sociology, 14*(2), 313–330. https://doi.org/10.14254/2071-789X.2021/14-2/17

Vasilyeva, T., Ziółko, A., Kuzmenko, O., Kapinos, A., & Humenna, Y. (2021b). Impact of digitalization and the Covid-19 pandemic on the AML scenario: Data mining analysis for good governance. *Economics and Sociology, 14*(4), 326–354. https://doi.org/10.14254/2071-789X.2021/14-4/19

Vasylieva, T., Jurgilewicz, O., Poliakh, S., Tvaronavičienė, M., & Hydzik, P. (2020). Problems of measuring country's financial security. *Journal of International Studies, 13*(2), 329–346. https://doi.org/10.14254/2071-8330.2020/13-2/22

Voronkova, O., Hordei, O., Barusman, A. R. P., & Ghani, E. K. (2019). Social integration as a direction for humanization of economic relations and improvement of social welfare. *SocioEconomic Challenges, 3*(4), 52–62. https://doi.org/10.21272/sec.3(4).52-62.2019

Zenchenko, S., Strielkowski, W., Smutka, L., Vacek, T., Radyukova, Y., & Sutyagin, V. (2022). Monetization of the economies as a priority of the new monetary policy in the face of economic sanctions. *Journal of Risk and Financial Management, 15*(3), 140. https://doi.org/10.3390/jrfm15030140

Zolkover, A., & Georgiev, M. (2020). Shadow investment activity as a factor of macroeconomic instability. *Financial Markets, Institutions and Risks, 4*(4), 83–90. https://doi.org/10.21272/fmir.4(4).83-90.2020

Chapter 14
Leading Information and Communication Technologies in Tourism and Hospitality Within the Context of the COVID-19 Pandemic

Natalia Alekseeva, Zhanna Kevorkova, and Elena Chernikina

Abstract This contribution tackles the issue of the leading information and communication technologies in tourism and hospitality within the context of the COVID-19 pandemic. We examine the use of information and communication technologies (ICTs) for removing various discrimination in tourism destinations, as well as for revitalizing tourism industries, during and right after the COVID-19 pandemic. This study also highlights the role of information technology in the course of the pandemic, especially when helping the authorities with the health aspects. This chapter also brings an overview that discusses the benefits of emerging state-of-the-art technologies for stemming the COVID-19 pandemic, as well as the ways various sectors have been adapting in short timeframes. The COVID-19 pandemic has made the missions and goals of the tourism and hospitality industry more urgent, changing organizations perspectives of work, leadership, and enterprise interactions with customers through radical transformations in digital capabilities.

Keywords Information and communication technologies · Tourism and hospitality · COVID-19

N. Alekseeva (✉)
Volga Region State University of Service, Togliatti, Russian Federation

Z. Kevorkova
Department of Audit and Corporate Reporting, Faculty of Taxes, Audit, and Business Analysis, Financial University under the Government of the Russian Federation, Moscow, Russian Federation
e-mail: zhkevorkova@fa.ru

E. Chernikina
Department of Finance and Credit, Russian State University for the Humanities, Moscow, Russian Federation
e-mail: chernikina.e@rggu.ru

W. Strielkowski (ed.), *Leadership, Entrepreneurship and Sustainable Development Post COVID-19*, Springer Proceedings in Business and Economics, https://doi.org/10.1007/978-3-031-28131-0_14

JEL Classification D38 · L38 · L60

14.1 Introduction

In the challenging times of the COVID-19 pandemic, information and communication technology (ICT) has emerged as the key vehicle for addressing both pandemic-induced challenges and responding to new realities in everyday life (Angell & Potoglou, 2022; Becken & Hughey, 2021). In some ways, the pandemic has increased demand from businesses for 5G technological solutions, a fifth-generation technological standard for mobile networks (Strielkowski et al., 2021). When developing a tourism strategy after the COVID-19 pandemic, it will be essential to take into account not just the technological issue but also the problems with discrimination and xenophobia at tourism destinations with greater care (Ide, 2021).

Generally, COVID-19 shutdowns and travel bans have affected nearly every industry, including tourism, hospitality, and education. Data show that the tourism industry has been heavily affected by the COVID-19 pandemic (Liang et al., 2021; Varzaru et al., 2021). As many countries implemented national lockdowns and heavily restricted border controls during the COVID-19 pandemic, the tourism and hospitality industry has been struggling to survive. With regard to tourism, the governments' strategy during the COVID-19 pandemic has not been effective at minimizing the economic losses, whether it is national or regional lockdowns (Chen et al., 2021). Although aviation was heavily damaged due to the COVID-19 pandemic, hotel industries continued to host a number of business travelers moving around within the countries' borders (Herédia-Colaço & Rodrigues, 2021). Whereas other sectors such as healthcare, education, mass transit, the media, and the hospitality sector are beginning to reopen after the sop from the pandemic, the travel and tourism sector is still in shutdown because of its vulnerability to COVID-19's capriciousness. In this study, we examine the tourism and hospitality sector under the hypothesis that many business activities were suspended because of the COVID-19 pandemic. Previous literature has shown that the pandemic has had significant impacts on all sectors, including tourism, considered to be the most affected sector (Moreno-Luna et al., 2021). The healthcare crisis is by no means a new one, but in terms of economy, COVID-19 has had more disruptive effects compared to the previous crises of recent history. The pandemic has been a major blow to the global economy, leaving no sector unaffected (Goldstein et al., 2021). The present research objective is to assess the negative effects of the pandemic on different sectors, such as education, healthcare, utilities, and tourism. One of the most interesting scenarios of COVID-19 was recorded in India, the world's largest vaccine push that employed the innovative approaches for solving India's vaccine hesitancy, cross-sectoral coordination, and innovations to tackle coronavirus crisis (Ahmad et al., 2021). The role of the World Health Organization (WHO) and national governments worldwide to prevent dissemination of fake news, increase awareness, and mitigate severity of COVID-19 pandemic is also quite important within this context.

All in all, the tourism sector had to undergo significant changes in the post-COVID era. Thence, researchers and engineers need to present their innovative ideas that should be implemented in the tourism sector in order to maintain them either during pandemic or in the time after COVID-19 (Toubes et al., 2021). Hotel management companies and developers must adopt interesting new technology ideas if they are to remain relevant, as well as if they are to function effectively in post-COVID-19 times. Hotel technologies can be used in all departments to enhance the guest experience, optimize operations, handle protocols after COVID-19, and enhance staff-guest interactions. Smarter capabilities and services, which were previously considered additions to guest experiences, will rapidly become necessities in the hospitality post-pandemic landscape (Xiang et al., 2022). During the pandemic, hotels have been employing AI technology and robotics to enhance the quality of service. Robots are also being used at hotels in various countries to fill roles hotel employees cannot do during the pandemic. During the pandemic, remote meetings and live-streaming have become increasingly common, for hotels as well as for their guests. Therefore, the pandemic is driving different industries toward implementing contactless options. Smart technologies hold the potential to enable a surmounting of barriers hoteliers are now facing when trying to manage an influx of complications from COVID-19 (Kim et al., 2021). As hotels reopen and strive for full occupancy (while following local, state, and national guidelines for COVID-19 reopening), technology has never been more critical, and it will be critical in the recovery. Those who make the right investments and use digital technologies to enhance their guest experience will be best placed to survive the COVID-19 pandemic and to thrive afterward.

14.2 Tourism Business and Data

As data collection and analytics technologies get more complex, companies are going to be finding new, more efficient ways to capture and contextualize data about everything, including consumers and their behavior (Guliyeva & Rzayeva, 2018; Gruzina et al., 2021). For companies capturing vast amounts of data, collecting the insights and then selling them represent opportunities for new revenue streams. Companies that capture data have the potential to make money off it. A data broker, or data services provider who buys and sells customer information, has emerged as a new industry along with Big Data (Shah et al., 2021). The sharing economy is changing how jobs are organized and services are delivered, as well as providing opportunities to leverage digital advances in transactions, capture and manage tourism supply and demand information and data, and enhance and link operations across the value chains and tourism ecosystems. It lists strong customer-facing bases as a key requirement for employees within the tourism and hospitality industries (Xu et al., 2021).

Now that we have an good post-pandemic knowledge of customer service basics, maintaining a positive relationship with guests and striving to surpass their expectations let us take a look at a few organizations supporting tourism and hospitality staff training and development. In order to do so, one needs to identify and discuss the possible ways that tourism and hospitality companies can build a lasting relationship with their guests. The development of a tourism businesses ecosystem involves taking a strategic approach to the establishment of networks and relationships through which products and services are delivered, customer relationship management occurring throughout a customers' travel experience including research, booking, preparing, traveling, arriving, staying at a destination, and post-trip interaction (Barbulescu et al., 2021). The focus forces tourism businesses to consider customer relationships in a longer-term perspective, or the customer lifetime value (CLV) cycle, instead of looking only at individual transactions (Riyanto & Azis, 2021). With the continued growth of competition among tourism destinations and businesses, organizations are becoming more focused on keeping existing customers, often less costly than attracting new ones. Various hotspots for economic downturns in tourism experienced business gains in times of global economic crises, thanks to relatively lower costs of living and the slower global labor market, suggesting travelers are procrastinating their trips, with more money traveling farther (Czerny et al., 2021).

In general, there are many new emerging types of tourism that are gathering their pace after the COVID-19 pandemic. Recession tourism is a tourism trend that has developed due to the global economic crisis. Recession tourism is defined by lower-cost, higher-value experiences taking place in formerly popular shared retreats (Plzáková & Smeral, 2022). Experiential tourism (or immersion tourism) is one of the key market trends of tourism (Guo et al., 2021). DNA tourism, also called ancestral tourism or heritage tourism, is tourism that is based on DNA testing (Mehtiyeva & Prince, 2020). Within this context, travel has a definition similar to tourism, but implies a more focused travel. Tourism represents a form of traveling, either for pleasure or business; it is also tourism theory and practice, tourism as a business for the attraction, accommodation, and entertainment of tourists, as well as tourism as the operation of tours (Miao et al., 2021). Digital businesses, such as online travel agents and accommodations platforms, have transformed tourism, connecting tourism products and services to customers everywhere in the world in real-time, as well as dramatically increasing market visibility for tourism-related small- and medium-sized businesses. According to many researchers, digitalization presents tourism SMEs with unprecedented opportunities to tap new markets, develop new tourism products and services, adopt new business models and processes, improve their positions on the global tourism value chain, and integrate in digital ecosystems (Alkier et al., 2022). In our increasingly digitalized world, businesses and consumers are becoming dramatically more dependent on a range of powerful platforms. The transformative effects of using interactive systems arise from the human-centered opportunities provided by efforts at Tourism 4.0, combined with users' desire to participate in long-term body-mind transformations. Tourism providers may also play an important role in designing experiences for Tourism 4.0, which mindfully mitigates possible harms through the selective implementation of technologies, keeping human users, not profits, in mind (Ioannides et al., 2021). Here, the utility of the interactive system becomes irrelevant as tourist experience goals and the quest for well-being are severely compromised. If using an interactive system which requires too much attention from a tourist, by demanding that they pay attention either to the system (e.g., there are too many tasks, complicated devices, unfamiliar or unresponsive interfaces) or the digital content, instead of enjoying their tourism experience, we refer to this as the distraction effect. For instance, the presence of voice-activated interfaces in busy airports or the necessity of downloading and using different tourism providers mobile app can frustrate users and divert their attention away from enjoying a touristic experience (Uzir et al., 2021). The biggest risk with these kinds of initiatives, of course, is that switching to a new platform might leave users behind with an established one. Clearly, cholera and its many siblings, waterborne diseases, are real, existential hazards in a globalized world where millions travel for work, tourism, and migration. The years before the COVID-19 pandemic seem to become a benchmark in anti-tourism sentiment, with the rise of the new social movement against an economic model of development that relies on the profits of mass tourism, following a series of high-profile attacks targeting both international tourists and local business interests. Data analysts' duties and responsibilities generally involve mining insights from a wide range of data sets, performing statistical analyses to predict and identify and suggest new ways of improving business, all informed by the data. Both at the national and the international level, tourism authorities and stakeholders are responsible for best practices research, training, and other support for professional development on behalf of the tourism businesses and millions of individuals employed in tourism-related professions worldwide.

14.3 Recent Changes in Tourism and Hospitality Industry

The evolution and applications of digital technologies are dramatically changing how people live, work, travel, and conduct business in the twenty-first century, and they are also changing and restructuring the tourist industry along the way. Digital innovations and technical breakthroughs are drivers of development and are showing their effects all over, particularly in the fields of manufacturing, information technology, and other service industries (Akhtar et al., 2021). Indeed, in all sectors, there is clear evidence of a significant business impact from technologies at the heart of the Fourth Industrial Revolution. Overall, the relentless move away from merely digitizing things (the Third Industrial Revolution) toward innovations that are built upon combinations of technologies (the Fourth Industrial Revolution) is forcing companies to reconsider how they are doing business. Almost every industry, including airlines, retail, hospitality, healthcare, education, construction, and technology, will need to rapidly transform nearly every sector's business models in order to stay viable. Similarly, food, transportation, health, education, and tourism sectors still require significant innovation to food sector IT infrastructure. Technological developments cannot be stopped; however, by being flexible enough and open, tourist businesses can be prepared to face upcoming challenges (Scott & Gössling, 2021). For that reason, it is essential for hotels, airlines, restaurants, and others to stay on top of the latest technological trends in tourism and travel.

Technology is an integral part of the tourism and travel industry, helping businesses in their daily operations, as well as improving customer experiences. Smartphones, smart speakers, and AI assistants all have helped increase the relevancy of voice searches, in terms of tech trends for the tourism industry. As a result of the digital revolution, the international tourism development trends opened up space for new solutions such as cloud-based reservation sites or sharing of information and experiences through digital platforms (Mavragani & Dionysios, 2022). Policymakers have a major role to play to help tourism businesses of all sizes, including more traditional, smaller businesses, engage with the digital revolution and prosper in response to these emerging technologies. As the technology-enabled transformation is essential, the renewed mentality for people will be a driving force for seeing the successful adoption that propels businesses forward. While digital business transformation is not as much about digital technologies per se, clearly adopting and taking advantage of technologies, in the umbrella of social business, cloud, mobility, Big Data analytics, cognitive computing, IoT, etc., will accelerate changes throughout society at all times. In that sense, digital transformation is a deep transformation of activities, processes, competencies, and business and organizational models in order to take full advantage of the changes and opportunities from the digital portfolio and its accelerating impacts on society, strategically and prioritized, taking into account both current and future transformations (Hai et al., 2021).

Digital trends driving changes in tourism digitalization are a process by which technology and data-driven governance are changing our social and economic systems and lives. The advancements in digitalization have finally reached the point that can really underpin cost-effectiveness and resilience of industrial food production, leading to a new path for the tourism and hospitality industries (Gupta et al., 2021). To date, the people who have benefitted most have been consumers able to afford and access a digital world; technology has made possible new products and services that enhance the efficiency and enjoyment of our personal lives. Needless to say, in 2020, the trend toward digital, touchless services has gained further traction due to the COVID-19 pandemic with its shutdowns and disruptions (Kesavan et al., 2022). On the supply side, a number of industries are seeing new technologies being introduced which are creating completely new ways of serving existing needs, as well as dramatically disrupting existing industrial value chains. Technology will continue to fuel innovation in various industry business models, enabling new businesses to come into the market and seriously disrupt existing businesses. The development and adoption of new technologies is expected to continue apace in the future, driven by the cumulative nature and exponential pace of technology change, convergence of technologies in novel combinations, significant reductions in costs, emergence of new digital business models, and declining costs to enter. The ability of tourism businesses of all sizes to develop their business models, to embrace digital technologies in order to effectively engage with global value ecosystems, and to embrace new ways of data-driven work will shape future productivity and social and economic well-being (Nicoletti et al., 2020). The downturn of restaurants and the food sector in times of pandemic has also led stakeholders to embrace technologies and digitalization in order to more efficiently respond.

14.4 Pandemic-Induced Changes to the Economy

The current pandemic, COVID-19, has led to worldwide challenges, economic crises, and health crises, as well as spillover impacts for the worlds industries, including tourism and travel, which are the largest global service sector participants (Maqsood et al., 2021). Tourism is one of the sectors that has been hit hardest by the pandemic, impacting economies, livelihoods, government services, and opportunities across continents. Tourism has seen sharp declines in revenues and is one of the economic sectors among the hardest hit by the COVID-19 outbreak. Ten months after an initial wave of closures because of COVID-19, workers in the leisure and hospitality industry continued to experience the highest rates of job loss amid the pandemic; more than a quarter of the sectors workforce was out of work (McCartney, 2021). While hotels are present in every metro area, only a handful have staked their economies on them.

The pandemic has devastated the tourist sector in such countries as the United States, pushing leisure and hospitality into the highest unemployment rates of any major sector. To show the broad effects of COVID-19 on economies such as those of Las Vegas or Orlando, we compared metropolitan areas' concentration of employment in hospitality and leisure prior to the pandemic to its changes in unemployment. Many researchers aimed at exploring the economic impacts of a disease outbreak such as the coronavirus pandemic, which is different depending on different capitalist structures (Sonmez et al., 2020).

With regard to all of the above, it is probably a good thing that the effects of different conditions of the pandemic, like SARS (Cov-1 and Cov-2), influenza H1N1, or swine flu, and current case of COVID-19, on capitalist countries economic growth (comparative analyses for the countries with CMEs and CMEs) were examined, and recommendations were made to restore financial health (Shang et al., 2021). As the new coronavirus COVID-19 was declared as a pandemic disease on the 11th of March 2020 by the WHO, in addition to human suffering involved, widespread outbreaks and pandemics also resulted in economic consequences.

However, there were also positive and climate-friendly effects of the pandemic on tourism and hospitality. This is due to the fact that the pandemic is a watershed moment for bringing efforts for sustainable livelihoods reliant on tourism into line with sustainable development goals, and to ensuring a more sustainable, inclusive, carbon-neutral, and resource-efficient future. The pandemic is an opportunity to redefine the way tourism engages with our societies, other economic sectors, and our natural resources and ecosystems; better measure and manage it; ensure equitable distribution of its benefits; and promote a shift toward a tourism economy that is carbon neutral and sustainable (Everingham & Chassagne, 2020). These socio-economic impacts the pandemic had on tourism, including on the millions of livelihoods that it supports, were beneficial for its development. Thence, recommendations to governments and other stakeholders, and to the United Nations system, for supporting safe reopening for sporting events and supporting physical activity throughout the COVID-19 outbreak and beyond can also be made using this positive experience.

The economic impacts of COVID-19 represent the substantial uncertainties facing the economic recovery of the world economy, which will depend heavily on the future course of the COVID-19 pandemic, relief efforts, and vaccine deployments. This reflects multiple factors, including challenges faced by many lower-income countries, relative to their high-income counterparts, to deliver fiscal assistance to deal with COVID-19 internal effects, and variations in severity of pandemic-related interruptions, and now, access to vaccines. This should also be linked to the situation on the labor markers of the developing countries when mostly young people are disproportionately out of work, and those that are employed are typically employed in informal economies or the gig economy, under temporary contracts, or the service sectors of the economy, which are likely to be heavily affected by COVID-19.

14.5 Novel Leading Trends in the Post-Pandemic Tourism

The COVID-19 pandemic introduced a plethora of novel trends in the use of e-tourism technologies and had an impact on the functioning of the tourism structures, organizations, and functions. Information and communication technologies applied in tourism constitute the core of an e-tourism industry, which will generate new tourist products, skills, and jobs (Hamid et al., 2021). The diffusion of technology advances and improved access through wireless connectivity will assist tourism sectors in satisfying the demand for participative consumption, as well as increasing awareness-raising tools to reach a larger number of prospective visitors. Mobile technologies and wireless connectivity impact a number of aspects of tourism at the global level. The key technological trends that are impacting tourism and hospitality today and in the near future are mobile technologies and connectivity. For years, technology has been heavily tied to tourism, as the tourism industry has sought to capitalize on developments and changes, opening up destinations and providing new products and services (Chen et al., 2020). Moving products and services around to suit the needs and wants of travelers worldwide will demand flexibility and creativity from BC tourism sector going forward. Like most industrial sectors, tourism is affected by global economic trends. Technological advocates speculate that the traditional ways in which information is provided about tourism and hospitality will be gone, with mobile technologies becoming more ubiquitous (Lehto & Lehto, 2019). Destination management systems are an ideal application domain for semantic web and P2P (peer-to-peer) technologies, since the distribution and sharing of tourist information is a fundamental basis for the destination management of tourism. The management and interoperation of semantically diverse tourist information systems is made easier with semantic web technologies, which provides methods and standards allowing for precise information access, and also the flexibility to meet the needs of tourist information systems users and administrator. Some researchers review ontologies, semantic modeling and querying, semantic portals, and semantically based e-markets, which are related to exploitation of semantic web technologies and applications to tourist information systems (Navío-Marco et al., 2018). Tourism technologies involve a number of processes (e.g., dynamic packing) that offer new useful options to consumers. It is clear that digital technologies increased respondents' interest in visiting cultural heritage sites, as users would like to use digital display devices and technologies while visiting other cultural heritage sites.

Regarding information and knowledge, the contributions of digital technologies to the recreation of historical scenes and presentation of historical information are broadly recognized by the potential users. Thinking about the future, please predict a single trend within each of the tourism-related domains – political, economic, environmental, social, and technological – that are likely to have the lasting effects on tourism and hospitality.

14.6 Conclusions

Overall, at a sectoral level, tourism and travel-related industries would be some of the most affected industries by the pandemic, in a situation when the authorities encouraged social distancing and consumers staying inside. Up to 100 million direct jobs in tourism were threatened, as well as sectors related to tourism, such as labor-intensive hospitality and food services industries, which employed around 200 million workers globally.

Key technological trends emerging in travel and tourism is a handy guide explaining the top technology trends within travel and tourism. The evolution and applications of digital technologies are dramatically changing how people live, work, travel, and conduct business, and they are changing and reshaping travel and tourism along the way. Digitalization, before and after the COVID-19 pandemic, has also opened new opportunities for tourism businesses to compete in the global marketplace. Just as travel needs and behaviors have changed among tourists, the ways tourism businesses sell their services has also changed. Travel agencies, tour operators, destination management organizations, and other legacy tourism businesses typically operate on a double-track model, maintaining the travel agencies proprietary web pages and social media marketing, while simultaneously offering their products and services across global platforms. Quick-service restaurants are comprised of an intricate network of relationships among various suppliers, tourist products, destination marketing organizations, tour operators, and travel agents, among many others. The retail and tourist industries are tightly intertwined, and most large tourism destinations will provide retail facilities as well. Coach and bus services are a significant component of tourism, providing travel over distances, helping passengers from the airport get to their accommodations, taking groups of tourists on day trips to local attractions or popular tourism destinations, and providing access to other neighboring towns and cities. The transport industry plays a fascinating role in the tourism industry, providing tourists with necessary refreshment throughout every phase of the tourism experience, including while traveling, while spending time at the chosen accommodation, and while they are on location exploring the destination to which they have traveled. Tourism services support the growth of the industry and guest experience. Tourism also provides large numbers of jobs to individuals working in transportation and the hospitality sector, among others. Working in tourism administration opens a world of new opportunities, which are rarer in other sectors. Technology has helped the tourism and hospitality industries to replace costly human labor with technology-based work. Technology is helping not just the big chains of hotels; it is useful for bed and breakfasts and other smaller businesses within the hospitality and tourism industries. The use of technology in hospitality and tourism has helped to accelerate operations and make the travel experience far more pleasant and effective.

In many ways, the tourism and hospitality industry is leading the way when it comes to adopting smart business technology in the post-pandemic era. From operations and guest experiences to marketing, smart hotel technology offers numerous opportunities for cost savings and increased revenue, enabling hotel owners to achieve new levels of profitability. The advancements in digitalization have finally reached the point that can really underpin cost efficiency and resilience in the production of food products in industry, leading the way for future hospitality and tourism businesses. Digitalization is providing tourism operators and companies with an unprecedented opportunity to tap into new markets, develop new tourism products and services, adopt new business models and processes, improve their positions on the global tourism value chain, and integrate in the digital ecosystem. As a consequence of the digital revolution, the international tourism development trends opened up new solutions such as cloud-based reservation sites or the sharing of information and experiences through digital platforms. Although the new trends such as trip packages (including rental cars) or taking the reviews of prior travelers (Lonely Planet) are, from many aspects, opposed to the former business models, the fast-growing popularity of online offerings requires quick, easy-to-use development of tourist products by the tourist industry. Technological developments cannot be stopped. However, with enough flexibility and openness, tourism businesses could be prepared to face the challenges ahead. The scale and income-generating power of technology companies in the tourism industry, such as Booking.com, AirBnB, and Expedia, allows such companies to develop new technologies and enhance their offerings for travelers, and the highly competitive nature of tourism will fuel continued developments.

In addition, the high-quality developments in 3D printing technology present an excellent opportunity to the tourism and hospitality sectors. Using 3D food printing for these purposes could improve the economic viability, efficiency, and sustainability, thereby supporting both the tourism industry and the hospitality industry. The Internet of Things (IoT) could power the tourism industry with rich data, supporting intelligent tourism while making cities more efficient. Thence, the Smart Tourism Platform can be built using data extraction technologies, which could facilitate the smart tourism development, providing the foundation of the sustainable growth of the tourism sector.

Thanks to the Big Data collection and assessment technologies, tourism stakeholders in the post-COVID-19 era are receiving information about tourists and data about the market operations in real-time, providing public services on smart tourism, which improves the time-sensitivity and the effectiveness of the administration. Those working in entertainment tourism know that providing a welcoming, friendly experience is the critical ingredient in maintaining any tourist destination. Whether it is festivals, events, attractions, or the emerging virtual components, the meetings and events industry rely on entertainment to round out packages and make sure guests, both business and vacation travelers, are increasing their spending and enjoying themselves. The impacts felt across the hospitality industry are significant, driven by factors including the migration of hospitality workforces and a global slowdown in tourism, travel, and hospitality services.

However, travel and tourism may be increasing with global levels of prosperity, but skilled hospitality workers are scarce throughout the sector, and ever-evolving market dynamics are creating entirely new types of jobs within the industry. The increasing complexity of the global hospitality industry and evolution in its operating models – due to increased usage of technology and data, evolving business models that separate leadership from operations, or the increasingly common trend of brands becoming publishers – are leading to the creation of new job profiles, such as asset managers, data scientists, or content marketing specialists.

In short, all of the factors suggested and discussed above represent the ways how to restore the leading position of the tourism and hospitality sector in our turbulent times that follow the COVID-19 pandemic and to help it not only to regain its former position but also to thrive and face the new challenges presented by the economic and (geo)political situation in the over-globalized world of the twenty-first century.

References

Ahmad, N., Hasan, M. G., & Barbhuiya, R. K. (2021). Identification and prioritization of strategies to tackle COVID-19 outbreak: A group-BWM based MCDM approach. *Applied soft Computing, 111*, 107642. https://doi.org/10.1016/j.asoc.2021.107642

Akhtar, N., Khan, N., Mahroof Khan, M., Ashraf, S., Hashmi, M. S., Khan, M. M., & Hishan, S. S. (2021). Post-COVID 19 tourism: Will digital tourism replace mass tourism? *Sustainability, 13*(10), 5352. https://doi.org/10.3390/su13105352

Alkier, R., Milojica, V., & Roblek, V. (2022). The complexity of the tourism supply chain in the 21st century: A bibliometric analysis. *Kybernetes, ahead-of-print*, ahead-of-print. https://doi.org/10.1108/K-03-2022-0430

Angell, C., & Potoglou, D. (2022). An insight into the impacts of COVID-19 on work-related travel behaviours in the Cardiff Capital Region and following the UK's first national lockdown. *Cities, 124*, 103602. https://doi.org/10.1016/j.cities.2022.103602

Barbulescu, O., Nicolau, C., & Munteanu, D. (2021). Within the entrepreneurship ecosystem: Is innovation clusters' strategic approach boosting businesses' sustainable development? *Sustainability, 13*(21), 11762. https://doi.org/10.3390/su132111762

Becken, S., & Hughey, K. F. (2021). Impacts of changes to business travel practices in response to the COVID-19 lockdown in New Zealand. *Journal of Sustainable Tourism, 30*(1), 108–127. https://doi.org/10.1080/09669582.2021.1894160

Chen, S., Law, R., Xu, S., & Zhang, M. (2020). Bibliometric and visualized analysis of mobile technology in tourism. *Sustainability, 12*(19), 7975. https://doi.org/10.3390/su12197975

Chen, H., Shi, L., Zhang, Y., Wang, X., & Sun, G. (2021). A cross-country core strategy comparison in China, Japan, Singapore and South Korea during the early COVID-19 pandemic. *Globalization and Health, 17*(1), 1–10. https://doi.org/10.1186/s12992-021-00672-w

Czerny, A. I., Fu, X., Lei, Z., & Oum, T. H. (2021). Post pandemic aviation market recovery: Experience and lessons from China. *Journal of Air Transport Management, 90*, 101971. https://doi.org/10.1016/j.jairtraman.2020.101971

Everingham, P., & Chassagne, N. (2020). Post COVID-19 ecological and social reset: Moving away from capitalist growth models towards tourism as Buen Vivir. *Tourism Geographies, 22*(3), 555–566. https://doi.org/10.1080/14616688.2020.1762119

Goldstein, I., Koijen, R. S., & Mueller, H. M. (2021). COVID-19 and its impact on financial markets and the real economy. *The Review of Financial Studies, 34*(11), 5135–5148. https://doi.org/10.1093/rfs/hhab085

Gruzina, Y., Firsova, I., & Strielkowski, W. (2021). Dynamics of human capital development in economic development cycles. *Economies, 9*(2), 67. https://doi.org/10.3390/economies9020067

Guliyeva, A., & Rzayeva, U. (2018). The role of E-government in reduction of information asymmetry in developing countries on the example of Azerbaijan. *Advances in Economics and Business, 6*(4), 209–217. https://doi.org/10.13189/aeb.2018.060402

Guo, K., Fan, A., Lehto, X., & Day, J. (2021). Immersive digital tourism: The role of multisensory cues in digital museum experiences. *Journal of Hospitality & Tourism Research*, 10963480211030319. https://doi.org/10.1177/10963480211030319

Gupta, S., Justy, T., Kamboj, S., Kumar, A., & Kristoffersen, E. (2021). Big data and firm marketing performance: Findings from knowledge-based view. *Technological Forecasting and Social Change, 171*, 120986. https://doi.org/10.1016/j.techfore.2021.120986

Hai, T. N., Van, Q. N., & Thi Tuyet, M. N. (2021). Digital transformation: Opportunities and challenges for leaders in the emerging countries in response to COVID-19 pandemic. *Emerging Science Journal, 5*, 21–36. https://doi.org/10.28991/esj-2021-SPER-03

Hamid, R. A., Albahri, A. S., Alwan, J. K., Al-Qaysi, Z. T., Albahri, O. S., Zaidan, A. A., & Zaidan, B. B. (2021). How smart is e-tourism? A systematic review of smart tourism recommendation system applying data management. *Computer Science Review, 39*, 100337. https://doi.org/10.1016/j.cosrev.2020.100337

Herédia-Colaço, V., & Rodrigues, H. (2021). Hosting in turbulent times: Hoteliers' perceptions and strategies to recover from the Covid-19 pandemic. *International Journal of Hospitality Management, 94*, 102835. https://doi.org/10.1016/j.ijhm.2020.102835

Ide, A. (2021). Tourism and ICT solutions in the COVID-19 era: A comparison between Japan and Sweden. *The Review of Socionetwork Strategies, 15*(1), 195–211. https://doi.org/10.1007/s12626-021-00072-x

Ioannides, D., Gyimóthy, S., & James, L. (2021). From liminal labor to decent work: A human-centered perspective on sustainable tourism employment. *Sustainability, 13*(2), 851. https://doi.org/10.3390/su13020851

Kesavan, N., Seenuvasaragavan, S., Burnette, K., Heim, A. B., Patel, R. J., Johnston, J., Cao, X., Sakandar, H. A., Zhuang, Y., Oi, K. K., Zhi, Y., Diao, Y., & Strielkowski, W. (2022). Pandemic-inspired policies. *Science, 377*(6601), 22–24. https://doi.org/10.1126/science.add2046

Kim, J. J., Kim, I., & Hwang, J. (2021). A change of perceived innovativeness for contactless food delivery services using drones after the outbreak of COVID-19. *International Journal of Hospitality Management, 93*, 102758. https://doi.org/10.1016/j.ijhm.2020.102758

Lehto, X. Y., & Lehto, M. R. (2019). Vacation as a public health resource: Toward a wellness-centered tourism design approach. *Journal of Hospitality & Tourism Research, 43*(7), 935–960. https://doi.org/10.1177/1096348019849684

Liang, S., Leng, H., Yuan, Q., & Yuan, C. (2021). Impact of the COVID-19 pandemic: Insights from vacation rentals in twelve mega cities. *Sustainable Cities and Society, 74*, 103121. https://doi.org/10.1016/j.scs.2021.103121

Maqsood, A., Abbas, J., Rehman, G., & Mubeen, R. (2021). The paradigm shift for educational system continuance in the advent of COVID-19 pandemic: Mental health challenges and reflections. *Current Research in Behavioral Sciences, 2*, 100011. https://doi.org/10.1016/j.crbeha.2020.100011

Mavragani, E., & Dionysios, P. (2022). Gen "Z" and tourism destination: A tourism perspective of augmented reality gaming technology. *International Journal of Innovation and Technology Management, 19*, 2241001. https://doi.org/10.1142/S0219877022410012

McCartney, G. (2021). The impact of the coronavirus outbreak on Macao. From tourism lockdown to tourism recovery. *Current Issues in Tourism, 24*(19), 2683–2692. https://doi.org/10.1080/13683500.2020.1762549

Mehtiyeva, A., & Prince, S. (2020). Journeys of research, emotions and belonging: An exploratory analysis of the motivations and experience of ancestral tourists. *Scandinavian Journal of Hospitality and Tourism, 20*(1), 85–103. https://doi.org/10.1080/15022250.2020.1728376

Miao, L., Im, J., Fu, X., Kim, H., & Zhang, Y. E. (2021). Proximal and distal post-COVID travel behavior. *Annals of Tourism Research, 88*, 103159. https://doi.org/10.1016/j.annals.2021.103159

Moreno-Luna, L., Robina-Ramírez, R., Sánchez, M. S. O., & Castro-Serrano, J. (2021). Tourism and sustainability in times of COVID-19: The case of Spain. *International Journal of Environmental Research and Public Health, 18*(4), 1859. https://doi.org/10.3390/ijerph18041859

Navío-Marco, J., Ruiz-Gómez, L. M., & Sevilla-Sevilla, C. (2018). Progress in information technology and tourism management: 30 years on and 20 years after the internet-Revisiting Buhalis & Law's landmark study about eTourism. *Tourism Management, 69*, 460–470. https://doi.org/10.1016/j.tourman.2018.06.002

Nicoletti, G., von Rueden, C., & Andrews, D. (2020). Digital technology diffusion: A matter of capabilities, incentives or both? *European Economic Review, 128*, 103513. https://doi.org/10.1016/j.euroecorev.2020.103513

Plzáková, L., & Smeral, E. (2022). Impact of the COVID-19 crisis on European tourism. *Tourism Economics, 28*(1), 91–109. https://doi.org/10.1177/13548166211031113

Riyanto, R., & Azis, A. (2021). Modelling customers lifetime value for non-contractual business. *International Journal of Informatics and Information Systems, 4*(1), 55–62. https://doi.org/10.47738/ijiis.v4i1.77

Scott, D., & Gössling, S. (2021). Destination net-zero: What does the international energy agency roadmap mean for tourism? *Journal of Sustainable Tourism, 30*(1), 14–31. https://doi.org/10.1080/09669582.2021.1962890

Shah, S. I. H., Peristeras, V., & Magnisalis, I. (2021). Government big data ecosystem: Definitions, types of data, actors, and roles and the impact in public administrations. *ACM Journal of Data and Information Quality, 13*(2), 1–25. https://doi.org/10.1145/3425709

Shang, Y., Li, H., & Zhang, R. (2021). Effects of pandemic outbreak on economies: Evidence from business history context. *Frontiers in Public Health, 9*, 632043. https://doi.org/10.3389/fpubh.2021.632043

Sonmez, S., Apostolopoulos, Y., Lemke, M. K., & Hsieh, Y. C. J. (2020). Understanding the effects of COVID-19 on the health and safety of immigrant hospitality workers in the United States. *Tourism Management Perspectives, 35*, 100717. https://doi.org/10.1016/j.tmp.2020.100717

Strielkowski, W., Dvořák, M., Rovný, P., Tarkhanova, E., & Baburina, N. (2021). 5G wireless networks in the future renewable energy systems. *Frontiers in Energy Research, 9*, 714803. https://doi.org/10.3389/fenrg.2021.714803

Toubes, D. R., Araújo Vila, N., & Fraiz Brea, J. A. (2021). Changes in consumption patterns and tourist promotion after the COVID-19 pandemic. *Journal of Theoretical and Applied Electronic Commerce Research, 16*(5), 1332–1352. https://doi.org/10.3390/jtaer16050075

Uzir, M. U. H., Al Halbusi, H., Lim, R., Jerin, I., Hamid, A. B. A., Ramayah, T., & Haque, A. (2021). Applied Artificial Intelligence and user satisfaction: Smartwatch usage for healthcare in Bangladesh during COVID-19. *Technology in society, 67*, 101780. https://doi.org/10.1016/j.techsoc.2021.101780

Varzaru, A. A., Bocean, C. G., & Cazacu, M. (2021). Rethinking tourism industry in pandemic COVID-19 period. *Sustainability, 13*(12), 6956. https://doi.org/10.3390/su13126956

Xiang, K., Huang, W. J., Gao, F., & Lai, Q. (2022). COVID-19 prevention in hotels: Ritualized host-guest interactions. *Annals of Tourism Research, 93*, 103376. https://doi.org/10.1016/j.annals.2022.103376

Xu, Y., Hazée, S., So, K. K. F., Li, K. D., & Malthouse, E. C. (2021). An evolutionary perspective on the dynamics of service platform ecosystems for the sharing economy. *Journal of Business Research, 135*, 127–136. https://doi.org/10.1016/j.jbusres.2021.05.056

Chapter 15
Informal Employment, Digitalization, and Economic Sustainability: Lessons from the COVID-19 Crisis

Natalia Kunitsyna

Abstract The exacerbation of social problems during the pandemic is due not only to medical and psychological issue but also to the social factors such as the loss of work and sources of labor income. The hidden sector of the labor market is transformed during epidemics and pandemics, negatively affecting the economic sustainability and leadership in business and economics as a whole. This chapter is devoted to the analysis of the depth of the leading role of the informal employment on the structure of the Russian economy during the COVID-19 pandemic. The dynamics of the informal sector in total employment, the digitalization of the economy, the structure of income of the officially and unofficially employed, and the effectiveness of public spending on financing social policies were assessed and analyzed. This chapter presents the results of a study of the relationship between quarantine measures, shadow employment, the life quality, and well-being. The analysis concluded that social policy plays a leading role in reducing informal employment and, consequently, the strengthening of economic sustainability. Governmental policy measures identified in course of this research might help to stimulate the development of the formal labor market. The leading and most relevant policies appear to be the development of social programs, investing into the human capital and ICT, supporting the poor and families with children, and improving the life quality of vulnerable social groups.

Keywords Informal employment · Digitalization · Sustainable development · Leadership · Labor market

JEL Classification G14 · E24 · L80

N. Kunitsyna (✉)
North-Caucasus Federal University, Stavropol, Russian Federation
e-mail: nkunitcyna@ncfu.ru

W. Strielkowski (ed.), *Leadership, Entrepreneurship and Sustainable Development Post COVID-19*, Springer Proceedings in Business and Economics, https://doi.org/10.1007/978-3-031-28131-0_15

15.1 Introduction

The theory and practice of human resource management in the context of the pandemic of the new coronavirus infection has become a response to the challenges of the modern economy, bringing to the fore health, quality of life, digitalization, and social well-being of the population along with preserving and strengthening the sustainability of the economy.

The following leading Russian scientists have devoted their research to the study of the features of informal employment, the formation of state socioeconomic policy, their theoretical and applied aspects: Nureev and Akhmadeev (2019), Bobkov et al. (2015), Bobkov and Loktyuhina (2020), and Bobkov and Gulagina (2020), Aivazyan (2016), Kleyner et al. (2015, 2018), Gimpelson and Kapelyushnikov (2007, 2014), Grinberg (2017), and Volchik et al. (2021) and specialists from other countries, such as Hussmann (2001), Besley and Burgess (2004), Fields (2005), Günther and Launov (2006), Perry et al. (2007), J.-P. Cling et al. (2014), Chen and Carré (2020), Williams and Windebank (1998), Williams and Lansky (2013), Jhabvala et al. (2003), Packard et al. (2012) or Strielkowski et al. (2020), Rodriguez-Loureiro et al. (2020), Simionescu et al. (2020), and Utzet et al. (2021).

Since the shock state of the economy and the psychology of the population under the spread of COVID-19 are concomitant phenomena of modern life, there has been an increase in the interest in this problem in the scientific world. Scientists are investigating the impact of COVID-19 on the sustainability of economic sectors and assessing the results of work and labor in 2020 (Kesavan et al., 2022).

This chapter is organized as follows: Section 15.2 focuses on the aggravation of the problems of informal employment. Section 15.3 looks into the economic sustainability, social tension and losses of the budget system. Section 15.4 describes the financing of social policy. Section 15.5 hints the ways to overcome the socioeconomic consequences of quarantine. Finally, Sect. 15.6 comes with the overall conclusions and policy implications.

15.2 Aggravation of the Problems of Informal Employment

Informal employment has become a leading trend, a notorious issue, and the trademark of the busy and over-globalized twenty-first century. As the digitalization of economy is continuing, the information and communication technologies (ICT) are changing the labor market by offering new possibilities how to engage into the employer-employee relations that include home office, digital contract, blockchain technologies, and cryptocurrencies. In addition, today the issue of precarization has become quite an important social issue since the future of work and labor might be in freelancing as the so-called uberization of economy is becoming the new normal. “Uber economy” or “gig economy” might be useful for many people in quite a number of professions including information technologies (IT), creative sector (marketing, social networking, information systems), and entertainment industry; however, we can also see the whole array of professions and jobs that do not fall into that sphere and cannot be done remotely. Some people engaged in these "non-gig economy" suffer from upheaval on the labor market, such as the recent COVID-19 pandemic.

The problem of informal employment has become particularly acute in the context of the pandemic. The analysis of publications shows that more than 20% of the employed population of Russia are involved in shadow business to some extent: about 10% (about 7–8 million people) are involved exclusively in the shadow economy, and another 10–12% earn extra money informally: they work at their main job without official registration, have signed an employment contract with an employer but receive part of their salary "under the table" by mutual agreement, and earn extra money in addition to their main employment without registering labor relations.

The coronavirus pandemic has exposed and exacerbated the abovementioned problems. Problems in the migration sphere have also been actualized which resulted in a sharp increase in informal employment with salary dumping. Unemployment and illegal employment, as a rule, contribute to instability in economic sectors and the growth of offenses, pose a threat to public safety, and entail the development of unsanitary conditions in everyday life as well as social and psychological tension.

According to the Federal State Statistics Service, one in five employed Russians works in the informal sector, i.e., they carry out activities without state control; 56% of companies do not officially employ their employees at all but recruit them on an informal basis. In 22 regions, the share of informally employed people exceeds 30%, for which there are two reasons: a significant part of the residents moves to other regions or work informally.

At the same time, according to a study by the audit and consulting network FinExpertiza, the number of people employed in the informal sector decreased in 2020, as quarantine and restrictions narrowed employment opportunities, and also led to the need to reduce the number of employees with low wages and low-productivity cheap labor (see Fig. 15.1).

15.3 Economic Sustainability, Social Tension, and Losses of the Budget System

Another problem that has become more acute during the pandemic that should be emphasized is the growth of poverty in general and the increased share of low-wage workers (below 2/3 of the median hourly earnings), which increased from 24.7% in 2019 to 36.4% in 2020. In the structure of monetary income of the Russian population, the largest share is occupied by wages (58.1% in 2019); social benefits, including pensions and scholarships (19%); and other cash receipts, which include untaxed

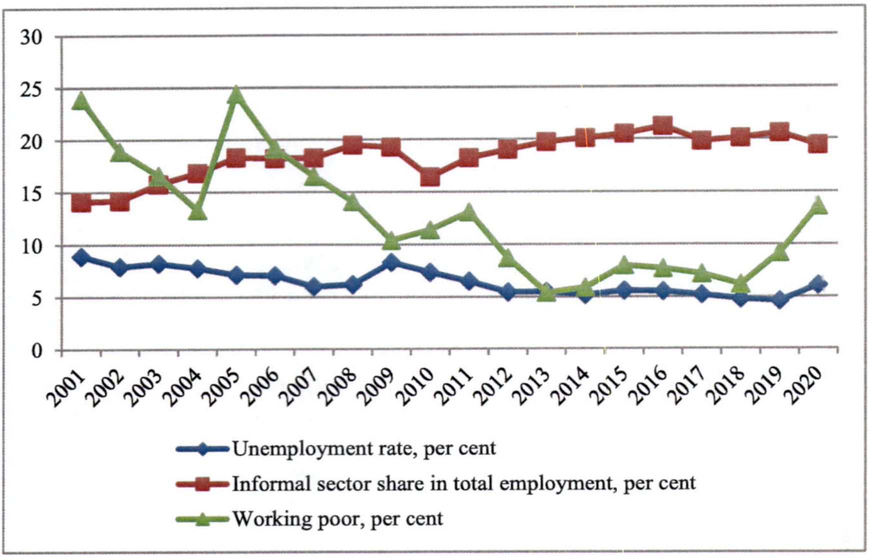

Fig. 15.1 Dynamics of labor market indicators in Russia. (Source: Own results)

income (12.4%). The share of income from entrepreneurial activity accounts for 6.1%. Under the restrictive measures, the reduction in income in 2020, according to various estimates, was about 15–20% compared to 2019. A significant decrease is due to strict quarantine measures for COVID-19 and the withdrawal of many small- and medium-sized businesses from the market. On the basis of the data on the collection of personal income tax to the budgets of the regions, the analysts of the National Credit Ratings agency calculated that due to the mass shutdown of small businesses, the wage fund in the economy has decreased by 10%. Russians lost almost a trillion rubles of salaries due to the pandemic.

The hidden wage fund is 12.6% of GDP. The losses of the budget system due to informal employment reach about 3 trillion rubles annually, or about 3% of GDP. The total number of informally employed people who escape tax statistics is estimated at 13 million people, or 18% of the labor force. Hidden income amounts to about 11 trillion rubles which is why social funds lose about 2.3 trillion rubles and budgets – 0.9 trillion rubles.

Our study of the relationship between quarantine measures, informal employment, quality of life, and well-being of the population correlates with the work of foreign scientists. Kanniainen et al. (2004) or Cabelkova and Strielkowski (2013) proved that unemployment, along with taxes, social security, and tax morality, has a significant impact on the sustainability of the economy and the size of its informal sector. Wanjie et al. (2011) evaluated the role of the government in regulating taxation, the social sphere, and economic issues as the most vital factors affecting the informal economy. Neck et al. (2012), Halla (2012), and Luttmer and Singhal

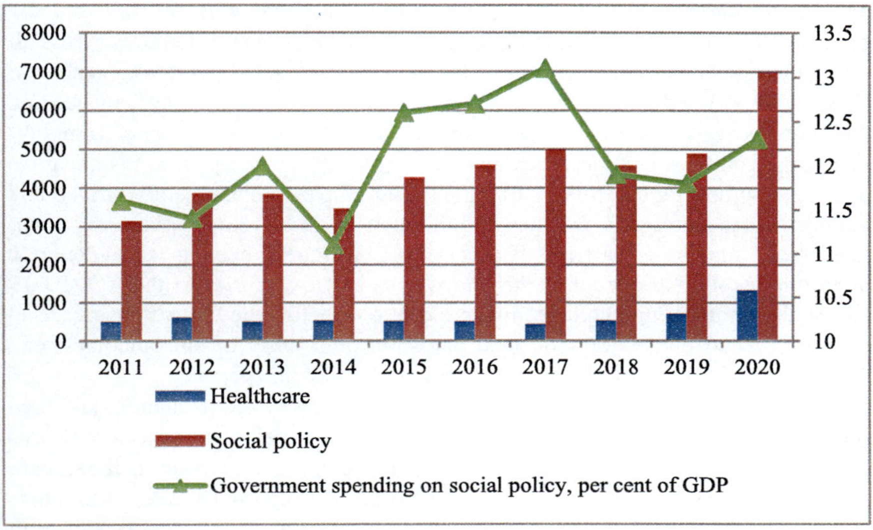

Fig. 15.2 Dynamics of state budget expenditures on social policy and healthcare, billion rubles. (Source: Own results)

(2014) illustrated how tax avoidance and tax morality affect the informal economy and informal employment.

Since government revenues during the pandemic are significantly lower than they could have been, this affects the ability to provide public services and maintain a healthy national balance. At the same time, the Government of the Russian Federation has taken a number of priority measures to finance social policy, increasing the number of deductions for these purposes to 6990.3 billion rubles in 2020 (4882.8 billion rubles in 2019), despite the reduction in GDP and federal budget revenues (Fig. 15.2). Due to additional costs, the share of social policy expenditures in total federal budget expenditures in 2020 amounted to a record 30.6%.

15.4 Financing Social Policy

Taking into account the opinion of an international team of researchers (Matytsin et al., 2019), we would like to emphasize that the tax and social policy measures taken before and after the outbreak of the pandemic are more focused on the poor and families with children.

In general terms, families with children tend to be more heavily reliant on social networks and public supports, and that dependence on social safety nets surely persisted, while the authorities worldwide tried to contain the COVID-19 pandemic and boost the economy, which was a very-slow-moving one. The economic crisis caused by COVID-19 is likely to be most acutely felt by children and families not only in Russia but also all around the world. The economic fallout from the COVID-19 pandemic is threatening the health and well-being of families across all countries, and is especially impactful for the children aged 18 and younger. Now, the COVID-19 pandemic's economic fallout is serving to further unravel and worsen childhood poverty, with higher rates among children because of widespread employment losses and lost income, meaning that more children can risk lasting damage to their healthy development. Prior to the COVID-19 pandemic, recently arrived and undocumented immigrant families and children in the developed Western economies faced numerous barriers to accessing government benefits and economic opportunities that supported health and well-being. Even prior to the COVID-19 pandemic, families with children living close to or below the federal poverty level historically had higher rates of food hardship, poor child health outcomes, and lower educational outcomes compared with higher-income peers.

Challenges for all families peaked at the beginning of the pandemic, and then improved through the summer of 2021 when vaccines against the new coronavirus disease were developed and started to be administered. However, adults living with children in some poorer countries have consistently reported greater difficulties paying their household expenses each day, providing their families with sufficient food and being less secure about housing payments than do other families. The improvements in measures of hardship in the second year of the pandemic are likely due to improved economic conditions and multiple governmental policies implemented during the COVID-19 pandemic that provided financial relief for individuals and families. For instance, in the United State, various stakeholders and policymakers took several actions to alleviate the economic effects of the pandemic on families and children, providing fiscal relief to states, increasing food aid, funding Medicaid, and the Child Tax Credit (CTC) and Earned Income Tax Credit (EITC) child tax credit. However, some state tax assistance has already expired or will expire soon. The improvements in the Child Tax Credit (CTC) are responsible for providing most low-income families experiencing the greatest levels of hardship with increased amounts of assistance in the form of monetary aid ($300 per month for children up to age 6/$250 for children ages 6–17 for the first time in CTC's history with increased amounts (up to $300 per month for children age 6–17) through increased amounts of cash assistance. Without additional action, improvements to the aid are set to expire soon, potentially leading to doubled child poverty, and increased taxes for more than 60 million families with children. Child tax credits would function as not just a short-term tool for alleviating child poverty, but a longer-term instrument to improve the social mobility of families. The results indicate that new measures would likely improve social mobility both for families and for their children. With regard to childhood poverty and racial/ethnic equity, researchers estimated the new policies may reduce childhood poverty by 45%, with its greatest effects for children of color and Blacks. Because child poverty is not distributed equally across the United States, so is its growth: poverty rates are growing most for Latino children (4.2%), Black children (2.8%), and children from women-headed households (4.1%), while remaining unchanged for White and Asian children. Despite the income losses many families with children experienced last year, child poverty has declined when accounting for the effects of the stimulus and other aid, including the first two rounds of economic impact payments, expansions to unemployment insurance benefits, nutritional help from the Supplemental Nutrition Assistance Program and the Pandemic-EBT Program, and others.

With the COVID-19 disruptions in employment, child care, and education, it is not surprising that child poverty increased significantly in 2020. About 1.2 million more children lived in poverty in 2020 than 2019 (15.7% to 17.5%).

All in all, one can see that the COVID-19 pandemic is harming children's health, social, and material well-being around the world, with poor children, including homeless children and children in custody, most affected. The COVID-19 outbreak is challenging the well-being of children with disabilities, on aspects including education, health, and social and family life. The pandemic challenges vulnerable children as resilience, because it increases in the children as environments the number of already existing risks (e.g., reduced access to healthy foods, higher household stress, and lack of contact with supportive adults) and decreases the number of protective factors (e.g., placement at school, access to playgrounds and after-school activities, and robust child protective systems). The concentration of risks associated with COVID-19 falls among children living in low-income families and those who have extra needs or are vulnerable. The urgent priority for any policy aimed at responding to the challenges faced by children in the COVID-19 pandemic should be minimizing risks to the health and psychological well-being of children; providing access to good food and nutrition, as well as education support; and providing care and protection for vulnerable children who are in need. Child care providers of all types will have to secure PPE, install new ventilation systems, increase the spacing between children and teachers, enact new health and safety standards, procure additional cleaning supplies, and develop strategies and plans for mitigating the COVID-19 pandemic. Schools may use the funds provided by the relevant authorities to enhance staffing capabilities, including hiring additional teachers, engaging parent coordinators and community liaisons to communicate with families prior to and during any time that students are temporarily physically unable to attend school because of COVID-19 cases, and implementing strategies listed throughout this document and the other resources provided by the educational departments. The novel coronavirus has also had significant impacts on child care costs for families with young children, both wealthy and poor, since even in the wealthier families, the pressure to work from home during the shutdowns meant that the parents had to look after their children and do their jobs in the same time, which constituted enormous psychological and logistical difficulties and complications for all the parties involved into this process.

Due to the reduction or loss of the main sources of income, the population that has bank deposits is forced to spend them, which affects the decrease in income from property. Two-thirds of Russians (63.6%) say that they have no savings at all in the case of job loss, and the majority of those who have savings will fully spend them in 2 to 6 months (Gilenko & Chernova, 2021).

In addition, the Central Bank of Russia recorded an increase in demand for cash as increased uncertainty and quarantine measures could push the population to accumulate paper money for settlements during the isolation period. According to some experts, the demand for cash may grow in the informal business.

We believe that the full recovery to the previous level and the structure of consumption will not occur immediately after the end of the pandemic as income and asset inequality (especially in real terms) will increase due to income losses during the period of unemployment and low economic activity. There may be a crisis in the management of housing and communal services since citizens were allowed to pay for services with a delay without penalties.

Moreover, fears of infection that may persist after the end of the first and possible subsequent outbreaks of the epidemic will also lead to a conservative correction of people's economic and social behavior. These facts lead to the conclusion that these problems will become worse during the consequences of coronavirus infection.

Unfortunately, it is impossible to make the informally employed register and pay taxes due to the specifics of their activities; the state, according to analysts, has neither incentive for the voluntary legalization of this kind of income nor measures of administrative influence on this category of citizens, and any attempts to establish control over payments through the banking system will lead to the cashing out of all settlements in this area.

15.5 Overcoming Socioeconomic Consequences of Quarantine

The entire scientific world is concerned about finding a "cure" for overcoming the COVID-19 pandemic with the least possible losses. For example, Auzan (2020) describes the civilizational consequences of the modern crisis emphasizing that the pandemic is the price for globalization. Buklemishev (2020) conducts a systematic analysis of the features of the modern crisis in terms of its main transmission channels through demand, supply, the financial sector, and expectations. Shastitko (2020) mentions three important areas of crisis research: the reflection of the features of socioeconomic systems in the complex of sanitary and epidemiological indicators, the comparison of discrete structural alternatives in the light of the decision-making in the fight against the pandemic and its consequences, and adaptive effectiveness of institutions.

In addition, Grigoryev (2020) writes about the pandemic as a global social drama for the entire society and economy. Denisova (2020) offers solutions for the Russian labor market which will make it easier for the government to overcome the economic consequences of forced quarantine measures. Among them are the introduction of an unemployment insurance system, employment subsidy programs with reduced working hours and the introduction of partial unemployment benefits based on the experience of OECD countries. The main objective of such measures is to avoid unnecessary layoffs during periods that involve rapid stabilization of the economic situation and job creation.

In addition, the solution to the COVID-19 crisis and its severe impacts on the labor markets might be in transforming these markets by using digitalization and information systems. Flexibility and systematic solutions need to be made and implemented.

ICT can empower the population, provide ample opportunities for personal and professional development, and make public services more accessible. Consequently, ICTs can contribute to economic growth and development by shrinking the informal economy. Similarly, it can be argued that those consumers who shop in the informal economy make a rational decision to turn to this area due to a lack of supply or lack in the formal market, leaving the consumer with no choice but to participate in transactions in the formal market and the informal economy. On the one hand, those who work informally, representing the supply side of the informal economy, are usually described as rational economic agents who engage in informal work, motivated by financial benefits, and on the other hand, consumers or those who produce that purchases in the informal economy are made solely in order to have a lower cost, which compensates them for any inconvenience. Any representative consumer living in Central and Eastern Europe or Russia is more likely to be forced to move to the informal economy for reasons of convenience (better price and/or faster and better-quality goods or services) or because of the lack of a necessary product or service on the ordinary market. Most citizens in emerging economies do not choose the informal economy of their choice: it is a by-product of citizens' need for survival, which supports themselves and their families; it is very important that everyone has access to the main income-generating activity. Thus, the variety of terms, interpretations, and characteristics related to this phenomenon is quite large, which determines the need to define the exact concept of the digital shadow economy from the point of view of consumers as active participants in this area. While informality is widespread in sub-Saharan Africa, creating pathways to full formalization in the region has so far met with limited success. In many developed countries, automation is fueling fears of losing jobs for less skilled workers. What used to be called traditional crimes has changed and adapted to the digital age.

Digitalization offers ways how to work from home or using the "hybrid" mode in order to save the money for renting office spaces. Digital technologies offer ways on how to sign labor contracts online and pay employees using digital currencies or cryptocurrencies. It might be time that employers in Russia and all around the world stop focusing on the employee's attendance at work and physical presence at the workplaces and start looking at the delivery of work results at the best possible terms.

15.6 Conclusions

Summing up our results, one can say that during the COVID-19 crisis, the leading problems of the labor market worsened manifesting themselves, in addition to the growth of the unemployment rate and the number of poor people, in the expansion of the scale of informal employment and the strengthening of its negative impact on the sustainability of the economy and its shadow or hidden (sometimes also dubbed as "underground" sector).

It stems from our research that the task of state institutions should be to become the leaders in creating favorable conditions for economic activity subject to taxation, reduce unaccounted rental income, the scale of appropriation of part of the profit by the owners of enterprises, the number of barter transactions, and unaccounted income from household production. The support measures taken by the Russian government during and after the quarantine contributed to smoothing the situation in the social sphere and in the labor market and made it possible to overcome and stabilize the shock situation in the society. At the same time, informal employment remains a serious problem that requires the development of adequate measures in the future. However, this might be just a trend of the current times, so that gig economy and uberization are going to take their toll. Working part-time or on some specific projects or contracts might become the new social and economic norm, but, in this case, new rules and legislative frameworks should be prepared and set into place. Our society needs clear and transparent labor rules and regulations that would take into the account the new trends on the labor market connected to the wide-scale deployment of ICT and precarization of the labor force. The COVID-19 pandemic and the crisis it caused showed us the problems we might be facing in the nearest future, but they also hinted the possible solutions for these issues that can be made by the stakeholders and policymakers not only in the Russian Federation but in all countries around the world. As digitalization and globalization continue, new technological and leading logistic solutions emerge that might help us to optimize the labor market and focus its hidden and shadow sector. Only by tackling these issues the new problems of the labor market of the twenty-first century can be effectively solved and sorted out in the post-COVID era.

References

Aivazyan, S. A. (2016). *Quality of life and living standards analysis, an econometric approach* (1st ed.). De Gruyter.

Auzan, A. A. (2020). The economy under the pandemic and afterwards. *Population and Economics, 4*(2), 4–12. https://doi.org/10.3897/popecon.4.e53403

Besley, T., & Burgess, R. (2004). Can labor regulation hinder economic performance? Evidence from India. *Quarterly Journal of Economics, 119*(1), 91–134. https://doi.org/10.1162/003355304772839533

Bobkov, V. N., & Gulugina, A. A. (2020). 5 years of monitoring income and living standards of the Russian population. *Living Standards of the Population in the Regions of Russia, 16*(3), 9–23. https://doi.org/10.19181/lsprr.2020.16.3.1

Bobkov, V. N., & Loktyuhina, N. V. (2020). Directions of social and economic policy in the context of the development of non-standard forms of employment in the Russian Federation. *Level of Life of the Population of the Regions of Russia, 16*(1), 21–32. https://doi.org/10.19181/lsprr.2020.16.1.2

Bobkov, V. N., Loktyuhina, N. V., Rozhkov, V. D., & Chernyshova, M. I. (2015). The informal employment: Ways of reduction and priorities in research. *Megapolis Management, 1*(43), 29–43.

Buklemishev, O. V. (2020). Coronavirus crisis and its effects on the economy. *Population and Economic, 4*(2), 13–17. https://doi.org/10.3897/popecon.4.e53295

Cabelkova, I., & Strielkowski, W. (2013). Is the level of taxation a product of culture? A cultural economics approach. *Society and Economy, 35*(4), 513–529. https://doi.org/10.1556/SocEc.2013.0007

Chen, M., & Carré, F. (2020). *The informal economy: Revisited, examining the past, envisioning the future* (1st ed.). Routledge, 326 p.

Cling, J.-P., Lagrée, S., Razafindrakoto, F., & Roubaud, M. (2014). *The informal economy in developing countries* (1st ed.). Routledge, 366 p.

Denisova, I. A. (2020). Lessons from the ongoing crisis for labour market institutions in Russia. *Population and Economics, 4*(2), 65–71. https://doi.org/10.3897/popecon.4.e53552

Denisova, I. A. In Gimpelson, V. E., & Kapelyushnikov R. I. (Eds.). (2007). Non-standard employment in the Russian economy. *Voprosy ekonomiki, 9*, 154–156.

Fields, G. S. (2005). *A guide to multisector labour market models* (Social protection discussion paper series, No. 0505) (1st ed.). World Bank, 49 p.

Gilenko, E., & Chernova, A. (2021). Saving behavior and financial literacy of Russian high school students: An application of a copula-based bivariate probit-regression approach. *Children and Youth Services Review, 127*, 106122. https://doi.org/10.1016/j.childyouth.2021.106122

Gimpelson, V. E., & Kapelyushnikov, R. I. (2014). *In the shadow of regulation: Informality in the Russian labor market* (1st ed.). HSE, 535 p.

Grigoryev, L. M. (2020). Global social drama of pandemic and recession. *Population and Economics, 4*(2), 18–25. https://doi.org/10.3897/popecon.4.e53325

Grinberg, R. S. (2017). *Looking for a balance* (1st ed.). Master, 160 p.

Günther, I., & Launov, A. (2006). *Competitive and segmented informal labor markets* (IZA discussion paper No. 2349) (1st ed.). Institute for the Study of Labor, 31 p.

Halla, M. (2012). Tax morale and compliance behavior: First evidence on a causal link. *Journal of Economic Analysis and Policy, 12*(1), 1–27. https://doi.org/10.1515/1935-1682.3165

Hussmann, R. (2001). *Informal sector and informal employment: Elements of a conceptual framework.* https://www.wiego.org/sites/default/files/publications/files/Hussmanns-IEpaper_lac_event.pdf. Accessed 22 Sept 2022.

Jhabvala, R., Sudarshan, M., & Unni, J. (2003). *Informal economy centrestage: New structures of employment* (1st ed.). Sage Publications, 290 p.

Kanniainen, V., Paakkonen, J., & Schneider, F. (2004). *Fiscal and ethical determinants of shadow economy: Theory and evidence discussion* (1st ed.). Helsinki Center of Economic Research.

Kesavan, N., Seenuvasaragavan, S., Burnette, K., Heim, A. B., Patel, R. J., Johnston, J., Cao, X., Sakandar, H. A., Zhuang, Y., Oi, K. K., Zhi, Y., Diao, Y., & Strielkowski, W. (2022). Pandemic-inspired policies. *Science, 377*(6601), 22–24. https://doi.org/10.1126/science.add2046

Kleyner, G. B., Dementyev, V. E., & Erznkyan, B. A. (2015). *Institutional economy and modern governance* (1st ed.). Aleteiia, 150 p.

Kleyner, G. B., Korablev, Y. A., & Shchepetova, S. E. (2018). Human in the digital economy. *Economics of Contemporary Russia, 2*(81), 169–175.

Luttmer, E., & Singhal, M. (2014). Tax morale. *Journal of Economic Perspectives, 28*(4), 149–168. https://doi.org/10.1257/jep.28.4.149

Matytsin, M., Popova, D., Freije, S., & Rusmod, A. (2019). *Tool for distributional analysis in the Russian Federation* (Policy research working paper No. 8994) (1st ed.). World Bank Group, 46 p.

Neck, R., Uwewachter, J., & Chneider, F. (2012). Tax avoidance versus tax evasion: On some determinants of the shadow economy. *International Tax and Public Finance, 19*(1), 104–117. https://doi.org/10.1007/s10797-011-9197-5

Nureev, R. M., & Akhmadeev, D. R. (2019). *Informal employment: Origins, current status and development prospects (institutional analysis experience)* (1st ed.). Knorus.

Packard, T., Koettl, J., & Montenegro, C. E. (2012). *From the shadow: Integrating Europe's informal labor* (1st ed.). Directions in Development: Human Development, 143 p.

Perry, G. E., Maloney, W. F., Arias, O., Fajnzylber, P., Mason, A. D., & Saavedra-Chanduvi, J. (2007). *Informality: Exit and exclusion* (No. 6730) (1st ed.). The World Bank.

Rodriguez-Loureiro, L., Vives, A., Martínez Franzoni, J., & López-Ruiz, M. (2020). Health inequalities related to informal employment: Gender and welfare state variations in the Central American region. *Critical Public Health, 30*(3), 306–318. https://doi.org/10.1080/09581596.2018.1559923

Shastitko, A. E. (2020). COVID-19: Moments of truth and sources of controversy. *Population and Economics, 4*(2), 34–38. https://doi.org/10.3897/popecon.4.e53285

Simionescu, M., Streimikiene, D., & Strielkowski, W. (2020). What does Google Trends tell us about the impact of Brexit on the unemployment rate in the UK? *Sustainability, 12*(3), 1011. https://doi.org/10.3390/su12031011

Strielkowski, W., Volchik, V., Maskaev, A., & Savko, P. (2020). Leadership and effective institutional economics design in the context of education reforms. *Economies, 8*(2), 27. https://doi.org/10.3390/economies8020027

Utzet, M., Botías, F., Silva-Peñaherrera, M., Tobías, A., & Benavides, F. G. (2021). Informal employment and poor self-perceived health in Latin America and the Caribbean: A gender-based comparison between countries and welfare states in a pooed analysis of 176,786 workers. *Globalization and Health, 17*(1), 1–11. https://doi.org/10.1186/s12992-021-00792-3

Volchik, V., Posukhova, O., & Strielkowski, W. (2021). Digitalization and sustainable higher education: Constructive and destructive potential of professional dynasties. *Transformations in Business & Economics, 20*(3), 21–43.

Wanjie, S., Hontat, H., & Rasli, A. (2011). Underground economy: Definition and causes. *Business and Management Review, 1*(2), 14–24.

Williams, C. C., & Lansky, M. A. (2013). Informal employment in developed and developing economies: Perspectives and policy responses. *International Labour Review, 152*(3–4), 355–380. https://doi.org/10.1111/j.1564-913x.2013.00196.x

Williams, C. C., & Windebank, J. (1998). *Informal employment in advanced economies* (1st ed.). Routledge.

Chapter 16
Development of Leading Digital Technologies in Business: New Risks and Prospects in the Post-Coronavirus Era

Elena Dombrovskaya, Alexander Neshcheret, and Tatiana Freze

Abstract This chapter studies the new risks and prospects in the post-coronavirus era with respect to the development of leading digital technologies in business by entrepreneurs and innovators. This chapter is about innovations in development and humanitarian efforts within the context of the coronavirus pandemic. The current spread of the COVID-19 in the past 2 years has highlighted the need for leadership by governments to expedite evaluation and implementation of digital technologies. The unprecedented humanitarian and economic needs presented by the COVID-19 pandemic are driving the development and deployment of new digital technologies at a large and rapid pace. Many researchers and policymakers alike share the view that the pandemic has precipitated changes to digital technologies already under way. Furthermore, it takes a deeper look at the informality situation as well as the impact that digital technologies and digital business models, especially digital platforms, are having on this situation. Several topics are addressed, providing support for COVID-19 control efforts and pandemics to come, using the artificial intelligence (AI) to identify outbreaks and to identify fake or adulterated medicines, while the new special group on digital technologies for coronavirus health emergencies sought to establish best practices for using AI during every phase of the life cycle of public health emergencies. Our results help to compare and balance new risks and

E. Dombrovskaya (✉)
Audit and Corporate Reporting Department, Faculty of Taxes, Audit and Business Analysis, Financial University under the Government of the Russian Federation, Moscow, Russian Federation
e-mail: ENDombrovskaya@fa.ru

A. Neshcheret
North-West Institute of Management, Saint Petersburg, Russian Federation
e-mail: neshcheret-ak@ranepa.ru

T. Freze
Institute of Engineering and Environmental Security, Togliatti State University, Togliatti, Russian Federation
e-mail: freze@tltsu.ru

W. Strielkowski (ed.), *Leadership, Entrepreneurship and Sustainable Development Post COVID-19*, Springer Proceedings in Business and Economics, https://doi.org/10.1007/978-3-031-28131-0_16

prospects in the post-coronavirus era and can provide politicians, academics, and business leaders valuable food for thought.

Keywords Digital technologies · Business leadership · Post-COVID-19

JEL Classification L86 · L96 · M10

16.1 Introduction

Now, in our globalized and interconnected world in which both developed and developing economy are fully using computers and benefit from the Internet, information and communication technologies (ICTs), the digital industries worldwide are slowly becoming major enablers of economic development. A tectonic shift is underway with information and communication technology expenditures cutting across sectors, from e-commerce and finance to manufacturing and agriculture (Haldar & Sethi, 2022). It becomes clear that we are approaching the biggest inversion of ICT investments into the physical and digital sectors, driven by the emerging technology suites for physical and digital industries such as 5G+ and others (Bojkovic et al., 2021). This indicates that the development of emerging technologies related to digital economies, such as the Internet and mobile communications, has contributed significantly to economic growth in countries both in the developed West and in the emerging countries. Currently, mobile technologies and services are widespread across the globe. On the African continent alone, they generated 1.7 million direct jobs in recent times (both formal and informal), contributed $144 billion of economic value (8.5% of GDP of Sub-Saharan Africa), and contributed $15.6 billion in revenue to the public sector via taxes (De Angelis et al., 2021).

Moreover, digitization has also solved problems related to the information asymmetries of the financial system and labor markets, thus increasing efficiency, reliability, and security, in a context in which information flows are crucial to economic growth and job creation (Kuteyi & Winkler, 2022). Digitization has affected economic growth through inclusive finance, by enabling unbanked people to transition into formality via electronic payment platforms at the retail level, as well as through technology-enabled savings and loan provision (Mpofu & Mhlanga, 2022). More generally, it is important to enable entrepreneurs and businesses to redefine their business models to become more influential, resilient, and connected. The rise of computing technology and digitization has transformed economic governance processes (Stroiko et al., 2021). It is worth noting that, on one hand, thanks to information technologies, efficiency has increased in most sectors of economics and government; opportunities to interact with subjects in legal relations are expanding thanks to new digital technologies from end-to-end, such as AI and robotics; and the pace at which decisions are developed and made by managers has increased

(Klochan et al., 2021). While cross-sectoral cooperation is necessary to make digital developments themselves successful, it also opens up opportunities for integration of cutting-edge digital solutions across sectors – be it using geospatial analytics to enhance risk-management of natural disasters, the deployment of Wi-Fi-enabled sensors to remotely monitor crops, or using big data from mobile devices to develop better public transportation networks. Beyond expertise in telecoms, building the five bases for the digital economy requires expertise across multiple other fields, including finance, private-sector development, education, labor, social protection, and more. It is comprised of not just underlying 5G fundamental networks but key technologies that will work in tandem with 5G to digitize all aspects of business operations (Strielkowski et al., 2021). According to research literature, a narrow-scope digital economy refers to an information technology or ICT sector producing fundamental digital goods and services, while the wide-spectrum digital economy encompasses all economic activities that are oriented toward the digital domain, that embrace ICT digitalization as the critical driver of production, that utilizes contemporary ICT infrastructures as the medium, and that delivers products or services using digital technologies (Williams, 2021). The key driver here is the continued increase of the variety, differentiation, and fragmentation of the target audiences and communications channels, which is once again caused by the growth of digital technologies. The authors' motivation for conducting this research is related to a wish to codify and perhaps to anticipate the evolution of digital governance and communications solutions under new conditions following the global COVID-19 pandemic, where a majority of working populations are experiencing re-thinking the paradigms of their economic and social-communication activities. If asked to think about how life would be in 2025 after the global pandemic outbreak in 2020 and the other crises, innovators and developers in technology would have probably articulated a variety of broader themes around how individuals and groups are adapting to a global crisis, outlining the more likely opportunities and challenges that arise as people rapidly increase the use and applications of digital technologies to cope. During the COVID-19 crisis, leaders in business and economics were the first to bring innovations to market and were the first company in their sector to experiment with new digital technologies (Farooq et al., 2022).

The idea of tipping points in technological adoption or digital disruption is not new, but the results from this survey indicate that the COVID-19 crisis is a tipping point – and one that is expected to prompt further changes as economic and human conditions develop (Srivastava et al., 2021). The infrastructure, the skilled labor, ICT capabilities, and the environment of innovation around digital technologies needed to develop a digital economy are severely lacking in most of these countries.

16.2 COVID-19 and Its Impact on Innovative Approaches

The COVID-19 pandemic has led to a predictable spike in digital tech usage, because of the norms of social distancing and nationwide quarantine. Internet access, as well as the digital divide information and communication technologies, will remain critical to a post-pandemic era, with innovations driving a spike in usage. Some behaviors developed during the crisis, including widespread adoption of the digital domain, will persist beyond the pandemic, long after restrictions on activities are lifted (Pandey & Pal, 2020). Detecting these signals of change sooner rather than later will be critical for optimizing the customer experience and revising customer value propositions in light of evolving preferences and needs. Digital transformation is needed even more in a pandemic, not less. Just as many companies had to rebuild the risk and financial models that failed in the financial meltdown in 2008, models will need to be replaced in similar ways due to massive economic and structural changes caused by the COVID-19 crisis (Berry et al., 2022).

Quickly reorienting business agendas to meet those changes will be essential for successful post-pandemic recovery. Continuously measuring performance across digital channels is crucial, allowing companies to adapt rapidly as more is learned. Changes that involve more foundational work, such as setting up a new e-commerce channel, typically will take more time. Accelerating the journey of digital transformation is going to be the number one priority for many companies, in order to guarantee the business continuity, increase the productivity, and roll out new business models in order to stay competitive. The concept of digital transformation is quite advanced today, with many organizations working to transform their businesses in the past couple years, if not longer. Digital transformation is a process where companies implement technology throughout their businesses in order to bring about a fundamental shift (Kraus et al., 2021). Digitalization is using digital technologies to transform business processes and projects, for example, training employees to work on a new software platform designed to help ship products more quickly. Digital transformation technologies, such as cloud, Internet of Things (IoT), blockchain (BoC), artificial intelligence (AI), and machine learning (ML), make up a large portion of what organizations are embracing as part of their transformation efforts. Digital technologies are joining the long-standing family of innovations in public health, which has been the focus of prevention-and-containment strategies in disease for centuries (Yörük & Gençer, 2022). The potential for digital technologies to underpin epidemiological intelligence through online data sets, identify cases and clusters of infections, quickly track contacts, track travel patterns in times of lockdown, and allow public-health messages to reach large audiences is well-known. With the significant use of technologies to access essential needs such as healthcare and education, it is critical that we understand the effects of digital disruptions on social equity (Kesavan et al., 2022). Following a vast body of literature of the overall role of innovation in the COVID-19 response, this chapter reviews innovation efforts that are under way within international development and humanitarian responses to the pandemic, how these efforts are working, and how they may need to be strengthened in order to meet pressing health, social, and economic challenges and ensure society's long-term resilience (Meagher et al., 2021). Much of this work is focused on policy-level innovations so the world's economies can re-establish themselves on more sustainable foundations, and on the associated organizational and technical innovations that could underpin these larger changes in societal and economic life. One part of this should examine the conditions that will lead to mass adoption of various types of digital assets, and the risks and opportunities such a rise could pose for the consumers, investors, and businesses, including an emphasis on the ways that technological innovations may affect those efforts, and focusing on those who are more vulnerable to the dispersed impacts (Hendershott et al., 2021; Lu & Strielkowski, 2022). We can see how the individuals and groups are adapting to a global crisis, outlining the more likely opportunities and challenges that arise as people rapidly increase the use and applications of digital technologies to cope. Our study suggests some useful recommendations for leaders about the business landscape after the pandemic, offering fresh perspectives on digital transformation, the future of work, transparency, and resilience.

As the dust settles after the pandemic, technology will continue to fuel innovation in various business models in different industries, enabling new companies to come into the market and seriously disrupt the incumbents. As transformations enabled by technology become critical, a new way of thinking about people will be a driving force for seeing implementation succeed to drive businesses forward (Zahoor et al., 2022). The new normal regarding the role digital technologies play in individuals personal and professional lives will be about ushering in, and learning how to navigate, an emerging metaverse. Human-centric technology adoption will be essential in order to make sure that communication and streams of knowledge are maintained. It is clear that it is not technology that is lacking to advance in many areas of development and humanitarian work, it is an enabled institutional context, the associated levels of political energy, and a willingness to promote and facilitate adoption (Brunetti et al., 2022). While quick and furious is the name of the game in terms of digital innovation, quick and frantic can result in mistakes.

16.3 Pandemic and Information Technology Implementations

As the blockchain technologies are already being applied to address tensions and trust issues in the context of maintaining confidentiality while also meeting the needs of the population, enabling tracking infected patients in combating the pandemic control and tracking the movement of patients can also bring important gains (Guo et al., 2021). The current pandemic expansion has highlighted the need for public leadership to expedite evaluation and implementation of digital technologies such as the blockchain. The unprecedented humanitarian and economic needs presented by COVID-19 are driving the development and deployment of new digital technologies at a large and rapid pace. Step-care, the practice of providing the most effective, lowest-resource-burdening treatments for patients who require them, then moving to higher-resource-burdening treatments depending on patient needs, is one helpful approach. Step-care would require systems to be both well-designed and well-equipped to provide such care for patients, from screening through to overflowing with the mental health needs inevitably emerging from this pandemic.

Furthermore, and generalized, effects from technology usage in the COVID-19 pandemic are transforming formerly offline tasks to activities that are now done online. In addition to these effects, technology can be used to reveal common attitudes and perceptions in the general public, which led to the effects of digital technologies used during the pandemic (Vargo et al., 2021). Additionally, the coronavirus pandemic forced education worldwide to work remotely, and most literature about digital technologies used in education estimates the effects of this drastic change. Now, a fundamental question to be answered is how much literature has been amassed, and how much is known, about digital technologies used during the COVID-19. Evidence about the efficacy of any new technology is essential to widespread adoption, but because the ongoing pandemic, many digital technologies have not been peer-reviewed, integrated into the public health system, undergone rigorous testing, or evaluated under evidence-based digital health frameworks, such as the evidence standard framework for digital health technologies. Digital technologies cannot work in isolation and must be integrated into existing public-health systems (Parra et al., 2021).

The technology advances at issue hold great promise for societal and economic benefits, increased efficiencies, and increased productivity in a range of sectors. Yet risks inherent in how these technologies are researched, designed, and developed, and in how they may be used, could equally easily undercut those benefits. There is growing concern that the technology advances at issue, and how they are used, will create major challenges, including workforce disruptions and other market disruptions, worsening inequalities, and new risks to public safety and national security (Haque et al., 2022). A widespread, almost ubiquitous sentiment among innovators and tech developers is that human relationships with technology will become deeper, with larger segments of the population coming to rely more heavily on digital connections for jobs, education, healthcare, everyday commerce, and basic social interactions. They articulated a number of broader themes on how individuals and groups are adapting to a global crisis, outlining the more likely opportunities and challenges to arise as humans speed up their use and applications of digital technologies in response. Gewin (2021) was able to offer an interpersonal perspective on how education technologies have engaged, identified, and supported struggling students during COVID-19, while Sun et al. (2020) shed light on more structural impacts arising from the digital technologies used during the COVID-19 pandemic.

16.4 COVID-19 and Remote Working

The relevance of telework in the context of the COVID-19 pandemic can influence the negative (work stress) and positive (work engagement and job performance) outcomes of telecommuting employees. Because trends toward remote work are expected to continue even following the COVID-19 pandemic, this study can provide useful insights into the individual- and work-related consequences of telecommuting during and following the pandemic (Parent-Lamarche, 2022). We recognize that the COVID-19 pandemic has upended the work situations for many students. Therefore, we have collected these resources to assist students in finding part-time, remote, and home-based opportunities. The continuing health and economic crises associated with the COVID-19 pandemic, as well as necessary physical distancing measures, have forced many businesses to implement remote working (work-from-home) at a larger scale. Due to the COVID-19 pandemic, many workers and employers had to make a fairly sudden transition, to telework, for the first time, and with no training (Kamal, 2020).

The outbreak of COVID-19 led many employers to switch to the remote working model for as many employees as possible, to try and limit the spread of coronavirus. The pandemic has increased the speed with which employers are coming up with ways to make sure employees have all of the necessary gear they need to be productive, even while working remotely. Employers are enjoying a boost in productivity when employees are working from home, but some corporate leaders are concerned about cultivating a highly innovative workplace in which employees are safe working on-site and remote (Ford et al., 2021). There are compelling reasons many companies and employees are excited by this blend of on-site and remote work – and equally compelling reasons many are wary about this transition. It presents an opportunity for companies to reimagine the future of work as a mix of in-office and remote arrangements. While 100% remote work is not going to be around forever, most employers are looking at having a hybrid model, including some in-office work on occasion and a work-from-anywhere (WFA) model (Choudhury et al., 2020). One way to mitigate a culture gap with remote employees is by rotating which employees are in the office and which are working from home. Many other companies have dabbled with a remote workforce, creating a policy of working from home for 1 to 2 days per week, or as an exception for certain employees. Even when it is safe to do so, companies may decide to keep a portion of the workforce remote. Of course, many jobs cannot be done remotely, requiring workers to physically be present at the workplace. If the remote worker is part of a team or a group, managers should make sure the work being done at a physical office is clearly defined and accessible for him or her. Remote workers who are clear about how their work contributes to the strategy and mission of the business choose working from home nearly twice as much as those who are not (Klerk et al., 2021). In other words, people new to working from home are about twice as likely as long-time telecommuters to say that they are now less productive. Framed differently, recently remote workers are almost twice as likely as their experienced counterparts to say their sense of belonging has suffered since starting to work from home. While these statistics might motivate both employees and employers to adopt work-from-home programs, remote workers also report higher levels of stress and greater difficulties in finding work-life balance compared with employees who are in an office environment. Teleworking has allowed some workers to keep working, but social distance persists. Many companies have been resistant to teleworking for various reasons. Now, with a pandemic emergency, this is the first time they are working remotely, and it has been successful in a significant way. The results are causing employers and employees alike to look at alternative working arrangements, particularly in a high-speed metropolis. Employers will be developing new work models through 2021, choosing from remote work policies in conjunction with on-site policies. As hybrid models arise, employers will determine how workers work together in order to stay safe but still engage. This means developing new hybrid working models that allow employees to move between on-site and remote work with ease, as well as thinking through appropriate physical spaces – both in terms of size and form for a hybrid office (Zajac et al., 2022). For every new challenge posed by the world of remote and hybrid work, employers will have to identify specific tools and technologies that will work within their company cultures and work patterns. To attract talent, companies will not just have to be accommodating to the employee's choice; they will have to improve that choice – whether it is working remotely or at the office – for both the experience and the performance. To meet employee's desires for flexibility, companies will need to close gaps in culture, work processes, socialization, and individual needs in order to improve both the remote work and office arrangements (Barath & Schmidt, 2022). All companies will also need to ensure the transition between on-site and remote work is as seamless as possible, providing employees with a seamless experience that feels designed rather than accidental and allows them to work as efficiently whether they are on-site or working remotely. Perhaps the most important thing that stems from our research is that leaders investing in getting remote work right can help employees cope with big changes in uncertain times, regardless of whether workers are at an office or at home. Many had to navigate technical difficulties the first time they switched to full remote work, but even switching to hybrid working can be hard. This effort needs to look at employee's career development journeys (such as being physically in an office early on, then remote work later on) as well as the various stages of projects (such as being physically onsite to do early planning, then working remotely for implementation).

16.5 Leading Role of Digital Technologies for Fighting COVID-19

The current spread of the pandemic has highlighted the need of leadership from government agencies to expedite evaluation and implementation of digital technologies. Many people stated the pandemic has precipitated changes to digital technologies already under way. Several recent studies discussed how expanding digital services could help mitigate the effects of the COVID-19 pandemic on economies now, and promote growth afterward (Ranjbari et al., 2021). Digital development could help the world to combat the virus. Digital platforms helped to create more resilient societies against COVID-19 disease, because now people can access official information, sign up for online courses, do online jobs, send mobile money, and even get telemedicine – regardless of where they live. Digital technologies now allow for online symptom checks, video consultations, and electronic prescribing and enable triage forward, rather than overwhelming hospitals and increasing the spread of COVID-19. In medicine and healthcare, the Internet of Things (IoT) together with few cutting-edge technologies form a more concrete technology system (called Internet of Medical Things (IoMT)) that has already made valuable contributions in managing coronavirus harm, to minimizing burden on overwhelmed hospital staff, thereby mitigating risk to transmitting the virus from the infected person to healthcare workers (Singh et al., 2020). In the past emergencies like the outbreak of Ebola virus, hurricanes and other natural disasters caused by the climate change, and now in COVID-19 pandemic, telemedicine has been explored for supporting the communication between patients and doctors, thus improving the healthcare outcomes (Kludacz-Alessandri et al., 2021). Wireless technologies are among the commonly implemented approaches during the modern-day COVID-19 pandemic in order to contain virus transmission and enhance public safety. Satellite communications are also playing an important role in monitoring, modeling, and responding to the COVID-19 pandemic. In addition, digital data and artificial intelligence (AI) may assist with diagnosis and surveillance of COVID-19. With them, the public health community can track the spread of the coronavirus and rapidly develop an effective response strategy (Malik et al., 2021).

Thence, the coronavirus pandemic has created a pressing need for coordinating mechanisms for responding to outbreaks in different sectors, with digital health solutions identified as a promising approach for solving the problem. Tech companies, governments, and international agencies have announced measures to help curb the spread of COVID-19. European Commission highlighted significant actions for combating the spread of coronavirus misinformation and established a program to track actions platforms are taking to limit the spread of COVID-19 misinformation (Di Mascio et al., 2021). Since the beginning of the COVID-19 pandemic, misinformation, fake news, and conspiracy theories regarding the origin, spread, and treatment of COVID-19 have been presented online and in public communications, social media, and text messages. One of the COVID-19 aftershocks would be a recognition that children and students need Internet access to succeed in school and that many families lack that. The world will rely more heavily on digital technologies to help us survive the pandemic. Digital data to model the spread, evolution, and perception of COVID-19 monitoring activity on the Internet or social media for public health purposes has been studied since the earliest days of the field of digital epidemiology, with the goal of collecting health-related trends and modeling disease outbreaks (Kostkova et al., 2021). The potential for digital technologies to support epidemiological intelligence through online data sets, case identification, infection clusters, rapid contact tracking, tracking travel patterns in times of lockdown, and public health message delivery at scale should not be underestimated. The addressed issues highlight the ways emerging state-of-the-art technologies could assist healthcare (involving contact tracing, live monitoring and screening for quarantine, disinfection, quarantine implementation, syndromic surveillance, and mental health), communications (involving remote care, information exchange, and communications support), logistics, tourism, and hospitality (Van Veldhoven & Vanthienen, 2022). There is the possibility to implement wireless technologies and telehealth services at hospitals in order to ensure uninterrupted healthcare facilities to the general public, including during pandemics, and minimize risk for infection by viruses. Another benefit of using robot services as a part of wireless technologies is that in times of pandemic, healthcare workers typically require specialized equipment, to minimize the risk of being infected with viruses and affected. The usage of 5G technologies could make it easier for the medical robots to provide drogues, monitor the temperature, pulse, respiratory rate, and respiratory rates of an infected patient, sanitize the hospital, and perform appropriate actions according to doctors' recommendations, which minimizes the risk of medical personnel being exposed to coronaviruses (Bokolo, 2021). Some findings conclude wireless technologies can minimize hospital visits, protect health resources, and reduce the virus transmission through remote treatment of patients during and after the COVID-19 pandemic (Sharifi et al., 2021). Therefore, it can be concluded that technologies such as contact tracing to fight COVID-19 outbreaks using mobile technologies like GPS, cell phone towers, and artificial intelligence-powered Big Data analytics can also help governments understand and control COVID-19 transmission in their communities.

16.6 Conclusions

To summarize and conclude, we can state that if there were some doubts and hesitations over the need for digital transformation for the long-term viability of businesses, then the COVID-19 pandemic has put those to rest. The present events have catalyzed a paradigm shift, evidenced by a marked shift in spending toward digital businesses. And it is only the beginning of the digital surge. The pandemic represents a reality check to businesses who were unwilling to adopt digital transformation, and now find themselves woefully underprepared. Businesses who were unwilling to adopt digital transformation now find themselves woefully underprepared. The opportunities afforded by digital technologies and the COVID-19 pandemic for organizations to reinvent their business models are being constrained by impediments like organizational inflexibility, digital disruption, and the unequal impact on worker well-being.

In catalyzing digital transformation, the COVID-19 outbreak has created a framework that will continue to drive innovation and technological adoption going forward. The long-term effects of the pandemic will strengthen businesses' ability to connect via digital means while continuing to run their businesses effectively. The outbreak of COVID-19 has forced companies to rapidly shift toward digital deployment, drastically speeding up trends that were developing at a much slower rate prior to the crisis. This may be explained in part by the capacity of certain companies to transition to digital technologies more rapidly, that is, those who had begun this transition prior to the pandemic; many therefore adapted to the pandemic by speeding up that transition, creating technology solutions as they reduced in-person customer interactions. One indication that technology is becoming increasingly important in both strategy and operations is the fact that companies have dedicated greater resources to their digital and technological capabilities during the COVID-19 crisis, even while cutting resources elsewhere in their businesses. Having seen the way that the pandemic has precipitated the acceleration of the deployment of digital technologies over a number of years, we have taken a closer look at how companies are reconsidering digital technologies' role in their overall business strategies and how they can run businesses at the faster speeds they must operate now. One year after the COVID-19 crisis, it becomes apparent that pandemic has increased business speeds, and technology capabilities will be crucial for companies COVID-19 exit strategies but also for what comes after. While workforce security and resiliency, cost management, and organizational flexibility emerge as the highest priorities in the near- to long-term, post-COVID-19 has amplified older business fears and introduced new ones. The post-pandemic world can be seen in novel industries ranging from healthcare to education, from finance to manufacturing, the pandemic has forced companies to rethink almost every aspect of their operations using technology. Executives are becoming increasingly confident about what technology can do and are pushing forward with digital transformation. Most companies are switching to new ways of doing business, and going digital, within a very short time, all the while taking care of the health and safety of their employees. In order to avoid a disastrous revenue loss, some organizations had little choice but to either switch focus to the existing digital channels in their company or to pivot completely toward the digital business model. Talent is an ongoing challenge for companies transforming their businesses with digital and technology – as several of our respondents said that their companies are trying to do with a lot of their respondents. Across economies, participation is unequal to new opportunities created by digital transformation. It is for this reason that it has become increasingly evident for governments to focus on developing and scaling up technological infrastructure in order to increase access and also connect rural communities with the new digital economy.

From a practical perspective, our analysis reinforces the case that the COVID-19 pandemic presents opportunities for firms to adopt at least some aspects of digitalization. Our analysis illuminates digitalized practices of the business models that have been brought about by the COVID-19 pandemic, including a move toward remote working and remote operations, as well as paperless offices and organizations. Our earlier investigation found that in key areas of business models, companies have seen their overall digital technology adoption accelerate 3 to 7 years within months. Technological drivers are also speeding up and disrupting the business landscape. As technology has proven its value for senior communities in times of pandemic, digital infrastructure is likely to be a constant in the provision of elder care, whether at home or at assisted living facilities. It is a good time for companies to invest in the kinds of bold investments in technologies and capabilities that will enable their businesses to overcome others. There should be new initiatives drawing on the new expertise and best practices, along with those of other digitally advanced countries, to create data ecosystems, drive eco-friendly digital innovations, and capitalize on the significant economic and societal opportunities that come with digital transformations driven by data. All of this would help to foster digitalization and solidify the position of leaders in business and education who would be the champions of the post-COVID era.

References

Barath, M., & Schmidt, D. A. (2022). Offices after the COVID-19 pandemic and changes in perception of flexible office space. *Sustainability, 14*(18), 11158. https://doi.org/10.3390/su141811158

Berry, C., Bailey, D., Beel, D., & O'Donovan, N. (2022). Building back before: Fiscal and monetary support for the economy in Britain amid the COVID-19 crisis. *Cambridge Journal of Regions, Economy and Society*, rsac024. https://doi.org/10.1093/cjres/rsac024

Bojkovic, Z., Milovanovic, D., Fowdur, T. P., & Hosany, M. A. (2021). What 5G has been and what should 5G+ be? *Athens Journal of Technology & Engineering, 8*(1), 27–38. https://doi.org/10.30958/ajte.8-1-2

Bokolo, A. J. (2021). Application of telemedicine and eHealth technology for clinical services in response to COVID-19 pandemic. *Health and Technology, 11*(2), 359–366. https://doi.org/10.1007/s12553-020-00516-4

Brunetti, D., Gena, C., & Vernero, F. (2022). Smart interactive technologies in the human-centric factory 5.0: A survey. *Applied Sciences, 12*(16), 7965. https://doi.org/10.3390/app12167965

Choudhury, P., Crowston, K., Dahlander, L., Minervini, M. S., & Raghuram, S. (2020). GitLab: Work where you want, when you want. *Journal of Organization Design, 9*(1), 1–17. https://doi.org/10.1186/s41469-020-00087-8

De Angelis, P., Tuninetti, M., Bergamasco, L., Calianno, L., Asinari, P., Laio, F., & Fasano, M. (2021). Data-driven appraisal of renewable energy potentials for sustainable freshwater production in Africa. *Renewable and Sustainable Energy Reviews, 149*, 111414. https://doi.org/10.1016/j.rser.2021.111414

Di Mascio, F., Barbieri, M., Natalini, A., & Selva, D. (2021). COVID-19 and the information crisis of liberal democracies: Insights from anti-disinformation action in Italy and EU. *Partecipazione e conflitto, 14*(1), 221–240.

Farooq, M. S., Ahmed, M., & Emran, M. (2022). A survey on blockchain acquainted software requirements engineering: Model, opportunities, challenges, and future directions. *IEEE Access, 10*, 48193–48228. https://doi.org/10.1109/ACCESS.2022.3171408

Ford, D., Storey, M. A., Zimmermann, T., Bird, C., Jaffe, S., Maddila, C., & Nagappan, N. (2021). A tale of two cities: Software developers working from home during the covid-19 pandemic. *ACM Transactions on Software Engineering and Methodology (TOSEM), 31*(2), 1–37. https://doi.org/10.1145/3487567

Gewin, V. (2021). Pandemic burnout is rampant in academia. *Nature, 591*(7850), 489–492. https://doi.org/10.1038/d41586-021-00663-2

Guo, X., Khalid, M. A., Domingos, I., Michala, A. L., Adriko, M., Rowel, C., & Cooper, J. M. (2021). Smartphone-based DNA diagnostics for malaria detection using deep learning for local decision support and blockchain technology for security. *Nature Electronics, 4*(8), 615–624. https://doi.org/10.1038/s41928-021-00612-x

Haldar, A., & Sethi, N. (2022). Environmental effects of Information and Communication Technology-Exploring the roles of renewable energy, innovation, trade and financial development. *Renewable and Sustainable Energy Reviews, 153*, 111754. https://doi.org/10.1016/j.rser.2021.111754

Haque, A. B., Bhushan, B., & Dhiman, G. (2022). Conceptualizing smart city applications: Requirements, architecture, security issues, and emerging trends. *Expert Systems, 39*(5), e12753. https://doi.org/10.1111/exsy.12753

Hendershott, T., Zhang, X., Zhao, J. L., & Zheng, Z. (2021). FinTech as a game changer: Overview of research frontiers. *Information Systems Research, 32*(1), 1–17. https://doi.org/10.1287/isre.2021.0997

Kamal, M. M. (2020). The triple-edged sword of COVID-19: Understanding the use of digital technologies and the impact of productive, disruptive, and destructive nature of the pandemic. *Information Systems Management, 37*(4), 310–317. https://doi.org/10.1080/10580530.2020.1820634

Kesavan, N., Seenuvasaragavan, S., Burnette, K., Heim, A. B., Patel, R. J., Johnston, J., Cao, X., Sakandar, H. A., Zhuang, Y., Oi, K. K., Zhi, Y., Diao, Y., & Strielkowski, W. (2022). Pandemic-inspired policies. *Science, 377*(6601), 22–24. https://doi.org/10.1126/science.add2046

Klerk, J. J. D., Joubert, M., & Mosca, H. F. (2021). Is working from home the new workplace panacea? Lessons from the COVID-19 pandemic for the future world of work. *SA Journal of Industrial Psychology, 47*(1), 1–14. https://doi.org/10.4102/sajip.v47i0.1883

Klochan, V., Piliaiev, I., Sydorenko, T., Khomutenko, V., Solomko, A., & Tkachuk, A. (2021). Digital Platforms as a tool for the transformation of strategic Consulting in Public Administration. *Journal of Information Technology Management, 13*(Special Issue: Role of ICT in Advancing Business and Management), 42–61. https://doi.org/10.1002/smj.3178

Kludacz-Alessandri, M., Walczak, R., Hawrysz, L., & Korneta, P. (2021). The quality of medical care in the conditions of the COVID-19 pandemic, with particular emphasis on the access to primary healthcare and the effectiveness of treatment in Poland. *Journal of Clinical Medicine, 10*(16), 3502. https://doi.org/10.3390/jcm10163502

Kostkova, P., Saigí-Rubió, F., Eguia, H., Borbolla, D., Verschuuren, M., Hamilton, C., & Novillo-Ortiz, D. (2021). Data and digital solutions to support surveillance strategies in the context of the COVID-19 pandemic. *Frontiers in Digital Health, 3*, 89. https://doi.org/10.3389/fdgth.2021.707902

Kraus, S., Jones, P., Kailer, N., Weinmann, A., Chaparro-Banegas, N., & Roig-Tierno, N. (2021). Digital transformation: An overview of the current state of the art of research. *Sage Open, 11*(3), 21582440211047576. https://doi.org/10.1177/21582440211047576

Kuteyi, D., & Winkler, H. (2022). Logistics challenges in Sub-Saharan Africa and opportunities for digitalization. *Sustainability, 14*(4), 2399. https://doi.org/10.3390/su14042399

Lu, J., & Strielkowski, W. (2022). Russia: Could cryptocurrency mining exploit unsold fuel? *Nature, 605*(7909), 225–225. https://doi.org/10.1038/d41586-022-01267-0

Malik, Y. S., Sircar, S., Bhat, S., Ansari, M. I., Pande, T., Kumar, P., & Dhama, K. (2021). How artificial intelligence may help the Covid-19 pandemic: Pitfalls and lessons for the future. *Reviews in Medical Virology, 31*(5), 1–11. https://doi.org/10.1002/rmv.2205

Meagher, K., El Achi, N., Bowsher, G., Ekzayez, A., & Patel, P. (2021). Exploring the role of City Networks in supporting urban resilience to COVID-19 in conflict-affected settings. *Open Health, 2*(1), 1–20. https://doi.org/10.1515/openhe-2021-0001

Mpofu, F. Y., & Mhlanga, D. (2022). Digital financial inclusion, digital financial services tax and financial inclusion in the fourth industrial revolution era in Africa. *Economies, 10*(8), 184. https://doi.org/10.3390/economies10080184

Pandey, N., & Pal, A. (2020). Impact of digital surge during COVID-19 pandemic: A viewpoint on research and practice. *International journal of information management, 55*, 102171. https://doi.org/10.1016/j.ijinfomgt.2020.102171

Parent-Lamarche, A. (2022). Teleworking, work engagement, and intention to quit during the COVID-19 pandemic: Same storm, different boats? *International Journal of Environmental Research and Public Health, 19*(3), 1267. https://doi.org/10.3390/ijerph19031267

Parra, C. M., Gupta, M., & Mikalef, P. (2021). Information and communication technologies (ICT)-enabled severe moral communities and how the (Covid19) pandemic might bring new ones. *International Journal of Information Management, 57*, 102271. https://doi.org/10.1016/j.ijinfomgt.2020.102271

Ranjbari, M., Esfandabadi, Z. S., Zanetti, M. C., Scagnelli, S. D., Siebers, P. O., Aghbashlo, M., & Tabatabaei, M. (2021). Three pillars of sustainability in the wake of COVID-19: A systematic review and future research agenda for sustainable development. *Journal of Cleaner Production, 297*, 126660. https://doi.org/10.1016/j.jclepro.2021.126660

Sharifi, A., Khavarian-Garmsir, A. R., & Kummitha, R. K. R. (2021). Contributions of smart city solutions and technologies to resilience against the COVID-19 pandemic: A literature review. *Sustainability, 13*(14), 8018. https://doi.org/10.3390/su13148018

Singh, R. P., Javaid, M., Haleem, A., Vaishya, R., & Ali, S. (2020). Internet of Medical Things (IoMT) for orthopaedic in COVID-19 pandemic: Roles, challenges, and applications. *Journal of Clinical Orthopaedics and Trauma, 11*(4), 713–717. https://doi.org/10.1016/j.jcot.2020.05.011

Srivastava, S., Heim, A. B., Yoho, R., Moore, E. B., Richter, W. E., Strielkowski, W., Li, K., Penders, B., Sojli, E., & Cao, B. (2021). Imagining a teaching utopia. *Science, 374*(6563), 31–32. https://doi.org/10.1126/science.abm1361

Strielkowski, W., Dvořák, M., Rovný, P., Tarkhanova, E., & Baburina, N. (2021). 5G wireless networks in the future renewable energy systems. *Frontiers in Energy Research, 9*, 714803. https://doi.org/10.3389/fenrg.2021.714803

Stroiko, T., Nazarova, L., & Danik, N. (2021). Transformation of economic processes on the basis of digitalisation. *Baltic Journal of Economic Studies, 7*(1), 2256–0742. https://doi.org/10.30525/2256-0742/2021-7-1-102-106

Sun, L., Tang, Y., & Zuo, W. (2020). Coronavirus pushes education online. *Nature Materials, 19*(6), 687–687. https://doi.org/10.1038/s41563-020-0678-8

Van Veldhoven, Z., & Vanthienen, J. (2022). Digital transformation as an interaction-driven perspective between business, society, and technology. *Electronic Markets, 32*(2), 629–644. https://doi.org/10.1007/s12525-021-00464-5

Vargo, D., Zhu, L., Benwell, B., & Yan, Z. (2021). Digital technology use during COVID-19 pandemic: A rapid review. *Human Behavior and Emerging Technologies, 3*(1), 13–24. https://doi.org/10.1002/hbe2.242

Williams, L. D. (2021). Concepts of Digital Economy and Industry 4.0 in Intelligent and information systems. *International Journal of Intelligent Networks, 2*, 122–129. https://doi.org/10.1016/j.ijin.2021.09.002

Yörük, E., & Gençer, A. Ş. (2022). The dynamics of welfare state regime development in the global south: Structures, institutions, and political agency. *Journal of Comparative Policy Analysis: Research and Practice, 24*(5), 452–472. https://doi.org/10.1080/13876988.2021.2004541

Zahoor, N., Golgeci, I., Haapanen, L., Ali, I., & Arslan, A. (2022). The role of dynamic capabilities and strategic agility of B2B high-tech small and medium-sized enterprises during COVID-19 pandemic: Exploratory case studies from Finland. *Industrial Marketing Management, 105*, 502–514. https://doi.org/10.1016/j.indmarman.2022.07.006

Zajac, S., Randall, J., & Holladay, C. (2022). Promoting virtual, informal learning now to thrive in a post-pandemic world. *Business and Society Review, 127*, 283–298. https://doi.org/10.1111/basr.12260

Part III
Leadership in Academia and Higher Education After COVID-19 Pandemic

Chapter 17
Digitalization and Innovation Transfer as a Leadership Trend in Education: Bibliometric Analysis and Social Analytics

Vitaliia Koibichuk, Anastasiia Samoilikova, and Tetiana Vasylieva

Abstract Today, digitalization is both an important direction and a significant result of the transfer of innovation. It has covered all spheres of modern life, and education is no exception. Moreover, the COVID-19 pandemic has significantly affected the transformation of learning around the world. Thus, the main purpose of this chapter is to determine key trends in learning transformation in the context of digitalization and innovation transfer as a leadership trend in education based on bibliometric and cluster analysis and social analytics. To achieve this target, the bibliometric analysis of indexed scientific papers by the Scopus database, the comparative and quantitative analysis of statistical metrics were made using ScientoPy software and Python language. The number of publications devoted to research of development and features of the e-learning market has increased rapidly over the past 2 years (2020–2021), almost three times compared to 2016–2020. The growing use of artificial intelligence and machine learning in the e-learning system stimulates the mastery of digitalization tools in education. This chapter presents the dynamics of the market of electronic educational services in accordance with statistical data and the results of the forecast of the leading research and consulting company, the advantages of e-learning, and the most used of them, and disadvantages and problems of distance learning. Comparative characteristics of indicators of infrastructure, access, opportunities, and barriers are studied for Poland, Ukraine, Germany, and Cyprus. Also, there is the cluster analysis of the median values of countries' web behavior in relation to online learning. It was performed using the Ward's method and Agglomeration scheme in Statgraphics 19 software based on

V. Koibichuk · A. Samoilikova (✉)
Sumy State University, Sumy, Ukraine
e-mail: v.koibichuk@biem.sumdu.edu.ua; a.samoilikova@biem.sumdu.edu.ua

T. Vasylieva
Sumy State University, Sumy, Ukraine

Silesian University of Technology, Gliwice, Poland

The London Academy of Science and Business, London, UK
e-mail: t.vasylieva@biem.sumdu.edu.ua

W. Strielkowski (ed.), *Leadership, Entrepreneurship and Sustainable Development Post COVID-19*, Springer Proceedings in Business and Economics, https://doi.org/10.1007/978-3-031-28131-0_17

IEEE data set from 20 countries and Google Trends analytics for the 18 years period. Six clusters of countries that have similar web behavior related to online learning were determined. The received results can be used by state and local authorities and scientific and educational institutions for the further development and improvement of e-learning technologies to strengthen leadership in education.

Keywords Artificial intelligence · Distance learning · e-learning · Leadership in education · Web behavior

JEL Classification I21 · O31 · D83

17.1 Introduction

The rapid development of digitalization as a result of innovation transfer and at the same time its dominant direction has covered all spheres of casual life. It, of course, requires knowledge of the tools of modern information technology, the ability to use them to obtain effective solutions to problems (Russ, 2021). The global surge in COVID-19 has led to the closure of some schools, colleges, and universities in 2020 in almost all parts of the world and the transition to online or distance learning, which has affected student learning in different ways (Guliyeva et al., 2021; Kesavan et al., 2022). In such conditions rapid development and efficient use of digital technologies of e-learning have become one of the most significant leadership trends in education. Also, this issue is significantly actual during other pandemics, military actions, and other global challenges in the context of sustainable development and inclusiveness.

The main purpose of this chapter is to determine key trends in learning transformation in the context of digitalization and innovation transfer as a leadership trend in education based on bibliometric and cluster analysis and social analytics.

The research methods and methodology include bibliometric, quantitative, and comparative analysis using the Scopus database tools, ScientoPy software, and Python language. The cluster analysis of the median values of web behavior in relation to online learning was performed using the Ward's method and Agglomeration scheme in Statgraphics 19 software on the basis of IEEE data set from 20 countries and Google Trends analytics for the 18 years period.

The information base was formed by the data of the World Bank, the Institute of Electrical and Electronics Engineers, the Statista data, etc.

The research novelty is the development of methodological approach to determine key trends of learning transformation and cluster analysis of countries' web behavior related to education.

The practical significance is that the research results can be used by state and local authorities and scientific and educational institutions for the further development and improvement of e-learning technologies to strengthen leadership in education.

17.2 Literature Review

The issue of digital transformation and innovations in education, e-learning services, tools and platforms, its strengths and weaknesses, and other actual aspects in this research area were investigated by a wide road of scholars (Ivanov et al., 2018; Artyukhov & Omelyanenko, 2021; Gruzina et al., 2021; Onopriienko et al., 2021; Artyukhov et al., 2021; Shvindina et al., 2020; Iatsyshyn et al., 2020; Strielkowski et al., 2022). Popova (2021) proved the necessity of education to develop closely with advancing digitization. Kuzior and Kwilinski (2022) paid attention on artificial intelligence and cognitive technologies. The role of open science was analyzed by Bukhtiarova et al. (2021), the influence of digital innovations by Lagodiienko and Yakushko (2021); the impact of digitalization on innovation development and entrepreneurship by Simionescu et al. (2020), Sadigov (2022), and Melnyk et al. (2022); or the influence of digital transformation generally on macro stability by Tiutiunyk et al. (2021); and the innovations efficiency and its factors by Wang et al. (2022).

Contemporary approach to innovation was proposed by Kuzior and Zozul'ak (2019). Kuzior et al. (2022) positioned the digitalization as a factor of sustainability and ethic. The issue of informational impact on education and people protection was raised up by Dutta et al. (2020), the problem of information security in the context of innovations by Petroye et al. (2020), the issue of "information bubbles" by Kuzmenko et al. (2021).

Buchynska et al. (2020) analyzed the problem of dual education in the context of innovations transfer in education. The coopetition of education, science, and business to reach innovation transfer and development was investigated by Samoilikova et al. (2022). Bilan et al. (2020) analyzed business models toward innovation development. Kyrychenko et al. (2021) characterized universities as a great driver of society development and educational and scientific progress. Vardanyan (2017) considered institutional capacity and other domains of education quality. Liuta et al. (2021) put attention on providing quality in education and its tools.

The important factor of management and leadership in the sphere of educational technologies was described by Yarovenko et al. (2021), Vorontsova et al. (2020a, b), Khanin et al. (2019), etc. Yu Sing Ong (2018) emphasized leadership in education based on a humanistic approach and principles of Confucianism and Islamism. Polyakov et al. (2020) identified leadership factors of universities. Beyi (2018) explained the influence of digital communication and technical interaction on social behavior. Other social behavioral aspects were shown by Didenko et al. (2020).

Lyeonov and Liuta (2016) and Jikia et al. (2017) studied the financial component in education development, Samoilikova (2020) financing innovation development on the country level in general, Frederick and Kasztelnik (2020) financial policy for higher education, and Kotenko and Bohnhardt (2021) financing of digital projects and its strengths and weaknesses. Brimah et al. (2020) determined interconnection between knowledge management and financial state. The knowledge marketing was investigated by Polyakov et al. (2019).

The number of publications devoted to research of development and features of the market of electronic educational services has increased rapidly over the past 2 years (2020–2021), almost three times compared to 2016–2020. Such conclusions can be drawn from bibliometric analysis of indexed scientific papers by the Scopus database. The bibliometric analysis was made in ScientoPy software based on 150 keywords in 1022 scientific papers in the Scopus database for the period from 2016 to 2021 (Fig. 17.1).

The Scopus database search engine found 1022 documents for the last 5 years, according to the "e-learning market" query. Quantitative comparison of the found publications on the ten most used keywords until 2020 and for 2020–2021 was carried out using the modern software ScientoPy (Fig. 17.2).

Thus, in the period from 2020 to 2021, the use of the keyword "artificial intelligence" is 88% of the total number of publications in the period from 2016 to 2021 (Fig. 17.2). The use of the keyword "machine learning" is 63%, "virtual reality" 62%, the keywords "online education" and "deep learning" 59%, "higher education" 52%, "online learning" 51%, "training" 50%, "education" 45%, and "e-learning" 38%.

The average growth rate (AGR) of keywords compared to the other ten top keywords in the period from 2020 to 2021, the average number of publications (ADY), and the value of the Hirsch index are shown in Fig. 17.3.

Thus, the statistics of Fig. 17.3 for the period from 2020 to 2021 show a list of the ten most used keywords, their total number in the sample of bibliometric research for the "e-learning market" query, the value of the Hirsch index, the average keyword growth among other most used keywords in the list of found publications, and the average number of publications with the corresponding keyword for the last 2 years.

Fig. 17.1 Bibliometric analysis of publications on "e-learning market". (Source: Compiled by the authors based on the results of the Scopus database analysis and ScientoPy software tools)

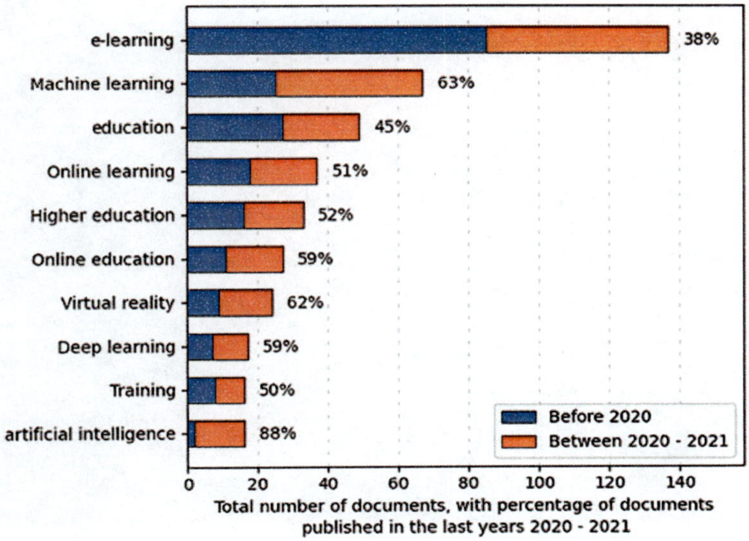

Fig. 17.2 Quantitative comparison of the found publications on the ten most used keywords until 2020 and for 2020–2021. (Source: Compiled by the authors based on the results of the Scopus database analysis and ScientoPy software tools)

For example, the highest value of the Hirsch index is 10 for the keyword "e-learning," the average growth rate (AGR) relative to other keywords decreased by 4 units, and the average annual number of publications for 2020–2021 is 26 units (ADY), which corresponds to 38% compared of the number of publications in the period from 2016 to 2019 (Fig. 17.3).

17.3 Results

The dynamics of the market of electronic educational services in accordance with statistical data (Statista, 2022) and the results of the forecast of the leading research and consulting company "Allied Market Research" (E-Learning Market, 2022) are shown in Fig. 17.4.

The inner circle (Fig. 17.4) reflects the quantitative comparison of the types of e-learning services in 2019, and the outer the projected amount of profit from e-learning services in 2026 (E-Learning Market, 2022).

The advantages of e-learning with the use of video and audio materials of e-books on various online platforms include increased productivity of employees of enterprises and organizations due to the convenience of choosing the time to study the course at a convenient time. Conditions of distance learning promote self-education. During training any skills are honed; there is a comprehension of best practices.

```
Calculating papers statistics...
Calculating accumulative ...
Calculating Average Growth Rate (AGR)...
Calculating Average Documents per Year (ADY)...
Calculating h-index...

Top topics:
Average Growth Rate (AGR) and Average Documents per Year (ADY) period: 2020 - 2021

--------------------------------------------------------------------------------------------
Pos authorKeywords                          Total      AGR       ADY      PDLY      h-index
--------------------------------------------------------------------------------------------
1   e-learning                                137     -4.0      26.0      38.0           10
2   Machine learning                           67      8.5      21.0      62.7            8
3   education                                  49      0.0      11.0      44.9            5
4   Online learning                            37      2.0       9.5      51.4            5
5   Higher education                           33      4.0       8.5      51.5            6
6   Online education                           27      2.5       8.0      59.3            5
7   Virtual reality                            24      2.5       7.5      62.5            5
8   Deep learning                              17      0.5       5.0      58.8            5
9   Training                                   16      2.0       4.0      50.0            2
10  artificial intelligence                    16      6.0       7.0      87.5            1
--------------------------------------------------------------------------------------------

Saved top results on: results\AuthorKeywords.csv
Saved extended top results on: results\AuthorKeywords_extended.csv
Saving results on: results\papersPreprocessed.csv, with Scopus fields

Analysis finished.
```

Fig. 17.3 Comparative analysis of statistical metrics of keywords for the "e-learning market" query. (Source: Calculated by the authors based on the results of the Scopus database analysis, ScientoPy software tools, and Python language)

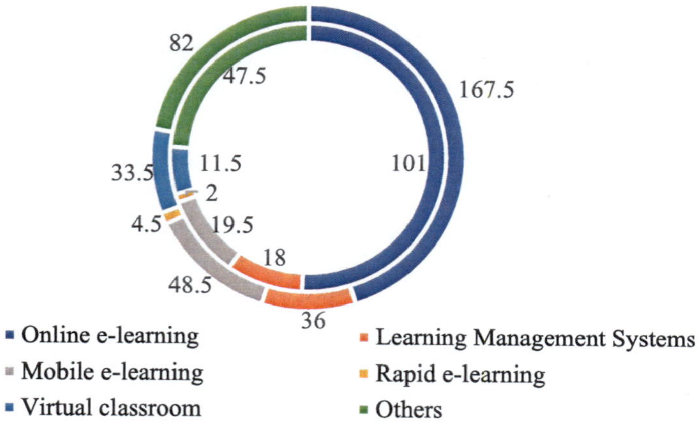

Fig. 17.4 Size of the global e-learning market, million US dollars. (Source: Compiled by the authors on the basis of E-Learning Market, 2022)

And for students of all levels (junior bachelor, bachelor, master, PhD student), e-learning is the only opportunity to get an education in a global pandemic.

The COVID-19 pandemic is boosting the e-learning market by increasing the spread of digital technology among various schools, colleges, and universities around the world and growing government support to improve the e-learning platform in different countries. According to the World Bank (The World Bank, 2022), in March 2020, 1.6 million students in 161 countries were deprived of the opportunity to attend school; the number of students who did not have the opportunity to attend classes in March 2020 was 80% and in June 61.8%.

Google software, Zoom telecommunications products and Skype, Telegram, WhatsApp, and Viber messengers are most actively used in quarantine training in the world. Thus, the share price of Zoom actively grew during the 2019–2020 school year; in April 2020, there were about 200 million active users; and between November 2020 and January 2021, the number of users increased 20 times (Zoom, 2022). This fact is especially important given that Zoom's stock prices have risen sharply over the past year (Fig. 17.5).

Also, particularly popular tools in the digital education market are tools from companies and corporations such as Adobe, Aptara Inc., Articulate Global, LLC, CERTPOINT, Cisco Systems Inc., Citrix Systems Inc., D2L Corporation, Microsoft Corporation, Oracle Corporation, and SAP SE.

To strengthen the development of digital literacy in Ukraine, a national online platform Diya.Tsyfrova Osvita (Action.Digital Education) (Diya, 2022) was created, which contains categories of mastering digital learning tools for parents, beginners, teachers, civil servants, youth, business, and startups, as well as digital opportunities for people with disabilities. In addition, the growing use of artificial intelligence and machine learning in the e-learning system stimulates the mastery of digitalization tools. So, in our opinion, the key advantages of distance learning, without being tied to a particular program or company, are the following: accessibility (having an Internet connection, the student has the opportunity to communicate

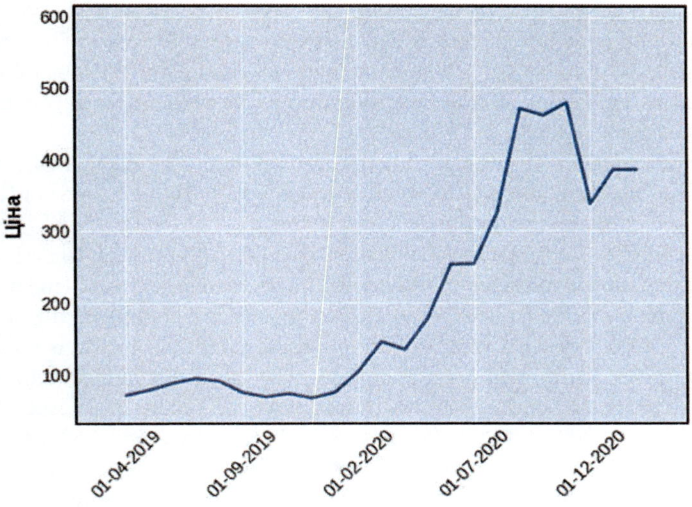

Fig. 17.5 Zoom's share price in 2019–2020. (Source: Compiled by the authors based on Zoom, 2022)

with the teacher); flexibility (the ability to master the material at a pace convenient for everyone); and saving money and time when moving to the physical location of the educational institution.

The current level of technological progress has greatly simplified human life, but the positive impact and the effectiveness of distance learning leaves much to be desired. Disadvantages of distance learning include the lack of face-to-face interaction in e-learning systems and the need for higher self-motivation. Also, the disadvantages of distance learning are that not everything can be digitalized (some things are simply impossible to transmit through the monitor screen); communication, or rather its distorted form (the problem of communication in the Internet epoch has been discussed for a long time, and the main thesis of the problem of communication is its lack); and lack of control, visible freedom of action gradually and imperceptibly grows into laziness and lack of motivation. It should be emphasized that distance learning creates inequality in exams, as pupils and students of large cities do not always have the opportunity to study offline, while in some suburban educational institutions, the educational process almost does not stop. A particularly acute problem is the interference of pranksters in the educational process, who demonstrate obscene material, candid videos. This, of course, has a negative effect on the psyche of children and adolescents, and it is virtually impossible for a teacher to find the "author" of criminal acts.

For example, learning restrictions in Romania because of negative effects of COVID-19 were personal support for students with special educational needs (63.80%), monitoring of the educational process (51.70%), feedback for immediate corrections (45.30%), explanations for understanding basic concepts (45.10%), individual counseling 36.30%), and others (3.2%) (Education Statistics, 2022).

Another major obstacle to distance learning is the level of technical equipment and the ability to organize a quality Internet connection. Thus, if we compare the digital development infrastructure and network access features of all countries (ITU, 2022), the dynamics for some of them differs significantly. Comparative characteristics of indicators of infrastructure, access, opportunities, and barriers for Cyprus, Germany, Poland, and Ukraine are given in Table 17.1.

Of course, the COVID-19 pandemic has caused many unprecedented challenges for students' learning and education around the world. Both teachers and students began to spend more time online than ever before.

Open-ended data collected by the Institute of Electrical and Electronics Engineers (IEEE, International Association of Engineering Specialists, world leader in areas of development of standards of radio electronics, electrical engineering, and hardware of computer systems and networks) were used as the research information base of behavior of different population categories to get acquainted with information, tools, programs, and structures for adaptation to online or distance learning (IEEE DataPort, 2022). The data set covers 20 countries and characterizes the behavior on the Internet in the form of search interests in online (distance) learning, recorded in the search engine Google Trends. The time period covers 18 years,

Table 17.1 Indicators of infrastructure, access, stimuli, and barriers to digital development

Indicator/country	Poland	Ukraine	Germany	Cyprus
Network coverage	100%	100%	100%	100%
Population covered by at least the 3G mobile network	100%	89%	98%	100%
Population covered by at least 4G mobile network	100%	87%	100%	100%
Access to ICT at home (households with a computer at home)	90%	66%	92%	93%
Active subscriptions to mobile broadband per 100 inhabitants	197	89	91	118
Fixed broadband subscriptions per 100 inhabitants	22	19	43	37
Fixed broadband access (% of total volume): 256 kbps–<2 Mbps	0%	1%	0%	0%
Fixed broadband access (% of total): 2–10 Mbps	9%	4%	5%	2%
Fixed broadband access (% of total):> 10 Mbps	79%	94%	93%	97%
Total number of fixed broadband subscriptions	8,212,601	7,769,401	36,040,739	332,080
Mobile data and voice shopping cart (high consumption, % of GNI per capita)	0,9%	1,8%	0,9%	1,4%
Mobile data and voice shopping cart (low consumption, % of GNI per capita)	0,8%	1,6%	0,9%	0,9%
Fixed broadband basket, % GNI per capita	1,3%	1,6%	1,0%	0,9%
Mobile fixed broadband shopping cart, % of GNI per capita	0,2%	1,5%	0,4%	0,9%
People with advanced skills	5%	1%	5%	4%

Source: Built by the authors based on statistics from the International Telecommunication Union's Digital Development Panel (ITU, 2022)

namely, from January 1, 2004, to October 1, 2021. Although the first case of COVID-19 in humans was recorded in 2019, the sample characterizes web behavior since 2004, on a scale from 0 to 100 points (0, the lowest value of search for online learning; 100, the highest value) (IEEE DataPort, 2022). Therefore, we have the opportunity to analyze in detail the degree of change in web behavior associated with online learning and to analyze trends in these changes.

The cluster analysis of the median values of web behavior in relation to online learning was performed using the procedure Describe/Multivariate methods/Cluster Analysis in Statgraphics 19 to form groups of countries that have similar web behavior related to online learning. To implement cluster analysis, the median values of online learning indicators in the studied countries were used, because they are insensitive to emissions. The Ward's method was used for the formation of clusters that reflect the search interests, because, unlike other methods of cluster analysis (single connection method, full connection method, centroid method, median method), it bases on methods of analysis of variance. The distance between clusters is the increment of the sum of the squares of the distances of the objects to the center of the cluster resulting from their merging:

$$V_I = \sum_i \sum_j \left(x_{ij} - \overline{x}_{jI}\right)^2, \tag{17.1}$$

where I is the cluster number, i is the object number ($i = 1, 2,\ldots, nI$), nI is the number of objects in the I cluster, j is the feature number ($j = 1, 2,\ldots, k$), and k is the number of features that characterize each object.

At each step of the algorithm, the following two clusters are grouped, combining which leads to a minimal increase in variance: the smaller the value of intragroup variance and the greater the value of intergroup variance – the better the characteristic characterizes the membership of a particular cluster and the better the results of clustering (Klebanova et al., 2018). That is, those objects (clusters) that have the smallest increment VI are grouped together.

The most stable is the division of the studied indicators of student mobility into six clusters. This is primarily due to the similarity of objects within the cluster. Second, the optimal number of clusters is confirmed by the results of the analysis of the union protocol, namely, the values of agglomeration coefficients. The calculation is based on formula below,

$$\text{K} \quad N - n_i \tag{17.2}$$

where K is the number of clusters, N is the number of steps, n_i is the number of the and i-th step on which the jump took place; there it was a sharp change in the values of the agglomeration coefficient (Table 17.2).

Table 17.2 shows the scheme of agglomeration of observations at each stage of the clustering procedure. For example, in the first step of the agglomeration scheme, observation 1 is combined with observation 7 (Table 17.2); the distance between the combined groups is zero. The next step in which the merger took place is step 7. Therefore, the optimal number of clusters is 6.

The results of cluster analysis are presented in Table 17.3.

The cluster analysis allowed to form six groups of countries that have similar web behavior related to online learning, in particular the following:

1. Argentina, Colombia, Indonesia, Iran, Peru, Russia, Turkey, and Ukraine
2. Brazil, France, Mexico, and the Philippines
3. Germany and the United Kingdom
4. India and Spain
5. Italy, Poland, and South Africa
6. The United States

A similar division into clusters was received on the basis of using the centroid method and the median method, which also confirms the correctness and accuracy of the results of the clustering.

Table 17.2 Agglomeration scheme (distance metric: Euclidean metric square)

Step	Combined cluster 1	Combined cluster 2	Distance	Previous step, cluster 1	Previous step, cluster 2	The next step
1	1	7	0,0	0	0	7
2	2	10	0,0	0	0	10
3	3	8	0,0	0	0	4
4	3	17	0,0	3	0	11
5	5	19	0,0	0	0	0
6	13	15	0,0	0	0	14
7	1	18	0,00077925	1	0	11
8	6	16	0,003117	0	0	0
9	11	14	0,00545475	0	0	13
10	2	12	0,00857175	2	0	12
11	1	3	0,0134421	7	4	13
12	2	4	0,0325337	10	0	0
13	1	11	0,0629731	11	9	0
14	9	13	0,101156	0	6	0

Source: Built by the authors using Statgraphics 19

Table 17.3 The results of cluster analysis by the Ward's method

Cluster 1	Cluster 2	Cluster 3	Cluster 4	Cluster 5	Cluster 6
Argentina	Brazil	Germany	India	Italy	United States
Colombia	France	United Kingdom	Spain	Poland	
Indonesia	Mexico			South Africa	
Iran	Philippines				
Peru					
Russia					
Turkey					
Ukraine					

Source: Built by the authors using Statgraphics 19

17.4 Conclusions

It was justified that the development and use of digital technologies of e-learning have become one of the most significant leadership trends in education, especially during the COVID-19 pandemic, military actions, and other global challenges in the context of sustainable development and inclusiveness.

The results of bibliometric analysis of indexed scientific papers by the Scopus database show that the number of publications devoted to the research of development and features of the e-learning market has increased rapidly over the past 2 years (2020–2021), almost three times compared to 2016–2020. Quantitative comparison of the found publications on the ten most used keywords for the "e-learning market" query makes it possible to argue that the use of the keyword "artificial intelligence" in 2020–2021 is 88% of the total number of publications in the period from 2016 to 2021, "machine learning" 63%, "virtual reality" 62%, "online education" and "deep learning" 59%, "higher education" 52%, "online learning" 51%, "training" 50%, "education" 45%, and "e-learning" 38% (based on the results of the Scopus database analysis and ScientoPy software tools and Python language). The growing use of artificial intelligence and machine learning in the e-learning system stimulates the mastery of digitalization tools.

Comparative characteristics of indicators of infrastructure, access, opportunities, and barriers were given for Poland, Ukraine, Germany, and Cyprus and showed significant differences. The cluster analysis of the median values of countries' web behavior (the sample of 20 countries based on IEEE data) in relation to online learning was performed using Ward's method and Agglomeration scheme in Statgraphics 19 software to form groups of countries that have similar web behavior related to online learning, in particular the following: (1) cluster 1: Argentina, Colombia, Indonesia, Iran, Peru, Russia, Turkey, and Ukraine; (2) cluster 2: Brazil, France, Mexico, and the Philippines; (3) cluster 3: Germany and the United Kingdom; (4) cluster 4: India and Spain; (5) cluster 5: Italy, Poland, and South Africa; and (6) cluster 6: the United States.

The received results can be used by state and local authorities and scientific and educational institutions for the further development and improvement of e-learning technologies to strengthen leadership in education.

Acknowledgments This research was funded by the Ministry of Education and Science of Ukraine (projects No 0122 U000772, No 0122 U000780, No 0121 U109553, No 0120 U102001).

References

Artyukhov, A., & Omelyanenko, V. (2021). Methodical aspects of innovation cooperation processes educations and R&D effects estimations. *Proceedings of 11th international conference on Advanced Computer Information Technologies* (*ACIT* 2021), pp. 250–254. https://doi.org/10.1109/ACIT52158.2021.9548433.

Artyukhov, A., Omelyanenko, V., & Prokopenko, O. (2021). University technology transfer network structure development: Education and research quality issues. *TEM Journal, 10*(2), 607–619. https://doi.org/10.18421/tem102-16

Beyi, W. A. (2018). The trilogy of a digital communication between the real man, his digital individual and the market of the digital economy. *SocioEconomic Challenges, 2*(2), 66–74. https://doi.org/10.21272/sec.2(2).66-74.2018

Bilan, Y., Pimonenko, T., & Starchenko, L. (2020). Sustainable business models for innovation and success: Bibliometric analysis. *E3S Web of Conferences, 159*, 04037. https://doi.org/10.1051/e3sconf/202015904037

Brimah, B. A., Olanipekun, W. D., Bamidele, A. G., & Ibrahim, M. (2020). Knowledge management and its effects on financial performance: Evidence from Dangote Flour Mills, Ilorin. *Financial Markets, Institutions and Risks, 4*(2), 34–42. https://doi.org/10.21272/fmir.4(2).34-42.2020

Buchynska, O., Davlikanova, O., Hofstetter, H., & Lylyk, L. (2020). The Ukraine-based employers' awareness of dual studies and willingness to engage into the implementation of education innovations. *Business Ethics and Leadership, 4*(3), 137–144. https://doi.org/10.21272/bel.4(3).137-144.2020

Bukhtiarova, A., Shcherbachenko, V., Artyukhov, A., Krmela, J., & Volk, I. (2021). Open science investigation of the health economics sector. *Health Economics and Management Review, 2*(4), 27–37. https://doi.org/10.21272/hem.2021.4-03

Didenko, I., Paucz-Olszewska, J., Lyeonov, S., Ostrowska-Dankiewicz, A., & Ciekanowski, Z. (2020). Social safety and behavioral aspects of populations financial inclusion: A multicountry analysis. *Journal of International Studies, 13*(2), 347–359. https://doi.org/10.14254/2071-8330.2020/13-2/23

Diya. (2022). *Diya.Tsyfrova Osvita. Action.Digital Education*. Available at: https://osvita.diia.gov.ua/

Dutta, P., Dutta, U., Hasan, S., Sarkar, S., & Sana, T. (2020). Educate, empower and protect people through timely authentic information: Explore how Bangladeshi newspapers response to COVID-19 pandemic. *SocioEconomic Challenges, 4*(3), 93–103. https://doi.org/10.21272/sec.4(3).93-103.2020

Education Statistics. (2022). *55 education statistics: Global data, latest trends & predictions*. Available at: https://www.guide2research.com/research/education-statistics#:~:text=A%20total%20of%20258.4%20million,school%20population%20at%20148.9%20million

E-Learning Market. (2022). *Allied market research*. Available at: https://www.alliedmarketresearch.com/e-learning-market-A06253

Frederick, D. T., & Kasztelnik, K. (2020). An analytical study of impact of international merger and acquisitions on the financial performance for Higher Education Institution in the United States. *Financial Markets, Institutions and Risks, 4*(4), 5–30. https://doi.org/10.21272/fmir.4(4).5-30.2020

Gruzina, Y., Firsova, I., & Strielkowski, W. (2021). Dynamics of human capital development in economic development cycles. *Economies, 9*(2), 67. https://doi.org/10.3390/economies9020067

Guliyeva, A., Aliyeva, T., Rzayeva, U., & Huseynova, R. (2021). Lifelong education as a challenge of digital era. *Turkish Journal of Computer and Mathematics Education (TURCOMAT), 12*(6), 1282–1289. https://doi.org/10.17762/turcomat.v12i6.2468

Iatsyshyn, A., Kovach, V., Lyubchak, V., Zuban, Y., Piven, A., Sokolyuk, O., Iatsyshyn, A., Popov, O., Artemchuk, V., & Shyshkina, M. (2020). Application of augmented reality technologies for education projects preparation. *Proceedings of CEUR Workshop, 2643*, 134–160. https://doi.org/10.55056/cte.318

IEEE DataPort. (2022). *Online learning global queries dataset: A comprehensive dataset of what people from different countries ask Google about online learning*. Available at: https://ieee-dataport.org/documents/online-learning-global-queries-dataset-comprehensive-dataset-what-people-different

ITU. (2022). *Digital development dashboard*. Available at: https://www.itu.int/en/ITU-D/Statistics/Dashboards/Pages/Digital-Development.aspx

Ivanov, V., Pavlenko, I., Trojanowska, J., Zuban, Y., Samokhvalov, D., & Bun, P. (2018). Using the augmented reality for training engineering students. *Proceedings of the international conference of the Virtual and Augmented Reality in Education (VARE 2018)*, 17-19.09.2018, pp. 57–64.

Jikia, G., Vorontsova, A., & Petrushenko, Y. (2017). Measuring efficiency of financial support in lifelong learning system: A case study of Ukrainian regions. *Business Ethics and Leadership, 1*(4), 84–92. https://doi.org/10.21272/bel.1(4).84-92.2017

Kesavan, N., Seenuvasaragavan, S., Burnette, K., Heim, A. B., Patel, R. J., Johnston, J., Cao, X., Sakandar, H. A., Zhuang, Y., Oi, K. K., Zhi, Y., Diao, Y., & Strielkowski, W. (2022). Pandemic-inspired policies. *Science, 377*(6601), 22–24. https://doi.org/10.1126/science.add2046

Khanin, I., Shevchenko, G., Bilozubenko, V., & Korneyev, M. (2019). A cognitive model for managing the national innovation system parameters based on international comparisons (the case of the EU countries). *Problems and Perspectives in Management*, (4), 17, 153–162. https://doi.org/10.21511/ppm.17(4).2019.13

Klebanova, T. S., Guryanova, L. S., Chagovets, L. O., et al. (2018). *Biznes-analityka bahatovymirnykh protsesiv [Business analytics of multidimensional processes]: A textbook* (1st ed., p. 2018). KhNEU named after S. Kuznets.

Kotenko, N., & Bohnhardt, V. (2021). Digital health projects financing: Challenges and opportunities. *Health Economics and Management Review, 2*(1), 100–107. https://doi.org/10.21272/hem.2021.1-10

Kuzior, A., & Kwilinski, A. (2022). Cognitive technologies and artificial intelligence in social perception. *Management Systems in Production Engineering, 30*(2), 109–115. https://doi.org/10.2478/mspe-2022-0014

Kuzior, A., & Zozul'ak, J. (2019). Adaptation of the idea of phronesis in contemporary approach to innovation. *Management Systems in Production Engineering, 27*(2), 84–87. https://doi.org/10.1515/mspe-2019-0014

Kuzior, A., Kettler, K., & Rąb, Ł. (2022). Digitalization of work and human resources processes as a way to create a sustainable and ethical organization. *Energies, 15*(1). https://doi.org/10.3390/en15010172

Kuzmenko, O., Cyburt, A., Yarovenko, H., Yersh, V., & Humenna, Y. (2021). Modeling of "information bubbles" in the global information space. *Journal of International Studies, 14*(4), 270–285. https://doi.org/10.14254/2071-8330.2021/14-4/18

Kyrychenko, K., Laznenko, D., & Reshetniak, Y. (2021). Green university as an element of forming a sustainable public health system. *Health Economics and Management Review, 2*(4), 21–26. https://doi.org/10.21272/hem.2021.4-02

Lagodiienko, N., & Yakushko, I. (2021). Digital innovations in taxation: Bibliometric analysis. *Marketing and Management of Innovations, 3*, 66–77. https://doi.org/10.21272/mmi.2021.3-06

Liuta, O., Lieonov, S., Artyukhov, A., Sushko-Bezdenezhnykh, M., & Dluhopolskyi, O. (2021). Student survey as a tool for quality assurance in higher education: The case of Ukrainian University. *Naukovyi Visnyk Natsionalnoho Hirnychoho Universytetu, 2021*(4), 158–164. https://doi.org/10.33271/nvngu/2021-4/158

Lyeonov, S., & Liuta, O. (2016). Actual problems of finance teaching in Ukraine in the post-crisis period. *The Financial Crisis*, 145–152. https://doi.org/10.1007/978-3-319-20588-5_8

Melnyk, L., Matsenko, O., Kubatko, O., Korneyev, M., & Tulyakov, O. (2022). Additive economy and new horizons of innovative business development. *Problems and Perspectives in Management, 20*(2), 175–185. https://doi.org/10.21511/ppm.20(2).2022.15

Onopriienko, K., Onopriienko, V., Petrushenko, Y., & Onopriienko, I. (2021). Environmental education for youth and adults: A bibliometric analysis of research. *E3S Web of Conferences, 234*, 00002. https://doi.org/10.1051/e3sconf/202123400002

Petroye, O., Lyulyov, O., Lytvynchuk, I., Paida, Y., & Pakhomov, V. (2020). Effects of information security and innovations on Country's image: Governance aspect. *International Journal of Safety and Security Engineering, 10*(4), 459–466. https://doi.org/10.18280/ijsse.100404

Polyakov, M., Bilozubenko, V., Korneyev, M., & Shevchenko, G. (2019). Selection of parameters for multifactor model in the knowledge economy marketing (country level). *Innovative Marketing, 15*(1), 89–99. https://doi.org/10.21511/im.15(1).2019.08

Polyakov, M., Bilozubenko, V., Korneyev, M., & Nebaba, N. (2020). Analysis of key university leadership factors based on their international rankings (QS world university rankings and times higher education). *Problems and Perspectives in Management, 18*(4), 142–152. https://doi.org/10.21511/ppm.18(4).2020.13

Popova, M. (2021). Wage differentials and educational attainment in Germany. How do job profiles affect earnings? *SocioEconomic Challenges, 5*(2), 15–25. https://doi.org/10.21272/sec.5(2).15-25.2021

Russ, M. (2021). Knowledge management for sustainable development in the era of continuously accelerating technological revolutions: A framework and models. *Sustainability, 13*(6), 3353. https://doi.org/10.3390/su13063353

Sadigov, R. (2022). Impact of digitalization on entrepreneurship development in the context of business innovation management. *Marketing and Management of Innovations, 1*, 167–175. https://doi.org/10.21272/mmi.2022.1-12

Samoilikova, A. (2020). Financial policy of innovation development providing: The impact formalization. *Financial Markets, Institutions and Risks, 4*(2), 5–15. https://doi.org/10.21272/fmir.4(2).5-15.2020

Samoilikova, A., Zhylinska, O., Pal, Z., & Kuttor, D. (2022). "Business-education-science" coopetition and innovation transfer for sustainable development. *Marketing and Management of Innovations, 2*, 220–230. https://doi.org/10.21272/mmi.2022.2-20

Shvindina, H., Kotenko, S., Vorontsova, A., Gordienko, V., Petrushenko, Y., & Jha, D. (2020). Reforming the education system to prevent labor migration. *Knowledge and Performance Management, 4*(1), 52–66. https://doi.org/10.21511/kpm.04(1).2020.05

Simionescu, M., Strielkowski, W., & Tvaronavičienė, M. (2020). Renewable energy in final energy consumption and income in the EU-28 countries. *Energies, 13*(9), 2280. https://doi.org/10.3390/en13092280

Statista. (2022). *Size of the global e-learning market in 2019 and 2026, by segment*. Available at: https://www.statista.com/statistics/1130331/e-learning-market-size-segment-worldwide/#statisticContainer

Strielkowski, W., Firsova, I., Azarova, S., & Shatskaya, I. (2022). Novel insights in the leadership in business and economics: A post-coronavirus update. *Economies, 10*(2), 48. https://doi.org/10.3390/economies10020048

The World Bank. (2022). *The World Bank Group data*. Available at: https://databank.worldbank.org/reports.aspx?source=Education%20Statistics#

Tiutiunyk, I., Drabek, J., Antoniuk, N., Navickas, V., & Rubanov, P. (2021). The impact of digital transformation on macroeconomic stability: Evidence from EU countries. *Journal of International Studies, 14*(3), 220–234. https://doi.org/10.14254/2071-8330.2021/14-3/14

Vardanyan, N. (2017). Education quality Assessment from the perspective of stakeholders on the example of Armenian higher education institutions. *Business Ethics and Leadership, 1*(3), 93–97. https://doi.org/10.21272/bel.1(3).93-97.2017

Vorontsova, A., Mayboroda, T., & Lieonov, H. (2020a). Innovation management in education: Impact on socio-labour relations in the national economy. *Marketing and Management of Innovations, 3*, 346–357. https://doi.org/10.21272/mmi.2020.3-25

Vorontsova, A., Shvindina, H., Mayboroda, T., Mishenina, H., & Heiets, I. (2020b). The impact of state regulation in a sphere of education on sustainable development of national economy. *Problems and Perspectives in Management, 18*, 275–288. https://doi.org/10.21511/ppm.18(4).2020.23

Wang, Q., Chen, Y., Guan, H., Lyulyov, O., & Pimonenko, T. (2022). Technological innovation efficiency in China: Dynamic evaluation and driving factors. *Sustainability (Switzerland), 14*(14). https://doi.org/10.3390/su14148321

Yarovenko, H., Bilan, Y., Lyeonov, S., & Mentel, G. (2021). Methodology for assessing the risk associated with information and knowledge loss management. *Journal of Business Economics and Management, 22*(2), 369–387. https://doi.org/10.3846/jbem.2021.13925

Yu Sing Ong. (2018). A humanistic model for the higher education excellence: Embedding Confucianism and Islamic principles. *Business Ethics and Leadership, 2*(3), 103–122. https://doi.org/10.21272/bel.2(3).103-122.2018

Zoom. (2022). *A message to our users*. Available at: https://blog.zoom.us/a-message-to-our-users/

Chapter 18
Leading Quality Issues of Higher Education in Tourism and Hospitality: The Case of Azerbaijan

Yasaman Abasova and Bahar Ganiyeva-Dadashova

Abstract This chapter analyzes the leading quality issues of the higher education in tourism and hospitality as one of the flexible and rapidly developing career paths that has the potential to generate future leaders. Quality criteria, tourism industry requirements, students' competences, required skills, curriculums, stakeholders' interests, and many factors have been touched based this research and in course of its processing. In turn, to reveal the current state of quality fitness of graduates' higher education to the needs of industry and identification of important criteria determining quality and competitiveness of higher education in the field of tourism and hospitality, a survey was held both among the students and the industry representatives. The survey's results revealed some gaps existing between higher education and the genuine requirement of tourism industry employment in Azerbaijan. To positively address the problems identified in the result of the research and maintain the quality of higher education in this field, the authors considered several practical decisions.

Keywords Tourism · Education · Leadership · Hospitality

JEL Classification L8 · J2 · D2

Y. Abasova (✉)
Azerbaijan Tourism and Management University, Baku, Azerbaijan
e-mail: yasaman.abasova@atmu.edu.az

B. Ganiyeva-Dadashova
Ministry of Youth and Sport of Azerbaijan Republic, AzerİdmanServise LLc, Baku, Azerbaijan
e-mail: B.dadashova@azeridmanservis.com

W. Strielkowski (ed.), *Leadership, Entrepreneurship and Sustainable Development Post COVID-19*, Springer Proceedings in Business and Economics, https://doi.org/10.1007/978-3-031-28131-0_18

18.1 Introduction

For decades, tourism, as the generator of economic growth and activity, has been occupying a considerable and important space in the economies of the majority of countries worldwide (Akan et al., 2008; Radovic et al., 2017; Grecu et al., 2019; Zhang & Yang, 2021). Accordingly, tourism as a priority sector of the development of non-oil industries of the country and ensure the sustainability of economy in Azerbaijan has been one of the fastest-growing field (Madina, 2019; Aliyeva & Rzayeva, 2020; Akhundova, 2022). Of course, for the past few decades, lots of successful work has been done both at international and local levels in the comprehensive development of this area. However, obtaining an advantage in an increasingly competitive environment for international tourism and hospitality mostly depends on the quality of the service which requires the training of more skilled and professional specialists who would become the leaders in the field (Sharifovich, 2021), since it is well recognized from the experience of developed countries that only countries with high-level staff win the competition on the international market (Guliyeva et al., 2020). Therefore, the development of tourism and hospitality industry in a country is substantially determined with overall level of education and qualification of work force (Lei et al., 2021). Consequently, the quality fitness of the graduates to the needs of industry and the identification of important criteria determining the quality and competitiveness of higher education in tourism and hospitality become the essential factors.

But a quite reasonable question pops out: "Does the management and quality of higher education meet the quality needs of the tourist and hospitality industry?" If we study, in general, the current state of labor market in Azerbaijan, a variety of concerns may be identified as well. The major reasons for these issues are the lack of permanent jobs, the quality of the labor market, the insufficient workforce in the economy, and the high level of youth migration. Employment issues are mostly seen among the young people. The increased unemployment among them has resulted in growth the number of youths leaving the country. Moreover, lots of engaged youth earn a pay near to the minimum living standards. Thus, according to 2020 data from the labor force sample statistical survey, 1,378,000 youth aged 15–29 were economically engaged, and 154,100 of them were jobless (comprising men – 69,900, women – 84,200, persons). If by the end of 2019 there were 251,600 (4.8%) of the unemployed young people aged 15–29 years, this indicator rose to 375,900 (7.2%) by 2020 (State Statistical Committee of the Republic of Azerbaijan, 2021). 10.6% of 154,100 of unemployed youth aged 15–29 have higher education, 11.9% secondary specialized, 4.4% primary vocational, 59.5% complete secondary, 13.4% general secondary education, and 0.2% primary education. In turn, referring to the information obtained from Azerbaijan Tourism and Management University (ATMU) on the employment of the graduates, it was stated that for the period of 2016–2018, 50% of them were engaged in the field of tourism, 22% in other fields not related to tourism, 13% in continued education, and 15% were unemployed. In addition, we would like to note that the personnel policy in the country is not sufficiently thought out and, in many cases, does not meet the requirements of the day. Thus, despite not needed in some specialties, there is an overstaffing, while in others there is a shortage of staff despite the demand. At the same time, in general, there is a lack of proportionality in the distribution of human resources in the country. As far as Baku, the capital, and several large cities are more developed, so in the conditions of internal migration, skilled human resources are concentrated there, while there is a shortage of labor resources in the regions.

In turn, as it was mentioned above currently, there are a few problems in the level of knowledge and professionalism of personnel trained in this field. In 2019, the "Meridian Integrated Business Solutions Inc." conducted surveys to highlight such problems with the request and funding of the State Tourism Agency of the Republic of Azerbaijan. The surveys were held in 25 cities and found a total of 289 facilities, out of which 170 were hotels, 112 were restaurants, and 7 were wineries. Based on the results of surveys, a report was prepared as an outcome of the project "Selection and Needs Analysis of Restaurants, Hotels and Wineries for enhancing Business Administration and Capacity Building" (Meridian Integrated Business Solutions Inc., 2019). Overall, in terms of training needs, problems and challenges were identified as follows:

- Limited number of the qualified and customer-oriented workforce (require outsource agency or company to arrange necessary trainings for the personnel as well as its funding)
- Limited number of talents with language proficiency to eliminate the gap in communication with the travelers and guests
- Insufficient level of communication among the peers
- Overall lack of knowledge and skills in tourism as industry and field

Apart from the aforesaid, low awareness of hygiene and sanitary requirements, design of menus, and service techniques were observed in the restaurants. Thus, summarizing the above points, we come to the validation of the research problem – how relevant the education in the field of tourism and hospitality is to the industry requirements and what alignment is to be implemented to eliminate or minimize the existing problems in higher education. In this regard, we have set several questions:

- What is the higher education quality criteria in tourism and hospitality?
- What are the realities and the quality requirements of the industry to the graduates in this field?
- How does the education in tourism and hospitality prepares qualified graduates for the industry and how aligned should it be?

Thus, the aim of the research is to find the answers to the above given questions, diagnose the current state, and assess the quality and competitiveness of higher education on tourism and hospitality in the Republic of Azerbaijan in the light of the leadership formation within this industry in the post-COVID era.

18.2 Literature Review

Nowadays, higher education is developing in conditions of vigorous competition, and consequently a crucial factor determining competitiveness of any university is the quality of education (Beykan & Kadir, 2015; Čábelková et al., 2015). Quality represents an exceptional characteristic that determines the socioeconomic significance of higher education in tourism and hospitality. However, it is worth to point out that there is no precise and simple definition for the term "quality of higher education" and its indicators as various stakeholders' groups place varying emphasis on different aspects of it. Tsinidou et al. (2010) noted that education services are often impalpable and hard to measure, since the effect is the returned transformation of individuals in their knowledge, comprehension, skills, and behavior. Consequently, there is no generally accepted definition of quality applicable specifically to higher education in tourism and hospitality (Kim et al., 2021). In turn, the difference in nature and the extent of higher education systems may have various mechanisms of quality assurance. The particular importance has also the fact that quality changes due to socioeconomic changes and there is continuing concern that higher education does not keep pace with changing needs (Mishra et al., 2020; Nureev et al., 2020).

A review of the relevant literature in this area shows that there is a diversity and fragmentation of viewpoints that characterize the quality of higher education in the field of tourism and hospitality (Mínguez et al., 2021; Bavik & Kuo, 2022). Some associate it with the quality of teaching (Goh & King, 2020); others consider this concept as the quality of teaching and upbringings as a whole (Kazmina et al., 2020); third relates it with the degree of personality development; fourth assumes it as the readiness of the individual for life as such (Sandybayev, 2020). From the early 1980s to the present day, plenty of books and articles have been published on the quality of higher education; however, there is no generally accepted definition, and moreover, the concept is still often distorted by many scientists (Doherty, 2008). The complication is related with the fact that the set of quality parameters to be assessed and their relative weight are not constant; moreover, they are distinct due to different points of view of the stakeholders (Tsinidou et al., 2010). In this regard, it is pertinent to analyze the scientific interpretations of the concept of "quality of higher education."

For instance, Barnett (1992) considers two main concepts of higher education where the first bases on intellectual property and cooperation within the participants of higher education. The second is the concept of a performance, in which the quality of higher education is seen as a result measured by performance reflected in performance indicators. Akvazba et al. (2015) in determining the quality of higher education propose the so-called process approach, where quality should be considered as the quality of process, quality of result, and the quality of education system. According to Abdullayev (2015), the chief adviser in the Ministry of Education, the quality of higher education is determined by the competitiveness of graduates both in the national and international labor market and their role in the socioeconomic development of the country.

The analysis of a certain list of viewpoints and approaches to characterize the quality of higher education shows a lack of unity, which is associated with the differences in the respective methodological approaches. However, due to the multidimensionality of a concept, the quality of higher education, determination given by Koreneva (2010), seems in our opinion more acceptable. She considers the concept of quality in education as the correspondence of education (as an outcome, process, and social system) to the different needs and interests of the individual, society, and state. In other words, the institution provides services to three subjects: person who receive education, to industry by providing tertiary educated workforce, and the state which acts as guarantor of educational process. In turn, as she assumes, the quality of higher education has to be determined not only by the goal of matching the acquired knowledge to the needs and expectations of society but also by the tasks of adherence to professional standards put forward and controlled by the state.

The competitiveness of higher education institutions within the economic system has now been increasingly appreciated. In recent years, a complete transformation has been established in the operating framework of the higher education systems. The more competitive an educational institution, the more chances for students to continue their education in it, which means that this institution has more survival opportunities in future in comparison with those lagging behind competitors.

Nowadays, social scientists should face a very big challenge, and globalization can be nothing. This may be an incredibly fast phenomenon, a multidimensional and multi-actor process that affects all elements of the social formation (Dallen, 2019). The main features include the diffusion of diverse types of competitiveness and cooperation, the technological link system, the economy and the social subsystems, and their reciprocal dependence. In addition, fast and radical changes have a direct impact on higher education.

The most important consideration in the establishment of a competitive internal economy is the direct share of viable development in the country regarding quality of education. Today, the growing importance of human life education transforms the idea of transferring knowledge over time to a skills training ideology. The quality of education in the world does not depend on the duration of the study, but on the effectiveness of the learning and its effect on the life of society. The concept of lifelong education is highly applicable to changing countries and to the continuous use of sustainable education systems, quickly changing the scope of their operations. The concept of lifelong learning and the continuing use of sustainable education systems is very much in place for countries that undergo changes, quickly changing the scope of Azerbaijani culture's operations and overall living environment, renewed job market, and data conditions.

Quality assurance of higher education is the elaboration and implementation of a coherent actions that make it possible to effectively provide educational services of such quality that ensures the result of education has met the requirements of consumers. However, here we ought to keep in mind that the education system provides services to students and products to the rest of consumers – organizations, enterprises, and the state as a whole. Since the sphere of application of knowledge, skills, and abilities of each graduate is society represented by state and private organizations and enterprises, then they are deceiving customers of qualified specialists. In addition, these two groups of consumers have varied requirements for the education system in terms of the criteria and characteristics of the quality higher education. Consequently, based to aforementioned, we will analyze the quality requirements of each group.

18.3 Methodology

The analysis is conducted to disclose the Azerbaijani case of quality of higher education in tourism and hospitality. As the higher educational institutions provides services to three main stakeholders (a person who receive education, industry for which a tertiary educated workforce are training, and state which acts as the guarantor of the educational process), so in research we used an approach that considers the interests of stakeholders in educational process, since in our opinion it seems as most suitable and reliable. The research draws on the description of general concept, study of state policy, and inclusive growth and criteria that guarantee the quality of higher education on tourism and hospitality. The online survey, consisting of 15 questions, was designed to identify the student's point of view related to the criteria and the satisfaction level of the quality of education they receive. Overall, 161 students and alumni responded to the survey questions.

In turn, to reveal the quality requirements and needs of the industry to the tertiary-educated workforce, an online survey consisting of 15 questions was conducted with representatives of tourism and hospitality. In total, 148 respondents of the industry participated in the online survey. Later in this chapter, we broadly overviewed the results of online surveys. We conclude this chapter by highlighting the reasonable and essential measures which may maintain the quality of higher education in tourism and hospitality.

All in all, the state quality assurance of higher education in tourism and hospitality is important. Higher education, probably under the general financial and social framework, is the largest subsystem for all nations. The importance of the system for higher education should be taken into consideration by each nation for a sustainable and robust turn. Advanced higher education is essential to the nation's improvement since it has profoundly positive results both for individual work positions and in socioeconomic development in the country as whole. Consequently, the state interests in high education are that universities prepare highly qualified specialists who can develop and put the country's economy on a par with the economies of developed countries and thereby justify the budgetary funds spent on their training. For long-term economic growth, a fruitful and effective functioning of higher education system plays crucial role, and "its internationally recognized performance can greatly improve the picture of economy and society" (Bonanno & Pozzoli, 2009). At the same time, graduates and the quality of their education must meet the requirements of state educational standards. Certainly, if the state intents to create durable education system behind which education standards are competitive, then it must take time to implement the conditions as transformation of higher education is a time-consuming process.

The demand for higher education on tourism and hospitality in Azerbaijan began to grow from the 2000s. Subsequently, in 2006, the Azerbaijan Tourism Institute (in 2015, renamed to Azerbaijan Tourism and Management University – ATMU) was established as a specialized institution promoting tourism and hospitality education at all levels. Up until now, there are several universities providing higher education on tourism and hospitality vocations apart from ATMU.

In turn, having integrated into the Bologna Process in 2005, Azerbaijan began reforming the whole system of higher education, and consequently a new Law on Education was adopted in 2009. This law formally allowed the transition to the European Credit Transfer System, covering three-cycle system of education: first level, bachelor studies; second level, master studies; and the third level, Ph.D. studies (The Ministry of Science and Education of the Republic of Azerbaijan, 2022).

The national level of quality assurance in Azerbaijan is regulated by the educational standards consisting of internal and external quality arrangements. In addition, the external quality arrangements are regulated by the Law on Education, the Accreditation Rules of Educational Institutions approved by Executive Order No.167 of the Cabinet of Ministers as of the 28th of September 2010, and other regulatory documents (ENIC-NARIC, 2022).

The central executive body ensuring the development of state policy and implementation of state control quality of education is the Ministry of Education of the Republic of Azerbaijan. In recent years, significant efforts have been made to develop education, notably in the establishment and development of lifelong learning and lifelong learning in Azerbaijan. The fourth strategy guideline in the Republic of Azerbaijan on the state education adopted in October 2013 is an educational infrastructure that fully meets the requirements of the modern world and offers a long-term education. The Cabinet of Ministers of Azerbaijan approved the National Framework for Qualifications for Lifelong Education in the Republic of Azerbaijan with the purpose of implementing that strategy in the scope of its Action Plan.

Furthermore, it ought to be noted that as a rule, in every country, the accreditation agencies are in charge and authorized to control the quality given by the educational institutions through assessing and accrediting their standards and the educational work offered. In this context, based on the Decree signed by the president of Azerbaijan on 29 December 2019, the Education Quality Assurance Agency (EQAA) was established which determines the quality levels of educational programs and ensures the implementation of requirements (TKTA, 2022). The activities of the EQAA are as follows:

- To carry out accreditation of educational institutions
- To recognize the qualifications of foreign countries related to higher education
- To evaluate and recognize knowledge, skills, competencies, and experience acquired through nonformal and informal forms of vocational education
- To prepare recommendations and organize trainings for the implementation of accreditation of educational institutions

- To organize consulting and methodological support services tied up with the establishment and development of internal quality assurance systems in educational institutions
- To promote the sustainable development national education system and state standards in this area, increase the prestige of educational institutions and improve the quality of specialist training
- Participation in the issuance of reviews for the purpose of licensing educational activities
- To study and apply best international practices in the relevant field

However, the case of Azerbaijan shows that various higher education institutions preparing students for tourism and hospitality industry are not certified in phrases of first-class teachers, centers, and curricula. Accordingly, there is a great need to tighten standards and create a more stringent accreditation mechanism in this direction. In addition, higher education institutions and colleges ought to cooperate with a broad range of nongovernmental and governmental organizations, including UNESCO, to enhance and develop their control mechanism.

When it comes to the quality criteria of high school institutions in tourism and hospitality, it becomes apparent that the highly skilled employees are the key to global development and innovation (Kofler et al., 2020). The higher the education standard, the higher the trend toward intellectual evolution. It is therefore important for high school institutions to play a deceiving role in our personal and social development directly, as indirectly as well. The main task of a high school institution is not only the formation of academic knowledge among students but above all the ability to work, live, and adapt in a rapidly changing world. From the second half of the twentieth century, universities have been involved in a powerful movement of educational expansion and democratization of educational opportunities. The massive spread of higher education has come to be seen as a guarantee of the competitiveness of the state in the new global economy.

The fast growth of science and technology, the expanding role of knowledge and information over the past decades, contributed to the fact that universities have always undergone significant transformation. In addition, with purpose to identify the role of higher educational institutions in the socioeconomic growth of society, theories were formed that consider universities as follows:

- Learning and personal development communities of their members, particularly students (this might be called the "liberal" theory)
- Sources of professional identity (professional education' theory)
- Creators, experts, and locations for evaluation and application of new knowledge (the theory of the research engine)
- Creators, testers, and new knowledge appraisal sites (the theory of "research engine" and an important corollary – theory of "business and industrial services")
- Significant social and national contributors (the "civic and social engagement" theory)

Quality has no absolute standards. An educational institution depends on its definition, its portfolio program, and the position it creates in the minds of its students and the public (Rinehart, 2016). However, it is ought to note that the higher education quality concept is multidimensional and cannot be measured by a single specific indicator. According to Rinehart (2016), the quality concept of higher education should be considered in terms of input, process, and output elements, as well as services provided to its internal and external stakeholders. Apart from this, the relative importance of each quality elements may differ from one institution to another. Some universities focus mainly on the quality of inputs to education systems, while others concentrate on the quality of the results.

Accordingly, considering the category "quality of higher education" from the base of high school institutions, it is a multidimensional concept that encompasses all aspects of the university's activities: the quality of curriculums and academic programs; teaching staff and their motivation; the state of material and technical base; scientific and research activities; faculties and specialties; finance and investments to education; administrative management and implementation of innovative activity; educational technologies; contingent of students; and demand for graduates in the labor market and their achievements. These criteria make it possible to determine the level of education quality and in accordance with this build the pricing policy of a higher educational institutions.

The curriculum as one of the major successes determining the quality of education in the educational institution has to be taken seriously. It is believed as "being a curriculum that includes the choice of knowledge, the link between tourist knowledge and the curriculum should be understood" (Tribe, 2005). It is essential to understand the significance of the curriculum program on tourism and hospitality management and its impact on sphere. In turn, it is crucial to introduce undergraduates the relationship-building and team-working skills and logical, analytical, language, managerial, and leadership skills via the education they receive. Consequently, to reap a higher first-class of hospitality and tourism program, efforts have to be made to improve curriculum layout. Currently, curriculum for tourism and hospitality specialties is evolved with the aid of using educating officers and academia without thinking about the requirement for workforces inside the enterprises, and alumnus cannot meet the industry requirements. Therefore, the related experts and specialists from this industry should be involved in the process of curricula layout. In addition, realistic education system should base more on instilling practical knowledge than theoretical lectures in classrooms, with the aim of equipping students with real abilities and skills before they become members of any enterprise after graduation.

Teaching staff is ranked as vital element, which embraces instructors' satisfaction and their coaching overall performance. Furthermore, it also comprises whether or not teacher's specializations correspond with the program purposes and whether or not the publications are appropriate to instructors in shaping their experience. Lecturer's overall performance is gauged based on the participation in the projects he/she obtains from the institute and active engagement in professional activities.

Moreover, it is inevitable to properly spur the teaching staff, because without this the proper development of competitiveness becomes almost unworkable.

For 800 years the aim of the modern universities and institutions has been to provide a competitive account. In addition, this has a strong contemporary resonance, and today universities generally reflect a balance between the various strands, be it explicit or implicit. Nowadays the following main fundamental factors (indicators) defining the competitiveness and quality of education in higher school institutions are as follows:

- Qualification of teaching staff and intellectual potential in general
- Teaching and methodological support, methods, and technology of teaching
- Appropriate educational and academic standards
- Material and technical base of the university
- Funding and use of financial resources
- Internationality and international acknowledgment
- Students as the main subject of the educational process, by the state of which one can judge the quality of education
- Graduates as an assessment of the relationship of the university with various industries and their career

The Wall Street Journal found that 89% of executives have difficulty in finding candidates with the required set of social and psychological skills such as teamwork, communication, and adaptability (Vedomosti, 2020). In turn, creating scholarly information is significant for the alumni professions, yet reality encounters should be shared during their schooling by building up an all the more impressive participation between the upper training establishments and the business (Koc et al., 2014). Consequently, the higher education institutions ought to understand that they operate in competition and in gap with the industry requirements, so just the "diploma manufacturing" process shall be ceased. They ought to strength the partnership and relationship with the enterprises. Vocational institutions which offer tourism and hospitality program have to evolve nigh cooperation with enterprise in this sphere with the aid of using providing tailored education applications and collaborating in consultancy and control of tourism and hospitality (Beykan & Kadir, 2015).

18.4 Results

In general, the quality requirements of the industry to the tertiary-educated workforce in tourism and hospitality are important for securing workplaces and regulating the workforce. Uncontrollable rapid changes in world economy, fast growth of tourism industry, and keen competition in tourism market require highly qualified workforce. Some scientists state this as follows: "Employees require more acknowledged and 'portable' professional competencies across the borders and tourism qualifications which are accessible" (Strietska & Tessaring, 2005).

The key task in assuring the quality of higher education in tourism and hospitality is to find a balance between general knowledge and specific skills and competencies required by industry. As we have noted, the practical use of knowledge and skills of graduates happen in various organizations and enterprises; thence the requirements that they claim on qualifications and quality should be the main components of education system of higher institutions. Thus, the "quality higher education" is the ability of an educational product or service to comply with the requirements of state standard and social order.

In this regard, a preferred educational program ought to base on standards agreed by representatives from education institutions and hospitality industry and include the following:

- Transition of industry quality requirements into the system of education
- Adjustment to the industry quality requirements
- Evaluation of the end product (professional skills and competencies of graduates)

To expose the competitiveness and quality of the graduates, we distributed the survey among 148 senior-level industry representatives as follows:

- Department heads (front of house, back of house)
- Division heads (rooms operation, sales and marketing, food and beverage, finance)
- General managers and hotel managers
- Recruitment specialists
- Human resources and personnel managers
- Incoming and outgoing travel agency managers

The participated leaders manage overall 8942 employees, and out of them, only 502 (5%) represent the graduates of ATMU. Mainly the industry leader's competency expectations and skills requirements to the candidates are as follows:

- Higher education in the field of tourism and hospitality
- Mindset, skillfulness, and desire to learn
- Effective communication
- Personal qualities (attentiveness, honesty, truthfulness, accuracy)
- Language competencies
- Responsibility
- Innovative and creative
- Experienced
- Flexibility
- Teamwork ability
- Hardworking

The survey led on students and graduates on the competency expectations and skill requirements from the industry stakeholders. The opinion of the industry representatives was acquired to find out what the main disadvantages and weaknesses of the candidates are, and the graph describes the overview of the responses:

- Lack of knowledge on the field of tourism and hospitality
- Do not develop themselves and unaware of new technologies
- Lack of experience
- Low language skills
- Low desire to work and business acumen
- Low self-esteem
- Low initiative to work
- Low introduction abilities of themselves
- Irresponsibleness, undisciplined, and unethical

The main criteria for election process during recruitment was responded as following by the industry representatives which makes the university/ties enhance the given qualification and competencies during education process (work experience: 46, personal qualities: 25, educational credentials: 12, professional certification: 4, areas of knowledge: 12, all of them: 9).

The main criteria for election process is given in percentage (%). The responses given to the question "What channels/sources do you refer for the recruitment process?" show that 19.8% of them mainly apply to the job websites, 32.1% to website, and 2.7% to recruitment agencies, and only 4.1% of them use posting at university boards. Low percentage of postings at university boards means that majority of industry representatives refers to the job websites rather than university, and the weakest contribution between industry and university is in place. To prove the given statement of having less contribution, we received only 14 out of 146 responses on any job announcement or posting shared within the university education platform/intranet; nevertheless, 72 had negative responses and the remainder was left as blank.

For instance, UNEC (Azerbaijan State Economic University) has collaboration programs with industry and government agencies to prepare candidates/students based on the industry requirements (banking, finance, audit, consultancy, and taxation) which is a good example of how local university can contribute in depth with the industry and support local talents improvement and eliminates brain flow. The partnership with the industry is to support education, science programs, application of innovative achievements in education (science) and information technologies in business processes, joint implementation of innovation activities, use of modern technologies and scientific developments in education and industry, and organization of joint seminars, trainings, and conferences, students to have further internship accordingly.

Referring to the practice of UNEC, the same can be applied to tourism and hospitality specialty-led universities where chain and local hotels can be involved to the process and develop industry oriented to narrow down specific qualities and competencies. During the research process, it was also highlighted that ATMU being the specialized education unit in this field could not improve its market position so that other universities also have the faculties on tourism, social cultural services, and hospitality management (UNEC: Zaqatala Branch, Baku State University; Qerbi Kaspi University; Khazar University; and Odlar Yurdu University).

Referring to the student responses on the knowledge of languages and the skills, we received the same impactful responses (71 responded requires, 7 not important, and the remainder is blank or N/A) from the industry representatives where the language skills is a significant competency every graduate as well as candidate in the recruitment process should possess. Thus, based on the data of students and industry representatives, it is strongly recommended to exclude irrelevant/not impactful subjects out of curriculum and have language classes for the duration of studies and provide international-accepted knowledge certification upon successful completion. This will support the recruitment and hiring process during the work interview of the candidates to the given position as well as strengthen self-confidence and presentation skills of the applying party.

Further on the survey outcome made among students on the lack of opportunity to have the practical classes and gain real work experience, we would recommend implementing management trainee program that would contribute to long-term partnership with industry and university, as well as provide students with guaranteed commitment relevant to their skills and competencies. Alongside with the gained experience, the nominees would achieve Junior Management (assistant manager) level positions after successful completion of the program. There are 6-, 12-, and 18-month training programs for new managers, called management trainee programs, which take in fresh graduates after evaluating their skills and then put them through intensive training in every department and division of the company. Under the guidance of senior managers and executives, a management trainee performs many roles. It is their goal to learn everything they can, so they can go on to be future managers, frequently in specific sectors like marketing or sales.

With the projected changes in the industry for upcoming years and current post-COVID-19 era, having remote classes and industry down period, we have requested the industry representatives to advise on the skills, competencies, and knowledge expected from the future graduates/students which will be significant data to import to the education system in general, specifically tourism and hospitality industry specialization units. The responses are as follows:

- International experience
- Digitalization
- Creativity and novation
- Integration to new technology
- Change of education system and curriculum
- Responsibility at all times
- Flexible mindset
- Improvement of the soft skills, competencies
- Communication skills
- Organizational and management skills
- Agile approach
- Customer orientation

The quality assurance of higher education can also be assessed from the student's point of view. The purpose of the survey was to evaluate the quality and satisfaction level of the education student receive, identify the factors impacted the choice of tourism and hospitality specialty, and assess whether the knowledge and skills obtained at the university meet the industry requirements. Although the survey was sent approximately to 650 students, only 161 responded. The share of students by language division was 61.9% (100 students) Azerbaijani, 24.8% (40 students) English, and 13.3% (21 students) Russian. In turn, the grading of respondents by course was as follows: I: 22.4%; II: 18.6%; III: 33.5%; and IV: 15.5% and 9.9%, graduates. To the question "Who decided or advised to make the decision for the specialty you study/studied at?", 67.1% of students responded myself, where 8.7% answered parents, 14.9% defined based on the entrance score, 5% the teacher's advice, and 4.3% of respondents had "Other" as the answer.

The reason for choosing the tourism specialty (%) were the following: We may observe the reasons why did the students choose the tourism specialty and as show the results – 60.4% of them made decisions by themselves, and 24.5% plan to launch a business in this area. In our opinion, the independence of choice of tourism specialty by the majority of students and having business plans for future in this field show confidence and intentions of students for further actions.

According to the satisfaction rate of respondents to the quality of education they receive at the university, we had the results. Based on the survey 50.3% of the students agree, 18% strongly agree, 15.6% had neutral responses, 11.8% disagree and 4.3% strongly disagree with the education quality at the universities. However, to the question "What would be preferred to get provision of the qualified education at the University you study at?", the personal opinions and claims of students were related mainly with the lack of practical classes, less interactivity and creativeness of teachers, insufficient infrastructure and equipment of auditoriums, imperfect library and lack of relevant literature, incorrect administration decisions, etc. For instance, the comments were as follows:

- "I would like to see more practical classes and teachers working in the field of tourism."
- "New teaching methods should be used instead of traditional methods."
- "Decisions made by the administration of university must be correct."
- "Take into consideration the views of the students in organization and making changes in the processes of exams and classes."
- "Teachers should be more creative in the teaching process."
- "Reduce the hours of non-specialty classes."
- "Enrich the university library."
- "Organization of various individual development or specialty workshops and trainings."
- "Fully equipped, spacious, modern auditoriums."
- "Continuous teaching of foreign language subjects until the end of the fourth year and preference for specialty-oriented subjects" and so forth confirm the aforesaid.

Another comment was like: "Academic knowledge the students gained during the whole education process is not relevant with the work life. More improvements must be dispensed and more must be done by the teacher during the education process. No one teaches me what to try to improve the skills and competencies. For the customer relationships, improper behavior of the guests and the psychology of the guest are not the cases that were experienced or studied during the University classes."

For the university and industry integration process where the students have more involvement and engagement to the practical sessions, the participants of the survey had the following statements: "We are entitled to only 22 weeks for internship period, and this at best allows us to be involved in a group of travel agencies or five-star hotels. I did my internship both in a hotel and in a travel agency. And if you ask me what I learned there, my response would be that I learned how to take a photocopy, polish silver, glassware, cutleries. Unfortunately, in terms of gaining managerial and industrial experience, this practice couldn't meet my expectations. I do know that several of my groupmates had the same experience." "Internship is the backbone of the tourism and hospitality education. Without experience, you can't cope with hard-working conditions of this sphere. Therefore, we ourselves try to work hard and communicate with many guests in the process of work. Although the university invited industry representatives several times and they gave us seminars and workshops, but I do not think they're collaborating academic activities. It doesn't provide you with additional hard skills. They have to figure together to develop programs in line with the industry needs. Thence, after we select internship the industry as well as stakeholders also would help us improve our skills."

The views of students to the question "Do you think that the education you receive meets the requirements of the industry?", the correspondence of the received education to the requirements of the industry (%) and the satisfaction rate of students to the skills and methodology of teaching staff (%). The satisfaction rate of students to the skills and methodology of teaching staff is one of the essential criteria determining the quality of higher education. The opinions of respondents on what factors depend more on the quality of education they receive were as follows: 39.6%, university management and decisions; 9.4%, university internal infrastructure; 37.1%, the knowledge and skills of teaching staff; 6.9%, on decisions of the Ministry of Education; and 7%, indicated Other as the answer. The independence of choice selecting the tourism specialty by most of students shows that they relate plans for future with this sector. Since the answers to the question "What direction (specification, branch) of tourism and hospitality do you see your future at?" were as follows: travel agency: 33.1%, hotel industry: 33.1%, guide: 1.3%, restaurant business: 12.3%, teacher: 4.5%, and Other: 15.7% where the answers like Blogger, Marketer, and Startup tourism business were given as options. In turn, for the variety of responses regarding the future work position in tourism and hospitality, the respondents mainly see themselves as general manager (56.6%), food and beverage manager (16.4%), restaurant manager (11.2%), front office associate (3.9%), housekeeping (0.7%), and in other direction (11.2%).

The wishes of students about continuous teaching of foreign language subjects until the end of the fourth year is quite reasonable as the tourism industry is closely related to foreign languages. The variation of language knowledge of respondents to express independently thoughts besides the mother tongue were 24.1% for one language, 49.7% for two, 20.1% for three, and 3.1%–4% and 2.5% for no one.

In order to identify the students' opinion on the quality of higher education within the online survey, they were directed on to question "How should qualified and competitive education be?" And the most common thought was as follows:

- Preparation for the business world – organization classes through working in hotels.
- Individual approach to every student and new teaching methods should be applied instead of traditional methods.
- Prefer quality over quantity. Graduation of students in the required quality, not in the quantity required by the industry.
- It is impossible achieve high quality and competitiveness in educational without critical thinking and academic freedom.
- Get internships in foreign countries at the expense of the state and elaboration of new academic curriculums.
- Replacement of non-specialty subjects with specialty-related subjects or application the principle of free choice of subjects.
- It depends on teachers to increase students' interest in learning and not make difficult to study.
- Organization of practical classes, seminars, excursions to hotels, and meetings with managers of hotels or tourism companies that will increase students' worldview.
- Encourage and create interest among students; competitive classes should be held regularly among students, and the winners should be awarded, which in turn will create a competitive environment and increase the quality of education.

What approach the students/graduates should choose to be aligned with the ethics and not to harm the image of the industry unit? In practice, we see that our current system of higher education in most cases does not arrange students for thinking, and mainly it works in an examination vein, which makes the long-term employment of graduates uncompetitive. After entering to the "labor market," graduates are not able to solve appropriately their work responsibilities, since within the framework of upper education, they did not encounter every reasonable exam case that could be a phenomenon which will appear in life. They are less "capable" to think, so this case will remain unsolved for an extended time.

Thence, according to the questionnaire answers, we may summarize that the quality of higher education is influenced more by the systems and a series of practical classes, professional competencies and the methods of teaching staff, student motivation to learn, academic evaluation and university administrative management, library resources, university infrastructure, and appropriateness of classrooms. The respondents mainly feel that these factors are the most important and have a direct impact on the quality assurance process in higher education. Thus, as for the increasing the competitiveness and quality of higher education on tourism and hospitality, it is ought not only to organize a simple process of issuing diplomas but also induce essential knowledge and taking into consideration; these factors make structural changes are required within the education system.

18.5 Conclusions

Overall, we can note that the powerful coordinated effort between the training organizations and hence the industry would inevitably result to a proficient and useful input framework where the researchers and area connections will be fortified moreover. Thus, referring to all of the mentioned above, it seems reasonable to take the following actions to maintain the quality of higher education in the tourism and hospitality sphere.

First of all, sufficient practical and theoretical knowledge and skills are required in order to ensure high labor productivity, reducing the nonproductive costs of enterprise.

Future employees at the Azerbaijani tourism industry should develop professional skills necessary for work (language skills, initiative, responsiveness, organizational skills, sociability, leadership and teamwork skills, customer service skills, networking skills, communication skills, flexibility skills, commitment, can-do attitude, multitasking skills, cultural awareness, problem solving skills, empathy, creativity, problem solving, and compassion).

Furthermore, Azerbaijani universities need to adjust the curriculum and arrange classes to include the development of organizational skills, soft skills, self-introduction ability, presentation skills, eagerness to learn, and flexible mindset to respond to the Global Business changes. In addition, some management trainee program implementation should be done between the universities and industry in order to improve the ratio of skilled and customer-oriented workforce.

The extension of language classes at the universities for the whole duration of Bachelor degree needs to be conducted, and certification should be made automatic, since the industry requires to recruit candidates with improved language skills in order to maintain and develop communication level with international travelers and guests. There should be the practical recommendation of replacement of irrelevant subjects to industry required skills (mainly soft skills).

Moreover, some specialization per discipline (finance and hotel development, rooms operation management, guest experience management, customer satisfaction management, etc.) need to be narrowed down.

Awareness and distribution of the information about the latest trends, innovations, projects, and government programs in the field of tourism and hospitality also need to be fostered.

All in all, with the profound changes in the economy and business, the necessities to launch centralized talent development hub at universities based on some industry representatives' specific requirements are required for helping to create the future leaders in tourism and hospitality industry of tomorrow.

Acknowledgments This contribution was undertaken as a part of the BERCA research project of ADA University and the Embassy of the USA in Azerbaijan.

References

Abdullayev, A. (2015). Higher education, quality, and labor market. *525th Newspaper, 1*, 5.

Akan, Y., Arslan, İ., & Işık, C. (2008). The impact of tourism on economic growth: The case of Turkey. *Journal of Tourism, 9*, 1–24.

Akhundova, N. (2022). The impact of tourism on the intersectoral development in Azerbaijan. *WSEAS Transactions on Environment and Development, 18*, 725–736. https://doi.org/10.37394/232015.2022.18.68

Akvazba, E., Uhabina, T., & Cheremisina, E. (2015). The quality of education in a modern educational organization: Problems and prospects. *Modern Problems of Science and Education, 5*(3), 382.

Aliyeva, L. Z., & Rzayeva, S. M. (2020). The main directions of the development of tourism in the economy of Azerbaijan. *Economic and Social Development: Book of Proceedings, 2*, 663–673.

Barnett, R. (1992). *Improving higher education: Total quality care* (1st ed.). Open University Press.

Bavik, A., & Kuo, C. F. (2022). A systematic review of creativity in tourism and hospitality. *The Service Industries Journal, 42*(5–6), 321–359. https://doi.org/10.1080/02642069.2022.2041605

Beykan, C., & Kadir, C. (2015). The higher education system in the field of hospitality tourism management: A comparative analysis between Turkey, Switzerland and Germany. *Mediterranean Journal of Humanities, 5*(2), 165. https://doi.org/10.13114/mjh.2015214565

Bonanno, P., & Pozzoli, D. (2009). Early labour market returns to college subject. *Labor, 23*(4), 559–588. https://doi.org/10.1111/j.1467-9914.2009.00466.x

Čábelková, I., Abrhám, J., & Strielkowski, W. (2015). Factors influencing job satisfaction in post-transition economies: The case of The Czech Republic. *International Journal of Occupational Safety and Ergonomics, 21*(4), 448–456. https://doi.org/10.1080/10803548.2015.1073007

Dallen, T. (2019). *Handbook of globalisation and tourism* (1st ed.). Edward Elgar Publishing.

Doherty, G. (2008). On quality in education. *Quality Assurance in Education, 16*(3), 255–265. https://doi.org/10.1108/09684880810886268

ENIC-NARIC. (2022). *Higher education in Azerbaijan*. Available at: https://www.enic-naric.net/azerbaijan.aspx#anc05_5

Goh, E., & King, B. (2020). Four decades (1980-2020) of hospitality and tourism higher education in Australia: Developments and future prospects. *Journal of Hospitality & Tourism Education, 32*(4), 266–272. https://doi.org/10.1080/10963758.2019.1685892

Grecu, A., Gruia, A. K., Marin, M., Bănuță, M., Olteanu, C., Constantin, I., & Drăghici, C. C. (2019). Specificity of sustainable structural dynamics of local economy in Romanian tourist resorts. *Sustainability, 11*(24), 7155. https://doi.org/10.3390/su11247155

Guliyeva, A., Rzayeva, U., & Abdulova, A. (2020). Impact of information technologies on HR effectiveness: A case of Azerbaijan. *International Journal of Advanced Computer Science and Applications, 11*(2), 81–89. https://doi.org/10.14569/IJACSA.2020.0110212

Kazmina, L., Provotorina, V., Makarenko, V., & Petrenko, A. (2020). Innovative trends of further education in the tourism and hospitality industry. *E3S Web of Conferences, 210*, 12001. https://doi.org/10.1051/e3sconf/202021012001

Kim, E. J., Kim, J. J., & Han, S. H. (2021). Understanding student acceptance of online learning systems in higher education: Application of social psychology theories with consideration of user innovativeness. *Sustainability, 13*(2), 896. https://doi.org/10.3390/su13020896

Koc, E., Yumusak, S., Ulukoy, M., & Totas, A. (2014). Are internship programs encouraging or discouraging?, A viewpoint of tourism and hospitality students in Turkey. *Journal of*

Hospitality, Leisure, Sports, and Tourism Education, 15, 135–142. https://doi.org/10.1016/j.jhlste.2014.10.001

Kofler, I., Innerhofer, E., Marcher, A., Gruber, M., & Pechlaner, H. (2020). *The future of high-skilled workers: Regional problems and global challenges* (1st ed.). Palgrave.

Koreneva, M. (2010). The quality of education is the key to success in modernizing Russian education. *Vestnik MGOU, 2*(2), 79–83.

Lei, C., Hossain, M. S., Mostafiz, M. I., & Khalifa, G. S. (2021). Factors determining employee career success in the Chinese hotel industry: A perspective of job-demand resources theory. *Journal of Hospitality and Tourism Management, 48*, 301–311. https://doi.org/10.1016/j.jhtm.2021.07.001

Madina, H. (2019). *Azerbaijan's tourism sector – Opportunities and obstacles* (1st ed.). CESD Press.

Meridian Integrated Business Solutions Inc. (2019). *Analysis of regional tourism in Azerbaijan, restaurants, hotels & wineries*. Available at: https://azeriobserver.az/2020/05/08/azerbaijani-tourism-a-success-story-that-continues-to-unfold

Mínguez, C., Martínez-Hernández, C., & Yubero, C. (2021). Higher education and the sustainable tourism pedagogy: Are tourism students ready to lead change in the post pandemic era? *Journal of Hospitality, Leisure, Sport & Tourism Education, 29*, 100329. https://doi.org/10.1016/j.jhlste.2021.100329

Mishra, L., Gupta, T., & Shree, A. (2020). Online teaching-learning in higher education during lockdown period of COVID-19 pandemic. *International Journal of Educational Research Open, 1*, 100012. https://doi.org/10.1016/j.ijedro.2020.100012

Nureev, R., Volchik, V., & Strielkowski, W. (2020). Neoliberal reforms in higher education and the import of institutions. *Social Sciences, 9*(5), 79. https://doi.org/10.3390/socsci9050079

Radovic, D., Strielkowski, W., Wang, J., Cepel, M., & Rausser, G. (2017). Economic analysis of sustainable tourism: A case study of Nottingham. *Transformations in Business & Economics, 16*(2B), 703–714.

Rinehart, G. (2016). *Quality education: Why it matters, and how to structure the system to sustain it* (1st ed.). Stormwatch.

Sandybayev, A. (2020). The impact of e-learning technologies on student's motivation: Student centered interaction in business education. *International Journal of Research in Tourism and Hospitality (IJRTH), 6*(1), 16–24. https://doi.org/10.20431/2455-0043.0601002

Sharifovich, Y. K. (2021). Ways to improve information and consulting services in the field of tourism. *International Journal of Business, Technology and Organizational Behavior (IJBTOB), 1*(2), 98–104. https://doi.org/10.52218/ijbtob.v1i2.15

State Statistical Committee of the Republic of Azerbaijan. (2021). *Statistical yearbook*. Available at: www.stat.gov.az

Strietska, I., & Tessaring, M. (Eds.). (2005). *Trends and skill needs in tourism. The European Centre for the Development of Vocational Training* (1st ed.). Office for Official Publications of the European Communities.

The Ministry of Science and Education of the Republic of Azerbaijan. (2022). *Study in Azerbaijan*. Available at: https://studyinazerbaijan.edu.az/web/overview-on-education-system

TKTA. (2022). *Our activities*. Available at: https://www.tkta.edu.az/p/how-we-work

Tribe, J. (2005). *An international handbook of tourism education* (1st ed.). Elsevier Ltd.

Tsinidou, M., Gerogiannis, V., & Fitsilis, P. (2010). Evaluation of the factors that determine quality in higher education: An empirical study. *Quality Assurance in Education, 18*(3), 227–244. https://doi.org/10.1108/09684881011058669

Vedomosti. (2020). *What do the students need from universities in the 21st century?* Available at: https://www.vedomosti.ru/management/articles/2020/01/12/820388-nuzhno-studentam

Zhang, H., & Yang, Y. (2021). Does tourism contribute to real estate boom? A DSGE modeling in small open economy. *Journal of Hospitality & Tourism Research, 45*(1), 257–279. https://doi.org/10.1177/1096348020932993

Chapter 19
Leadership in Business and Economics Through the Development of Popular Scientific Research and Practice-Oriented Education After COVID-19

Liudmila Kashirskaya, Nadezhda Nikolaeva, and Aleksandr Platitzyn

Abstract Our chapter focuses on the contemporary issues of the integration of leadership in economics and business through the development of popular scientific research and practice-oriented education in the post-COVID-19 era. The pandemic brought new threats and challenges for the world's economies, in particular effecting the labor markets. Thence, the emphasis on the popular scientific research and practice-oriented education that is based on social partnership between the society, entrepreneurs, and organizations becomes a must for regaining leadership for business and economics in the new conditions. Our results demonstrate that COVID-19 pandemic contributed to making inequalities between the developing and developed countries worse around all the world by altering the balance on the economic relations and labor markets. Therefore, there is a pressing need for the scientific and technological systems around the world to be fostered and supported by the governments enabling them to offer important answers to challenges that were presented by the pandemic. Digitalization of education can also become an answer to these issues and help to regain the leadership paths for business and economics in the post-pandemic world.

Keywords Economic integration · Practice-oriented education · COVID-19 · Leadership

---

L. Kashirskaya (✉)
Audit and Corporate Reporting Department, Faculty of Taxes, Audit and Business Analysis, Financial University under the Government of the Russian Federation, Moscow, Russian Federation
e-mail: lvkashirskaya@fa.ru

N. Nikolaeva
Volga Region State University of Service, Togliatti, Russian Federation

A. Platitzyn
Togliatti State University, Togliatti, Russian Federation

W. Strielkowski (ed.), *Leadership, Entrepreneurship and Sustainable Development Post COVID-19*, Springer Proceedings in Business and Economics, https://doi.org/10.1007/978-3-031-28131-0_19

JEL Classification I20 · F02 · J40

19.1 Introduction

The recent COVID-19 pandemic made the integration of economy and business into the development of leading and popular scientific research and practice-oriented education especially relevant (He et al., 2021; Guliyeva et al., 2021; Csáki et al., 2022). This can be found through the direct comparisons between the performance of students compared with a year before the pandemic emerged (when in-person instruction was in place) with the changes in teaching and learning introduced during the pandemic (partial/complete substitution of in-person instruction by online instruction) (Kumar et al., 2021). The surveys of teachers and students conducted by many researchers worldwide explore students and teachers' previous experiences, preparations, interests, supports received, and challenges encountered with remote teaching, learning, and teaching during the COVID-19 pandemic (Zou et al., 2021). For example, the teachers' surveys ask participants to describe whether or not they had prior online/remote teaching experiences before the COVID-19 pandemic and how those experiences prepared them to teach during the pandemic (Walsh et al., 2021). Initial qualitative research focuses on experiences in teaching and learning during COVID-19 and its impacts and outcomes for students (Suliman et al., 2021).

Let us consider underscored the importance of identifying resources to teach remotely and assess learning. In the face of COVID-19 shutdowns, Ministries of Education worldwide implemented three key interventions to support continuity of learning throughout COVID-19 using different digital e-learning platforms made available for students who were benefiting from in order to continue learning during COVID-19-based school lockdowns (Meira et al., 2022). The COVID-19 response plan included such measures as developing TV programs at primary education levels, strengthening the e-learning platforms at lower primary and lower middle primary levels, and supporting the use of Zoom, Microsoft Teams, Skype and other popular platforms for video-conferencing lessons at higher primary and higher middle education levels (Yeap et al., 2021). These novel learning opportunities surely presented overwhelming challenges for educators, schools, institutions, and governments regarding e-learning in a variety of ways (Singh et al., 2021). Generally, there are several pathways created by the COVID-19 pandemic for those who were not prepared for, and far off plans to adopt, e-learning systems (Strielkowski, 2022). Online teaching, remote learning, and continuous learning have been the positive externality of the COVID-19 pandemic, in spite of challenges faced by both educators and students (Roman & Plopeanu, 2021).

When assessing the impacts of COVID-19 pandemic on teaching and learning worldwide, it can be concluded that while several studies were conducted, suitable pedagogies and platforms in the context of the various grade levels in higher, intermediate, and elementary education should be explored further, especially for developing countries (Pokhrel & Chhetri, 2021). In the face of the sudden changes to practical training for students caused by the COVID-19 pandemic, it is essential to understand students' experiences related to participation in practically oriented classes and draw conclusions which would enable implementation of innovations that would, in turn, enable optimal development of required vocational competencies in an education process, despite the limitations of the pandemic (Dziurka et al., 2022). This is one of many directions for change following experiences of primary and higher education during the COVID-19 pandemic, including increased utilization of current technologies in nursing education, such as online classes and the use of game-based computer learning applications requiring the students to address problems via model cases, virtual simulations, or such aspects as e-health solutions, and telenursing (Poudevigne et al., 2022). The landscape for education after the pandemic has changed in ways that would promote teaching of new skills to employees, regardless of where they might be (Mok et al., 2021). Professional development training uses an expertise-based approach, both in classrooms and in clinical settings. Students fulfill an extensive clinical training requirement, which is embedded in clinical integration competencies (in the sense of the professional, practice-oriented outcomes). Today's educational requirements include acquiring knowledge, skills, and clinical proficiency, as well as a wide range of fundamental behaviors for the profession. Continued education requirements should emphasize increasing knowledge, skills, and abilities related to the sport of athletic training (Hussein et al., 2021). As instructional leaders, teachers and lecturers need to move beyond routine practices for wellness in the post-COVID-19 pandemic world and adopt a multimodal approach for tracking trainee and faculty health within the context of both physical and social distancing (Pathak et al., 2021).

It is essential that today's educators are prepared to navigate this interrupted training time with a plan that helps and supports residents as they strive to become competent, independent practitioners. Thanks to the experience gained from the pandemic, experts from the fields of law, economics, nursing, medicine, and dentistry meet virtually to discuss current gaps in the scope of practice for healthcare practitioners, what can be learned from the COVID-19 disaster, and how scopes of practice can be redefined to better serve the needs of communities and public health (Vargo et al., 2021). During the global COVID-19 pandemic, work for the professionals at various organizations has changed dramatically with an abrupt transition to remote work, as training and education typically occurs in in-person settings (Kraft et al., 2021). Future work should examine how the organizations in life sciences are structuring their remote work and remote learning in the post-COVID-19 world. Although many employees learned by doing in the early phases of the COVID-19 crisis, or received fast-and-dirty training, continuing to work remotely is likely to remain a skill-building challenge (Taufiq-Hail et al., 2021). These findings also offer insights on how a COVID-19 global pandemic requires practitioners learning flexibility both to start learning themselves and to support learning flexibility for others within an organization. Students explicitly highlighted how, although unable to provide clinical education at hospitals during the COVID-19 pandemic, instructors attempted to provide them with the highest levels of theory and practice skills by seeking alternate methods of instruction, in order for students in future jobs to be able to deliver the highest levels of services.

19.2 Post-Pandemic Integration of Business and Economics

After the COVID-19 pandemic, innovations are developing at an unprecedented pace, and the characteristics and evidence exist that these are showing positive results both for the business companies and for the economy (Kesavan et al., 2022). This study examines the problems associated with how innovation research develops in the field of business and economics, and what characteristics they change, following the coronavirus disease pandemic. To investigate the changes in characteristics of the innovation literature prior and following the COVID-19 pandemic, the vast amount of academic and professional literature on the subject can be analyzed using the academic papers, briefs, policy reports, and industry analyses (Klein et al., 2022).

Additionally, one of the aims of this chapter is to explore the shifting emphasis of the literature about innovation in the fields of business and economics following the COVID-19 pandemic with the aid of bibliometric analysis, as well as explore the informative review on the innovation studies, which is the fast-developing research orientation as technology advances and dynamic environments. While efforts are underway to develop drug interventions for COVID-19, the social and behavioral sciences may offer valuable insights for managing a global pandemic and its impacts (Abbas et al., 2021). It is critical that academic research is conducted to assess the impacts of a COVID-19 pandemic in terms of quality of life as well as in terms of the economy, psychology, and environment (Epifanio et al., 2021). The COVID-19 pandemic could have serious impacts on the world from a perspective of quality of life, sustainable development in politics, environment, and economy, as well as on the world economy. Not only the economic losses but also social parameters such as changes in psychological sustainability are affected by the COVID-19 pandemic (Faulks et al., 2021). In addition to the direct human suffering caused by the COVID-19 crisis itself, as well as millions losing their livelihoods, the COVID-19 pandemic has highlighted several critical vulnerabilities in our social and economic systems (Shittu et al., 2022). Left unchecked, global ecological emergencies, such as climate change and the loss of biodiversity, have the potential to inflict far greater social and economic harms than COVID-19. The vulnerabilities exposed by the COVID-19 crisis could serve to underscore why ecological issues were becoming the world's highest policy priorities prior to the COVID-19 attacks (Watson et al., 2022). In this time of pandemic, we need to consider how we might respond to both the present and the post-COVID-19 scenario. In the longer run, the challenges are reviving the economy and helping the people who are the worst affected by COVID-19. The challenges in re-igniting the world economy after an economic downturn caused by COVID-19 are, of course, different. To rescue economies from the COVID-19 pandemic, as well as their citizens' mental health, a number of countries have developed strategies, but it will take years of struggles to restore economic stability. More generally, the pace and depth of the COVID-19 crisis has demonstrated that the central principle of the global economy prioritizing short-term economic growth and efficiency over long-term sustainability can come at enormous social costs. According to Ahmad et al. (2020), there is a need for global economic integration in order to address the consequences of the COVID-19 pandemic. Global interconnectedness has helped create enormous economic and social benefits over decades, although unevenly, but it has also contributed to the rapid spread of the COVID-19 crisis. Nundy et al. (2021) mention how the COVID-19 pandemic has changed the way humans live on a global scale, by impacting on their social, economic, and ecological environments. This explains how COVID-19 has devastated communities and economies worldwide, profoundly changing how we live and work.

In general, it has become clear that leaders of economies and businesses across the globe must take advantage of the COVID-19 pandemic crisis to rethink their supply networks, take steps to understand their vulnerabilities, and then act to create resilience (Strielkowski et al., 2022). In the face of the COVID-19 pandemic, it is also imperative that the leadership of business and the economy sector be redefined, in order to redraft this topic and situate it within the context of the Sustainable Development Goals (SDGs) defined and outlined by the United Nations and widely used to assess the sustainable development perspectives.

Another important question related to the recovery of business and economics is whether the lessons from the COVID-19 pandemic in 2020–2021 offer us a template or way forward to facilitate green transition, meaning the move toward sustainable, low-carbon economies, relative to the pre-crisis scenario. Many researchers unanimously agree that COVID-19 would ultimately be bad for the environment, contending that post-pandemic carbon emissions could rise above pre-pandemic levels, while the world engaged in dirty rebuilding (Corfee-Morlot et al., 2021; Scott, 2021). In assessing whether research on COVID-19 and air pollution is focused on informed action, glib attributions of changes in air quality during the pandemic itself reflects the larger problem of air pollution literature regarding causality assignment for complex, interrelated phenomena (Fos et al., 2021). Research on air pollution responses to the COVID-19 pandemic can meet the three characteristics of systems-oriented resilience research described earlier to the extent that it will capture physical and social components influencing both pollution concentrations and impacts, take into account the dynamic nonlinearities and feedbacks occurring in both atmospheric chemistry and in individual and institutional responses, and resolve the causation of changes in the atmosphere and multiple potential physical and social drivers, as well as resolve the underlying causes and effects in ways that could inform effective action (Selin, 2021). In addition, this research fulfills the three characteristics of systems-oriented sustainability research described above to the extent that it captures the physical and societal components that influence both pollution concentrations and impacts, accounts for dynamic nonlinearities and feedbacks that occur both in the chemistry of the atmosphere and in individual and institutional responses, and addresses causality of air pollution changes and their multiple potential physical and societal drivers in ways that can inform effective action. These are the systemic components that are relevant for both the environmental properties of the atmosphere, as well as for the environment itself (Payne-Sturges et al., 2021). The systems that are relevant for both the ecological properties of the atmosphere, as well as environmental properties. The systems that produce emissions, as well as environmental properties. The systems that produce emissions, including knowledge and technologies, which produce emissions, as well as in terms of environmental properties. Emission changes also reflect changes in the local and global demand for industrial goods. Many system components related to the pandemic itself and air pollution system involve not only the environmental properties of the atmosphere but also knowledge and technologies that create emissions and human and institutional factors that affect exposure and impacts.

19.3 Practice-Oriented Education Based on Social Partnership

Within the context of the post-pandemic recovery, looking at social innovation through an educational research lens illuminates the strong relationship between education, creativity, and innovation. By sharing experiences from work, lessons learned, solutions to problems in the field, and the creative ideas needed to address those problems, students are able to acquire a broad range of knowledge in the context of a social learning environment (Chankseliani & McCowan, 2021). Social learning promotes collaboration organically, particularly within a learning online environment where insights are shared and valued throughout an entire organization. As more training opportunities arise within a workplace, a collaborative social learning environment creates opportunities to troubleshoot, solve problems, implement projects, conduct research, and innovate in order to come up with answers, even when there is no coach or mentor (Valk & Planojevic, 2021). While, a formal training environment remains necessary to achieve particular training outcomes, it is critical that organizations in the post-pandemic world take advantage of platforms that allow social and informal learning, in which learners connect, share, collaborate, and share ideas for solving problems. This shift in training practices can happen only in environments that place high value on collaboration and discussion (Zajac et al., 2022). In universities and schools, changing a teaching theory in the post-COVID era is a far deeper transformation. The approach rejects the traditional learning models, preferring more practical approaches to learning. In this sense, a learner-centered approach is more oriented toward evidence-based teaching and ongoing evaluation. Standardized curricula designed to reproduce outcomes typically remove any opportunity for spontaneous inquiry, taking away potentially teachable moments from students and teachers alike with cookie-cutter approaches to education in the classroom (Bennett & McWhorter, 2021). Given that a lecture-based approach is not making sense anymore in traditional education, it makes even less sense in online education models. Online education cannot be focused on delivering content that replicates the lecturing-based, faculty-centric, face-to-face, transferrable models. From that standpoint, the school is not a preferred learning focus.

Community-based learning after the coronavirus shutdowns is also motivated by a belief that all communities possess inherent educational assets and resources which educators can draw upon to improve students' learning experiences. The ability for students to bring reflective knowledge and ideas from their own observations and social interactions into the classroom topic and academic discussion adds depth to learning experiences for individuals as well as to the content of courses. Teachers are supported by tools and professional development ICT courses, and by a teaching community, as well as by other supporting materials including a section on best practices in teaching (Teras et al., 2020).

The combination seems promising for future efforts supporting social innovations in the social and educational sectors. Parts of this milieu are the regional networks for social innovation (including advocates, initiators, investors, knowledge providers, facilitators, and entrepreneurial actors) and also a domain-specific milieu, that is, national, regional, and local educational systems. There are collaborative efforts across nearly all aspects of healthcare, such as patient-care and healthcare collaborations, shared learning, interprofessional cooperation in practice and education, healthcare value cooperation, enterprise cooperation, collaborative efforts on research, and financing. Collaboration is a term that is often used in research, clinical practice, and education in the healthcare professions. The fields of research, education, and clinical practice are related; research informs education, which, in turn, influences clinical practice as well as business practice (Roux et al., 2021).

In addition, the research on practices that may help educators to address individual variation, manage adversity, and foster resilience so that schools are capable of supporting all children in learning and finding affirmative pathways into adulthood can also be examined and yield interesting results (Stockinger et al., 2021). It is a child's or a student's relationships with parents, teachers, and environment that spark learning. If we are going to foster the hunger to learn, we need to create environments where students and teachers feel safe stepping outside of what is already known – environments that reflect our values and celebrate students and teachers as unique, creative individuals. We suggest looking to initiatives in related, intertwined areas of social services provision, which are heavily regulated in most European countries and reliant on state financing, to analyze the occurrence, emergence, and first success stories, and to identify lessons for the field of education. In the field of cross-organizational structures and processes, the top-right quadrant contains innovation learning labs, where companies try to develop new products and services, working in collaboration with universities and building upon the findings of research.

19.4 Labor Market Growth After the Pandemic

The labor market revival after COVID-19 is likely to be slow and will take significant time to bring key indicators of the labor world back to levels seen prior to the health crisis, with more needed time for them to meet targets that the Sustainable Development Goals (SDGs), the European Central Bank (ECB), and International Labor Organization (ILO) have all indicated (Dokic, 2022). In 2020, the health crisis dealt unprecedented blows to the world's economies and labor markets, leading to the largest contractions in the past dozens of years, at significant costs, affecting the economy, labor, and society, with significant costs, in terms of economy, labor, society, and the overall economic output (Nayak et al., 2021). In response to the COVID-19 pandemic, the socioeconomic impacts of the new COVID-19 on different aspects of the global economy can be summarized. It might be relevant to explore how sectors including healthcare, airline travel, consumer retail, and asset management may be planning for the longer-term, even as sectors including healthcare face the immediate devastation from the COVID-19 pandemic. This would require the researchers to focus on the analysis related to the pandemic this week examines a possible new era of global prosperity, one of higher rates of economic growth and increased innovation in healthcare (Sheng et al., 2021). Our study suggests several major takeaways for business owners and entrepreneurs about the business landscape after the pandemic, offering fresh perspectives on digital transformation, the future of work, transparency, and resilience. Beyond the immediate crisis response, the COVID-19 pandemic must serve as an impetus for sustained gains and the acceleration of implementation of long-overdue measures that put the world on a path toward greater sustainability and that make the worlds' economies more resilient to future shocks (Ambrogio et al., 2022). For emerging markets and developing countries, many of which are facing daunting vulnerabilities, strengthening public health systems, countering challenges from informality, and implementing reforms will be essential to sustaining robust, sustained growth after the health crisis subsides. Achieving better standards of living, jobs, education, healthcare, food, and consumer goods will depend on changes in the world people now confront.

This requires additional green policies to spur jobs and growth, underpinned by an active fiscal policy to promote jobs, with labor-intensive investment projects and a focus on ecological sustainability, that should be accompanied by industrial and technical policies to build the nations productive capacity and boost its competitiveness (Morea, 2020). Part of a big reboot, as companies and countries imagine the post-COVID-19 future, is to redefine how technology serves our needs. Many businesses navigating the pandemic have already announced plans to boost productivity through investments in automation and accelerate the effects of technological megatrends, all of which will present further challenges for workers who are already vulnerable, such as the lower-skilled, women, labor market newcomers, and migrants (Yang et al., 2021). In particular, evidence presented in many papers and policy briefs suggests the lower-skilled are likely to beat particularly risk from negative impacts from the COVID-19 crisis, because they are most vulnerable to job losses and high risks associated with automation (Collings et al., 2021). Regarding employment among youth in a time of health crisis, evidence shows that the pandemic has dramatically affected the employment of youth aged between 15 and 24, especially those entering the labor market for the first time. Moreover, the pandemic has had more negative effects on demand for jobs that require less qualifications, but this is not the case for the five OECD countries (Papadakis et al., 2022). The number of jobs opened has declined for many occupations, but has increased in core service occupations. In line with earlier work by OECD countries examining the first months of the pandemic, results from this policy brief indicate that the pandemic has also had a heterogeneous effect on the number of jobs advertised online by occupation. The pandemic has devastated the tourism sector, pushing the hotel and hospitality sector toward the highest level of job losses of any large sector (Barbour et al., 2021). In many countries (e.g., in the United States), Hispanic or Latino workers who lost jobs during the pandemic could not find jobs for months to years after the final COVID-19 assistance was distributed. The effects of COVID-19 economic downturn also will not go away in the coming years (Riley et al., 2021). The COVID-19 pandemic has also raised fears about a coming economic crisis and economic downturn. That is how much experts expect another pandemic such as coronavirus-related disease, which has already caused potentially $16 trillion in worldwide economic damage, to cost (Assaad & El-adaway, 2021). The pandemic has intensified a trend toward employers playing a larger role in the financial, physical, and psychological well-being of their employees. Even before the pandemic, organizations were increasingly using unconventional tools to monitor employees, but this trend will accelerate with the new tracking of telecommuters and collection of data on employees' health and safety. The results stemming from the related research suggest that organizations will continue expanding the use of contingent workers in order to retain greater workforce flexibility after COVID-19, and they will also look at adopting other job models that they saw through the pandemic, such as talent sharing and compensation for work.

19.5 Conclusions

In summary, it becomes quite clear that the COVID-19 pandemic might have altered the world's labor market by increasing inequalities across the board, halting progress in, for example, poverty alleviation and global clean energy. Whether it is causing lost jobs or shuttered schools, the pandemic is threatening to reverse decades of progress in achieving great economic and social equality. The COVID-19 pandemic is obviously making inequalities worse around the world, and because the worst effects of the pandemic are not being reported to developed countries, many will have issues with their development. The pandemic is never far from the forefront in terms of the catastrophic inequalities worldwide with respect to COVID-19 vaccine access. Now, just in a short period of time, the COVID-19 pandemic has unleashed unprecedented crises, further disrupting the progress in sustainable economic development and tackling the climate change, with the poorest and most vulnerable people around the world being hit hardest. Using updated data and assessments, this annual snapshot of progress across the Sustainable Development Goals (SDGs), it become apparent that the poorest and most vulnerable citizens – including children, older people, people with disabilities, migrants, and refugees – are the ones who were hit by the effects of the COVID-19 pandemic. The COVID-19 pandemic holds a mirror up to society, revealing existing flaws in our digital infrastructure, as well as gross inequality in the distribution of resources and opportunities for using them.

In general terms, the crisis is symptomatic of capitalisms failure to evolve into an economically and socially resilient model. For European Union (EU), the situation created by the coronavirus disease-induced crisis is seen as an opportunity to tackle education inequality as an outstanding policy challenge, not a phenomenon that is exclusively related to the COVID-19 crisis. Moreover, the impacts of the COVID-19 pandemic on educational systems globally are enormous and can reflect on the potential role of the characteristics of national education systems in mediating countries responses to common external challenges created by the crisis. In different educational systems, the crisis caused by coronavirus has generated debates about the limits and potentialities of learning platforms and online learning, the need for revisiting the role of exams and external assessment, and/or about how effective policies can be developed to tackle the new forms of education inequalities. In addition, the crisis has also highlighted the importance of developing students' digital skills as a key strategy for dealing with a post-COVID world and mitigating the economic impacts of the pandemic. In addition to the direct effects on education from the coronavirus disease crisis, economic and social crises stemming from the pandemic are expected to raise the levels of poverty among children and young people in many countries, decisively impacting educational inequalities. As the number of children infected by COVID-19 increases, a growing number of them will be living with the chronic health conditions which it is increasingly becoming clear are associated with the survival of the virus. COVID-19 is the great leveler, because anyone can get it and be affected, but it is also the great inequality-breaker, because it will hit poor people in overcrowded slums and the informal economy, who have little to no access to healthcare and, for many, little food, for the near future. The pandemic is also halting progress in the field of clean energy, curbing investments and delaying expansions in clean energy technologies. The global effort, which has been in place for many years to improve peoples' lives around the world through the implementation of 17 Sustainable Development Goals (SDGs) by 2030, is already falling behind schedule. According to the 2020 SDG report, the world had made progress – though uneven and not enough to achieve the goals – in areas like improving maternal and child health, expanding access to electricity, and increasing female representation in governance. Even before the COVID-19 pandemic took hold, the targets embedded in the SDGs were ambitious and broad-ranging: from eliminating extreme hardships to significantly reducing inequality to changing the trajectory toward protecting nature. The list was headed by the twin, both green and digital, transformations, and COVID-19 has obviously affected both.

The twin pressures to prosper from the current situation and to bounce back after the pandemics economic failures are providing a fresh impetus for the complete transformations into "go digital." We contend that over the next decade, it is the public relations departments of go digital multinational enterprises who can, and should, be doing the most to ensure that investments in the digitalization are converging with the SDGs progress. Global communication strategies are also needed to inform communities, policymakers, and investors about the value of the digitalization and the technologies required for it, for example, for participative digital healthcare and data-gathering tools to prepare for and respond to the pandemic (proposed by the SDG number 3). We think that digital technologies have the potential to advance and improve people's lives and standards of living globally; facilitate access to government services, quality education, and better and new jobs; and to lower barriers, decrease inequality, and foster peace. If we look at what happened during the COVID-19 pandemic, we see our scientific and technological systems being mobilized and reinvigorated, offering answers to challenges that were presented, even as they worked in extremely difficult circumstances. For instance, online shopping has been developed in recent decades, and was ready to go when the crisis struck.

We are living in crucial times, facing all kinds of climate, pandemic, and demographic threats, yet we also have greater technological capabilities than we have ever had in trying to address all kinds of climates. While developed countries had means of early intervention to protect people and businesses – with emergency stimulus packages and social protections – these measures are not long-term. Additionally, it has to be remembered that some groups of population – like women or migrants – were disproportionately more severely affected by the pandemic, because they are a greater share of workers in sectors heavily hit by the COVID-19, including accommodation and food services, as well as frontline professions, such as the health and social care sectors. These aspects also need to be studied carefully and taken into account when planning for the strategies needed to face new disasters and unexpected events in the future.

References

Abbas, J., Wang, D., Su, Z., & Ziapour, A. (2021). The role of social media in the advent of COVID-19 pandemic: Crisis management, mental health challenges and implications. *Risk Management and Healthcare Policy, 14*, 1917. https://doi.org/10.2147/RMHP.S284313

Ahmad, T., Baig, M., & Hui, J. (2020). Coronavirus disease 2019 (COVID-19) pandemic and economic impact. *Pakistan Journal of Medical Sciences, 36*(COVID19-S4), S73. https://doi.org/10.12669/pjms.36.COVID19-S4.2638

Ambrogio, G., Filice, L., Longo, F., & Padovano, A. (2022). Workforce and supply chain disruption as a digital and technological innovation opportunity for resilient manufacturing systems in the COVID-19 pandemic. *Computers & Industrial Engineering, 169*, 108158. https://doi.org/10.1016/j.cie.2022.108158

Assaad, R., & El-adaway, I. H. (2021). Guidelines for responding to COVID-19 pandemic: Best practices, impacts, and future research directions. *Journal of Management in Engineering, 37*(3), 06021001. https://doi.org/10.1061/(ASCE)ME.1943-5479.0000906

Barbour, N., Menon, N., & Mannering, F. (2021). A statistical assessment of work-from-home participation during different stages of the COVID-19 pandemic. *Transportation Research Interdisciplinary Perspectives, 11*, 100441. https://doi.org/10.1016/j.trip.2021.100441

Bennett, E. E., & McWhorter, R. R. (2021). Virtual HRD's role in crisis and the post COVID-19 professional lifeworld: Accelerating skills for digital transformation. *Advances in Developing Human Resources, 23*(1), 5–25. https://doi.org/10.1177/1523422320973288

Chankseliani, M., & McCowan, T. (2021). Higher education and the sustainable development goals. *Higher Education, 81*(1), 1–8. https://doi.org/10.1007/s10734-020-00652-w

Collings, D. G., Nyberg, A. J., Wright, P. M., & McMackin, J. (2021). Leading through paradox in a COVID-19 world: Human resources comes of age. *Human Resource Management Journal, 31*(4), 819–833. https://doi.org/10.1111/1748-8583.12343

Corfee-Morlot, J., Depledge, J., & Winkler, H. (2021). COVID-19 recovery and climate policy. *Climate Policy, 21*(10), 1249–1256. https://doi.org/10.1080/14693062.2021.2001148

Csáki, C., Szabó, I. B., Szabó, Z., Csillik, O., & Gábor, A. (2022). Reconsidering the challenges of BIS education in light of the COVID pandemic. In W. Abramowicz, S. Auer, & M. Stróżyna (Eds.), *Business information systems workshops. BIS 2021. Lecture notes in business information processing* (Vol. 444). Springer. https://doi.org/10.1007/978-3-031-04216-4_6

Dokic, M. (2022). World economy in the time of pandemic: Consequences of COVID-19 on world output, trade and employment. *Economics of Sustainable Development, 6*(1), 57–72. https://doi.org/10.5937/ESD2201057D

Dziurka, M., Machul, M., Ozdoba, P., Obuchowska, A., Kotowski, M., Grzegorczyk, A., & Dobrowolska, B. (2022). Clinical training during the COVID-19 pandemic: Experiences of nursing students and implications for education. *International Journal of Environmental Research and Public Health, 19*(10), 6352. https://doi.org/10.3390/ijerph19106352

Epifanio, M. S., Andrei, F., Mancini, G., Agostini, F., Piombo, M. A., Spicuzza, V., & La Grutta, S. (2021). The impact of COVID-19 pandemic and lockdown measures on quality of life among Italian general population. *Journal of Clinical Medicine, 10*(2), 289. https://doi.org/10.3390/jcm10020289

Faulks, B., Song, Y., Waiganjo, M., Obrenovic, B., & Godinic, D. (2021). Impact of empowering leadership, innovative work, and organizational learning readiness on sustainable economic performance: An empirical study of companies in Russia during the COVID-19 pandemic. *Sustainability, 13*(22), 12465. https://doi.org/10.3390/su132212465

Fos, P. J., Honore, P. A., & Honore, R. L. (2021). Air pollution and COVID-19: A comparison of Europe and the United States. *European Journal of Environment and Public Health, 5*(2), em0074. https://doi.org/10.21601/ejeph/9706

Guliyeva, A., Rzayeva, U., & Guliyeva, G. (2021). University life during pandemic COVID-19: Attitudes of staff and students towards distance education. *SHS Web of Conferences, 92*, 01011. https://doi.org/10.1051/shsconf/20219201011

He, W., Zhang, Z. J., & Li, W. (2021). Information technology solutions, challenges, and suggestions for tackling the COVID-19 pandemic. *International Journal of Information Management, 57*, 102287. https://doi.org/10.1016/j.ijinfomgt.2020.102287

Hussein, R., Lin, E. C., & Grindrod, K. (2021). Effects of computer-based education on health professionals' knowledge, skills, and behavior: A scoping review. *Journal of the American Pharmacists Association, 61*(3), e44–e68. https://doi.org/10.1016/j.japh.2021.01.036

Kesavan, N., Seenuvasaragavan, S., Burnette, K., Heim, A. B., Patel, R. J., Johnston, J., Cao, X., Sakandar, H. A., Zhuang, Y., Oi, K. K., Zhi, Y., Diao, Y., & Strielkowski, W. (2022). Pandemic-inspired policies. *Science, 377*(6601), 22–24. https://doi.org/10.1126/science.add2046

Klein, M., Gutowska, E., & Gutowski, P. (2022). Innovations in the t & l (transport and logistics) sector during the COVID-19 pandemic in Sweden, Germany and Poland. *Sustainability, 14*(6), 3323. https://doi.org/10.3390/su14063323

Kraft, M. A., Simon, N. S., & Lyon, M. A. (2021). Sustaining a sense of success: The protective role of teacher working conditions during the COVID-19 pandemic. *Journal of Research on Educational Effectiveness, 14*(4), 727–769. https://doi.org/10.1080/19345747.2021.1938314

Kumar, A., Sarkar, M., Davis, E., Morphet, J., Maloney, S., Ilic, D., & Palermo, C. (2021). Impact of the COVID-19 pandemic on teaching and learning in health professional education: A mixed methods study protocol. *BMC Medical Education, 21*(1), 1–7. https://doi.org/10.1186/s12909-021-02871-w

Meira, D., Azevedo, A., Castro, C., Tomé, B., Rodrigues, A. C., Bernardino, S., & Gomes, M. (2022). Portuguese social solidarity cooperatives between recovery and resilience in the context of COVID-19: Preliminary results of the COOPVID project. *CIRIEC España, 104*, 233–266. https://doi.org/10.7203/CIRIEC-E.104.21486

Mok, K. H., Xiong, W., Rahman, B. A., & H. N. (2021). COVID-19 pandemic's disruption on university teaching and learning and competence cultivation: Student evaluation of online learning experiences in Hong Kong. *International Journal of Chinese Education, 10*(1), 22125868211007011. https://doi.org/10.1177/22125868211007011

Morea, J. P. (2020). Post COVID-19 pandemic scenarios in an unequal world challenges for sustainable development in Latin America. *WORLD, 2*(1), 1–14. https://doi.org/10.3390/world2010001

Nayak, P., Mishra, V., Singh, M., & Tambuwala, M. M. (2021). Impacts and consequences of COVID-19 epidemic on global economy. *Coronaviruses, 2*(1), 77–88. https://doi.org/10.2174/2666796701999200905094151

Nundy, S., Ghosh, A., Mesloub, A., Albaqawy, G. A., & Alnaim, M. M. (2021). Impact of COVID-19 pandemic on socio-economic, energy-environment and transport sector globally and sustainable development goal (SDG). *Journal of Cleaner Production, 312*, 127705. https://doi.org/10.1016/j.jclepro.2021.127705

Papadakis, N., Drakaki, M., Saridaki, S., Amanaki, E., & Dimari, G. (2022). Educational capital/level and its association with precarious work and social vulnerability among youth, in EU and Greece. *International Journal of Educational Research, 112*, 101921. https://doi.org/10.1016/j.ijer.2021.101921

Pathak, N., Deb, P. K., Mukherjee, A., & Misra, S. (2021). IoT-to-the-rescue: A survey of IoT solutions for COVID-19-like pandemics. *IEEE Internet of Things Journal, 8*(17), 13145–13164. https://doi.org/10.1109/JIOT.2021.3082838

Payne-Sturges, D. C., Cory-Slechta, D. A., Puett, R. C., Thomas, S. B., Hammond, R., & Hovmand, P. S. (2021). Defining and intervening on cumulative environmental neurodevelopmental risks: Introducing a complex systems approach. *Environmental Health Perspectives, 129*(3), 035001. https://doi.org/10.1289/EHP7333

Pokhrel, S., & Chhetri, R. (2021). A literature review on impact of COVID-19 pandemic on teaching and learning. *Higher Education for the Future, 8*(1), 133–141. https://doi.org/10.1177/2347631120983481

Poudevigne, M., Armstrong, E. S., Mickey, M., Nelson, M. A., Obi, C. N., Scott, A., & Thompson, T. N. (2022). What's in your culture? Embracing stability and the new digital age in moving colleges of health professions virtually during the COVID-19 pandemic: An experiential narrative review. *Education Sciences, 12*(2), 137. https://doi.org/10.3390/educsci12020137

Riley, A. R., Chen, Y. H., Matthay, E. C., Glymour, M. M., Torres, J. M., Fernandez, A., & Bibbins-Domingo, K. (2021). Excess mortality among Latino people in California during the COVID-19 pandemic. *SSM-Population Health, 15*, 100860. https://doi.org/10.1016/j.ssmph.2021.100860

Roman, M., & Plopeanu, A. P. (2021). The effectiveness of the emergency eLearning during COVID-19 pandemic. The case of higher education in economics in Romania. *International Review of Economics Education, 37*, 100218. https://doi.org/10.1016/j.iree.2021.100218

Roux, T. L., Heinen, M. M., Murphy, S. P., & Buggy, C. J. (2021). A unified theoretical framework of learning theories to inform and guide public health continuing medical education research and practice. *The Journal of Continuing Education in the Health Professions, 41*(2), 130. https://doi.org/10.1097/CEH.0000000000000339

Scott, D. (2021). Sustainable tourism and the grand challenge of climate change. *Sustainability, 13*(4), 1966. https://doi.org/10.3390/su13041966

Selin, N. E. (2021). Lessons from a pandemic for systems-oriented sustainability research. *Science Advances, 7*(22), eabd8988. https://doi.org/10.1126/sciadv.abd8988

Sheng, J., Amankwah-Amoah, J., Khan, Z., & Wang, X. (2021). COVID-19 pandemic in the new era of big data analytics: Methodological innovations and future research directions. *British Journal of Management, 32*(4), 1164–1183. https://doi.org/10.1111/1467-8551.12441

Shittu, E., Kantamaneni, K., & Campos, L. C. (2022). Investigating how COVID-19 has challenged the Eurocentric concept of 'development': A case for sustainable food systems in the UK. *Food and Energy Security*, e416. https://doi.org/10.1002/fes3.416

Singh, M., Adebayo, S. O., Saini, M., & Singh, J. (2021). Indian government E-learning initiatives in response to COVID-19 crisis: A case study on online learning in Indian higher education system. *Education and Information Technologies, 26*(6), 7569–7607. https://doi.org/10.1007/s10639-021-10585-1

Stockinger, K., Rinas, R., & Daumiller, M. (2021). Student adaptability, emotions, and achievement: Navigating new academic terrains in a global crisis. *Learning and Individual Differences, 90*, 102046. https://doi.org/10.1016/j.lindif.2021.102046

Strielkowski, W. (2022). New trends in international education: Impact of COVID-19 and digitalization on higher education and student mobility. In B. Akgün & Y. Alpaydın (Eds.), *Education policies in the 21st century* (Maarif Global Education series). Palgrave Macmillan. https://doi.org/10.1007/978-981-19-1604-5_8

Strielkowski, W., Firsova, I., Azarova, S., & Shatskaya, I. (2022). Novel insights in the leadership in business and economics: A post-coronavirus update. *Economies, 10*(2), 48. https://doi.org/10.3390/economies10020048

Suliman, W. A., Abu-Moghli, F. A., Khalaf, I., Zumot, A. F., & Nabolsi, M. (2021). Experiences of nursing students under the unprecedented abrupt online learning format forced by the national curfew due to COVID-19: A qualitative research study. *Nurse Education Today, 100*, 104829. https://doi.org/10.1016/j.nedt.2021.104829

Taufiq-Hail, G. A. M., Sarea, A., & Hawaldar, I. T. (2021). The impact of self-efficacy on feelings and task performance of academic and teaching staff in Bahrain during COVID-19: Analysis by SEM and ANN. *Journal of Open Innovation: Technology, Market, and Complexity, 7*(4), 224. https://doi.org/10.3390/joitmc7040224

Teras, M., Suoranta, J., Teräs, H., & Curcher, M. (2020). Post-COVID-19 education and education technology 'solutionism': A seller's market. *Postdigital Science and Education, 2*(3), 863–878. https://doi.org/10.1007/s42438-020-00164-x

Valk, R., & Planojevic, G. (2021). Addressing the knowledge divide: Digital knowledge sharing and social learning of geographically dispersed employees during the COVID-19 pandemic. *Journal of Global Mobility, 9*(4), 591–621. https://doi.org/10.1108/JGM-02-2021-0019

Vargo, D., Zhu, L., Benwell, B., & Yan, Z. (2021). Digital technology use during COVID-19 pandemic: A rapid review. *Human Behavior and Emerging Technologies, 3*(1), 13–24. https://doi.org/10.1002/hbe2.242

Walsh, L. L., Arango-Caro, S., Wester, E. R., & Callis-Duehl, K. (2021). Training faculty as an institutional response to COVID-19 emergency remote teaching supported by data. *CBE-Life Sciences Education, 20*(3), ar34. https://doi.org/10.1187/cbe.20-12-0277

Watson, R., Kundzewicz, Z. W., & Borrell-Damián, L. (2022). COVID-19, and the climate change and biodiversity emergencies. *Science of the Total Environment, 844*, 157188. https://doi.org/10.1016/j.scitotenv.2022.157188

Yang, Y., Zhang, C. X., & Rickly, J. M. (2021). A review of early COVID-19 research in tourism: Launching the annals of tourism Research's curated collection on coronavirus and tourism. *Annals of Tourism Research, 91*, 103313. https://doi.org/10.1016/j.annals.2021.103313

Yeap, C. F., Suhaimi, N., & Nasir, M. K. M. (2021). Issues, challenges, and suggestions for empowering technical vocational education and training education during the COVID-19 pandemic in Malaysia. *Creative Education, 12*(8), 1818–1839. https://doi.org/10.4236/ce.2021.128138

Zajac, S., Randall, J., & Holladay, C. (2022). Promoting virtual, informal learning now to thrive in a post-pandemic world. *Business and Society Review, 127*, 283–298. https://doi.org/10.1111/basr.12260

Zou, B., Huang, L., Ma, W., & Qiu, Y. (2021). Evaluation of the effectiveness of EFL online teaching during the COVID-19 pandemic. *Sage Open, 11*(4), 21582440211054491. https://doi.org/10.1177/21582440211054491

Part IV
Leadership and the Digital Surge in Business and Education

Chapter 20
Information Openness as a Factor of Business Leadership in Today's Digital Environment

Serhii Lyeonov, Oleksii Zakharkin, and Yevhenii Okhrimchuk

Abstract The study aims to determine the role of information openness in ensuring business leadership in the digital business environment. Based on the analysis of European and Ukrainian statistics on information openness of business, the main advantages of entrepreneurship from such transparency, and the obstacles that arise are identified. The data was generated from EU Open Data Portal, Corporate Governance Professional Association (CGPA), CSR Ukraine, and Center for International Private Enterprise (CIPE) for EU countries and Ukraine for 2019–2021. The findings allowed identification that today's business is informationally open mainly due to the state policy of introducing transparent online services and procedures to ensure practical cooperation between the state and firm. However, the rating of information openness of the company itself shows that it remains low. The results of a comparative analysis of the level of transparency of large business enterprises in Europe and Ukraine demonstrate the need to ensure the practices of their transparency and openness to increase success in the context of digitization processes. The study showed that companies that implement the principles of transparency in business strategy improve their reputation, increase the loyalty of partners and consumers, and increase investment attractiveness.

Keywords Information openness · Digitalization and transparency · Business leadership · Digital environment

JEL Classification L21 · L86 · M21

S. Lyeonov
Sumy State University, Sumy, Ukraine

Silesian University of Technology, Gliwice, Poland

The London Academy of Science and Business, London, UK
e-mail: s.lieonov@uabs.sumdu.edu.ua

O. Zakharkin (✉) · Y. Okhrimchuk
Sumy State University, Sumy, Ukraine
e-mail: o.zakharkin@biem.sumdu.edu.ua

W. Strielkowski (ed.), *Leadership, Entrepreneurship and Sustainable Development Post COVID-19*, Springer Proceedings in Business and Economics, https://doi.org/10.1007/978-3-031-28131-0_20

20.1 Introduction

The demands of modern society to ensure publicity and the growth of economic relations have long been implemented in most developed countries at the level of public finances, where transparency is one of the main factors in overcoming the corruption component (Koudelková et al., 2015; Khanin et al., 2019; Lu et al., 2019; Moskalenko et al., 2022; Petroye et al., 2020; Pilia, 2017; Rekunenko et al., 2022; Tiutiunyk et al., 2021; Vasilyeva et al., 2016; Vasilyeva et al., 2021a, b). At the same time, at the level of the business environment and corporate finance, the principles of transparency are just beginning to actively spread to the activities of enterprises and corporations, significantly influencing the level of trust of stakeholders in them (Zakharkina, 2009; Kasych & Vochozka, 2017; Mercado & Vargas-Hernández, 2019; Meresa, 2019; Dutta et al., 2020; Kaya, 2020; Kuzior et al., 2021; Melnyk et al., 2022; Zenchenko et al., 2022; Navickas et al., 2021).

The trust of employees and external stakeholders in companies is one of the main factors influencing the reputation of the brand and ensuring its leading position in the market, which is especially relevant in the conditions of a tough competitive environment (Guliyeva et al., 2018; Brychko et al., 2021; Kuzmenko et al., 2021; Wang et al., 2022). Modern technological capabilities of the digital economy make it possible to increase the level of business transparency by using the latest Internet technologies, artificial intelligence, and data monetization strategies (Kolosok et al., 2020; Gruzina et al., 2021; Kotenko & Bohnhardt 2021; Kotenko et al. (2021); Kuzior et al., 2022; Midor et al., 2021; Semenova & Tarasova, 2017; Grant Thornton 2021).

In addition to the publication of purely financial results of business activity, information transparency of its direct and indirect impact on the environment and society in accordance with the principles of the ESG strategy is becoming more and more widespread (Antonyuk et al., 2021; Artyukhov et al., 2022; Hanić & Jevtić, 2020; Kotenko 2021; Voronkova et al., 2019). Effective management of these influences is possible through the implementation of a transparent system of corporate governance (Bilan et al., 2022; Company, 2021; Fila et al., 2020; Kobushko et al., 2021; Skrynnyk & Vasilyeva, 2020a, b. Thus, the question of researching the influence of information openness of business on its leadership positions and competitive advantages in the market becomes especially relevant.

20.2 Methods and Data

20.2.1 Information Openness and Transparency in the Modern Business Environment

The information openness of the economies of European countries can be analyzed on the basis of data provided on the official portal for European data. A series of indicators have been selected to measure Open Data maturity across Europe. These indicators cover the level of development of national policies promoting Open Data, an assessment of the features made available on national data portals as well as the expected impact of Open Data (Open Data, 2021). It should be noted that Ukraine also entered the top ten European countries in terms of data openness.

The methodological basis for determining the level of information openness of European countries is based on the calculations of a specially defined number of indicators that include publicly available data on the following dimensions of transparency: policy (the policy of data openness in the country is evaluated, the national strategy on open data, the number of state bodies that provide information about their activity), information portal (sites providing information on the status are evaluated), impact (this indicator includes an assessment of the social and economic impact of open data), and quality (maintenance of data relevance, breadth of the spectrum of provided data, improvement of the quality of open data deployment). Hence, according to the given data, the average result in European countries is 2106 points out of a possible 2600, and the leader in data openness is France, which has been in the first position for 7 years in a row. Ukraine received 2447 points in total and moved from 17th to 6th place in the overall ranking, which is a remarkable achievement.

Modern concepts of business transparency understand this concept as full visibility of all business operations. This may include, for example, sharing information and ideas with employees, stakeholders, suppliers, and customers regarding business performance, revenue, pricing, and other business functions. The conducted marketing research made it possible to identify the main advantages and disadvantages that a business receives by implementing the principles of transparency in its activities (Fig. 20.1).

The benefits of transparent business may vary slightly depending on the specific industry and sector. Services based on open data help to optimize the workflow and reduce operational costs, freeing company employees from large volumes of the same routine work (Kuzmenko et al., 2020; Smiianov et al., 2020). In general, transparency in any business promotes exponential progress and contributes to the overall satisfaction of employees and employers (Polyakov et al., 2019; Hinke et al., 2020). At the same time, the implementation of artificial intelligence and other forms of new technologies can either have a positive effect on the company's performance or be an expensive and risky business venture that ends in the disappointment of many stakeholders and a damaged brand reputation (Bilan et al., 2022; Bondarenko et al., 2020).

+ Creating positive advertising and building trust in the brand

- only 34% of consumers say they trust most of the brands they use; while 81% of consumers say that trust influences their purchasing decisions (Quality trumps 2021);
- Accenture reports that a lack of trust costs American brands $756 billion a year (Barton et al. 2022).

+ Ensuring better efficiency in the use of labor resources

- 92% of employees say they would work harder if their goals were transparent (Goals and Transparency 2021).

+ Increasing business profitability

- 73% of customers are willing to pay more for a product that they believe is fully transparent (Quality trumps 2021);
- Cisco's 2021 Data Privacy Benchmark Study found that, on average, companies saw nearly twice as much profit for every dollar spent on privacy (Forged by the Pandemic 2021);
- attracting a new client costs 9–11 times more than keeping an existing one (Customer 2022);
- an increase in brand loyalty of only 7% can increase the company's profit for the entire period by 85% (Customer 2022).

+ Stimulation of the involvement of qualified workers

- 59% of respondents called transparency very important for a potential employer (Could trust 2016);
- employees who feel their voice is heard are 4.6 times more likely to feel empowered to perform at their best (The Impact 2017; Hanić 2020).

- Data protection

- an Accenture study found that companies faced an average of 270 attacks in 2021, a 31% increase over 2020. And since the beginning of the war, there has been an exponential increase in the activity of cyber threats. The business impact is huge and the costs associated with a breach can last for years and include costs such as loss of data, customers, profits (Barton et al. 2022; Semenova 2017);
- in a 2020 McKinsey survey, 71 percent of customers said they would leave a company if it shared their sensitive data without permission (Anant et al. 2020).

- Mistrust of information

- according to an Edelman survey, 45% of consumers would never trust a brand again after it engaged in unethical behavior or was involved in a scandal, and 40% said they would stop buying from that brand altogether (Fertik 2019).

Fig. 20.1 The results of marketing research on the role of transparency in business. (Source: Own results)

20.2.2 The Role of Open Information Resources in Ensuring Business Success

The lack of publicity and transparency of business for a wide range of stakeholders is one of the factors increasing its shadowing (Brychko et al., 2021; Lyulyov et al., 2021; Moskalenko et al., 2022; Vasilyeva et al., 2016; Vasylieva et al., 2020; Zolkover & Georgiev, 2020). That is why the policy of introducing transparent online services and procedures to ensure effective cooperation between the state and business is being actively spread in Ukraine at the state level, which is aimed at the following:

- Automation and efficiency improvement of business processes. With the help of special services and applications, business owners can reduce the amount of paper document flow, significantly simplify interaction with control bodies, automate the filling of primary accounting documentation, and attract new customers. For example, the "Opendatabot" service provides access to tools for optimizing production business processes
- Business protection from economic crimes and fraud. Research by PwC (Global Economic 2020) on the level of economic crimes and fraud in Ukraine conducted in 2020 shows that 51% of surveyed business representatives suffered from fraud in the last 2years, and at the same time, almost half of the respondents did not conduct integrity checks on their counterparties. Currently, such checks are largely made possible by services based on open data, such as YouControl, Vkursi.pro, and Liga Contr Agent.
- Reduction of manifestations of corruption during business inspections. In total, there are about 80 spheres of state supervision in Ukraine, which every year appoint 150,000–215,000 scheduled and unscheduled inspections. And today, the availability of data on the Inspection Portal about planned inspections largely helps enterprises to prepare in advance and avoid fines and administrative liability. In turn, the SaveEcoBot chatbot was created for environmental inspections, which allows you to conveniently get the necessary information about the necessary planned inspections and special permits.
- Protection against raider attacks. The problem of raiding in Ukraine deserves special attention. According to the rating of the international property rights index in 2021, according to the level of raiding, Ukraine ranked 106th out of 129 studied countries (International property, 2021), but it is worth noting that the number of raiding attacks decreased by 18% in a year. This improvement was achieved with the help of the Opendatabot service, which allows monitoring changes in state registers, illegal actions of registrars, and timely response to such actions. In addition, services based on open data have become a new tool for countering corrupt manipulation of registries.

The implementation of such digital services helps not only to simplify the work of business but also helps to combat corruption at the state level (Kobushko et al., 2021; Tiutiunyk et al., 2021). Currently, special services have become widespread in Ukraine and provide transparency of information, among them the following are the most popular:

- YouControl. This is a special online service that allows you to get all available information about the company from more than 180 state registers using the code or the name of the owner. With the help of special tools, YouControl allows you to monitor changes in state registers, conduct market analysis, search for connections between counterparties, etc.
- Prozorro. An electronic system of public procurement, where state and utility customers announce tenders for the purchase of goods, works, and services and business representatives compete at auctions for the opportunity to supply it to the state.
- Diia.Business. The online portal was initiated by the Ministry of Digital Transformation of Ukraine for small- and medium-sized business entrepreneurs. As part of the Diia.Business project, an online portal for potential and active entrepreneurs and a network of support centers for entrepreneurs is functioning.
- Open Budget. Dissemination of information in an accessible form for the population and public organizations regarding the effectiveness of the use of tax funds within the framework of publicity and transparency of budget policy.
- "Under supervision." Inspection portal providing access to information on scheduled and unscheduled inspections carried out by state control bodies, as well as their results.

Ukraine has made significant progress in publicizing open data and building an ecosystem around this data to facilitate its productive use. Much of this development is a result of the work of the Ministry of Digital Transformation, so in recent years an unprecedented amount of public information has become open, including registers of companies, court decisions, and public transactions. This contributed to increased efficiency and cost savings, as well as the emergence of new and better services, new business models, and more efficient markets (Samoilikova, 2020).

20.3 The Role of Information Openness in Business Implementation of ESG Strategies

All over the world, companies are increasingly aligning their core business strategies with ESG factors (environmental, social, and governance factors) in order to achieve competitive advantage and ensure market leadership, which is in line with the Sustainable Development Goals (Bilan et al., 2020; Mercado & Vargas-Hernández, 2019; Navickas et al., 2021; Skrynnyk & Vasilyeva, 2020a; Samusevych et al., 2021). The goal of the business is to move from the usual profit-making to providing value for internal and external stakeholders. Ensuring the transparency of business implementation of ESG strategies plays a significant role in this process.

This criterion is aimed at taking into account the needs of environmental and social development and achieving management results during the implementation of investment processes along with ensuring financial results and achieving the required level of investment efficiency. The term "ESG" encompasses a wide range of complex and often interrelated issues such as resource management, community relations, and governance. Today, most of the world's leading investors undertake to include ESG issues in their investment analysis.

Constant monitoring of world trends in the field of application of the concept of "values-based investing" is carried out by the Global Sustainable Investment Alliance (GSIA), the goal of which is the international cooperation of sustainable investment organizations. According to the Global Alliance, the growth of investment assets continues around the world, based on the principles of "values-based investing," but the pace of such growth is different in different countries and regions.

The implementation of the principles of "values-based investing" does not include the possibility of using a wide range of strategic directions, both by types of investment assets and by subjects of investment activity. As the world practice shows, such subjects can be institutional participants, investment funds, and private investors. Depending on this, new forms and investment strategies are created and developed, providing for the achievement of the required level of return on investment, and the simultaneous formation of social value through the provision of positive social, environmental, or other effects.

The goal of the enterprise's investment activity should be an increase not only in the corporate value (the value of the company's share capital and profit) but also in the social value expressed through the net financial balance between positive and negative external economic, social, environmental, and other effects. Such investments provide an opportunity to introduce new or significantly improved products, technological processes, and management technologies, the use of which by the enterprise ensures an increase in corporate value for owners and management, as well as social value for other stakeholders in the form of an increase in economic, environmental, social, and other external effects and/or minimization of relevant external risks.

To assess the level of such transparency, the ESG Transparency Index is used, which is one of the important tools for investors to assess business sustainability. The evaluation is based on an international methodology developed by Beyond Business (Israel), the Center for CSR Development, and the Corporate Governance Professional Association with the support of the Center for International Private Entrepreneurship (CIPE). Information on corporate governance and corporate social responsibility is assessed in the following categories (Company, 2021):

A General information about the company and its CSR policy
B Corporate governance
C Human rights
D Employment relationships
E Environmental protection
F CSR integration in the supply chain

G Responsible consumption
H Community relations
I Management and company leadership
J Reporting
K Navigation
L Accessibility

Experts evaluate the informational openness of CSR policy and practice and the results of their implementation on such aspects as corporate governance, human rights, and labor relations, environmental protection, implementation of CSR in the "supply chain," responsible consumption, relations with the community, and leadership of management and the company. Management leadership was considered as the personal involvement of senior management in solving social and environmental problems of society (reflection of their public activity on the website and pages of such social networks as LinkedIn and Facebook), and company leadership as the company's membership in associations. This category also evaluates general information about the company (mission and values, company strategy, CSR strategy, commitment to achieving the Sustainable Development Goals). If the information on indicators was placed in the appropriate sections of the site, the company receives maximum points. If such information was contained in reports, policies, and documents, the company receives half as many points as provided by the methodology. Table 20.1 shows the level of growth of the average level of information disclosure according to the ESG transparency index compared to 2019 for Ukrainian companies and the level of growth of similar indicators in EU countries.

Table 20.1 Growth in the average level of disclosure by transparency index category

Categories	An increase in the level of the ESG Transparency Index compared to 2019	
	Ukraine	EU countries
General information about the company and its CSR policy	32%	11.9%
Corporate governance	17.3%	6.9%
Human rights	133.%	22.5%
Employment relationships	13.6%	8.8%
Environmental protection	46.7%	24.7%
CSR integration in the supply chain	0	12.5%
Responsible consumption	42.9%	15%
Community relations	45.4%	18.7%
Management and company leadership	100%	10.5%
Reporting	31.8%	3.6%
Navigation	−3.7%	3.2%
Accessibility	17.9%	2.5%

Source: Own results based on (Company, 2021) and (ESG, 2021)

According to the ESG Transparency Index, the overall level of transparency of Ukrainian companies in 2020 is 32%, which is 6.5% higher than in 2019. The leaders of the 2020 ESG Transparency Index were PJSC "VF Ukraine," JSC First Ukrainian International Bank, DTEK Group, JSB UKRGASBANK, Coca-Cola Beverages Ukraine Limited, NNEGC Energoatom, MHP, PJSC "Ukrhydroenergo," JSC Ukrainian Railways, and NJSC Naftogaz of Ukraine. All these enterprises and companies are leaders in their types of economic activity and for many years have demonstrated stable growth in production and financial results of their activities. The greatest increase in the level of transparency took place in information provision of relations with the community (supporting projects of local communities; posting information about the implementation of own social and educational projects; developing of charity) (12.5%), respect for human rights (disclosure of corporate documentation on human rights and labor relations; employment of people with special needs) (10%), and environmental protection (11.7%). To evaluate the general level of data disclosure by category, it is possible to note navigation and accessibility (convenience and availability of information on sites) at 65% and 41.8%.

It is also worth noting that the level of disclosure of information by state-owned companies is 39.4%, which is 11% higher than that of private companies, which is most likely the result of the corporate governance reform of state-owned enterprises. And in general, the highest level of information disclosure was demonstrated by companies in the financial sphere, transport, and electricity supply.

A company's transparency increases its reputation. Disclosure of information on ESG indicators is important not only for investors. It is becoming more and more important for consumers and company employees.

That is why the publication of CSR strategies and programs, the results of their implementation, information about corporate governance, and constant updating of such information should become the main points of reputation programs of companies.

Companies are recommended to develop a corporate evaluation and reporting system based on ESG indicators that will be consistent with the national objectives of the Sustainable Development Goals.

Companies are recommended not only to publish information on ESG indicators on corporate websites but also to constantly update it, as well as make it available to many groups of stakeholders.

Companies should pay attention to the coverage of those issues that are at the center of attention of the world community – respect for human rights, implementation of CSR in the supply chain, environmental protection, and the company's contribution to achieving Sustainable Development Goals.

State bodies are recommended to make changes to the existing legal framework (or develop a separate law) regarding the terms, methods, and place of publishing management reports and determination of state authorities, whose sphere of management includes issues of collection and analysis of management reports and control over compliance with legal norms regarding their preparation.

The regulatory and legal framework needs additions in terms of defining ESG indicators (general and sectoral), which will be consistent with the national objectives of the Sustainable Development Goals.

At the state level, it is recommended to introduce a policy of support and encouragement of companies that prepare nonfinancial reports (Company, 2021).

20.4 Conclusions

Overall, one can see that in today's digital information environment, business transparency is one of the key factors in its success and ensuring leadership positions in the market. The information openness of a business creates its competitive advantages through positive advertising and the formation of trust in the brand, increasing the efficiency of the use of company resources and increasing sales volumes, which in turn contributes to an increase in financial results. Modern digital online services based on open data are a significant function of ensuring business information openness, the use of which allows avoiding not only economic and financial losses but also reputational losses. Yes, open data has a powerful potential for optimizing business processes, but at the same time, the introduction of new technologies will increase the risks of cyber threats, which business often underestimates. Companies now find themselves in a world with persistent levels of disruption, increased inflationary pressures, and the specter of further uncertainty.

Another important factor in ensuring business leadership is the implementation of an ESG strategy by enterprises and companies that meets the Sustainable Development Goals. European experience shows that openness in the disclosure of information for ESG indicators is important not only for investors. It is becoming more and more important for consumers and company employees. That is why the publication of CSR strategies and programs, the results of their implementation, information on corporate governance, and constant updating of such information should become the main points of reputation programs of companies.

Acknowledgments The research was funded by grants from the Ministry of Education and Science of Ukraine: No 0122U000774, No 0121U109553, No 0121U109559, and No 0120U102001.

References

Anant, V., Donchak, L., Kaplan, J. & Soller, H. (2020). The consumer-data opportunity and the privacy imperative. Available at: https://www.mckinsey.com/business-functions/risk-and-resilience/our-insights/the-consumer-data-opportunity-and-the-privacy-imperative

Antonyuk, N., Plikus, I., & Jammal, M. (2021). Human capital quality assurance under the conditions of digital business transformation and COVID-19 impact. *Health Economics and Management Review, 2*(3), 39–47. https://doi.org/10.21272/hem.2021.3-04

Artyukhov, A. E., Volk, I. I., & Vasylieva, T. A. (2022). Agile methodology in higher education quality assurance system for SDGs 4, 8 and 9 achievement: National experience. *CTE Workshop Proceedings, 9*, 81–94. https://doi.org/10.55056/cte.105

Barton, R., Roark, C., & Delawalla, A. (2022). Operating through volatility: Five pillars to manage business continuity. https://www.accenture.com/hu-en/insights/strategy/ukraine-operating-through-volatility-business-continuity

Betterworks. (2021). Goals and transparency help people win at work, says employee sentiment survey. Available at: https://www.betterworks.com/magazine/goals-transparency-help-people-win-work-says-employee-sentiment-survey/

Bilan, Y., Pimonenko, T., & Starchenko, L. (2020). Sustainable business models for innovation and success: Bibliometric analysis. *E3S Web of Conferences, 159*, 04037. https://doi.org/10.1051/e3sconf/202015904037

Bilan, S., Šuleř, P., Skrynnyk, O., Krajňáková, E., & Vasilyeva, T. (2022). Systematic bibliometric review of artificial intelligence technology in organizational management, development, change and culture. *Business: Theory and Practice, 23*(1), 1–13. https://doi.org/10.3846/btp.2022.13204

Bondarenko, A. F., Zakharkina, L. S., Syhyda, L. O., & Saher, L. Y. (2020). The economic and marketing attractiveness of countries: Measurement and positioning in terms of economic security. *International Journal of Sustainable Development and Planning, 15*(4), 439–449. https://doi.org/10.18280/ijsdp.150404

Brand Keys. (2022). Customer loyalty engagement index identifies secret to lasting loyalty. Available at: https://brandkeys.com/wp-content/uploads/2019/02/Press-Release-2019-Customer-Loyalty-Engagement-Index-Top-10.pdf

Brychko, M., Savchenko, T., Vasylieva, T., & Piotrowski, P. (2021). Illegal activities of financial intermediaries: A burden of trust crisis. *Journal of International Studies, 14*(1), 172–189. https://doi.org/10.14254/2071-8330.2021/14-1/12

Cisco. (2021). Forged by the pandemic: The age of privacy. Available at: https://www.cisco.com/c/dam/en_us/about/doing_business/trust-center/docs/cisco-privacy-benchmark-study-2021.pdf

Clear Channel. (2021). Quality trumps price and social responsibility in fight for consumer trust. Available at: https://www.clearchannel.co.uk/latest/clear-channel-and-jcdecaux-unveil-the-moment-for-trust-research

Company transparency index 2020 – Ukraine. (2021). CGPA. Available at: https://index.cgpa.com.ua/

Data.europa.eu. (2021). Open data in Europe 2021. Available at: https://data.europa.eu/en/dashboard/2021

Dutta, P., Dutta, U., Hasan, S., Sarkar, S., & Sana, T. (2020). Educate, empower and protect people through timely authentic information: Explore how Bangladeshi newspapers response to COVID-19 pandemic. *SocioEconomic Challenges, 4*(3), 93–103. https://doi.org/10.21272/sec.4(3).93-103.2020

EY. (2016). Could trust cost you a generation of talent? Global generations 3.0: A global study on trust in the workplace. Available at: http://ccnews.pl/wp-content/uploads/2016/09/report_global_generations_3.0.pdf

Fertik, M. (2019). How to get customers to trust you. Available at: https://www.forbes.com/sites/michaelfertik/2019/11/26/how-to-get-customers-to-trust-you/?sh=7de7763ef8d6

Fila, M., Levicky, M., Mura, L., Maros, M., & Korenkova, M. (2020). Innovations for business management: Motivation and barriers. *Marketing and Management of Innovations, 4*, 266–278. https://doi.org/10.21272/mmi.2020.4-22

Grant Thornton. (2021). Transparency with advanced technologies drives trust. Available at: https://www.grantthornton.com/insights/whitepapers/audit/2021/transparency-in-using-advanced-tech-drives-trust

Gruzina, Y., Firsova, I., & Strielkowski, W. (2021). Dynamics of human capital development in economic development cycles. *Economies, 9*(2), 67. https://doi.org/10.3390/economies9020067

Guliyeva, A., Britchenko, I., & Rzayeva, U. (2018). Global security and economic asymmetry: A comparison of developed and developing countries. *Journal of Security and Sustainability Issues, 7*(4), 707–717. https://doi.org/10.9770/jssi.2018.7.4(8)

Hanić, A., & Jevtić, D. (2020). Human resource management between economy and ethics – Research of Serbia and Bosnia and Hercegovina. *Business Ethics and Leadership, 4*(3), 127–136. https://doi.org/10.21272/bel.4(3).127-136.2020

Hinke, J., Gezo, M., Smutka, L., & Strielkowski, W. (2020). Management of financial statements auditing in the Visegrad Group countries. *Problems and Perspectives in Management, 18*(1), 1–16. https://doi.org/10.21511/ppm.18(1).2020.01

International property rights index 2021. (2021). Property Rights Alliance. Available at: https://www.internationalpropertyrightsindex.org/.

Kasych, A., & Vochozka, M. (2017). Theoretical and methodical principles of managing enterprise sustainable development. *Marketing and Management of Innovations, 2*, 298–305. https://doi.org/10.21272/mmi.2017.2-28

Kaya, H. D. (2020). Business friendliness, firm performance and owner's optimism. *Financial Markets, Institutions and Risks, 4*(3), 13–23. https://doi.org/10.21272/fmir.4(4).109-118.2020

Khanin, I., Shevchenko, G., Bilozubenko, V., & Korneyev, M. (2019). A cognitive model for managing the national innovation system parameters based on international comparisons (the case of the EU countries). *Problems and Perspectives in Management, 17*(4), 153–162. https://doi.org/10.21511/ppm.17(4).2019.13

Kobushko, I., Tiutiunyk, I., Kobushko, I., Starinskyi, M., & Zavalna, Z. (2021). The triadic approach to cash management: Communication, advocacy, and legal aspects. *Estudios De Economia Aplicada, 39*(7). https://doi.org/10.25115/eea.v39i7.5071

Kolosok, S., Myroshnychenko, I., & Zakharkina, L. (2020). Open data in electrical energy balancing of Ukraine: Green deal and security aspects. *CEUR Workshop Proceedings, 2732*, 270–281.

Kotenko, N., & Bohnhardt, V. (2021). Digital health projects financing: Challenges and opportunities. *Health Economics and Management Review, 2*(1), 100–107. https://doi.org/10.21272/hem.2021.1-10

Kotenko, S., Kobushko, I., Heiets, I., & Rusanov, O. (2021). KPI model impact on employee motivation and competitiveness of private healthcare facilities. *Health Economics and Management Review, 2*, 31–42. https://doi.org/10.21272/hem.2021.2-04

Koudelková, P., Strielkowski, W., & Hejlová, D. (2015). Corruption and system change in the Czech Republic: Firm-level evidence. *Danube: Law, Economics and Social Issues Review, 6*(1), 25–46. https://doi.org/10.1515/danb-2015-0002

Kuzior, A., Mańka-Szulik, M., & Krawczyk, D. (2021). Zmiany w zarządzaniu elektronicznymi usługami publicznymi w metropolii podczas pandemii covid-19. *Polish Journal of Management Studies, 24*(2), 261–275. https://doi.org/10.17512/pjms.2021.24.2.16

Kuzior, A., Kettler, K., & Rąb, Ł. (2022). Digitalization of work and human resources processes as a way to create a sustainable and ethical organization. *Energies, 15*(1), 172. https://doi.org/10.3390/en15010172

Kuzmenko, O., Vasylieva, T., Vojtovič, S., Chygryn, O., & Snieška, V. (2020). Why do regions differ in vulnerability to covid-19? Spatial nonlinear modeling of social and economic patterns. *Economics and Sociology, 13*(4), 318–340. https://doi.org/10.14254/2071-789x.2020/13-4/20

Kuzmenko, O., Cyburt, A., Yarovenko, H., Yersh, V., & Humenna, Y. (2021). Modeling of "information bubbles" in the global information space. *Journal of International Studies, 14*(4), 270–285. https://doi.org/10.14254/2071-8330.2021/14-4/18

Lu, J., Ren, L., Qiao, J., Yao, S., Strielkowski, W., & Streimikis, J. (2019). Corporate social responsibility and corruption: Implications for the sustainable energy sector. *Sustainability, 11*(15), 4128. https://doi.org/10.3390/su11154128

Lyulyov, O., Paliienko, M., Prasol, L., Vasylieva, T., Kubatko, O., & Kubatko, V. (2021). Determinants of shadow economy in transition countries: Economic and environmental aspects. *International Journal of Global Energy Issues, 43*(2–3), 166–182. https://doi.org/10.1504/ijgei.2021.10037735

Melnyk, L., Matsenko, O., Kubatko, O., Korneyev, M., & Tulyakov, O. (2022). Additive economy and new horizons of innovative business development. *Problems and Perspectives in Management, 20*(2), 175–185. https://doi.org/10.21511/ppm.20(2).2022.15

Mercado, M. P. S. R., & Vargas-Hernández, J. G. (2019). Analysis of the determinants of social Capital in Organizations. *Business Ethics and Leadership, 3*(1), 124–133. https://doi.org/10.21272/bel.3(1).124-133.2019

Meresa, M. (2019). The effect of strategic management practices on the institutional performance; the case of Dedebit credit and saving institution in Eastern Tigray. *SocioEconomic Challenges, 3*(3), 80–97. https://doi.org/10.21272/sec.3(3).80-97.2019

Midor, K., Kuzior, A., Płaza, G., Molenda, M., & Krawczyk, D. (2021). Reception of the smart city concept in the opinion of local administration officials-a case study. *Management Systems in Production Engineering, 29*(4), 320–326. https://doi.org/10.2478/mspe-2021-0040

Moskalenko, B., Lyulyov, O., & Pimonenko, T. (2022). The investment attractiveness of countries: Coupling between core dimensions. *Forum Scientiae Oeconomia, 10*(2), 153–172. https://doi.org/10.23762/FSO_VOL10_NO2_8

Navickas, V., Petroke, I., Baciuliene, V., & Vasylieva, T. (2021). The impact of the sharing economy as an ecosystem on the tourism sector. *Journal of Tourism and Services, 12*(22), 66–88. https://doi.org/10.29036/jots.v12i22.241

Orbis. (2021). ESG in private equity: The ESG transparency index 202. Available at: https://www.orbisadvisory.com/latest-news/the-esg-transparency-index-2021-esg-in-private-equity-and-moving-forwards%20(2021)

Petroye, O., Lyulyov, O., Lytvynchuk, I., Paida, Y., & Pakhomov, V. (2020). Effects of information security and innovations on Country's image: Governance aspect. *International Journal of Safety and Security Engineering, 10*(4), 459–466. https://doi.org/10.18280/ijsse.100404

Pilia, G. (2017). Estonia and Lithuania in transition: A compared analysis of the change and its costs and benefits. *Business Ethics and Leadership, 1*(2), 12–19. https://doi.org/10.21272/bel.1(2).12-19.2017

Polyakov, M., Bilozubenko, V., Korneyev, M., & Shevchenko, G. (2019). Selection of parameters for multifactor model in the knowledge economy marketing (country level). *Innovative Marketing, 15*(1), 89–99. https://doi.org/10.21511/im.15(1).2019.08

PwC. (2020). Global economic crime and fraud survey 2020. Available at: https://www.pwc.com/ua/en/survey/2020/economic-crime-survey.html

Rekunenko, I., Zhuravka, F., Nebaba, N., Levkovych, O., & Chorna, S. (2022). Assessment and forecasting of Ukraine's financial security: Choice of alternatives. *Problems and Perspectives in Management, 20*(2), 117–134. https://doi.org/10.21511/ppm.20(2).2022.11

Salesforce. (2017). The impact of equality and values driven business. Available at: https://www.salesforce.com/content/dam/web/en_us/www/assets/pdf/datasheets/salesforce-research-2017-workplace-equality-and-values-report.pdf

Samoilikova, A. (2020). Financial policy of innovation development providing: The impact formalization. *Financial Markets, Institutions and Risks, 4*(2), 5–15. https://doi.org/10.21272/fmir.4(2).5-15.2020

Samusevych, Y. V., Novikov, V. V., Artyukhov, A. Y., & Vasylieva, T. A. (2021). Convergence trends in the "economy – education – digitalization – national security" chain. *Naukovyi Visnyk Natsionalnoho Hirnychoho Universytetu, 6*, 177–183. https://doi.org/10.33271/nvngu/2021-6/177

Semenova, K., & Tarasova, K. (2017). Establishment of the new digital world and issues of cyber-risks management. *Marketing and Management of Innovations, 3*, 236–244. https://doi.org/10.21272/mmi.2017.3-22

Skrynnyk, O., & Vasilyeva, T. (2020a). Comparison of open learning forms in organizational education. *CEUR Workshop Proceedings, 2732*, 1314–1328.

Skrynnyk, O., & Vasilyeva, T. (2020b). Neuro-genetic hybrid system for management of organizational development measures. *CEUR Workshop Proceedings, 2732*, 411–422.

Smiianov, V. A., Vasilyeva, T. A., Chygryn, O. Y., Rubanov, P. M., & Mayboroda, T. M. (2020). Socio-economic patterns of labor market functioning in the public health: Challenges connected with COVID-19. *Wiadomosci Lekarskie (Warsaw, Poland: 1960), 73*(10), 2181–2187. https://doi.org/10.36740/wlek202010114

Tiutiunyk, I., Drabek, J., Antoniuk, N., Navickas, V., & Rubanov, P. (2021). The impact of digital transformation on macroeconomic stability: Evidence from EU countries. *Journal of International Studies, 14*(3), 220–234. https://doi.org/10.14254/2071-8330.2021/14-3/14

Vasilyeva, T., Sysoyeva, L., & Vysochyna, A. (2016). Formalization of factors that are affecting stability of Ukraine banking system. *Risk Governance and Control: Financial Markets and Institutions, 6*(4), 7–11. https://doi.org/10.22495/rcgv6i4art1

Vasilyeva, T., Kuzmenko, O., Kuryłowicz, M., & Letunovska, N. (2021a). Neural network modeling of the economic and social development trajectory transformation due to quarantine restrictions during covid-19. *Economics and Sociology, 14*(2), 313–330. https://doi.org/10.14254/2071-789x.2021/14-2/17

Vasilyeva, T., Ziółko, A., Kuzmenko, O., Kapinos, A., & Humenna, Y. (2021b). Impact of digitalization and the COVID-19 pandemic on the AML scenario: Data mining analysis for good governance. *Economics and Sociology, 14*(4), 326–354. https://doi.org/10.14254/2071-789x.2021/14-4/19

Vasylieva, T., Jurgilewicz, O., Poliakh, S., Tvaronavičienė, M., & Hydzik, P. (2020). Problems of measuring country's financial security. *Journal of International Studies, 13*(2), 329–346. https://doi.org/10.14254/2071-8330.2020/13-2/22

Voronkova, O., Hordei, O., Barusman, A. R. P., & Ghani, E. K. (2019). Social integration as a direction for humanization of economic relations and improvement of social welfare. *SocioEconomic Challenges, 3*(4), 52–62. https://doi.org/10.21272/sec.3(4).52-62.2019

Wang, Q., Chen, Y., Guan, H., Lyulyov, O., & Pimonenko, T. (2022). Technological innovation efficiency in China: Dynamic evaluation and driving factors. *Sustainability, 14*(14), 8321. https://doi.org/10.3390/su14148321

Zakharkina, L. S. (2009). Balancing of innovative development of machinebuilding enterprises in strategic planning process. *Actual Problems of Economics, 3*, 88–95.

Zenchenko, S., Strielkowski, W., Smutka, L., Vacek, T., Radyukova, Y., & Sutyagin, V. (2022). Monetization of the economies as a priority of the new monetary policy in the face of economic sanctions. *Journal of Risk and Financial Management, 15*(3), 140. https://doi.org/10.3390/jrfm15030140

Zolkover, A., & Georgiev, M. (2020). Shadow investment activity as a factor of macroeconomic instability. *Financial Markets, Institutions and Risks, 4*(4), 83–90. https://doi.org/10.21272/fmir.4(4).83-90.2020

Chapter 21
Novel Leading Approaches to the Digitalization of Professional Education: Best World Practices

Larisa Gorina, Olga Naumova, and Marina Gordova

Abstract This chapter describes the novel leading approaches to the digitalization of professional education in the course of the COVID-19 pandemic as well as in the post-COVID era. Moreover, it also assesses the world's best practices applied in the field the digitalization of professional education. Our research represents the abundant resource for teachers looking for strategies, examples, and practical recommendations to deal with the challenges of teaching and lecturing in a digital and globalized age. Our results and outcomes can be used for managing classrooms, considering the race, improving literacy, teaching in the digital age, and much more. In particular, we argue that the leading approaches of the digitalization of professional education should be targeted to teachers and instructors who are eager to use technology for learning as best as they can. This chapter explores fundamental principles guiding effective teaching in an age where everybody, and especially the students who are being taught, is using modern digital information and communication technology. We issue a useful reminder to all teachers using digital tools in the education process which is not focusing on particular tools, but rather the ways these tools should be used to improve learning. Sometimes, this is exploring ways that digital technologies might improve the professional teaching practices for fostering sustainable education and sustainable economic development.

Keywords Leadership · Digitalization · Professional education · COVID-19

L. Gorina (✉)
Institute of Engineering and Environmental Security, Togliatti State University, Togliatti, Russian Federation
e-mail: gorina@tltsu.ru

O. Naumova
Volga Region State University of Service, Togliatti, Russian Federation

M. Gordova
Financial University under the Government of the Russian Federation, Moscow, Russian Federation
e-mail: MGordova@fa.ru

W. Strielkowski (ed.), *Leadership, Entrepreneurship and Sustainable Development Post COVID-19*, Springer Proceedings in Business and Economics, https://doi.org/10.1007/978-3-031-28131-0_21

JEL Classification H40 · I20 · L86

21.1 Introduction

Digital technologies in higher education constitute an important tool for its development and sophistication, making the process of learning dependent on the digital tools and approaches. In order to develop such digital-based teaching and learning, theoretically and practically, current media education and e-learning studies have faced challenges to open a transdisciplinary dialogue with other academic disciplines (Lohr et al., 2021; Núñez-Canal et al., 2022). A necessary issue arising is how to connect emerging technologies and education concepts in order to cultivate practical digital expertise, which can improve students learning and social belonging and, at the same time, reduce attrition rates (Guliyeva et al., 2021; Srivastava et al., 2021). While there is the potential for technology to support learning centered on students, there needs to be deliberate ways of strengthening and augmenting traditional teaching and learning practices (Gu et al., 2022). When used appropriately, classroom instruction is strengthened and given new dimensions. Technology is advancing so fast right now. The traditional ways of teaching and learning are not pushing students and teachers to reach their full potential (Li et al., 2021). Professional development for teachers is also becoming a priority, with many schools embracing new technologies, aiming to bring in new pedagogical foundations to their classroom instruction. For example, some proactive faculty members can develop Blended Learning Massive Open Online Courses (bMOOC) about digital teaching, and offered it to other teachers-trained faculty members as a resource for building their own digital professional skills (Dale et al., 2021).

At the beginning of any program leading to the digitalization of education, it is important to get teachers, students, and parents using the technology, as well as developing new ways to teach, something that was extremely uncomfortable for many of his teachers. The pedagogical uses of the new technologies for enhancing instruction, from the perspective of a longer-term view, are seldom mentioned in any digital strategies (Lampos et al., 2021). Paralleling conversations about technology at community level was one related to expanding digital access at schools, which led to an abundance of resources. Such dialogues allowed for addressing and discussing educations complexity in a digital era, drawing upon the insights and research that have emerged from diverse fields. There are some key changes forcing teachers and instructors to reconsider their goals and methods of instruction, specifically, identifies the critical knowledge and skills students need in a digital age, and how technology is changing everything, including the contexts within the teaching and learning process itself (Marouli, 2021). There are several aspects addressing the issues of pedagogy, collaboration, experience, methods for teaching students before they enter teaching, professional development, digital tools, and equity. They are crucial for documenting best practices, strategies, and efforts of teachers' educators, professional developers, researchers, and practitioners. These aspects focus on emerging and innovative pedagogical approaches to helping preservice teachers build the skills needed to teach in the discipline (Archambault et al., 2022). Equity issues (digital tools chapters) provide information and tools to preservice and in-service teachers for meeting the varied needs of all students. Digital tools include elements focusing on providing guidelines for using digital tools in an online instructional setting in a pedagogically sound manner. This can become a helpful resource to faculty at colleges and universities, and some K-12 teachers. Due to this fact, faculty at colleges and universities are considering the best way to engage students and support learning for the digital natives, as well as others who are seeking knowledge and instruction in this era based on information and knowledge. There are some necessary and crucial policy and operational supports needed by schools, colleges, and universities in order to provide high-quality, relevant teaching in a digital era (Huang et al., 2021).

In general, one can draw a lot of examples from post-secondary education, but many principles would apply equally to teachers in schools or in the K-12 sector, though one should be keenly aware that schools have much less resources and support for technology than colleges or universities (Rock & Curry, 2021). Although technology is the central goal, advocating the removal of the present, human-based education system and replacing it with a heavily computerized teaching model should not be done haphazardly. As the use of technology in classrooms advances, there will be an increasing number of students who might ask for its inclusion. Faculty members may be awarded grants to utilize technology in their classrooms and to develop new teaching technologies. Students are interested in doing new things with their writing, playing with technology, and being original. Students need to be taught to successfully negotiate, as a class, for them to become plentiful and to push their digital narratives farther and to embrace the digitalization of education as a leading element in its further development in the post-COVID era.

21.2 Digital Transformation of Educational Institutions

The leading factor behind education institutions digital transformation has certainly been its leaders' vision. The governance strategies related to digital technologies that facilitated a shift from vision to action, taking into consideration the heterogeneity of education institutions across the world, need to be properly formulated and implemented (Alenezi, 2021). It is thought that an examination of teachers' views and experiences on their school principals' digital leadership roles would contribute to relevant literature in an ongoing pandemic where ICTs are still heavily used in education practices. According to some studies, the roles of principals and school managers are highly relevant, since they are expected to serve as transformative leaders, transforming the education institutions that they are leading into a digital era. Being a principal and managing different implementation projects in the area of digital education, it is important to be accompanied by educational principals and school managers, when transforming an educational institution in the digital era (Bellibas et al., 2021). In the digital learning environment, it is essential that school leaders and teams of digital learning leaders think about how to use technology effectively, collaborate with parents, set protections, and create safe, supportive communities. Just as students who are learning using digital tools require time and support from schools and districts, so should do the teachers, who require professional learning investments by the digital learning leadership team. Supporting school administrators as digital leaders and developing opportunities for professional learning with digital tools can promote the self-development of teachers as well as students. Effective communications around digital learning will ensure parents and families are informed and involved, and they trust decisions made by educators and school leaders (Zahorodna et al., 2022). As the school or system leader, the person in change or the team of digital learning leaders have a critical role to play in advocating for digital access for all students and teachers, as well as communicating effectively to parents about access and resources that are available. The instructional leaders identified as part of the digital learning leadership team are critical to setting expectations for effective digital learning, equipping teachers to plan and organize powerful learning experiences, and identifying resources and tools that may improve student outcomes and engagement (Vilchez et al., 2021). To do so well, instructional leaders must consider digital transformation from an entire systems perspective and create a culture that promotes leadership everywhere and promotes innovation. Digital transformation goes far beyond technology integration and requires changes in the universities and schools' curriculum and the entire teaching-learning culture in order to prepare the next generation for a diverse labor market and society (Rodríguez-Abitia & Bribiesca-Correa, 2021). Thence, it can be stated that the digital transformation is a process of organizational change, which requires a unique view, perspective, and set of leadership skills.

All in all, the digital transformation in higher education is a process that uses digital technologies to enhance and improve student experiences, improving interactions with instruction services, support teams, and other key areas of campus operations. Digital transformation is not focused on a particular strategy or piece of software, but instead uses technologies to advance improvements to fundamental strategic and operational challenges. This digital transformation is not limited to digitizing business processes and adding technology portfolios; it requires adopting new ways of learning and leadership, driving a cultural change, and aligning strategies, talent, and resources throughout an organization (García-Peñalvo, 2021). Cultural or organizational change changes how campus leaders engage with one another and requires laser-like attention to progress on institutional goals, an increased focus on managing change, and greater institutional flexibility to respond to fast-moving needs – a shift toward digital-first needs to start with mapping processes and customers a journey. There should be a digital strategy that is focused on innovation that addresses problems in the service of the school or the university as mission. Having leaders who are committed to transformation is the key to successful cultural shifts required for becoming a truly digital-first organization (Monteiro et al., 2021). In the mid-term, organizations can enhance their pure digital capability by exposure of key leaders to digital experiences. The important aspect is ensuring that top leaders take on a meaningful role, working throughout campus to build shared understanding and define problems and opportunities, and being deliberate in requiring digital solutions. To supplement their digital preparedness, in-house leaders might need development of leadership skills, exposure to more digital experiences, broader perspectives, and opportunities for accomplishing results with others. Some of these leaders may have already developed expertise in digital strategies and applications; others may be keenly interested (Cohen et al., 2022). From an organization standpoint, most relevant actors are students, faculty, organic units, college administrators, corporate leaders, content providers, rectors, or schools. From the education dimension, most of the related actors are students, teachers, industries, organic units, digital platforms, governments, faculty preparation units, information systems, or communities. From a societal dimension, major technologies considered in digital transformation processes are digital technologies, social networks, learning management systems, big data, digital education technologies, software, machine learning, computers, and information systems (Elgohary, 2022). The districts' professional development, instructional technology support, content curation, learning management systems, and asset management approaches are integrated in our digital transformation. In this regard, using digital instructional resources is perceived to allow new roles for teachers and students, creating flexible, motivating ways to learn that are more self-directed and collaborative. The integration of technology into education is expected to respond to challenges from a wider digital transformation of society, as well as promote digital literacy and twenty-first-century skills. Some authors define digital leadership in education as accepting, adopting, and applying emerging technologies in order to transform schools into digital-age places of learning (Milenkova & Lendzhova, 2021). Therefore, the transformations occurring at digital and administrative density have altered the view of leadership, too, and have affected the education industry in terms of its revised conceptions of digital leadership. In this sense, it is the goal of school administrators to design and guide the overall vision of institutional excellence, create digital learning cultures, and promote innovation, the establishment of professional learning environments, and appropriate uses of IT.

21.3 Digital Leadership Roles and Technology Capabilities During the COVID-19 Pandemic

Even before the coronavirus pandemic, it has always been essential for the education systems to support teacher professional learning and development around digital citizenship in ways that tailor instruction to individual students' needs. Facilitating professional learning opportunities for employees is another role of teachers' leaders. Just as students who are learning through digital tools require time and support from schools and districts, so do your teachers who require professional learning investments by the leadership teams who are leading the charge in digital learning (Udeogalanya, 2021).

The rewards to leaders who innovate and work collaboratively with teachers at their schools and school systems will be stronger, safer, and more successful student populations that are better equipped to keep up digital learning, whether at school or home, possessing enduring resilience and yielding greater transition and workforce readiness. Creating schools and universities that work for students requires digital leaders who articulate a bold vision of change, one that does not just challenge the status quo embedded in an industrialized education model, but one that also embraces the inherent value of technology in augmenting the teaching and learning process (Lawson et al., 2021). For years, school leaders have been pushing for education technology implementations in ways that actually affect students learning. Now, it is time to go beyond this, where the digital technology becomes just one more thing educator must teach and students must learn. Educational leaders today must strategize about how they will deploy technology at their schools, taking the long-term view of a topic that is always changing. They are charged both with managing staffing and managing a shift toward teaching and learning differently, equipping radically different students to enter a radically different world. Teachers are playing into a classrooms' technology-enabled vision, encouraging their students to engage in learning processes using technology (Kesavan et al., 2022).

Additionally, in the nearest future, all administrators, teachers, and school or university personnel are going to be using technology in the blended learning approach in order to effectively help students achieve higher standards and be prepared for the workforce of tomorrow. Digital technology tools will provide records of students learning histories and ways to manage learning progress and activities. Technology tools providing feedback and data in real-time enable teachers to meet students where they are and to assist students in developing individualized learning plans (Singh et al., 2021). Teachers must also manage technology usage across digital platforms to make sure students are using it for its intended purpose. Beyond this basic level of device access and Internet, systems, supports, and practices need to be in place to ensure students develop the skills and competencies to best use these digital resources. While individual educators may be developing the skills needed to help students succeed in a new economy, if this capability is not distributed across a school, and if schools do not have the tools needed for teachers to teach effectively in the digital world, students ultimately will suffer. If the schools' leaders are not capable of creating a digital culture, then their schools will struggle. A vision developed by working collaboratively with a large school community – through conversations with students, parents, teachers, administrators, and business communities – will become your lighthouse in times of uncertainty. When shared widely across the many members of a schools' community, the foundational vision becomes woven effortlessly into a schools' culture to serve as a guide to all students' future learning (Tawfik et al., 2021).

The future-ready schools' culture should be developed through building a leadership team, creating a cohesive vision for change, developing a systemic action plan, modeling effective ways for leaders to use digital tools for increased impact, and modeling how teachers can use tools to support student learning. It can provide districts' resources and support to ensure local tech and digital learning plans are aligned with best practices in instruction, implemented by highly trained teachers, and result in customized learning experiences for all students, especially those from traditionally underserved communities (Domina et al., 2021). Districts are embracing college and career readiness as the goal for all students, and they are acknowledging the potential for digital tools to help teachers customize instruction. As online and blended learning reach more and more K-12 students across the world, there is an excellent opportunity to learn from practitioners, study what is fostering success, and share best practices. Working with school leaders, both in a research capacity and a consultancy capacity, is key for understanding more about, and lead their schools and districts in developing and implementing digital learning programs that are transformative and sustainable (Saboowala & Manghirmalani Mishra, 2021). When integrated into the curriculum, technology allows all stakeholders – students, teachers, and parents – to focus on the process of learning and growth, rather than the product. From the outset, a leadership team for digital learning can plan for how they will intentionally solicit feedback from stakeholders during implementation and convey to teachers, parents, families, and students the progress they are making toward their goals. The leader collaboratively establishes an intent for the initiative and becomes a driving force for the other five components of the schools' leadership for implementing a digital learning initiative (Walker et al., 2021). This standard requires leaders to build their own and others learning capacity and to direct professional learning, to champion it, provide a support system, and distribute leadership and accountability for its effectiveness and outcomes.

21.4 Digital Leadership and Professional Education After the COVID-19

The unprecedented uses of technology during the COVID-19 pandemic and the lockdowns and restrictions it imposed have exposed educational systems shortcomings regarding availability and appropriateness of digital infrastructure, and, simultaneously, revealed significant digital gaps between students and teachers, as well as between schools and countries, leading to increased attention on education digitalization policies (Cheshmehzangi et al., 2021). Post-COVID-19 has precipitated a shift toward digitization by the majority of schools and universities.

After the COVID-19 pandemic, most of organizations worldwide accelerated digital transformation, with many acknowledging that they were able to execute initiatives they had previously encountered resistance. COVID-19 has helped to accelerate digital transformation, and it has created the kind of landscape that will continue to drive innovation and technology adoption going forward. Prior to the pandemic, more than half of companies believed a lack of prioritization was a leading cause for the failure to embrace digital transformation and capabilities (Klein & Todesco, 2021). However, COVID-19 dependence on technology has forced people to take a hard look at what tech investments they had, and how those aligned with what was needed to run an effective, digitally focused company. As businesses start to understand more about the capabilities of these types of cutting-edge technologies, they will also start to see the opportunities they will be presented with, even once COVID-19 is behind us. Because the pandemic has forced brands to remove many barriers that previously stood in the way of digital implementation, like network security, in order for employees to work remotely, companies will find it considerably easier to work on are mote basis (John & Thakur, 2021).

During the shutdowns, multiple ed-tech companies offered free access to certain digital supports for education institutions, teachers, and students for the continuation of instruction, garnering millions of new users in the process. There are new digital platforms featuring teacher-video classes and game-based learning activities, provided teachers with tools to support remote teaching of children who were temporarily withdrawn from school because of COVID-19. They are part of a resource that contributes to the development of education that provides curriculum support and teacher professional development in addition to temporary school closures (Chango et al., 2022). This is just one of many digital e-learning platforms provided by the stakeholders which students have been benefiting from in order to carry out learning while schools are closed based on COVID-19. Other tech-based learning platforms also used to assist students include Google Suite Education, Smart Class, Microsoft Teams, Zoom, and others.

Following the pandemic, the authorities identified critical areas that will assist students, including providing educational services via radio for elementary school students and digital technologies for secondary and higher education (Almaiah et al., 2020). School leaders could hold workshops for teachers, families, and students about digital literacy using remote learning platforms and psychology of adolescents. Key takeaways for students to experience successful digital learning environments, schools, and school systems play a critical role in promoting, modeling, and teaching digital citizenship, working collaboratively with parents as their children's primary educators (Cabelkova et al., 2022). In the digital learning environment, it is important that school leaders and teams leading the digital learning effort think about how to use technology effectively, work collaboratively with parents, set protections, and create safe, supportive communities. Just as students who are learning using digital tools require time and support from schools and districts, so do the teachers, who require professional learning investments by the digital learning leadership team. In addition, digital learning leadership teams need to think about how you will thoroughly communicate legal obligations and the school systems policies laid out in the last section with students, parents, families, and teachers (Zimmer & Matthews, 2022). As the schools' leaders, acting as the shared leader between the teachers' organizations and administrative delegated by the Ministry of Education, a principal's engagement with managing the pandemic situation along with the schools is critical. Many international scholars have highlighted the importance of community and family involvement to help school leaders navigate

COVID-19 challenges. We also need to focus on how school leaders engaged stakeholders in education (families and communities) throughout the pandemic. Harris and Jones (2022) highlight how a number of community-based initiatives helped those within schools, including school leaders, cope with the challenges of COVID-19. Froiland (2021) emphasizes how parents and careers are still important partners in education, and how their work with schools during the pandemic has been extremely influential in providing many young people with access to effective education. Unfortunately, students and teachers, including school authorities, are likely to have little technical capacity to handle pandemic-driven increases in remote instruction. In many ways, education has been redefined during the pandemic as an experience where the face-to-face bond between teachers and students has been broken. The lockdown experience and resulting experiment with online learning, viewed by many teachers, students, and families, as traumatizing, as well as by policymakers, may have exaggerated the significance of online education and digitization to have in the post-COVID-19 era. Many authors suggest that during the pandemic, a new model of school leadership has evolved – one that is distributed, digital, networked, and reliant on technology infrastructure. Even within schools, with students opting for remote classes in many school systems and universities, teachers are becoming more proficient in creating virtual lesson plans.

21.5 Conclusions

To sum this all up, it is essential that we take this time to innovate, to act proactively, and to adapt to the post-COVID-19 world; higher education needs to redefine its future, and to act on it. For certain stakeholders at the national and international levels (i.e., MoEs, European Commission, World Bank, and the OECD), the coronavirus crisis has marked a pivotal moment, which should be taken as an opportunity to promote digitization in educational systems. In different educational systems, the pandemic has spurred debates about the limits and potentialities of learning platforms and online learning, the need to reconsider the role of exams and external assessment, and/or how to design effective policies for dealing with new forms of education inequalities. This contribution provided an overview of the impacts of the coronavirus disease crisis in educational systems around the world and considers the potential role of the characteristics of national education systems in mediating countries responses to common external challenges created by the crisis – predictors of perceived job strain for teachers and guidance counselors in a transitional period for in-school e-learning in the COVID-19 pandemic. Our results illuminate challenges and opportunities for education systems in the post-pandemic world. Challenges and opportunities in education after COVID-19 includes developing skills among students, creating new learning methodologies, and flexibility in assessment. Teacher inoculations, more support, and resources were also features of countries responses to the pandemic. The Organization for Economic Cooperation and Developments survey confirmed that tracking responses to the pandemic from

30 education systems around the world. According to it, the COVID-19 crisis has brought an unprecedented level of educational disruption, with school closings impacting 1.2 billion students and young people around the globe. Even prior to school closings due to the COVID-19 pandemic, UNESCO estimated that there was a teacher shortage in order to reach education for all. The education crisis is real, and has been exacerbated even further by COVID-19, even as education ministries throughout the world are working on measures to reopen schools. In addition to the direct effects on education from the coronavirus disease crisis, economic and social crises stemming from the pandemic are expected to raise the levels of poverty among children and young people in many countries, with significant impacts on inequality of access to education. The COVID-19 pandemic, which has been experienced globally since the end of 2019, has had major impacts on health, economy, finance, society, culture, environment, politics, and education. The COVID-19 crisis has not only undermined confidence in the ability of public services like education and healthcare to respond to the crisis; it has raised more fundamental concerns about a potential erosion of human rights, both in the present moment and after the crisis. The COVID-19 pandemic has also challenged the fit, vitality, and resilience of the operating models, practices, and systems of universities. There are few silver linings in the global pandemic, but for higher education, the pandemic has been a chance to demonstrate that universities can quickly adapt. With each crisis comes deeper challenges and opportunities to transform – and the past crises in higher education show it is possible to rebuild in better ways. When COVID-19 dimmed the lights across universities and schools, it dramatically exposed the long-ignored cracks and limitations of its educational system. Without question, COVID-19 created new and difficult circumstances for the governments to support our scholars. It becomes apparent that learning opportunities and instructional quality varied widely – that there were clear disparities in access to quality instruction, whether it was online or in-person. In response to the challenges presented by COVID-19, curriculum affair teams in various countries developed and implemented a blended learning model that prioritizes safety while providing critical in-person learning experiences. One effort that the educational systems have been engaged in over the years is an attempt to foster greater, cross-professional learning. This can mean working with online educational platforms, or cultivating partnerships with international providers. In addition, universities can also be re-examined as a catalyst of growth and development, making sure it is possible to build ecosystems of learning – schools, vocational and higher education, apprenticeship systems, and universities that do not operate linearly, but rather that ease students' movement across structures. With the education and schools' ecosystems in place, our society can gain the necessary leverage to nest developments such as stacked credentials or micro-credentials and have them actually work for its benefits, with each playing its own role, with each respected not only as an institution but by colleagues and students. Online education is going to be a regular part of modern education. Education is a critical ally for the protection of rights and for defending the rule of law, and it is a vital life skill during times of crisis. UNESCO and the United Nations emphasize the importance of education, which can foster an understanding of human

rights, ultimately helping to create fairer, more sustainable, inclusive societies and economies, more resilient to crises. While the inclusion of remedial education, social-emotional learning, and digital technologies into the curriculum will be challenging for most countries, authorities and governments worldwide agree that remedial education is an aspirational goal; they should support countries in reaching this year and beyond, as education systems begin to recover from the current crisis. They offer perspectives from both within and outside of various governments, although their insights could arguably assist the many countries around the world struggling to keep their education going amid the deadly coronavirus pandemic. It has been inspiring to follow how students are being affected by the pandemic, and what the pandemic means for the education going forward. Regarding the process of sharing experiences with other teachers, it can be stressed that such opportunities offer space for intellectual, social, and scholarly oxygenation, since one loses at times the more global perspective on educational phenomena and the preparation of future professionals. Having an opportunity to share one's knowledge an experience with other teachers with common interests, but from different disciplines, becomes a very positive thing that can increase the leadership element of education in the post-pandemic world.

References

Alenezi, M. (2021). Deep dive into digital transformation in higher education institutions. *Education Sciences, 11*(12), 770. https://doi.org/10.3390/educsci11120770

Almaiah, M. A., Al-Khasawneh, A., & Althunibat, A. (2020). Exploring the critical challenges and factors influencing the E-learning system usage during COVID-19 pandemic. *Education and Information Technologies, 25*(6), 5261–5280. https://doi.org/10.1007/s10639-020-10219-y

Archambault, L., Leary, H., & Rice, K. (2022). Pillars of online pedagogy: A framework for teaching in online learning environments. *Educational Psychologist, 57*(3), 178–191. https://doi.org/10.1080/00461520.2022.2051513

Bellibas, M., Kılıncc, A., & Polatcan, M. (2021). The moderation role of transformational leadership in the effect of instructional leadership on teacher professional learning and instructional practice: An integrated leadership perspective. *Educational Administration Quarterly, 57*(5), 776–814. https://doi.org/10.1177/0013161X211035079

Cabelkova, I., Smutka, L., & Strielkowski, W. (2022). Public support for sustainable development and environmental policy: A case of The Czech Republic. *Sustainable Development, 30*(1), 110–126. https://doi.org/10.1002/sd.2232

Chango, W., Lara, J. A., Cerezo, R., & Romero, C. (2022). A review on data fusion in multimodal learning analytics and educational data mining. *Wiley Interdisciplinary Reviews: Data Mining and Knowledge Discovery*, e1458. https://doi.org/10.1002/widm.1458

Cheshmehzangi, A., Sedrez, M., Ren, J., Kong, D., Shen, Y., Bao, S., & Dawodu, A. (2021). The effect of mobility on the spread of COVID-19 in light of regional differences in the European Union. *Sustainability, 13*(10), 5395. https://doi.org/10.3390/su13105395

Cohen, A. B., Stump, L., Krumholz, H. M., Cartiera, M., Jain, S., Scott Sussman, L., & Navai, N. (2022). Aligning mission to digital health strategy in academic medical centers. *NPJ Digital Medicine, 5*(1), 1–8. https://doi.org/10.1038/s41746-022-00608-7

Dale, V., Kubincová, E., Kerr, J., & Murray, J. A. (2021). Lessons learned from being BOLD: Staff experiences of an institutional strategic project in blended and online learning development.

Journal of Perspectives in Applied Academic Practice, 9(2), 29–38. https://doi.org/10.14297/jpaap.v9i2.484

Domina, T., Renzulli, L., Murray, B., Garza, A. N., & Perez, L. (2021). Remote or removed: Predicting successful engagement with online learning during COVID-19. *Socius, 7*, 237802312098820. https://doi.org/10.1177/2378023120988200

Elgohary, E. (2022). The role of digital transformation in sustainable development in Egypt. *The International Journal of Informatics, Media and Communication Technology, 4*(1), 71–106. https://doi.org/10.21608/ijimct.2022.219953

Froiland, J. M. (2021). A comprehensive model of preschool through high school parent involvement with emphasis on the psychological facets. *School Psychology International, 42*(2), 103–131. https://doi.org/10.1177/0143034320981393

García-Peñalvo, F. J. (2021). Avoiding the dark side of digital transformation in teaching. An institutional reference framework for eLearning in higher education. *Sustainability, 13*(4), 2023. https://doi.org/10.3390/su13042023

Gu, C., Sun, J., Chen, T., Miao, W., Yang, Y., Lin, S., & Chen, J. (2022). Examining the influence of using first-person view drones as auxiliary devices in matte painting courses on college students' continuous learning intention. *Journal of Intelligence, 10*(3), 40. https://doi.org/10.3390/jintelligence10030040

Guliyeva, A., Aliyeva, T., Rzayeva, U., & Huseynova, R. (2021). Lifelong education as a challenge of digital era. *Turkish Journal of Computer and Mathematics Education (TURCOMAT), 12*(6), 1282–1289. https://doi.org/10.17762/turcomat.v12i6.2468

Harris, A., & Jones, M. (2022). Leading during a pandemic–what the evidence tells us. *School Leadership & Management, 42*(2), 105–109. https://doi.org/10.1080/13632434.2022.2064626

Huang, F., Teo, T., & He, J. (2021). Digital nativity of university teachers in China: Factor structure and measurement invariance of the Digital Native Assessment Scale (DNAS). *Interactive Learning Environments, 29*(3), 385–399. https://doi.org/10.1080/10494820.2019.1570278

John, J., & Thakur, R. (2021). Long term effects of service adaptations made under pandemic conditions: The new "post COVID-19" normal. *European Journal of Marketing, 55*(6), 1679–1700. https://doi.org/10.1108/EJM-08-2020-0607

Kesavan, N., Seenuvasaragavan, S., Burnette, K., Heim, A. B., Patel, R. J., Johnston, J., Cao, X., Sakandar, H. A., Zhuang, Y., Oi, K. K., Zhi, Y., Diao, Y., & Strielkowski, W. (2022). Pandemic-inspired policies. *Science, 377*(6601), 22–24. https://doi.org/10.1126/science.add2046

Klein, V. B., & Todesco, J. L. (2021). COVID-19 crisis and SMEs responses: The role of digital transformation. *Knowledge and Process Management, 28*(2), 117–133. https://doi.org/10.1002/kpm.1660

Lampos, V., Mintz, J., & Qu, X. (2021). An artificial intelligence approach for selecting effective teacher communication strategies in autism education. *NPJ Science of Learning, 6*(1), 1–10. https://doi.org/10.1038/s41539-021-00102-x

Lawson, H. A., Jones, E., Beddoes, Z., Estes, S., Morris, S. A., Mitchell, M. F., & Ward, P. (2021). Collective action for learning, improvement, and redesign. *Journal of Teaching in Physical Education, 40*(3), 412–422. https://doi.org/10.1123/jtpe.2020-0246

Li, Q., Li, Z., & Han, J. (2021). A hybrid learning pedagogy for surmounting the challenges of the COVID-19 pandemic in the performing arts education. *Education and Information Technologies, 26*(6), 7635–7655. https://doi.org/10.1007/s10639-021-10612-1

Lohr, A., Stadler, M., Schultz-Pernice, F., Chernikova, O., Sailer, M., Fischer, F., & Sailer, M. (2021). On powerpointers, clickerers, and digital pros: Investigating the initiation of digital learning activities by teachers in higher education. *Computers in Human Behavior, 119*, 106715. https://doi.org/10.1016/j.chb.2021.106715

Marouli, C. (2021). Sustainability education for the future? Challenges and implications for education and pedagogy in the 21st century. *Sustainability, 13*(5), 2901. https://doi.org/10.3390/su13052901

Milenkova, V., & Lendzhova, V. (2021). Digital citizenship and digital literacy in the conditions of social crisis. *Computers, 10*(4), 40. https://doi.org/10.3390/computers10040040

Monteiro, S., Isusi-Fagoaga, R., Almeida, L., & García-Aracil, A. (2021). Contribution of higher education institutions to social innovation: Practices in two southern European universities. *Sustainability, 13*(7), 3594. https://doi.org/10.3390/su13073594

Núñez-Canal, M., de Obesso, M. D. L. M., & Pérez-Rivero, C. A. (2022). New challenges in higher education: A study of the digital competence of educators in Covid times. *Technological Forecasting and Social Change, 174*, 12127. https://doi.org/10.1016/j.techfore.2021.121270

Rock, W. D., & Curry, J. R. (2021). Louisiana School Counselors' daily activities and the American School Counselor Association National Model: A complex history and a hopeful future. *International Journal of Education Policy and Leadership, 17*(5), n5. https://doi.org/10.22230/ijepl.2021v17n5a1089

Rodríguez-Abitia, G., & Bribiesca-Correa, G. (2021). Assessing digital transformation in universities. *Future Internet, 13*(2), 52. https://doi.org/10.3390/fi13020052

Saboowala, R., & Manghirmalani Mishra, P. (2021). Readiness of in-service teachers toward a blended learning approach as a learning pedagogy in the post-COVID-19 era. *Journal of Educational Technology Systems, 50*(1), 9–23. https://doi.org/10.1177/00472395211015232

Singh, J., Steele, K., & Singh, L. (2021). Combining the best of online and face-to-face learning: Hybrid and blended learning approach for COVID-19, post vaccine, & post-pandemic world. *Journal of Educational Technology Systems, 50*(2), 140–171. https://doi.org/10.1177/00472395211047865

Srivastava, S., Heim, A. B., Yoho, R., Moore, E. B., Richter, W. E., Strielkowski, W., Li, K., Penders, B., Sojli, E., & Cao, B. (2021). Imagining a teaching utopia. *Science, 374*(6563), 31–32. https://doi.org/10.1126/science.abm1361

Tawfik, A. A., Shepherd, C. E., Gatewood, J., & Gish-Lieberman, J. J. (2021). First and second order barriers to teaching in k-12 online learning. *TechTrends, 65*(6), 925–938. https://doi.org/10.1007/s11528-021-00648-y

Udeogalanya, V. (2021). Aligning digital literacy and student academic success: Lessons learned from COVID-19 pandemic. *The Business & Management Review, 12*(1), 275–284. https://doi.org/10.24052/BMR/V12NU01/ART-26

Vilchez, J. A., Kruse, J., Puffer, M., & Dudovitz, R. N. (2021). Teachers and school health leaders' perspectives on distance learning physical education during the COVID-19 pandemic. *Journal of School Health, 91*(7), 541–549. https://doi.org/10.1111/josh.13030

Walker, J. D., Johnson, K. M., & Randolph, K. M. (2021). Teacher self-advocacy for the shared responsibility of classroom and behavior management. *Teaching Exceptional Children, 53*(3), 216–225. https://doi.org/10.1177/0040059920980481

Zahorodna, O., Saienko, V., Tolchieva, H., Tymoshchuk, N., Kulinich, T., & Shvets, N. (2022). Developing communicative professional competence in future economic specialists in the conditions of postmodernism. *Postmodern Openings, 13*(2), 77–96. https://doi.org/10.18662/po/13.2/444

Zimmer, W. K., & Matthews, S. D. (2022). A virtual coaching model of professional development to increase teachers' digital learning competencies. *Teaching and Teacher Education, 109*, 103544. https://doi.org/10.1016/j.tate.2021.103544

Chapter 22
Sustainability of Economics and Education After the COVID-19 Pandemic Based on the Formation of Network Structures

Elena Korneeva, Raisa Krayneva, and Aizhan Omarova

Abstract This paper contemplates upon the sustainability of economics and education after the COVID-19 pandemic based on the formation of network structures. The primary objective of the present study is to assess the effects of the COVID-19 pandemic on sustainability psychology (which can be expressed by the quality of life), sustainability, as well as the global economic sustainable development. We show that the COVID-19 pandemic is likely to have serious impacts worldwide on the quality of life, on the sustainability of the world's politics, environment, and economics, as well as the world economy. The impacts of the COVID-19 pandemic are not limited to sustainable development because challenges related to the global trade governance are also present. There is a high level of uncertainty associated with COVID-19, which has affected the economic indicators, sustainable criteria, and development processes. Policy-makers and stakeholders worldwide need to consider these challenges and ensure remedies fostering the post-pandemic development of leadership in business and economics.

Keywords Sustainability · Economics · Education · COVID-19 · Network structures

---

E. Korneeva (✉)
Department of Mass Communications and Media Business, Financial University under the Government of the Russian Federation, Moscow, Russian Federation

Institute of Economics, Finance and Management, Togliatti State University, Togliatti, Russian Federation
e-mail: ENKorneeva@fa.ru

R. Krayneva
Department of Mass Communications and Media Business, Financial University under the Government of the Russian Federation, Moscow, Russian Federation
e-mail: RKKrayneva@fa.ru

A. Omarova
Yessenov University, Aktau, Kazakhstan
e-mail: Aizhan.omarova@yu.edu.kz

W. Strielkowski (ed.), *Leadership, Entrepreneurship and Sustainable Development Post COVID-19*, Springer Proceedings in Business and Economics, https://doi.org/10.1007/978-3-031-28131-0_22

JEL Classification I20 · L14 · Q01

22.1 Introduction

The COVID-19 pandemic has long-lasting and profound effects for the sustainability of economics and education. Not only the economic losses but also COVID-19's pandemic impacts social parameters such as changes in sustainable psychological development did matter. In addition to the direct human suffering caused by the COVID-19 crisis itself, as well as millions losing their livelihoods, the COVID-19 pandemic has highlighted several critical vulnerabilities in our social and economic systems (Newell & Dale, 2021; Štreimikienė et al., 2022). The government's top priorities in dealing with the COVID-19 pandemic were addressing the healthcare crisis and undertaking swift economic relief measures, the latter mostly focused on providing critical liquidity and protecting livelihoods amid sudden losses in revenue (Khalid et al., 2021). The vulnerabilities exposed in the COVID-19 crisis can serve as an underscore for why, prior to COVID-19, environmental issues were emerging as the top policy priorities globally (Strielkowski et al., 2022). What is quite concerning is that many people today did not think it necessary to refer to the COVID-19 emergency in their education on sustainability, regardless of the global ecological, social, and economic impacts of this pandemic, that is, reductions in CO2 emissions from major cities and reductions in consumption (Biffi et al., 2021). With regard to the above, there is a need for education to recover from COVID-19 (Guliyeva et al., 2021). In terms of analyzing research findings from included studies, it is quite evident that impacts from the pandemic were assessed across one or more factors (health, economics, the environment, resilience, or governance). There are reports and research papers currently separately discussing the impacts of rapid COVID-19 pandemic expansion on health systems, mental health, resilience, and global economy (Haldane et al., 2021; Schmuck et al., 2022). In this paper, we use literature reviews and individual insights to offer lessons on COVID-19 pandemic digitalization on the sustainability of various countries around the world from the perspectives of technology for the greater social good. This paper elaborates the foundations for supporting digital transformation after COVID-19 to sustainably develop and suggests practical and research implications for strengthening the success of business enterprise in the post-COVID-19 digital transformation. We further suggest additional areas for research that will improve successful digital transformation of economics and education after the coronavirus pandemic. While efforts are underway to develop political and economic interventions for COVID-19, the social and behavioral sciences may offer valuable insights for managing the global pandemic and its impacts. Responding to the coronavirus pandemic requires a comprehensive partnership between governments, international organizations, and the business community, a focus and the mission of the international organizations or the public-private collaboration.

In just weeks, coronavirus has changed how students are taught across the world and caused the digital surge in the educational systems worldwide bringing the "creative destruction" into this sector (Kesavan et al., 2022). While it is still too early to assess how responses to coronavirus will affect education systems across the globe, there are signs suggesting that it may have lasting effects on the trajectory of innovation in learning and the digitalization. COVID-19 has been the catalyst that has led educational institutions worldwide to seek out innovative solutions within a relatively short period of time (Maini et al., 2021). It is evident that COVID-19 affects all types of resilience, and its effects in higher education provide an opportunity to use universities, where faculty are employed, as living laboratories for teaching resilience, adaptability, fragility, organizational change management, marketing, accounting, and other resilience-related content. This evidence points to the missed opportunity for faculty in the field of sustainability in a variety of curriculums to leverage the tragedy of COVID-19 and its multiple impacts, including those on campus operations, as a gateway to discussions about resilience, the wicked problem, and sustainability (Oliveira et al., 2021). One expected COVID-19 impact for universities, suggested below, is that universities will invest more in developing faculty expertise in remote teaching so that they are better prepared to deal with future pandemics. A longitudinal study on the effects of COVID-19 on universities will be useful to clarify this potential dynamic of responses.

22.2 COVID-19 and Disruptions in the Global Equity

Following the pandemic, governments and researchers worldwide seek to provide guidance about areas where sovereign debt managers may have to react to challenges stemming from the COVID-19 crisis. There should be balanced actions and priorities that responsible supervisors should consider when planning medium- to long-term responses to the unfolding COVID-19 crisis and communicating about the crisis and regulatory adjustments. A broader guidance is needed for national authorities regarding potential responses by central banks to disturbances to money, securities, and currency markets likely to arise following financial disruptions, including a COVID-19 shock. In addition, there should be an appropriate central banks' policy response to the serious economic and financial impacts of the COVID-19 pandemic for emerging markets and developing economies (Beirne et al., 2021).

Furthermore, in the light of the recent events, it is important to focus on the disproportionately large impacts the ongoing economic disruption caused by the COVID-19-inspired war conflict in Ukraine that is having considerable impacts on the power sectors of emerging market and developing economies (Nezhyva & Mysiuk, 2022). The war conflict in Ukraine has not only caused massive humanitarian, migration, and refugee crises but has added a layer of downside risk for a global economy that is still dealing with the COVID-19 pandemic (Zakeri et al., 2022). The speed of change stemming from the pandemic is creating new forms of financial and operational risks, as companies struggle to cope with inflation, capacity constraints, and supply chain disruptions. This is a period of massive disruptions to global supply chains, with myriad challenges, from shifting consumer demands and reliability offshore to transport gridlock, anticompetitive practices, and geopolitical complications (Kang et al., 2021). Following the COVID-19 pandemic, the supply chains are now efficient, yet fragile – vulnerable to collapse in the face of pandemic, war, or other related disaster. The disruptions and shortages in the COVID-19 pandemic have exposed weaknesses in global supply chains that are already facing threats arising from trade wars. With most of the developed world moving toward offshore production and "just-in-time" supply chains, it did not take long for global supply chains to be stretched thin and strained following COVID-19's arrival (Magableh, 2021). Companies accustomed to producing critical products in the developed world, with fast turn-around times for global markets, found the pandemic to be devastating for components, production, distribution, public health, and labor. The explosion of demand for e-commerce and consumers, as well as the challenges of public health, in the COVID-19 pandemic has placed strains on production and distribution facilities worldwide. During the pandemic, as demand surged across multiple product categories, manufacturers were struggling to move from providing for one segment of the market to providing for another, or producing one type of product to producing another (Pu et al., 2021). The continued economic challenges from the COVID-19 pandemic made meeting these demands more fiscally difficult, and it was even more critical that they had access to additional markets. A decades-long focus on streamlining the supply chain – to keep costs down, inventory down, and assets utilization high – has removed buffers, and flexibility for absorbing disruptions and COVID-19 demonstrates how many companies are completely unaware of their supply chain vulnerabilities to global disruptions (Choksy et al., 2022). Moreover, there is little question that COVID-19 has tested supply chain leaders' global creativity, resilience, and flexibility in trying to keep critical operations running. Businesses and governments will have to redefine the meaning of global supply chain resilience. There is an ongoing debate about COVID-19's effect on global manufacturing structures and on global supply chains.

Fairness above that the level of the distribution of benefits at the national level and at the level of global competition is critical for maintaining public support for the trade and open markets needed to overcome and recover from COVID-19. Beyond immediate responses to the crisis, the COVID-19 pandemic must serve as an impetus for sustained gains and the acceleration of implementation of long-overdue measures that put the world on a path toward more sustainable development and that make the world's economies more resilient to future shocks. By 2025, the costs associated with coronavirus are expected to be \$16–\$35 trillion globally (Silva et al., 2021). These would be the expenses brought about by another disaster, just like COVID-19, or some other catastrophic even such as the recent war conflict in Ukraine.

All in all, with global living connected, as well as international travel and trade, companies need to plan for epidemics and periodic pandemics and to have public health systems ready to respond to large outbreaks. The supply shock starting with China, and the subsequent demand shock, with a shutdown of the world economy, exposed vulnerabilities in the production strategies and supply chains of firms almost everywhere. Even with government incentives, China took 20 years to develop a domestic base that could provide the overwhelming majority of electronics components, car parts, chemicals, and pharmaceutical ingredients needed to manufacture at home. This experience should not be forgotten and can be taken into account when planning the new strategies for the revitalization of business and economy in the post-COVID era.

22.3 Global Pandemic as an Element of Tensions in Supply Chains

The recent US–China trade wars, the ravages of the global pandemics, disruptive disruptions in the march of technology are all driving a radical reconsideration of the value of global, distributed, lean supply chains. Disruptions to the supply chains of fewer basic goods and shortages of essential medicines during the COVID-19 outbreak highlighted interconnectedness across countries via global value chains (GVCs) and renewed the debate about the costs and benefits of globalization (Barone et al., 2021). The policy debate over whether benefits of international specialization through global value chains outweigh associated risks for the spread of the outbreak has been heightened since the outbreak of COVID-19 and resulting supply chain disruptions in certain manufactured goods and medical products. The economic turmoil caused by the pandemic has exposed numerous vulnerabilities in supply chains and raised questions on globalization. The disruptions and shortages caused by the COVID-19 pandemic are exposing weaknesses in the global supply chains that are already facing threats of trade wars (Stevano et al., 2021). With most of the developed world moving toward offshore production and "just-in-time" supply chains, it did not take long for global supply chains to be stretched thin and strained following COVID-19's arrival. Value chain disruption was already a problem, with the COVID-19 pandemic being mostly contained in China. To be sure, challenges during the COVID-19 pandemic did not illuminate the interdependent or global nature of supply chains; instead, they highlighted how most organizations are ill-equipped to handle such interconnections when unfavorable impacts arise (Dirzka & Acciaro, 2022). Along with concerns about China, the COVID-19 pandemic exposed an inherent fragility in long, complex supply chains, which rely on unimpeded flows of ideas, designs, intermediates, and finished products for their operation. This is a moment of immense disruption for the world's supply chains, which have myriad challenges, from shifting consumer demands and off-shore reliability to transport gridlock, anticompetitive practices, and geopolitical complications. The supply chains are now efficient, yet fragile – vulnerable to collapse in the face of a pandemic, a war, or natural disaster (Nikookar et al., 2021). For companies competing globally, things can change rapidly – and not only because of an unexpected disruption to supply chains caused by a pandemic but also because of disruptive technologies or sudden shifts in markets. As chip and auto supply chains have shown, disruptions in one part of the world can have significant impacts on operations around the world. Managing supply chain disruptions is challenging, for global as well as local businesses, and requires risk-versus-economic-efficiency calculations by each firm. The situation has been particularly challenging for businesses with complex supply chains, since their output is susceptible to interruptions caused by the lack of inputs from other businesses (Khan et al., 2022). Most global manufacturers have some operations in China, and many businesses are reporting production and trade disruptions with that crucial partner. The supply shock starting in China in February 2020, and the subsequent demand shock, when the global economy stalled, exposed vulnerabilities in the manufacturing strategies and supply chains of firms almost everywhere. Firms used to produce critical products in the developing world, with fast turnarounds in global markets, found the pandemic was devastating components supplies, production, distribution, public health, and labor (Espitia et al., 2022). As borders began closing, nations' economies around the globe discovered the crippling lack of redundancy in supply chains needed to manufacture and distribute products critical to their economies, from vaccines and health supplies to critical components like semiconductors, and a broad range of technologies relying on semiconductors. Importantly, however, global medical shortages resulted from an unprecedented demand shock caused by the global spread of the pandemic rather than a supply side one (Ali & Liberty, 2022). This is not to mention the reports of shortages caused by the pandemic in sectors as diverse as construction, health, agriculture, and retail, which led to increased focus on supply chain organizations and the degree to which Australia-China is dependent on imports. The evaluation results demonstrated that, both for exports and imports, trade interruptions between Australia and the United States-China – even when offset through alternate supply chains – would cause both countries, as well as the world, economic losses and emissions increases (Barai & Dhar, 2021). Many studies show that although some economies might benefit from supply chain diversification, forced realignment would be counterproductive to global trade dynamics – thus incurring economic and environmental costs, and meaning there is deadweight loss to the global community (Colgan et al., 2021). Such an intervention into the supply chain may weaken global trade regimes and economic integration, especially in East Asia, which has significantly benefitted from integrated regional production networks. Impacts of COVID-19 pandemic have had significant impacts on the various businesses and activities, especially health care and food supplies. Phenomena of panic buying, interrupted supplies, a tendency toward localization, limitations in the availability of commodities, and transport problems, and political, social, and health-related issues related to COVID-19 are major phenomena stemming from the adverse effects of the pandemic. This is proved by the fact that the introduction of the global supply chain (GSC) disruptions began following the declaration by the

World Health Organization (WHO) that the outbreak of coronavirus disease was a global health emergency in the beginning of 2020. Thence, organizations need to revisit their supply chain strategies in light of everything they learned from COVID-19 and prioritize capabilities that they anticipate will be needed going forward in order to thrive in this new normal. Other studies examine the act of rebalancing that confronts many companies across commodity-producing value chains as they try to manage risks – not the constant challenges of doing business, but more deep-seated disruptions like financial crises, terrorism, extreme weather, and, yes, the pandemic. The devastation caused by the COVID-19 pandemic, erosions of global norms around innovation and trade, and the worsening geopolitical order have led to calls to redefine globalization.

22.4 Impact of COVID-19 on Socioeconomic Domains

It is obvious that COVID-19 has created a global health crisis in which untold numbers are dying, human suffering is spreading, and lives are being disrupted for individuals (Zerrusen, 2021). The situations created by the COVID-19 pandemic have been an emerging threat, which has increased social inequalities. The social crisis created by the outbreak of COVID-19 could also heighten inequalities, discrimination, and mid- to long-term unemployment, unless adequately addressed through adequate policies (Clouston et al., 2021).

Addressing the negative side, economic impacts of the pandemic and structural inequalities will require significant policy reforms across large economies. Some 2–3 years from now is far too early to expect the economic effects of the pandemic, as well as the fissures of inequality and healthcare inequalities that it has exposed, will recede. Many of the economic impacts of COVID-19 will still be felt in 5 years, from the inner cities never fully recovering their economic vitality to long-term wage depressions among those laid off or entering the labor force during the pandemic (Blundell et al., 2022). The global socioeconomic impacts of COVID-19 include higher rates of unemployment and poverty, lower oil prices, altered educational sectors, changes in the nature of work, lower GDP, and increased risks for healthcare workers. COVID-19 has increased human suffering, eroded economies, turned lives for billions around the world upside down, and has had significant impacts on health, the economy, environment, and society. In addition, COVID-19 represents a truly global pandemic, which can happen once in a century and which has disrupted economic activities and poses severe risks for general well-being. The coronavirus pandemic is impacting peoples' lives across the globe, but people in developing countries are likely to suffer most in this global crisis. COVID-19 is clearly increasing inequality globally, and the worst effects of the pandemic are not yet visible because of the lack of reporting from developing countries, many of which are expected to remain undeveloped (Clemente-Suárez et al., 2022).

Returning to the COVID-19 pandemic, it can be noted that developing countries and lower-income countries are being affected by COVID-19 in an outsized way, and that this exemplifies the importance of sustainable development goals to address all types of global crises (Bundervoet et al., 2022). Turning to the economic impact of COVID-19, it can be stated that the pandemic is severely undermining international efforts to meet the Sustainable Development Goals. Collective global efforts are also needed to coordinate the healthcare response to COVID-19 by strengthening WHOs capabilities crucial. The global COVID-19 response has demonstrated the highs that can be achieved by working in the spirit of friendship and solidarity, which should endure beyond the pandemic. Many developing countries welcomed the creation of a COVID-19 response and recovery fund by the United Nations noting universal access to appropriate technologies and medical products is crucial for global recovery from the pandemic (Hesselman et al., 2021). Despite the COVID-19 pandemic and ensuing economic crisis, even some low-income economies exhibit significant progress. In short, COVID-19 has not dramatically challenged those structural social realities. This highlights how economic inequalities, which are often translated directly into spatial inequalities, may be becoming more evident in the course of the COVID-19 crisis (Wethal et al., 2022). Populist approaches to containing COVID-19 often made matters worse. Several scientists have recommended that different countries indefinitely ban the wildlife trade to ensure human protection against novel viruses and global pandemics such as COVID-19 (Wyatt et al., 2022). While we had high hopes for technology, the pandemic has taken the digital inclusion/digital inequality gap to new heights with demand for homeschooling, remote working, and telemedicine. As digital transition is being precipitated by COVID-19, preexisting ethical, social, legal, political, and economic implications of artificial intelligence and big data are more crucial than ever. This handbook provides access to a curated list of resources for academic researchers in the social sciences and practitioners in the public policy field researching the coronavirus pandemic.

In many countries, classes are largely mediated by the coronavirus, and it impacts the social and economic impacts in terms of isolation from society and business shutdowns, as well as the availability of health care (Kabeer et al., 2021). The concept needs to be developed to analyze the experiences of households associated with clashing social practices in their household spaces, or ad hoc border traffic, in the COVID-19 crisis. There are some growing social practices that are connected with, and have been actively adapted by, other family practices during COVID-19 lockdown measures that prevented labor within a normal context and within time-space conditions. Working from home in a COVID-19 crisis introduced the new rules and norms that would probably stick and become the new normal. The COVID-19 crisis also forced households to reconstruct and manage social, spatial, temporal, and material boundaries among different practices within the household in order to create spaces for work and leisure in a balanced symbiosis. The burst of these practices in the lockdowns brought an array of new practices to the home and required the re-routinization of family life after the pandemic. Although the lockdown represented a remarkable situation, it may have had lasting effects on work organization. The demands of social segregation and fear put legal brakes on protests, even meetings, but the economic conditions faced by the most vulnerable are likely to be exacerbated under a system that does not take care of its most vulnerable populations.

22.5 Conclusions

All in all, in the current post-COVID times, relevant stakeholders need to find ways how to manage the economic impacts of the COVID-19 pandemic, saying that unity and innovation are how the societies worldwide are going to tackle coronavirus and move toward global recovery. Businesses are invited to collaborate with each other, working together with a number of multistakeholder organizations, to help tackle the coronavirus crisis. As many countries in the world prepare to face the effects of the coronavirus, and infections are rising, biotech firms are channeling millions of dollars in support for relief efforts that are meeting key needs of the communities affected by COVID-19. Organizations such as the International Monetary Fund (IMF) spend millions of dollars of fiscal support, which is also intended to assist various countries with its response to the COVID-19 pandemic. Some countries have also eased requirements to access unemployment benefits and rental assistance to individuals whose economic and social situations have deteriorated as a result of the health crisis. For the economic fallout, the law includes billions of dollars in benefits to unemployed people because of the COVID-19 pandemic, as well as in transfers to impoverished families. Other approved measures on revenues and spending in order to combat the COVID-19 pandemic and minimize its adverse economic impacts are being introduced. The pandemic and the containment measures had significant economic impacts, reducing the GDP by as high as two digits in percents. The analysis found that maintaining economic activity was critical: each week of no economic activity exponentially increased the adverse economic effects of COVID-19. Interestingly, although COVID-19 has led to very sharp decreases in air pollution in developed economies as a result of reduced economic activity caused by lockdowns, these positive impacts caused by the pandemic are only temporary, since they do not reflect changes in economic structures in the global economy. Indeed, during the early months of the novel coronavirus, it has become increasingly clear that a biological threat, whether naturally occurring, accidentally occurring, or intentionally introduced, in any nation, is now posing unquantifiable risks for global health and for the global economy. This continues to reinforce global human interconnections and interdependencies linking our natural environments, economies, and social systems. The United Nations Department of Economic and Social Affairs concluded that COVID-19 pandemics can also exacerbate global isolation, inequality, discrimination, and unemployment over the medium to long term, unless they are appropriately addressed with the most effective policy tools.

In order to help companies to continue navigating the impacts of COVID-19, task forces are created which highlight the critical public health, economic, and sociopolitical dimensions of the crisis. Other measures include the increased spending on the public health, economic, and sociopolitical dimensions of the crisis.

In addition, the spread of the virus has begun affecting financial markets, but how much further the novel coronavirus will stress the larger financial system is unclear. Reversal has made the new coronavirus COVID-19 a global economic shock unlike any since the Great Depression and has already outpaced the 2009 global financial crisis. This issue of trust has only grown more complex throughout the coronavirus outbreak, and this makes it even harder to estimate the effect of the new coronavirus on the world economy. This puts more pressure on governments to wield their budgetary powers in order to address the economic impact of coronavirus. In other words, all countries around the world need to undertake significant expenditures and investments in critical areas of the economy in order to boost economic activity, as well as to minimize the damage to health and prosperity for the public, and limit the impact on supply chains and business sectors.

References

Ali, M. A., & Liberty, F. S. (2022). Effects of Covid-19 pandemic on the Nigerian populace: Focusing short and long term effects on economy and health system. *Human Nature Journal of Social Sciences, 3*(1), 44–55.

Barai, M. K., & Dhar, S. (2021). COVID-19 pandemic: Inflicted costs and some emerging global issues. *Global Business Review*, 0972150921991499. https://doi.org/10.1177/0972150921991499

Barone, A. S., Matheus, J. R. V., de Souza, T. S. P., Moreira, R. F. A., & Fai, A. E. C. (2021). Green-based active packaging: Opportunities beyond COVID-19, food applications, and perspectives in circular economy-a brief review. *Comprehensive Reviews in Food Science and Food Safety, 20*(5), 4881–4905. https://doi.org/10.1111/1541-4337.12812

Beirne, J., Renzhi, N., Sugandi, E., & Volz, U. (2021). COVID-19, asset markets and capital flows. *Pacific Economic Review, 26*(4), 498–538. https://doi.org/10.1111/1468-0106.12368

Biffi, E., Gambacorti-Passerini, M. B., & Bianchi, D. (2021). Parents under lockdown: The impacts of the COVID-19 pandemic on families. *Rivista Italiana di Educazione Familiare, 18*(1), 97–111. https://doi.org/10.36253/rief-10332

Blundell, R., Costa Dias, M., Cribb, J., Joyce, R., Waters, T., Wernham, T., & Xu, X. (2022). Inequality and the COVID-19 crisis in the United Kingdom. *Annual Review of Economics, 14*, 607–636. https://doi.org/10.1146/annurev-economics-051520-030252

Bundervoet, T., Dávalos, M. E., & Garcia, N. (2022). The short-term impacts of COVID-19 on households in developing countries: An overview based on a harmonized dataset of high-frequency surveys. *World Development, 105844*. https://doi.org/10.1016/j.worlddev.2022.105844

Choksy, U. S., Ayaz, M., Al-Tabbaa, O., & Parast, M. (2022). Supplier resilience under the COVID-19 crisis in apparel global value chain (GVC): The role of GVC governance and supplier's upgrading. *Journal of Business Research, 150*, 249–267. https://doi.org/10.1016/j.jbusres.2022.05.068

Clemente-Suárez, V. J., Rodriguez-Besteiro, S., Cabello-Eras, J. J., Bustamante-Sanchez, A., Navarro-Jiménez, E., Donoso-Gonzalez, M., & Tornero-Aguilera, J. F. (2022). Sustainable development goals in the COVID-19 pandemic: A narrative review. *Sustainability, 14*(13), 7726. https://doi.org/10.3390/su14137726

Clouston, S. A., Natale, G., & Link, B. G. (2021). Socioeconomic inequalities in the spread of coronavirus-19 in the United States: A examination of the emergence of social inequalities. *Social Science & Medicine, 268*, 113554. https://doi.org/10.1016/j.socscimed.2020.113554

Colgan, J. D., Green, J. F., & Hale, T. N. (2021). Asset revaluation and the existential politics of climate change. *International Organization, 75*(2), 586–610. https://doi.org/10.1017/S0020818320000296

Dirzka, C., & Acciaro, M. (2022). Global shipping network dynamics during the COVID-19 pandemic's initial phases. *Journal of Transport Geography, 99*, 103265. https://doi.org/10.1016/j.jtrangeo.2021.103265

Espitia, A., Mattoo, A., Rocha, N., Ruta, M., & Winkler, D. (2022). Pandemic trade: COVID-19, remote work and global value chains. *The World Economy, 45*(2), 561–589. https://doi.org/10.1111/twec.13117

Guliyeva, A., Rzayeva, U., & Guliyeva, G. (2021). University life during pandemic COVID-19: Attitudes of staff and students towards distance education. *SHS Web of Conferences, 92*, 01011. https://doi.org/10.1051/shsconf/20219201011

Haldane, V., De Foo, C., Abdalla, S. M., Jung, A. S., Tan, M., Wu, S., et al. (2021). Health systems resilience in managing the COVID-19 pandemic: Lessons from 28 countries. *Nature Medicine, 27*(6), 964–980. https://doi.org/10.1038/s41591-021-01381-y

Hesselman, M., Varo, A., Guyet, R., & Thomson, H. (2021). Energy poverty in the COVID-19 era: Mapping global responses in light of momentum for the right to energy. *Energy Research & Social Science, 81*, 102246. https://doi.org/10.1016/j.erss.2021.102246

Kabeer, N., Razavi, S., & van der Meulen Rodgers, Y. (2021). Feminist economic perspectives on the COVID-19 pandemic. *Feminist Economics, 27*(1–2), 1–29. https://doi.org/10.1080/13545701.2021.1876906

Kang, H., An, J., Kim, H., Ji, C., Hong, T., & Lee, S. (2021). Changes in energy consumption according to building use type under COVID-19 pandemic in South Korea. *Renewable and Sustainable Energy Reviews, 148*, 111294. https://doi.org/10.1016/j.rser.2021.111294

Kesavan, N., Seenuvasaragavan, S., Burnette, K., Heim, A. B., Patel, R. J., Johnston, J., Cao, X., Sakandar, H. A., Zhuang, Y., Oi, K. K., Zhi, Y., Diao, Y., & Strielkowski, W. (2022). Pandemic-inspired policies. *Science, 377*(6601), 22–24. https://doi.org/10.1126/science.add2046

Khalid, U., Okafor, L. E., & Burzynska, K. (2021). Does the size of the tourism sector influence the economic policy response to the COVID-19 pandemic? *Current Issues in Tourism, 24*(19), 2801–2820. https://doi.org/10.1080/13683500.2021.1874311

Khan, S. A. R., Razzaq, A., Yu, Z., Shah, A., Sharif, A., & Janjua, L. (2022). Disruption in food supply chain and undernourishment challenges: An empirical study in the context of Asian countries. *Socio-Economic Planning Sciences, 82*, 101033. https://doi.org/10.1016/j.seps.2021.101033

Magableh, G. M. (2021). Supply chains and the COVID-19 pandemic: A comprehensive framework. *European Management Review, 18*(3), 363–382. https://doi.org/10.1111/emre.12449

Maini, R., Sehgal, S., & Agrawal, G. (2021). Todays' digital natives: An exploratory study on students' engagement and satisfaction towards virtual classes amid COVID-19 pandemic. *International Journal of Information and Learning Technology, 38*(5), 454–472. https://doi.org/10.1108/IJILT-03-2021-0055

Newell, R., & Dale, A. (2021). COVID-19 and climate change: An integrated perspective. *Cities & Health, 5*(sup1), S100–S104. https://doi.org/10.1080/23748834.2020.1778844

Nezhyva, M., & Mysiuk, V. (2022). War in Ukraine: Challenges for the global economy. *Foreign Trade: Economics, Finance, Law, 121*(2), 16–25. https://doi.org/10.31617/zt.knute.2022(121)02

Nikookar, E., Varsei, M., & Wieland, A. (2021). Gaining from disorder: Making the case for antifragility in purchasing and supply chain management. *Journal of Purchasing and Supply Management, 27*(3), 100699. https://doi.org/10.1016/j.pursup.2021.100699

Oliveira, G., Grenha Teixeira, J., Torres, A., & Morais, C. (2021). An exploratory study on the emergency remote education experience of higher education students and teachers during the COVID-19 pandemic. *British Journal of Educational Technology, 52*(4), 1357–1376. https://doi.org/10.1111/bjet.13112

Pu, M., Chen, X., & Zhong, Y. (2021). Overstocked agricultural produce and emergency supply system in the COVID-19 pandemic: Responses from China. *Food, 10*(12), 3027. https://doi.org/10.3390/foods10123027

Schmuck, J., Hiebel, N., Kriegsmann-Rabe, M., Schneider, J., Matthias, J. K., Erim, Y., & Geiser, F. (2022). Individual stress burden and mental health in health care workers during the COVID-19 pandemic: Moderating and mediating effects of resilience. *International Journal of Environmental Research and Public Health, 19*(11), 6545. https://doi.org/10.3390/ijerph19116545

Silva, S., Arinaminpathy, N., Atun, R., Goosby, E., & Reid, M. (2021). Economic impact of tuberculosis mortality in 120 countries and the cost of not achieving the sustainable development goals tuberculosis targets: A full-income analysis. *The Lancet Global Health, 9*(10), e1372–e1379. https://doi.org/10.1016/S2214-109X(21)00299-0

Stevano, S., Franz, T., Dafermos, Y., & Van Waeyenberge, E. (2021). COVID-19 and crises of capitalism: Intensifying inequalities and global responses. *Canadian Journal of Development Studies/Revue canadienne d'études du développement, 42*(1–2), 1–17. https://doi.org/10.1080/02255189.2021.1892606

Štreimikienė, D., Baležentis, T., Volkov, A., Ribašauskienė, E., Morkūnas, M., & Žičkienė, A. (2022). Negative effects of COVID-19 pandemic on agriculture: Systematic literature review in the frameworks of vulnerability, resilience and risks involved. *Economic Research-Ekonomska Istraživanja, 35*(1), 529–545. https://doi.org/10.1080/1331677X.2021.1919542

Strielkowski, W., Zenchenko, S., Tarasova, A., & Radyukova, Y. (2022). Management of smart and sustainable cities in the post-COVID-19 era: Lessons and implications. *Sustainability, 14*(12), 7267. https://doi.org/10.3390/su14127267

Wethal, U., Ellsworth-Krebs, K., Hansen, A., Changede, S., & Spaargaren, G. (2022). Reworking boundaries in the home-as-office: Boundary traffic during COVID-19 lockdown and the future of working from home. *Sustainability: Science, Practice and Policy, 18*(1), 325–343. https://doi.org/10.1080/15487733.2022.2063097

Wyatt, T., Maher, J., Allen, D., Clarke, N., & Rook, D. (2022). The welfare of wildlife: An interdisciplinary analysis of harm in the legal and illegal wildlife trades and possible ways forward. *Crime, Law and Social Change, 77*(1), 69–89. https://doi.org/10.1007/s10611-021-09984-9

Zakeri, B., Paulavets, K., Barreto-Gomez, L., Echeverri, L. G., Pachauri, S., Boza-Kiss, B., & Pouya, S. (2022). Pandemic, war, and global energy transitions. *Energies, 15*(17), 6114. https://doi.org/10.3390/en15176114

Zerrusen, L. (2021). The implications of COVID-19 on older adults: Challenges and opportunities. *Locus: The Seton Hall Journal of Undergraduate Research, 4*(1), 13. https://scholarship.shu.edu/locus/vol4/iss1/13

Part V
Leadership in Energy Policy: New Threats and Opportunities

Chapter 23
Digital Eco-Energy: Patterns of Achieving Economic Leadership, National Security, and Sustainability

Olena Chygryn, Svitlana Kolosok, and Vita Hordiienko

Abstract The solution to the priority world's scientific problem of ensuring sustainable development and reducing threats to national security within its components – economic, environmental, energy, and social – can be done by forming the foundation of digital eco-energy. For this purpose, the formalization of cross-sectoral and cross-border relationships in the chain "economy – energy – digitalization – national security" is relevant. The aim of this article is to analyze and study the patterns of digital eco-energy in the context of smart development of the energy sector of the national economy. This paper deals with digital eco-energy on improving the countries' national energy security. For that purpose, the authors conducted the bibliometric analysis using the VOSviewer and Scopus tools. The investigation involved 1224 studies focused on digitalization and 1331 studies on the energy sector. Thus, the obtained results demonstrated the strong links between investigating the digitalization trends, particularly the energy sector, smart technology, national security, and sustainable development. Given that, the authors formed the system of main determinants of the digitalization energy sector which will create the preconditions for its smart transformation.

Keywords Sustainable development · Energy security · Leadership · National economy

JEL Classification Q01 · O13 · P18

O. Chygryn (✉) · S. Kolosok · V. Hordiienko
Sumy State University, Sumy, Ukraine
e-mail: o.chygryn@biem.sumdu.edu.ua; s.kolosok@biem.sumdu.edu.ua; v.hordiienko@crkp.sumdu.edu.ua

W. Strielkowski (ed.), *Leadership, Entrepreneurship and Sustainable Development Post COVID-19*, Springer Proceedings in Business and Economics, https://doi.org/10.1007/978-3-031-28131-0_23

23.1 Introduction

As Ukraine has stepped on a track of tightening its connections with the European Union and establishing a closer cooperation with NATO in the light of the recent events marked by the Russian invasion in February 2022, sustaining and fostering the country's leadership position require updates in the directions and priorities of the energy sector. Ukrainian energy sector yields many negative traits of the Soviet past such as obsoleteness which calls for its modernization. In addition to that, Ukraine is seeking a new position and role in the world's economy which also imposes new challenges on its energy security as well as on its energy doctrine.

In general terms, the Russian-Ukrainian war offers yet another grim reminder that future progress in the field of climate change is increasingly reliant on the Chinese and Indian transitions to cleaner energy (Zakeri et al., 2022). As the Russian-Ukrainian war rages on, most of the European Union remains highly reliant on Russian oil and gas. It becomes apparent that Russia's undeterred invasion of Ukraine has shaken up the energy markets and geopolitics, driving up the price of oil and gas to its highest levels in almost a decade and forcing many countries to reconsider their energy supplies. Russia's invasion into Ukraine has made Russian energy resources, upon which European Union Member States depend, not just precarious, but unwanted.

Russia is the largest supplier of fossil fuels on the world market, including oil, gas, and coal. In addition, Russia is the largest oil exporter to the global market, while its natural gas powers the European economy. The current energy crisis brought about by the war in Ukraine is especially acute in Germany, which relies on Russia for about half of Germany's gas and coal, and more than one-third of its oil. Even though the European Union is especially reliant on Russian energy supplies, the implications are felt throughout the developing world, because sanctions, or just the expectation of sanctions, against Russian energy have strained markets, sending prices for all fossil fuels soaring. Oil and gas prices are rising, with a lot of volatility, as energy markets struggle to cope with the effects of sanctions and the prospect of a diminished supply from Russia.

Energy, in particular, has become the main concern of Moscow's war, resulting in oil and gas prices jumping sharply around the world. Global resources markets are still shaking from the fallout of Russia's incursion into Ukraine; both countries are large suppliers of energy, food, and fertilizer. Russia's undeterred invasion of Ukraine has led political leaders to reconsider energy plans, with potentially profound implications for a number of issues, from a growing food crisis to global efforts to curb greenhouse gas emissions.

Following the Russian invasion in February 2022, the European Commission revealed its plans on how Europe can shift away from Russian fossil fuels by 2027 and leverage a shift away from Moscow to accelerate a transition to green power. The plan is the European Commission's answer to the difficulties and global disruption of energy markets caused by Russia's invasion into Ukraine and includes moving toward more non-Russian natural gas imports, faster deployment of renewables, and greater efforts at energy conservation. Now that a new bill on the classification has passed through the European Parliament, including both gas and nuclear power, Ukraine can more easily assist the EU with its short- to medium-term transition. Mobilizing Eastern members, creating the core focus for a green European New Deal in Ukraine, can become a major accelerant to realizing the EU's ambitions of energy security and carbon neutrality. Creating a narrative around using green energy to build a more prosperous, independent Ukraine once the Russian full-scale invasion of Ukraine is over may prove valuable to future EU energy security. For countries seeking to limit Russian power, shifting away from Russian and other fossil fuels is a crucial step in diminishing Russian influence, making other countries energy-resilient, and keeping the goals of the Paris Agreement in place. There is a question what Russia is going to do with its unsold fuel, but some peculiar alternatives relating to mining cryptocurrencies might be arising (Lu & Strielkowski, 2022).

Generally, energy security is one of the priorities of national governments all around the world and the key element of building the national economy leadership (Guliyeva et al., 2018). Each state determines the goals of the policy in the field of energy security and the mechanisms for its implementation, taking into account its own peculiarities of the organization of society's life activities and the model of state management (Li et al., 2019; Bilan et al., 2020a; Didenko et al., 2020; George, 2020). Management activity in this area is not unambiguous and simple, as it requires taking into account various aspects of the country's life. These are resource, technical, economic, environmental, organizational-administrative, managerial, social, innovative, informational, digital, political, geopolitical, security, and other aspects, worldview aspects of energy supply of vital needs and functions of the state, society, and economy (Lisin et al., 2018; Lyeonov et al., 2018; Korneeva et al., 2021; Lyulyov et al., 2021b; Gong, 2022; Melnyk et al., 2022). In addition, activities in the field energy security can evolve significantly over time in one and the same country, depending on the dynamics of socioeconomic and scientific-technological development and transformation of energy market functioning models (Rausser et al., 2018; Pavlyk, 2020; Polcyn et al., 2022; Polyakov et al., 2021).

Recently, both in Ukraine and in the world, there have been significant changes in the principles of regulation of the energy sector. This is happening with the range of methods of using energy as a tool for achieving political goals expanding in its scope and implementation. Taking into account these circumstances which determine the variety of theoretical approaches and practices of identifying the object of research (which is defined by the sphere of management), its description and methods of assessing the state of energy security of the country, in this paper, we are focusing on the digital eco-energy which includes such essential patterns of achieving economic leadership as the national security as well as sustainability.

23.2 Literature Review

Energy security as a dominant factor in the leadership of the national economy is determined by a number of political, technological, socioeconomic, environmental, as well as the informational factors (Ziabina et al., 2020; Samusevych et al., 2021a; Strielkowski et al., 2021; Tiutiunyk et al., 2021; Štreimikienė et al., 2022). Reliable, economically efficient, and ecologically acceptable satisfaction of society's needs for energy resources will ensure the sustainable functioning of the national economy in normal and crisis conditions and protect the sovereignty of the state in the formation and implementation of the policy of protection of national interests.

The scientific community has made significant progress in determining the main determinants and components of national economic leadership. In particular, the scientists Bilan et al. (2020b) studied the links between the environmental pollution and the shadow economy taking into account the level of the carbon emissions on the gross domestic product. The authors employed a regression analysis aimed at eliciting the impact of individual factors and variables on the gross domestic product. The methodological approaches proposed in the papers by Cergibozan (2022), El Amri et al. (2020), and Yarovenko et al. (2021) evaluated the set of risks for the national economy connected with the possible threats to the energy security of the country. The scientists Lyulyov et al. (2021a, c) and Mentel et al. (2020) summarize the pros and cons of the global countries' cooperation's role in mitigating climate change and adaptation to its impacts. They emphasize renewable energy development and democracy support as key determinants for ensuring countries' energy security. At the same time, the set of authors separately are highlighting the necessity of supporting green energy technologies and implementing the smart energy management systems (Kuzior & Staszek, 2021; Kuzior et al., 2021a, 2021b; Kolosok et al., 2021b; Vakulenko et al., 2021; Us et al., 2021). The studies proved by Cebula and Pimonenko (2015), Didenko et al. (2021), and Elnikova and Barhaq (2020) confirm the hypothesis about the relevance of finding ways to adapt to climate changes and minimize the negative consequences to achieve sustainable development. Their studies found that providing energy security depends on the set of macro and micro determinants. The results of the papers by Kyrychenko et al. (2021) and Letunovska et al. (2021) who are assessing the influence of the social and economic determinants on the indicators of energy consumption are also relevant for this study. For hypothesis testing, the authors used a ranking correlation analysis where the key indicators were population health and the level of consumption of renewable and nonrenewable resources in some selected countries.

A number of scientific publications relate energy security to the sphere of national security, which in turn ensures the realization of fundamental national interests such as the secure development of the national economy alongside the principles of the sustainable development (United Nations' Sustainable Development Goals), the development of the state and society, as well as making sure that the population is growing and thriving (Burghard et al., 2022). However, it should not be forgotten that the energy security principles should also be embedded into the national agenda that features every country's national interest and priorities that can reflect, for example, the desire to join certain institutions, such as the European Union, or Mercosur (see Sahoo, 2017; Samusevych et al., 2021b; Sedmíková et al., 2021; Rekunenko et al., 2022; Kolosok et al., 2021a). The existing scientific background also characterizes the approaches for the assessment of the key factors on the countries' energy security levels (Vasylieva et al., 2021; Vysochyna et al., 2020; Vasudevan & Aslan, 2022; Us et al., 2020; Pimonenko et al., 2021; Khalatur & Dubovych, 2022). The scientist (Bhowmik, 2019) tested empirically the decoupling of carbon dioxide emissions per capita from the GDP per capita in the Nordic countries. That might be relevant to increasing a higher level of GDP per capita simultaneously with the decline CO_2 emission per capita. The set of scientific findings proved that the analysis of the green brand with the interconnection of the environmental management will determine the efficiency of the sustainable development of national economies. Entrepreneurship (Starchenko et al., 2021; Lyeonov et al., 2021; Vasylieva et al., 2020a, b; Ziabina & Kovalenko, 2021). Therefore, the objective of assessing and studying the potential and the development concept of the energy security for achieving the leading position of the national economy is an important task that needs to be pursued by the national governments and international organizations worldwide.

23.3 Data and Methods

The central hypothesis of the paper is the study of the increasing tendency on analyzing energy security strategies by modern scientists. For this purpose, were identified the quantity tendency of the scientific articles which are analyzing the issues connected with energy security strategies as determinants of the national economy leadership. At the same time, the paper is emphasizing the target subject areas under analyses of digital energy strategies was done. Also structured and clustered the scientific papers on the main areas for investigations.

In order to form a terminological basis and research the dominant trends of scientific publications on the subject of energy security and the development of smart technologies in the energy sector, the work carried out a bibliometric analysis in relation to the categories "economic leadership – energy – digitalization – national security" using the VOSViewer v. 1.6.13 tool. For this purpose, a sample of 2555 scientific articles on this issue were generated for the period 2000–2021 in publications indexed by Scopus scientometric databases.

23.4 Results and Discussion

The international database Scopus was chosen as the information base of the study. With the purpose of understanding the context of scientific development of the topic of energy security, we analyzed the connections between the key categories "energy

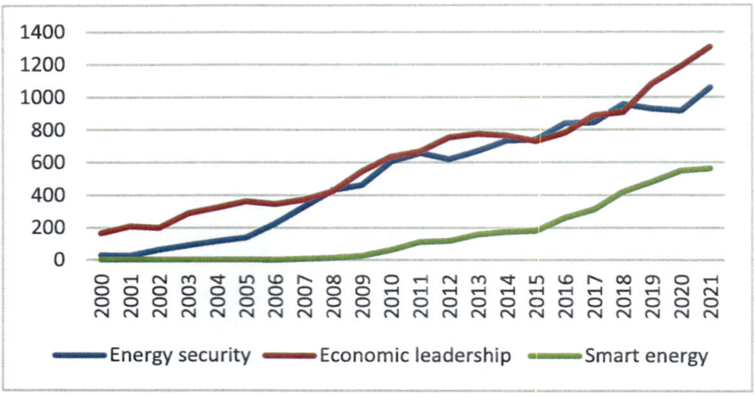

Fig. 23.1 Dynamics of publications in the field of energy security. (Source: Own results based on Scopus database)

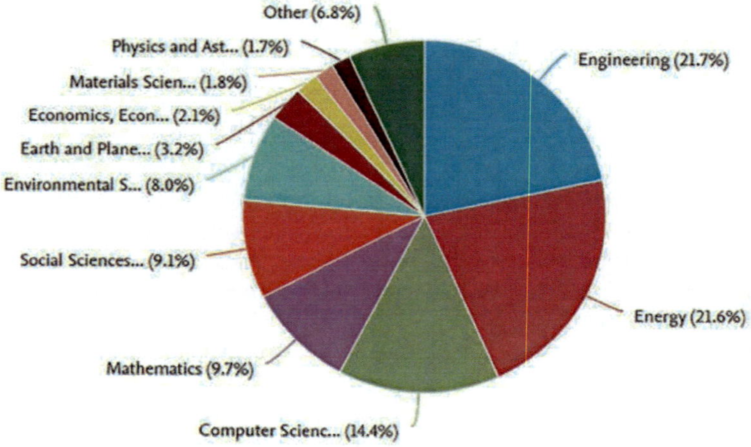

Fig. 23.2 The structure of subject areas in the field of energy security. (Source: Own results based on Scopus database)

security," "digitalization," "economic leadership." For the corresponding investigation were analyzed the scientific papers of the Scopus database from 2001 to 2020 (Fig. 23.1). More than 30,000 scientific articles were examined during such period.

Figure 23.1 illustrates the dynamics of publication activity, which characterizes the stable increasing of the scientific papers devoted to the chosen field.

Fig. 23.2 mirrored the cross-sector representation of the energy security scientific studying.

Thus, the main areas of research are energy, 21,6%; engineering, 21,7%; computer science, 14,4%; mathematics, 9,7%; social sciences, 9,1%; environmental sciences, 8%; etc. Such tendency talks about the huge development of the scientific technological background in the sphere of providing energy security for supporting the leadership of national economies. This thematic is especially relevant in today's condition of Russian war aggression. In the global level, the analyzed area is represented by such countries: India, China, the United States, the United Kingdom, Australia, Poland, Japan, and Germany. The most significant scientific investigations and researches concentrate on the next academic institutions: Siksha O Anusandhan Deemed to be University, C. V. Raman College of Engineering, Kalinga Institute of Industrial Technology, Bhubaneswar, National Institute of Technology Rourkela, International Islamic University Malaysia, Western Sydney University, or the University of Tokyo.

Table 23.1 represents the highly cited scientific articles in the field of smart energy security and development of the modern technologies in energy sector from 2013 to 2020 according to Scopus database.

The results revealed that the paper of Le (2017) is the most cited article (139 citations). Besides, other articles were highly cited in papers appeared in such journals as of "ISA Transactions" (121 citations), "Shiyou Xuebao/Acta Petrolei Sinica" (114 citations), "Journal of Network and Computer Applications" (90). These

Table 23.1 The Top-10 cited articles in the field of smart energy security

	Title	Authors, year	Citations
1	Key credit criteria among international green building rating tools	Le (2017)	139
2	Design and analysis of tilt integral derivative controller with filter for load frequency control of multi-area interconnected power systems	Kumar, Panda, Biswal, Chandra Sekhar (2016)	121
3	Shale oil and gas revolution and its impact	Zou, Pan, Jing, Gao, Yang, Wu, Zhao (2020)	114
4	Autonomic schemes for threat mitigation in Internet of Things	Hadi Habaebi (2015)	90
5	Hybrid BFOA-PSO approach for coordinated design of PSS- and SSSC-based controller considering time delays	Panda, Yegireddy, Mohapatra (2013)	67
6	A shade dispersion interconnection scheme for partially shaded modules in a solar PV array network	Satpathy, Sharma, Jena (2017)	64
7	Power and mismatch losses mitigation by a fixed electrical reconfiguration technique for partially shaded photovoltaic arrays	Satpathy, Priya, Sharma (2019)	61
8	Smart grids and the constitution of solar electricity conduct	Bulkeley, Powels, Bell (2016)	43
9	Energy monitoring prototype for internet of things: Preliminary results	Ashraf (2017)	24
10	Hybrid power systems frequency regulation using TID-based robust controller design and differential evolution (DE) algorithm	Behera, Biswal, Swain, Samantray (2018)	8

Source: Own results based on Scopus database entries

publishers devote their papers to questions connected with the development of smart technologies in the energy sector of national economy.

According to the type, the documents were structured as follows: conference papers, 49,6%; peer-reviewed articles, 36,6%; book chapters, 4,8%; and notes, editorials, and reviews.

The citation analysis provided with the VOSviewer tools revealed the most authoritative scientists as well as the six clusters of research teams which have the biggest influence on the development of energy security theories and approaches (Fig. 23.3).

The big blue cluster includes the categories connected with the energy security categories, such as alternative energy, biofuels, carbon emissions, carbon footprint, climate change mitigation, energy conservation, energy consumption, environmental impact, environmental management, energy market, information technology, innovation, international trade, national economy, etc. Considerable green cluster connected with the processes of the smart electric power transmission: battery management system, complex networks, demand side management, distributed power generation, energy management, electric vehicles, intelligent energies, etc. The significant orange cluster is chained with the smart cities and contains the next areas and categories: energy generation, optimization, power control, solar power generation, electric power management, etc.

The evolution of the energy security concept formation and development is presented in Fig. 23.4.

Therefore, the retrospective analysis in the evolutionary time dimension and the identification of the close relationship of key patterns of scientific research studies

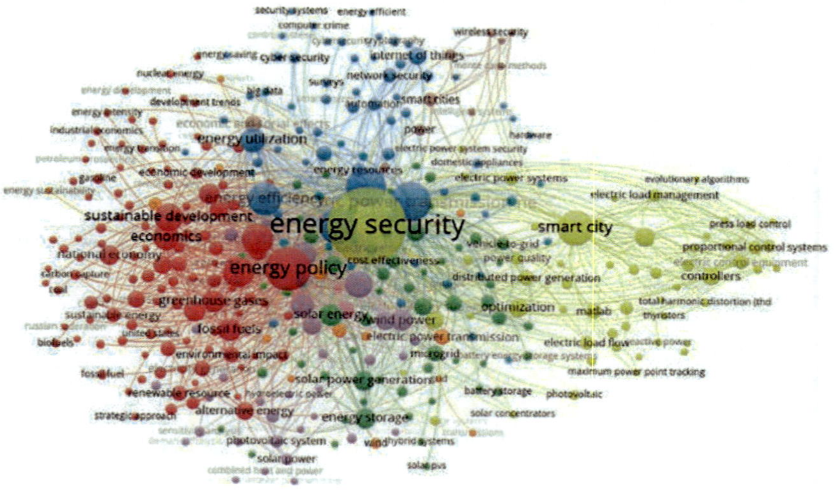

Fig. 23.3 The results of bibliometric analysis of the structural and functional environment of smart energy security theory formation and development. (Source: Own results based on the Scopus base and VOSviewer tool)

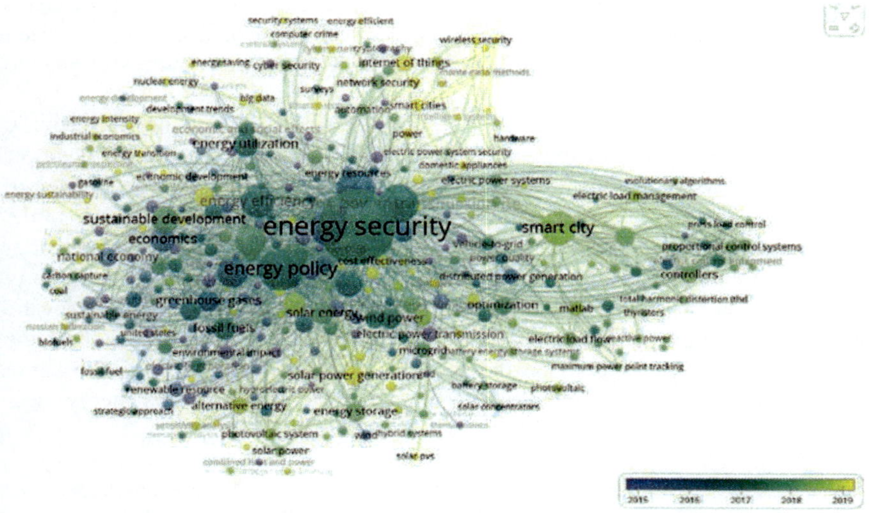

Fig. 23.4 The evolution of the smart energy security concept formation. (Source: Own results based on Scopus base and VOSviewer tool output)

from Scopus database made it possible for us to identify the most relevant stages of the theoretical development of the energy security concept in the scientific literature. These results can be useful for feeding other related studies and drawing conclusions as well as the inspiration regarding the energy security and economic development (as well as the correlation between those two).

23.5 Conclusions

All in all, in the today's situation when the fundamental changes are taking place in the geopolitical, military, geopolitical, and economic situation in the world, as well as taking into account the reorientation of the energy policy of both the Russian Federation and individual Member States of the European Union, there is currently a pressing need to rethink the issue of ensuring energy security as an important component of the national security. This step is essential for ensuring the leadership of the national economies not only in the EU but all around the world. Nowadays, energy security becomes one of the key priorities of the national governments and international organizations. Each country needs to determine its goals related to the energy policy targeted at ensuring its energy security and setting up the mechanisms for its implementation. This needs to be done while taking into account each country's specificities and peculiarities that feature the organization of society's life activities as well as the model of the public management and ways of running the state. Quite recently, both in Ukraine and in the world, significant changes in the principles of regulation of the energy sector have occurred, and the range of methods for using energy as a tool for achieving political and economic goals has been significantly expanding which marks the new normal in building and preserving the energy security concept.

In this paper, the identification of regularities in the formation of scientific approaches to energy security has been carried out using bibliometric (VOSviewer v. 1.6.13) analysis which enabled us to build the empirical model that allowed to obtain and analyze the dominant retrospective evolutionary-temporal trends in the development of this theory in relation to the concepts of sustainable digital development, environmental management, as well as the development of renewable energy sources. All of these findings might be of special interests and importance for the policy-makers and stakeholders both in Ukraine and in the world, especially in the current turbulent times marked by the post-COVID-19 complications and the war in Ukraine that has shaken the whole concept of energy cooperation and energy security established in the world since the 1970s.

Acknowledgments This research was funded by the grants from the Ministry of Education and Science of Ukraine: (grant numbers 0122 U000788, 0120 U102002, 0122 U000788).

References

Bhowmik, D. (2019). Decoupling CO2 emissions in Nordic countries: Panel data analysis. *SocioEconomic Challenges, 3*(2), 15–30. https://doi.org/10.21272/sec.3(2).15-30.2019

Bilan, Y., Pimonenko, T., & Starchenko, L. (2020a). Sustainable business models for innovation and success: Bibliometric analysis. *E3S Web of Conferences, 159*, 04037. https://doi.org/10.1051/e3sconf/202015904037

Bilan, Y., Srovnalã-KovÃi, P., Streimikis, J., Lyeonov, S., Tiutiunyk, I., & Humenna, Y. (2020b). From shadow economy to lower carbon intensity: Theory and evidence. *International Journal of Global Environmental Issues, 19*(1–3), 196–216. https://doi.org/10.1504/ijgenvi.2020.114874

Burghard, U., Dütschke, E., Caldes, N., & Oltra, C. (2022). Cross-border concentrated solar power projects – Opportunity or dead end? A study into actor views in Europe. *Energy Policy, 163*, 112833. https://doi.org/10.1016/j.enpol.2022.112833

Cebula, J., & Pimonenko, T. (2015). Comparison financing conditions of the development biogas sector in Poland and Ukraine. *International Journal of Ecology and Development, 30*(2), 20–30.

Cergibozan, R. (2022). Renewable energy sources as a solution for energy security risk: Empirical evidence from OECD countries. *Renewable Energy, 183*, 617–626. https://doi.org/10.1016/j.renene.2021.11.056

Didenko, I., Volik, K., Vasylieva, T., Lyeonov, S., & Antoniuk, N. (2020). Migration, environment, and country safety: Analysis of touchpoints. *E3S Web of Conferences, 202*, 03028. https://doi.org/10.1051/e3sconf/202020203028

Didenko, I., Volik, K., Vasylieva, T., Lyeonov, S., & Antoniuk, N. (2021). Environmental migration and country security: Theoretical analysis and empirical research. *E3S Web of Conferences, 234*, 00010. https://doi.org/10.1051/e3sconf/202123400010

El Amri, A., Boutti, R., Oulfarsi, S., Rodhain, F., & Bouzahir, B. (2020). Carbon financial markets underlying climate risk management, pricing and forecasting: Fundamental analysis. *Financial Markets, Institutions and Risks, 4*(4), 31–44. https://doi.org/10.21272/fmir.4(4).31-44.2020

Elnikova, J., & Barhaq, A. R. (2020). Transparency of responsible investment environment. *Business Ethics and Leadership, 4*(4), 68–75. https://doi.org/10.21272/bel.4(4).68-75.2020

George, B. (2020). Inclusive sustainable development in the Caribbean region: Social capital and the creation of competitive advantage in tourism networks. *Business Ethics and Leadership, 4*(3), 119–126. https://doi.org/10.21272/bel.4(3).119-126.2020

Gong, X. (2022). Energy security through a financial lens: Rethinking geopolitics, strategic investment, and governance in China's global energy expansion. *Energy Research and Social Science, 83*, 42–50. https://doi.org/10.1016/j.erss.2021.102341

Guliyeva, A., Britchenko, I., & Rzayeva, U. (2018). Global security and economic asymmetry: A comparison of developed and developing countries. *Journal of Security and Sustainability Issues, 7*(4), 707–717. https://doi.org/10.9770/jssi.2018.7.4(8)

Khalatur, S., & Dubovych, O. (2022). Financial engineering of green finance as an element of environmental innovation management. *Marketing and Management of Innovations, 1*, 232–246. https://doi.org/10.21272/mmi.2022.1-17

Kolosok, S., Bilan, Y., Vasylieva, T., Wojciechowski, A., & Morawski, M. (2021a). A scoping review of renewable energy, sustainability and the environment. *Energies, 14*(15). https://doi.org/10.3390/en14154490

Kolosok, S., Vasylieva, T., & Lyeonov, S. (2021b). Machine analysis of the UK electrical energy initiatives based on the e-petitions to the UK government and parliament. *Paper Presented at the CEUR Workshop Proceedings, 2870*, 1562–1573.

Korneeva, E., Olinder, N., & Strielkowski, W. (2021). Consumer attitudes to the smart home technologies and the Internet of Things (IoT). *Energies, 14*(23), 7913. https://doi.org/10.3390/en14237913

Kuzior, A., & Staszek, M. (2021). Energy management in the railway industry: A case study of rail freight carrier in Poland. *Energies, 14*(21), 6875. https://doi.org/10.3390/en14216875

Kuzior, A., Kwilinski, A., & Hroznyi, I. (2021a). The factorial-reflexive approach to diagnosing the executors' and contractors' attitude to achieving the objectives by energy supplying companies. *Energies, 14*(9), 2572. https://doi.org/10.3390/en14092572

Kuzior, A., Lobanova, A., & Kalashnikova, L. (2021b). Green energy in Ukraine: State, public demands, and trends. *Energies, 14*(22), 7745. https://doi.org/10.3390/en14227745

Kyrychenko, K., Laznenko, D., & Reshetniak, Y. (2021). Green university as an element of forming a sustainable public health system. *Health Economics and Management Review, 2*(4), 21–26. https://doi.org/10.21272/hem.2021.4-02

Letunovska, N., Saher, L., Vasylieva, T., & Lieonov, S. (2021). Dependence of public health on energy consumption: A cross-regional analysis. *E3S Web of Conferences, 250*, 04014. https://doi.org/10.1051/e3sconf/202125004014

Li, M., Li, L., & Strielkowski, W. (2019). The impact of urbanization and industrialization on energy security: A case study of China. *Energies, 12*(11), 2194. https://doi.org/10.3390/en12112194

Lisin, E., Strielkowski, W., Chernova, V., & Fomina, A. (2018). Assessment of the territorial energy security in the context of energy systems integration. *Energies, 11*(12), 3284. https://doi.org/10.3390/en11123284

Lu, J., & Strielkowski, W. (2022). Russia: Could cryptocurrency mining exploit unsold fuel? *Nature, 605*(7909), 225–225. https://doi.org/10.1038/d41586-022-01267-0

Lyeonov, S. V., Vasylieva, T. A., & Lyulyov, O. V. (2018). Macroeconomic stability evaluation in countries of lower-middle income economies. *Naukovyi Visnyk Natsionalnoho Hirnychoho Universytetu, 1*, 138–146. https://doi.org/10.29202/nvngu/2018-1/4

Lyeonov, S., Pimonenko, T., Chygryn, O., Reznik, O., & Gaynulina, R. (2021). Green brand as a marketing instrument: Principle, features and parameters. *International Journal of Global Energy Issues, 43*(2–3), 147–165. https://doi.org/10.1504/ijgei.2021.115141

Lyulyov, O., Lyeonov, S., Tiutiunyk, I., & Podgórska, J. (2021a). The impact of tax gap on macroeconomic stability: Assessment using panel VEC approach. *Journal of International Studies, 14*(1), 139–152. https://doi.org/10.14254/2071-8330.2021/14-1/10

Lyulyov, O., Paliienko, M., Prasol, L., Vasylieva, T., Kubatko, O., & Kubatko, V. (2021b). Determinants of shadow economy in transition countries: Economic and environmental aspects. *International Journal of Global Energy Issues, 43*(2–3), 166–182. https://doi.org/10.1504/ijgei.2021.10037735

Lyulyov, O., Pimonenko, T., Kwilinski, A., & Us, Y. (2021c). The heterogeneous effect of democracy, economic and political globalisation on renewable energy. *E3S Web of Conferences*, 25003006. https://doi.org/10.1051/e3sconf/202125003006

Melnyk, L., Matsenko, O., Kubatko, O., Korneyev, M., & Tulyakov, O. (2022). Additive economy and new horizons of innovative business development. *Problems and Perspectives in Management, 20*(2), 175–185. https://doi.org/10.21511/ppm.20(2).2022.15

Mentel, G., Vasilyeva, T., Samusevych, Y., Vysochyna, A., Karbach, R., & Streimikis, J. (2020). The evaluation of economic, environmental and energy security: Composite approach. *International Journal of Global Environmental Issues, 19*(1–3), 177–195. https://doi.org/10.1504/ijgenvi.2020.10037581

Pavlyk, V. (2020). Assessment of green investment impact on the energy efficiency gap of the national economy. *Financial Markets, Institutions and Risks, 4*(1), 117–123. https://doi.org/10.21272/fmir.4(1).117-123.2020

Pimonenko, T., Us, Y., Myroshnychenko, Y., Dubyna, O., & Vasylyna, T. (2021). Green financing for carbon-free growth: Role of banks marketing strategy. *Financial Markets, Institutions and Risks, 5*(3), 71–78. https://doi.org/10.21272/fmir.5(3).71-78.2021

Polcyn, J., Us, Y., Lyulyov, O., Pimonenko, T., & Kwilinski, A. (2022). Factors influencing the renewable energy consumption in selected European countries. *Energies, 15*(1), 10.3390/en15010108.

Polyakov, M., Khanin, I., Bilozubenko, V., Korneyev, M., & Shevchenko, G. (2021). Factors of uneven progress of the European union countries towards a circular economy. *Problems and Perspectives in Management, 19*(3), 332–344. https://doi.org/10.21511/ppm.19(3).2021.27

Rausser, G., Strielkowski, W., & Štreimikienė, D. (2018). Smart meters and household electricity consumption: A case study in Ireland. *Energy & Environment, 29*(1), 131–146. https://doi.org/10.1177/0958305X17741385

Rekunenko, I., Zhuravka, F., Nebaba, N., Levkovych, O., & Chorna, S. (2022). Assessment and forecasting of Ukraine's financial security: Choice of alternatives. *Problems and Perspectives in Management, 20*(2), 117–134. https://doi.org/10.21511/ppm.20(2).2022.11

Sahoo, K. K. (2017). Transform of ethical practices in globalized economy, a special focus on Ghana. *Business Ethics and Leadership, 1*(4), 58–65. https://doi.org/10.21272/bel.1(4).58-65.2017

Samusevych, Y., Maroušek, J., Kuzmenko, O., Streimikis, J., & Vysochyna, A. (2021a). Environmental taxes in ensuring national security: A structural optimization model. *Journal of International Studies, 14*(2), 292–312. https://doi.org/10.14254/2071-8330.2021/14-2/19

Samusevych, Y., Vysochyna, A., Vasylieva, T., Lyeonov, S., & Pokhylko, S. (2021b). Environmental, energy and economic security: Assessment and interaction. *E3S Web of Conferences, 234*, 00012. https://doi.org/10.1051/e3sconf/202123400012

Sedmíková, E., Vasylieva, T., Tiutiunyk, I., & Navickas, M. (2021). Energy consumption in assessment of shadow economy. *European Journal of Interdisciplinary Studies, 13*(2), 47–64. https://doi.org/10.24818/ejis.2021.12

Starchenko, L., Lyeonov, S., Vasylieva, T., Pimonenko, T., & Lyulyov, O. (2021). Environmental management and green brand for sustainable entrepreneurship. *E3S Web of Conferences, 234*, 00015. https://doi.org/10.1051/e3sconf/202123400015

Štreimikienė, D., Samusevych, Y., Bilan, Y., Vysochyna, A., & Sergi, B. S. (2022). Multiplexing efficiency of environmental taxes in ensuring environmental, energy, and economic security. *Environmental Science and Pollution Research, 29*(5), 7917–7935. https://doi.org/10.1007/s11356-021-16239-6

Strielkowski, W., Civín, L., Tarkhanova, E., Tvaronavičienė, M., & Petrenko, Y. (2021). Renewable energy in the sustainable development of electrical power sector: A review. *Energies, 14*(24), 8240. https://doi.org/10.3390/en14248240

Tiutiunyk, I., Drabek, J., Antoniuk, N., Navickas, V., & Rubanov, P. (2021). The impact of digital transformation on macroeconomic stability: Evidence from EU countries. *Journal of International Studies, 14*(3), 220–234. https://doi.org/10.14254/2071-8330.2021/14-3/14

Us, Y., Pimonenko, T., Tambovceva, T., & Segers, J.-P. (2020). Green transformations in the healthcare system: The covid-19 impact. *Health Economics and Management Review, 1*(1), 48–59. https://doi.org/10.21272/hem.2020.1-04

Us, Y., Pimonenko, T., & Lyulyov, O. (2021). Energy efficiency profiles in developing the free-carbon economy: On the example of Ukraine and the V4 countries. *Polityka Energetyczna, 23*(4), 49–66. https://doi.org/10.33223/epj/127397

Vakulenko, I., Saher, L., Syhyda, L., Kolosok, S., & Yevdokymova, A. (2021). The first step in removing communication and organizational barriers to stakeholders' interaction in smart grids: A theoretical approach. *E3S Web of Conferences, 234*, 00020. https://doi.org/10.1051/e3sconf/202123400020

Vasudevan, H., & Aslan, I. (2022). Developing organizational citizenship behaviour: Organizational climate and leader-member exchange effects. *Marketing and Management of Innovations, 1*, 94–107. https://doi.org/10.21272/mmi.2022.1-07

Vasylieva, T., Jurgilewicz, O., Poliakh, S., Tvaronavičienė, M., & Hydzik, P. (2020a). Problems of measuring country's financial security. *Journal of International Studies, 13*(2), 329–346. https://doi.org/10.14254/2071-8330.2020/13-2/22

Vasylieva, T., Machová, V., Vysochyna, A., Podgórska, J., & Samusevych, Y. (2020b). Setting up architecture for environmental tax system under certain socioeconomic conditions. *Journal of International Studies, 13*(4), 273–285. https://doi.org/10.14254/2071-8330.2020/13-4/19

Vasylieva, T., Pavlyk, V., Bilan, Y., Mentel, G., & Rabe, M. (2021). Assessment of energy efficiency gaps: The case for Ukraine. *Energies, 14*(5), 1323. https://doi.org/10.3390/en14051323

Vysochyna, A., Samusevych, Y., & Starchenko, L. (2020). Convergence trends of environmental taxation in European countries. *E3S Web of Conferences, 202*, 20303. https://doi.org/10.1051/e3sconf/20202020303

Yarovenko, H., Bilan, Y., Lyeonov, S., & Mentel, G. (2021). Methodology for assessing the risk associated with information and knowledge loss management. *Journal of Business Economics and Management, 22*(2), 369–387. https://doi.org/10.3846/jbem.2021.13925

Zakeri, B., Paulavets, K., Barreto-Gomez, L., Echeverri, L. G., Pachauri, S., Boza-Kiss, B., & Pouya, S. (2022). Pandemic, war, and global energy transitions. *Energies, 15*(17), 6114. https://doi.org/10.3390/en15176114

Ziabina, Y., & Kovalenko, Y. (2021). Regularities in the development of the theory of energy efficiency management. *SocioEconomic Challenges, 5*(1), 117–132. https://doi.org/10.21272/sec.5(1).117-132.2021

Ziabina, Y., Pimonenko, T., & Starchenko, L. (2020). Energy efficiency of national economy: Social, economic and ecological indicators. *SocioEconomic Challenges, 4*(4), 160–174. https://doi.org/10.21272/sec.4(4).160-174.2020

Chapter 24
Leading Aspects of Transition from the Traditional Energy to Renewable Energy Sources After the COVID-19 Pandemic

Khalida Sadigova, Farhad Aliyev, and Matin Firuzi

Abstract The growing demand for energy resources which represents the leading drivers of modern society tempts states and businesses to provide consumers with these resources. Energy resources are involved in many essential life processes of modern human beings. In fact, as much as energy resources are important in our life, still they are the main causes for pollution and climatic changes. Moreover, the depletion of the fossil fuels worldwide makes them scarce and costly. The idea of this research is to investigate the current situation in the world related to energy resources and present the ways how the alternative energy resources are applied and which policies related to expanding these sources are implemented in the post-pandemic era. We also discuss the role of the traditional sources of energy, like oil and gas; oil in modern society; and the advantages of using renewable energy sources after COVID-19. In addition, we show that the degree to which COVID-19 and its attendant disruptions to economic activity and reductions in demand can be expected to trigger a new wave of creative destruction and generate market-based incentives for the transition toward renewables remains to be observed.

A quantitative and qualitative data were used in this paper for comparison of different case studies and countries. Some criteria as actuality, credibility, and the accuracy of information and reliability of sources are applied. Following that, the inquiry turned to the cases of numerous nations at varying levels of economic growth. Our results outline the need of converting to alternative energy as a means of tackling the climate change and sustaining the leading position in the world of the future.

Keywords Renewable energy · Leadership · Energy transition

K. Sadigova (✉) · F. Aliyev · M. Firuzi
SABAH Center, Azerbaijan State University of Economics (UNEC), Baku, Azerbaijan
e-mail: Firuzi.matin.p@sabah.edu.az

W. Strielkowski (ed.), *Leadership, Entrepreneurship and Sustainable Development Post COVID-19*, Springer Proceedings in Business and Economics, https://doi.org/10.1007/978-3-031-28131-0_24

JEL Classification P50 · Q20 · Q42

24.1 Introduction

Even though prior to the Industrial Revolution, all energy sources appeared to be renewable, many people believe that the humanity first got acquainted with oil only in the nineteenth century. However, the first full-fledged industrial production of oil was carried out in 1847 at Apsheron (now in part of Azerbaijan) (Hajiyeva et al., 2018). However, humanity has been using oil for over 6000 years. The first oil was used almost 6000 years ago by Sumerians who mixed petroleum bitumen with clay and sand to make an extra strong brick (Bustillo Revuelta, 2021). The oil industry of the most ancient oil country in the world and Azerbaijan played a large part in it throughout its 150-year history.

Industrial oil production began in Azerbaijan in the middle of the nineteenth century. In 1848, an oil gusher in Baku, which hit from a borehole in the Bibiheybat field, laid the foundation for industrial production of "black gold" in Azerbaijan. In 1899, Azerbaijan came out on first in the world in the extraction and processing of oil, yielding half of the world's oil production and producing it industrially (Haiko & Biletsky, 2019).

In general terms, oil is an exhaustible and nonrenewable natural resource and at the same time very necessary and is in high demand (Ferreira da Cunha & Missemer, 2020). Since oil is a nonrenewable resource, it is inevitable that global production will peak someday. The Hubbert peak theory consists in the fact that the same calculations that successfully predicted the US peak oil production are also applicable to other cases, such as the peak of world oil production. Various estimates of the time of the world peak have been published, both by Hubbert peak theory and others, with some of these dates already in the past (Laherrère et al., 2022).

Despite the significant contribution to the world economy, the production of oil and oil products causes significant damage to the environment, both during the normal course of the process and as a result of various accidents. Thus, at all stages of production and transportation, more than 45 million tons of oil are lost annually (Lee, 2022). Having started the exploitation of oil and gas fields, a human being did not think about the consequences of intensive extraction of these natural resources. The great danger is fraught with the use of oil and gas as fuel. When these products are burned in the atmosphere, large quantities of carbon dioxide, various sulfur compounds, and nitrogen oxide are released. A decrease in oxygen and an increase in carbon dioxide in turn affect climate change. Molecules of carbon dioxide allow solar shortwave radiation to penetrate the Earth's atmosphere and trap infrared radiation emitted from the Earth's surface (Baldocchi & Penuelas, 2019).

Due to that threats, various projects for the creation of engines operating on other types of fuel appear. The way out of the situation of permanent oil and gas crisis is the use of alternative types of energy (Zakaria et al., 2021). Alternative energy sources are natural phenomena that are converted into thermal or electrical energy by means of conversion in special installations. They are solar electromagnetic radiation, kinetic energy of movement of air masses (wind), kinetic energy of water flow (rivers and streams), the energy of sea tides, and the thermal energy from hot springs (Guliyeva et al., 2021; Brożyna et al., 2020). In addition, in a world where energy resources are in short supply, the necessity for biofuels as an alternative source of energy is becoming increasingly apparent. Biofuel is a type of fuel made from plant or animal raw materials – from the products of vital activity of organisms or organic industrial waste. Biofuels are different in terms of their state of aggregation (Avtar et al., 2019).

The interplay between technological developments, regulations, and market conditions in the course of the COVID-19 pandemic triggered a rapid paradigm shift for the electricity industry (Strielkowski et al., 2021). The renewables sector, which is the focus of this shift, is suffering from the same uncertainties and impending job losses that have plagued so many other sectors. COVID-19 has occurred during a period in which falling renewable energy costs, continued low oil prices, rising debt within the power sector, and investors' concerns over the impacts of fossil fuels on carbon emissions and environmental regulations are already driving down capital investments into fossil fuel industries, making renewables one of the fastest growing industries (Nundy et al., 2021). According to the available data, renewable energy growth appears to have slowed due to the COVID-19 crisis (Karakosta et al., 2021). The growth in renewables power generation during the pandemic is not entirely a matter of conjecture. Global use of renewables across sectors increased 1.5%, and electricity generated from renewables increased nearly 3% during the first quarter of 2020, relative to the same period in the United States alone in 2019. In the same time, the solar power has experienced the largest percentage increase in all US power sources in the last decade (Khanna, 2021).

Most renewable energy growth has occurred in the electric power sector, driven by the rapid expansion of wind and solar, which has been enabled by continued policy support and falling costs. Wind, for example, has quadrupled its share of electricity generation, while solar has seen a 40-fold expansion, helping spur large job gains over this period (Lange & Santarius, 2020). The job total increases to millions of workplaces when considering impacts across the entire power industry, including jobs related to the transition, such as in infrastructure and grid flexibility, as well as traditional technologies including fossil fuels and nuclear energy. If the governments do not support maintaining and growing the clean energy industry as part of future stimulus or economic recovery packages, renewable industries may see a job loss of as much as 50 percent in certain subsectors (Barbier, 2020). In just months, fossil fuels have reached an unfavorable tipping point in electricity production, while electricity mixes are shifting toward renewables in all major regions. The proportion of renewables in the world's power mix is increasing only slowly, even though wind and solar are rapidly growing as sources of power. Global installations of solar power and energy storage are expected to fall nearly 20% compared with projections before COVID-19, and the installations of wind turbines are expected to fall 4.9 gigawatts (GW) (Pandey et al., 2022). For wind and solar, favorable policies and falling costs will continue, despite COVID-19, with renewables only expected to see a decline in capacity. The adjustment is made because of the continued transitions toward renewables and that this transition may accelerate with the promises to have a carbon-free electricity sector in place as early as 2035 or later. Gas demand has seen its biggest downgrade to 2030, because the pandemic has exposed long-standing structural weaknesses in gas demand (Valiyev & Valehli, 2022).

In short, alternative energy is becoming increasingly important nowadays as the fears of the global warming and climate change begin to prevail over the benefits or using fossil fuels for generating the world's energy. In this paper, we are focusing on the leading aspects of transition from the traditional energy to renewable energy sources after COVID-19 pandemic using a case study of Azerbaijan, an oil- and gas-abundant economy.

24.2 State of the Art

Apart from the sun and wind energy, biofuel and algae represent the new leading and very promising sources of renewable energy. In general terms, a biofuel is a form of fuel created from animal raw or from plant materials, such as the by-products obtained due to the essential activities of living organisms or organic industrial waste (Merlo et al., 2021). Biofuels are distinct in terms of their aggregation condition. Liquid biofuels include biodiesel, ethanol, and methanol, while solid biofuels are straw, firewood, and briquettes, and gaseous biofuels are hydrogen, biogas, and synthesis gas. As of today, bioethanol and biodiesel are the currently two most common biofuels (Zaky et al., 2022).

Another source of renewable energy is algae. As early as in the nineteenth century, chemists and microbiologists proposed the use of microalgae as an alternative source of energy (Phour et al., 2022). Green microalgae are able to convert carbon dioxide into organic compounds while having a cleansing effect on the atmosphere and hydrosphere. Such biofuel can be used in diesel engines: it is very close in composition to traditional motor fuels – refined petroleum products (Enamala et al., 2018). A number of countries have started serial production of special bioreactors for growing microalgae. Japan and the United States have already successfully tested aircraft and vehicles powered exclusively by algae biodiesel (Gray et al., 2021). Algae are the fastest-growing plants on earth (their mass doubles per day), and their growth requires readily available raw materials: sunlight, water, and carbon dioxide. According to the modern classification of biofuels, the fuel obtained using microalgae is called third-generation biofuel. At the same time, in this work, it is indicated that microalgae used to obtain biofuel should not be used to obtain a food or feed product and should not be a waste of any production. Algae production is also attractive because carbon dioxide is absorbed from the atmosphere during biosynthesis (Barati et al., 2021).

According to a number of authors, the possibility of using algae as a raw material for fuel production is determined by their high lipid content. Different types of fuel are made from algae rich in lipids – bio-oil and biodiesel. The most common is biodiesel, the production method of which is the transesterification of microalgae lipids with methyl alcohol carried out under mild conditions (Sadvakasova et al., 2019). There are several more good reasons to use seaweed more actively. Approximately the 70% of our planet is sea; we are already exploring the depths of the sea for the production of food and various materials. Algae need water, sunlight, and the presence of nutrients for the development. In such conditions, they can grow very quickly, and we just need such a stable and unpretentious raw material to achieve our environmental goals by 2050 (Chew et al., 2018). Ethanol and butanol currently are made from various sugars, and seaweed has a lot of sugar that can be converted to alcohol. It has been demonstrated that it is technically possible to meet the fuel needs of a car with algae. Like in the case of wind energy, reducing costs on the production of fuel from algae spread over 25–30 years. During this time, it is possible to reduce production costs by a factor of 100. So far, biofuels are very expensive. With the intensification of algae production, with access to a larger scale, it will be able to reduce the cost by the same hundred times (Patle et al., 2021).

For example, the US Department of Energy has researched high oil content algae under the Aquatic Species Program. The researchers concluded that California, Hawaii, and New Mexico are suitable for the industrial production of algae (Kandasamy et al., 2022). Biochemical engineering companies have started mass production of patented bioreactors for the cultivation of algae, suitable for immediate use, which includes fast-growing algae with high oil content. Recently, scientists have found one of the types of microalgae that are able to reproduce much faster than other biological counterparts under certain lighting conditions. In addition, ExxonMobil released news of a project aimed at the production of biofuels from algae. For 10 years, the company has been working with Synthetic Genomics to research a technology for the production of diesel fuel from algae with the high oil content. The goal is to create capacity by 2025 to produce 10,000 barrels of algal fuel per day (Pires da Mata Costa et al., 2021).

China is collectively perceived as the world chief interest in clean energy innovations. The country is investing heavily in the construction of plants for sustainable energy sources as well as the development of environmentally friendly energy developments such as batteries and electric cars. China built 77 gigawatts of solar farms and 149 gigawatts of wind farms in 2016 alone (Balakrishnan et al., 2020). A lot of the worldwide environmentally friendly power energy is relied upon to be 42% sunlight based, 35% hydropower, and 40% breeze power by 2022.

Denmark is a pioneer on the planet in the improvement of wind energy. In 2017, the creation of power from wind sets another precedent the wind ranches covered 43% of all power needs. Of all OECD nations, Denmark has the most noteworthy per capita power and wind creation over the most recent 15 years. Regardless of this, the Danes are creating by 2030 over half of power ought to be delivered by sustaining power plants, and by 2050 all 100% (Hrnčić et al., 2021). Yet, the speed of environmentally friendly power energy development shows that the nation will accomplish these objectives quicker. As indicated by specialists' conjectures, in 2020, environmentally friendly power plants (wind, sun based, and biomass) will cover over 80% of the power interest.

Iceland is prominent for its amazing regular scenes, yet it is additionally a country with perhaps the most noteworthy pace of sustainable power infiltration into the public energy balance on the planet and the most elevated among European nations. Presently, geothermal, hydro, and wind power supply 100% of Iceland's power needs. Practically hydroelectric force plants give 75% and the rest is given by geothermal sources. Over 90% of the nation's interest for high temp water and warmth is likewise met by geothermal energy (Spittler et al., 2020).

India is a fast-developing economy with 1.34 billion inhabitants. Furthermore, energy will be basic to the nation's turn of events. India is anticipated to require 1/4 of the world's energy. India appears to have chosen to cover the greater part of its requirements with clean energy. Coal actually represents 70% of the country's energy blend, yet India is attempting to build its sustainable power limit from the current 58 GW to 175 GW in 2022 (Chambon et al., 2020). With the presentation of green closeouts, India has made the most serious environmentally friendly power market on the planet. Because of cost reserve funds, the state is getting ready to decommission coal-terminated force plants, which they intend to supplant with sunlight-based and wind power plants. What's more, India intends to construct the world's first drifting sun-oriented force plant and in any event 10 GW of sustainable power in the following 3 years.

24.3 Azerbaijan's Leading Perspectives in Transition to the Renewable Energy

According to new energy initiatives, the Azerbaijani government is becoming increasingly interested in the development of "green" energy. Azerbaijan's renewable energy contribution to power generation now surpasses 8%, but the country's Ministry of Energy hopes to increase this to 30% by 2030 (Karatayev et al., 2021). Given Azerbaijan's geographical location, which generates excellent circumstances for the growth of the renewable energy sector, the aim is quite attainable. The country is actively participating in the commission established by the President of the Republic of Azerbaijan's order of December 5, 2019, "On measures to undertake pilot projects in the field of renewable energy sources." The first auction for the development of solar and wind power facilities in the Absheron region indicated a lot of international interest in this area. As a reminder, the Ministry of Energy inked contracts with ACWA Power and Masdar a year ago for the implementation of pilot projects in the field of renewable energy sources. A 200 MW solar power plant (Masdar) and a 240 MW wind power plant will be built as part of the project (Mustafayev et al., 2022). Most recently, Azerbaijan is counting on the areas of Karabakh region through green energy initiatives.

At the same time, the most contemporary technology, the most economically reasonable, energy-saving technologies, and green energy (water, wind, and sun), should be implemented throughout the whole country. The construction of a "green" zone idea has already begun in many areas of Azerbaijan which will involve the use of renewable energy sources, energy efficiency, and ecologically friendly technology, including the use of automobiles, and so on. Discussions with the International Renewable Energy Agency are proceeding in this direction. According to early estimations, these lands have a solar energy potential of over 4000 megawatts and the wind energy potential of up to 500 megawatts. As a result, eight prospective solar energy areas have been identified on the Azerbaijani territory, while the wind power also has many perspectives (Vidadili et al., 2017). All in all, following the changes brought by the COVID-19 pandemic with its restrictions, lockdowns, as well as its threats to energy security and stability, Azerbaijan embarked on to the worldwide trend of transitioning to a "green" economy, and despite its continued reliance on oil earnings, it is pleased to steer its future energy and economic policies toward green energy (Gulaliyev et al., 2020).

24.4 Formal Analysis

Figure 24.1 demonstrates the share of primary energy from renewable sources. At the first glance at Fig. 24.1, it should be noted that the world leadership is held by the United States, England, and Sweden. Wind, solar, biofuel, and hydropower in the United Kingdom generated a total of 29.5 TWh in Q3 2019, compared to 29.1 TWh from fossil fuels. The share of wind and solar in the United Kingdom's electricity generation in 2020 was 29.2%. The country also plans to increase the

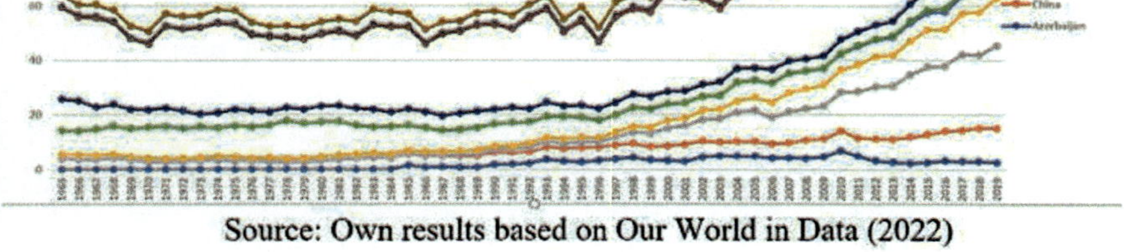

Source: Own results based on Our World in Data (2022)

Fig. 24.1 Share of primary energy from renewable sources (1985–2019). (Source: Own results based on Our World in Data, 2022)

capacity of wind farms from the current 8500 MW to about 20,000 MW by 2025 (Bulut, 2020).

And by 2030, the capacity of wind power plants, taking into account new projects, will be increased to 30,000 MW. By 2045, Sweden should be climate neutral, that is, the balance of greenhouse gas emissions should be zero or negative. By 2040, 100% of electricity should be generated from renewable energy sources (Lipiäinen et al., 2022).

After successfully navigating the COVID-19 pandemic, the renewables industry has a promising outlook for 2021 and beyond, as many drivers endorse continued rapid growth of wind, solar, and battery storage. There are currently 232 GW of wind and solar projects under construction in the United States. Over 21 GW of wind and nearly 15 GW of solar power are expected to be operational by the end of 2021. Energy storage capacity is expected to rise by 15,000 MW by 2024, compared to 1500 MW in 2020. Energy storage growth, which was projected to be 1500 MW in 2020, is expected to increase by 15,000 MW by 2024 (Jordaan et al., 2022).

Japan is rapidly increasing the share of renewable energy sources and strengthening its position. Japan is planning a large-scale expansion in offshore wind energy, which will lead to the creation of approximately 10 GW by 2030. By 2050, renewable energy sources will account for more than 75% of total electricity production, of which solar and wind generation will account for 32% each. For comparison, in 2019, the total share of RES is only 8% (Esteban et al., 2018).

According to Fig. 24.2, 114 GW of new wind capacity was added globally in 2020, an 82% increase over 2019. In 2030, wind farms will generate approximately 7300 TWh, which by then will account for approximately 22% of projected global electricity consumption. Considering a graph of the share of wind power in different ten countries, one can notice that the world leaders are China and the United States. Chinese wind energy had an installed capacity of more than 200 GW at the end of 2019, and it provided more than 5.5 percent of the country's electricity. According

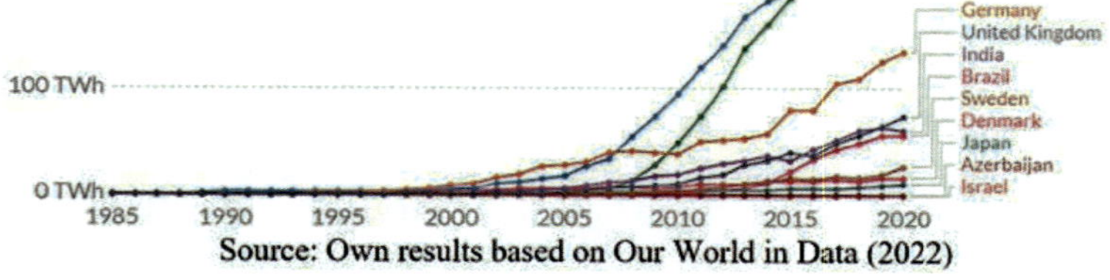

Fig. 24.2 Share of wind energy resources (1985–2022). (Source: Own results based on Our World in Data, 2022)

to the National Energy Administration, China's total wind energy potential was 281.5 GW at the end of 2020 (Zhang et al., 2022). The installed capacity of wind farms in the United States exceeded 90 GW at the end of the second quarter of 2018, indicating that wind has surpassed solar as the country's primary source of renewable energy. In December 2020, the United States set an hour record by delivering 82 gigawatts of renewable power. In 2021, the United States plans to commission 12.2 GW of new wind capacity (Alola, 2022).

While Germany lags far behind China and the United States in terms of renewable energy sources, as seen in the graph, it is working to increase the share of renewable energy sources in its energy balance. Germany produced 40% of its electricity from renewable energy sources in the first quarter of 2021. At the end of 2017, the installed capacity was 55.6 GW, with 5.2 GW from offshore installations. Wind power provided a quarter of the country's total electricity in 2019, compared to an estimated 9.3 percent in 2010 (Strobel et al., 2022).

Figure 24.3 depicts the share of solar energy in different ten countries describing the situation related to the usage of solar energy in the world. As one can see in the graph, all ten selected countries are rapidly increasing their share of solar power. It is apparent that the leaders of solar energy are China, the United States, Japan, India, and Germany.

By 2030, the number of solar installations in the United States is projected to quadruple. This sector would add 324 gigawatts (GW) of power over the next 10 years (Schoeneberger et al., 2020). In the same time, the solar energy installed capacity in China will exceed 500–530 GW by 2021–2025. That is, on average, the Chinese can build more than 50 GW of solar power plants (). Japan is now one of only five countries with a combined solar photovoltaic potential of 10 gigawatts. In August of this year, Japan's PV market surpassed the 10 GW mark, reaching 10.5 GW by the end of the month. Solar energy is a rapidly growing industry in India. As

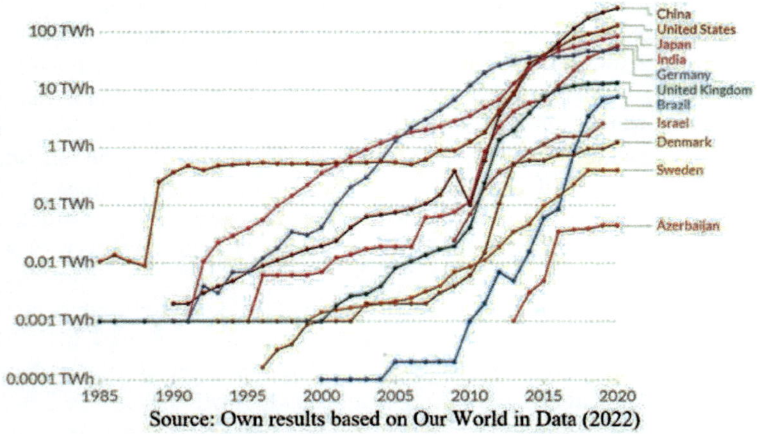

Fig. 24.3 Share of solar energy (1985–2020). (Source: Own results based on Our World in Data, 2022)

of November 30, 2020, the country's solar installed capacity was 36.9 GW. A total of 395.14 MW of solar energy capacity was added in February 2021. The Indian government sets a target of 20 gigawatts of power for 2022, which was met 4 years ahead of time. By the end of 2019, Germany will have installed solar energy capacity of 50 GW. The next group can be called Brazil, Israel, and Denmark, and they are also strengthening their positions and increasing the pace of development in this type of energy. According to some calculations, more than 4.9 gigawatts (GW) of installed power would be applied to the large plants and systems scattered around roofs, verandas, and small plots in Brazil. This represents a more than 68 percent increase over the country's total installed capacity of 7.5 GW. Around 887,000 grid-connected PV systems are expected to be deployed in Brazil by 2024. One can also note a sharp jump in Brazil in this energy industry in 10 years (Muhammed & Tekbiyik-Ersoy, 2020).

Azerbaijan, having all the necessary resources to increase the pace of development in this energy sector, is also striving to increase the supply of sun energy. According to the State Statistics Committee (SCS), power plants using solar panels generated 27.1 million kilowatt-hours of electricity in January–July 2020, up 5.4 percent from the same time last year (Asadi & Pourhossein, 2021).

24.5 Economic Potentials of the Renewable Energy in the Post-COVID Era

Our research is based on studying the role of the alternative energy sources in the world and assessing their importance for the sustainable economic development. In our paper, we already discussed the role of oil in modern society, what are the advantages of using renewable energy sources, and which states use more green energy, and which less. In our research, we used quantitative and also qualitative data for making analysis, and for collecting this data, we used secondary information. The information that we used in our research is taken from several main sources like books, articles, publications, news portals, and the electro-news portal. Also, we used some criteria when searching for a piece of information. These criteria were actuality, quality credibility, the accuracy of the information, and reliability of sources. After gathering all information, we made our prediction in this sphere and how it would develop. Then we began to analyze how to construct our article more properly for the reader and how it will be easier for the reviewer to accept information.

Renewable energy contributes to the reduction of the worldwide industrial pollution. By 2030, triple the renewable energy within the global energy mix will add 1.1%, or \$ 1.3 trillion, to global GDP. Which will be consistent with the analysis and this transformation will add 2030 the year, global GDP will increase by 0.6–1.1%, a rise of roughly \$ 700 billion to \$ 1.3 trillion compared to regular businesses (Mohsin et al., 2022).

Another important and dramatic advantage is the impact of investing in renewable energy on trade. For fuel importers, switching to a better source of renewable energy may bring trade benefits. Reducing fuel imports will help improve trade balances and GDP. Doubling the renewable and sustainable energy sources will increase the net energy exports by the $15 billion and can increase by US$21 billion as electrification progresses (Du Can et al., 2022).

According to a survey performed by the "Union of Concerned Scientists," the US renewable energy target is going to be 25% by 2025. Benefits include 263.4 billion (or $ 11.5 billion in wind land lease payments and extra land tax revenues for local governments) (Khan et al., 2021).

Furthermore, the algae-based fuel generation is still a relatively young technology. It exemplifies the fundamental shift in our attitude to combustibles that are required. Before this can be observed in practice, certain difficulties must be overcome. Fuel created from algae, like any novel concept, has both advantages and disadvantages. However, it appears that with sufficient research and development from scientists and engineers, the advantages will exceed the disadvantages. Because this form of fuel has all of the makings of game-changing technology, all we need now is time to improve the process and make it far more cost-effective.

Biofuels, particularly algal varieties, are not ideal, but they are vital to invest in since they can help us reduce our carbon footprint while scientists work on new, better, and more sustainable options. It may become mainstream with enough progress and enough goodwill and trust from the world's governments, as has been witnessed many times before, making our lives much better and more sustainable.

Moreover, it is obvious that many countries could build opportunities to recover economically, environmentally, and inclusively from the COVID-19 pandemic. Governments should define relevant strategies for responding to the pandemic, such that near-term policy objectives that support both the recovery efforts and sustainable energy development can be established (Rempel & Gupta, 2021). Governments and businesses alike should use any opportunities presented by the COVID-19 pandemic to promote a cleaner energy transition and increase access to energy, including developing COVID-19 recovery programs, accessing international funding, and taking advantage of renewables growing competitiveness (Rausser et al., 2021). While scholars and international agencies view the pandemic as presenting opportunities to make a transition to cleaner energy, the critical question for oil-rich developing countries is whether renewables can displace the amount of oil fuel used by the industrial sector and serve the needs of the unserved population. While making firm predictions and assessments about the total impact of the COVID-19 pandemic on the worlds renewable energy markets may be somewhat immature, there are strong indications from initiatives taken by selected countries in the promotion of sustainable, renewable energy development, as well as investments in clean energy sectors, as part of a green recovery effort (Hoang et al., 2021). In some countries, opportunities are increasing to develop high-capacity renewable energy projects, such as utility-scale solar, onshore, and offshore wind farms, as part of economic recovery plans following the pandemic. One of the key examples of the acceleration efforts is the investment in renewable energy, which has been the target of increased state support in certain countries, particularly in areas where the use of fossil fuel-based power is being phased out. Investment in renewables, particularly wind, has also been robust ahead of the grid-price parity implementation, which would eliminate subsidies to utility-scale solar and onshore wind projects (Aleluia et al., 2022). While that will increase demand for electric power, a rising proportion of renewables in power generation should lead to lower consumption of fossil fuels by the energy industry. Energy use drives economic growth, and therefore, access to electricity at all levels is critical to revitalizing economies and improving living conditions after a pandemic. In the European Union (EU), where the goal is to reach net-zero carbon economy by 2050, political support is greatly contributing to the EUs growth of renewables. Countries which are net fossil fuel importers and are major producers of low-carbon technologies (e.g., the EU, China) have seen an increase in GDP compared with pre-COVID-19 period, because reduced imports of fossil fuels are replaced with domestic activities related to energy efficiency and renewable sources of energy, leading to the creation of green jobs at home (Deutch, 2020).

Our evaluation shows that measures aimed at restoring the environment are likely to boost all countries growth rates of activities, driven by increased investments in low-carbon technological options (including solar PV, onshore and offshore wind, electric vehicles, heat pumps, refurbishments), which provide an incentive for growth of world economic output. Social benefits, such as job creation, economic growth, industry strength, and so forth, are not a primary goal of a Green Recovery scenario. However, our work captures the potential co-economic and employment benefits of such a Green Recovery as the result of increased investments in low-carbon, energy-efficient technologies and infrastructure, as opposed to a driver. We project that the post-pandemic recovery plans will require increased investment in infrastructure, which will ultimately contribute to lower energy intensity and carbon emissions. While upfront investments are higher, investments in renewables could help many counties all around the world to reduce electricity utility costs, fuel import bills, and subsidies costs; lower carbon emissions; and generate new jobs. In order to ensure the energy sector prospers, sound planning, favorable policies, and increased investments to sustain its growth and mitigate its risks from current uncertainties are required. Achieving a strong economic recovery without raising emissions at the pace seen after the global financial crisis in 2008 would require governments to pursue structural emissions reductions with sensible, sustainable, and ambitious policies that speed up development and deployment of the full spectrum of clean energy solutions (Lugo-Morin, 2021). To help drive these tough decisions, which are likely to shape countries infrastructures for decades, studies are conduced that will make explicit recommendations for how governments can make energy and sustainable energy issues central to stimulus plans for creating jobs and building more modern, resilient, and cleaner energy systems (Čábelková et al., 2022). The simultaneity of the COVID-19 crisis and climate challenges means the recovery from the pandemic will be a crucial time for climate action. A win-win scenario is possible through a post-COVID-19 pandemic recovery effort, where governments can leverage the major renewable energy initiatives to support a transition to cleaner energy, while creating appropriate conditions to rebuild the economy and create a stronger, responsive national healthcare system capable of withstanding a severe health crisis (Tian et al., 2022).

Nevertheless, while continued use of fossil fuels is undesirable in terms of decarbonization, many developing countries with an oil-producing sector cannot jump from a fossil-fuel-based economy to one powered entirely by renewables immediately following the COVID-19 pandemic. This is due to the fact that unlike other energy technologies, renewables showed a remarkable resilience. The main priorities and strategies, such as digitalization or indigenization of the renewable energy technologies could be deployed to speed up the renewables' growth during the post-COVID-19 global pandemic era. Within this context, countries worldwide should move toward sustainable energy futures, under the pretext of achieving sustainable development, access to energy, energy security, and economic growth and prosperity with low-carbon emissions, through partnerships at continental, subregional, and national levels. In short, in order to meet these goals, countries with large amounts of nonrenewable resources will need to establish targeted policies in areas including taxation policies, finance, energy and minerals sector regulations, and low-carbon technologies, with strong attention paid to equity aspects of the transition achieving a more sustainable future.

24.6 Conclusions

Overall, the conducted research was aimed at getting an answer to the question whether the world should switch to alternative energy as the leading source of power in the post-COVID-19 era. Through a long collection of information about alternative and traditional energy sources in the world, as well as analysis and processing of the collected data, graphs of the use of alternative sources in various countries the results were derived. According to the evaluation of results, the answer to the question whether the world should switch to alternative energy is unequivocal: it surely should do so in order to follow the leading principles of the sustainable development and tackling the threats of the global climate change.

Governments worldwide are using the opportunities presented by the COVID-19 pandemic to promote a cleaner energy transition and increase access to power, including developing COVID-19 restoration programs, accessing international funding, and taking advantage of renewables growing competitiveness. If all the world's leading leaders adopted such thinking, with investment in renewables as a core element of a recovery plan, then potential co-benefits from these policy decisions may soon become reality. While making predictions and assessments about the total impact of the COVID-19 pandemic on the world's renewable energy markets may be somewhat immature, there are strong indicators of initiatives taken by selected countries to foster the development and investments in sustainable, renewable energy sources within the clean energy sector within a green recovery effort. In some countries, opportunities are increasing to develop high-capacity renewable energy projects, such as utility-scale solar, onshore, and offshore wind farms, as part of economic recovery plans following the pandemic.

While the COVID-19 pandemic is seen by scholars and international agencies as presenting an opportunity to make the transition to cleaner energy, the critical question for oil-rich developing countries, like Azerbaijan, is whether renewables will be able to displace the amount of oil fuel used by the industrial sector and serve unserved populations. COVID-19 has confirmed oil and gas vulnerability to external shocks, while also underscoring their resilience relative to renewables, stimulating increased interest in shifting investments away from fossil-based power and toward renewables.

At the same time, it is common knowledge that traditional energy sources are running out with an astonishing speed, and in the near future, there will be no chance to choose between traditional and alternative energies. Therefore, those countries that will begin or have already begun the transition to renewable energy are very likely to benefit economically and, even more importantly, environmentally. From an ecological point of view, the transition to renewable energy will prevent the currently threatening climate apocalypse and improve the state of the hydrosphere, atmosphere, and biosphere. Therefore, we can conclude that by starting the transition now, countries can save themselves not only from future economic problems but also from environmental ones, which in turn generally lead to improved public health and well-being and also enable them to become the leaders of tomorrow.

References

Aleluia, J., Tharakan, P., Chikkatur, A. P., Shrimali, G., & Chen, X. (2022). Accelerating a clean energy transition in Southeast Asia: Role of governments and public policy. *Renewable and Sustainable Energy Reviews, 159*, 112226. https://doi.org/10.1016/j.rser.2022.112226

Alola, A. A. (2022). The nexus of renewable energy equity and agricultural commodities in the United States: Evidence of regime-switching and price bubbles. *Energy, 239*, 122377. https://doi.org/10.1016/j.energy.2021.122377

Asadi, M., & Pourhossein, K. (2021). Neural network-based modelling of wind/solar farm siting: A case study of East-Azerbaijan. *International Journal of Sustainable Energy, 40*(7), 616–637. https://doi.org/10.1080/14786451.2020.1833881

Avtar, R., Tripathi, S., Aggarwal, A. K., & Kumar, P. (2019). Population-urbanization-energy nexus: A review. *Resources, 8*(3), 136. https://doi.org/10.3390/resources8030136

Balakrishnan, P., Shabbir, M. S., Siddiqi, A. F., & Wang, X. (2020). Current status and future prospects of renewable energy: A case study. *Energy Sources, Part A: Recovery, Utilization, and Environmental Effects, 42*(21), 2698–2703. https://doi.org/10.1080/15567036.2019.1618983

Baldocchi, D., & Penuelas, J. (2019). The physics and ecology of mining carbon dioxide from the atmosphere by ecosystems. *Global Change Biology, 25*(4), 1191–1197. https://doi.org/10.1111/gcb.14559

Barati, B., Zeng, K., Baeyens, J., Wang, S., Addy, M., Gan, S. Y., & Abomohra, A. E. F. (2021). Recent progress in genetically modified microalgae for enhanced carbon dioxide sequestration. *Biomass and Bioenergy, 145*, 105927. https://doi.org/10.1016/j.biombioe.2020.105927

Barbier, E. B. (2020). Greening the post-pandemic recovery in the G20. *Environmental and Resource Economics, 76*(4), 685–703. https://doi.org/10.1007/s10640-020-00437-w

Brożyna, J., Strielkowski, W., Fomina, A., & Nikitina, N. (2020). Renewable energy and EU 2020 target for energy efficiency in The Czech Republic and Slovakia. *Energies, 13*(4), 965. https://doi.org/10.3390/en13040965

Bulut, M. (2020). Analysis of the COVID-19 impact on electricity consumption and production. *Sakarya University Journal of Computer and Information Sciences, 3*(3), 283–295. https://doi.org/10.35377/saucis.03.03.817595

Bustillo Revuelta, M. (2021). *Construction materials. Geology, production and applications* (Springer textbooks in earth sciences, geography and environment). Springer.

Čábelková, I., Smutka, L., & Strielkowski, W. (2022). Public support for sustainable development and environmental policy: A case of The Czech Republic. *Sustainable Development, 30*(1), 110–126. https://doi.org/10.1002/sd.2232

Chambon, C. L., Karia, T., Sandwell, P., & Hallett, J. P. (2020). Techno-economic assessment of biomass gasification-based mini-grids for productive energy applications: The case of rural India. *Renewable Energy, 154*, 432–444. https://doi.org/10.1016/j.renene.2020.03.002

Chew, K. W., Chia, S. R., Show, P. L., Yap, Y. J., Ling, T. C., & Chang, J. S. (2018). Effects of water culture medium, cultivation systems and growth modes for microalgae cultivation: A review. *Journal of the Taiwan Institute of Chemical Engineers, 91*, 332–344. https://doi.org/10.1016/j.jtice.2018.05.039

Deutch, J. (2020). Is net zero carbon 2050 possible? *Joule, 4*(11), 2237–2240. https://doi.org/10.1016/j.joule.2020.09.002

Du Can, S. D. L. R., Letschert, V., Agarwal, S., Park, W. Y., & Kaggwa, U. (2022). Energy efficiency improves energy access affordability. *Energy for Sustainable Development, 70*, 560–568. https://doi.org/10.1016/j.esd.2022.09.003

Enamala, M. K., Enamala, S., Chavali, M., Donepudi, J., Yadavalli, R., Kolapalli, B., & Kuppam, C. (2018). Production of biofuels from microalgae-a review on cultivation, harvesting, lipid extraction, and numerous applications of microalgae. *Renewable and Sustainable Energy Reviews, 94*, 49–68. https://doi.org/10.1016/j.rser.2018.05.012

Esteban, M., Portugal-Pereira, J., Mclellan, B. C., Bricker, J., Farzaneh, H., Djalilova, N., & Roeber, V. (2018). 100% renewable energy system in Japan: Smoothening and ancillary services. *Applied Energy, 224*, 698–707. https://doi.org/10.1016/j.apenergy.2018.04.067

Ferreira da Cunha, R., & Missemer, A. (2020). The Hotelling rule in non-renewable resource economics: A reassessment. *Canadian Journal of Economics/Revue canadienne d'économique, 53*(2), 800–820. https://doi.org/10.1111/caje.12444

Gray, N., McDonagh, S., O'Shea, R., Smyth, B., & Murphy, J. D. (2021). Decarbonising ships, planes and trucks: An analysis of suitable low-carbon fuels for the maritime, aviation and haulage sectors. *Advances in Applied Energy, 1*, 100008. https://doi.org/10.1016/j.adapen.2021.100008

Gulaliyev, M. G., Mustafayev, E. R., & Mehdiyeva, G. Y. (2020). Assessment of solar energy potential and its ecological-economic efficiency: Azerbaijan case. *Sustainability, 12*(3), 1116. https://doi.org/10.3390/su12031116

Guliyeva, A., Mammadzada, P., & Mamedov, T. (2021). Economic analysis of alternative energy development in Nagorno-Karabakh and surrounding regions. *E3S Web of Conferences, 250*, 03004. https://doi.org/10.1051/e3sconf/202125003004

Haiko, H., & Biletsky, V. (2019). The history of industrial oil extraction in Azerbaijan (the 19th-beginning of the 20th centuries). *Shid, 4*, 35–40. https://doi.org/10.21847/1728-9343.2019.4(162).176938

Hajiyeva, S. R., Veliyeva, Z. T., Samadova, A. A., Hajiyev, O. B., Frontasyeva, M. V., & Madadzada, A. I. (2018). Assessment of polycyclic aromatic hydrocarbon [pah] levels in air of Absheron peninsula, Azerbaijan, by gas chromatography of moss transplants. *Advances in Biology & Earth Sciences, 3*(3), 197–204.

Hoang, A. T., Nižetić, S., Olcer, A. I., Ong, H. C., Chen, W. H., Chong, C. T., & Nguyen, X. P. (2021). Impacts of COVID-19 pandemic on the global energy system and the shift progress to renewable energy: Opportunities, challenges, and policy implications. *Energy Policy, 154*, 112322. https://doi.org/10.1016/j.enpol.2021.112322

Hrnčić, B., Pfeifer, A., Jurić, F., Duić, N., Ivanović, V., & Vušanović, I. (2021). Different investment dynamics in energy transition towards a 100% renewable energy system. *Energy, 237*, 121526. https://doi.org/10.1016/j.energy.2021.121526

Jordaan, S. M., Park, J., & Rangarajan, S. (2022). Innovation in intermittent electricity and stationary energy storage in the United States and Canada: A review. *Renewable and Sustainable Energy Reviews, 158*, 112149. https://doi.org/10.1016/j.rser.2022.112149

Kandasamy, S., Zhang, B., He, Z., Bhuvanendran, N., EL-Seesy, A. I., Wang, Q., & Dar, M. A. (2022). Microalgae as a multipotential role in commercial applications: Current scenario and future perspectives. *Fuel, 308*, 122053. https://doi.org/10.1016/j.fuel.2021.122053

Karakosta, C., Mylona, Z., Karásek, J., Papapostolou, A., & Geiseler, E. (2021). Tackling covid-19 crisis through energy efficiency investments: Decision support tools for economic recovery. *Energy Strategy Reviews, 38*, 100764. https://doi.org/10.1016/j.esr.2021.100764

Karatayev, M., Lisiakiewicz, R., Gródek-Szostak, Z., Kotulewicz-Wisińska, K., & Nizamova, M. (2021). The promotion of renewable energy technologies in the former soviet bloc: Why, how, and with what prospects? *Energy Reports, 7*, 6983–6994. https://doi.org/10.1016/j.egyr.2021.10.068

Khan, I., Hou, F., & Le, H. P. (2021). The impact of natural resources, energy consumption, and population growth on environmental quality: Fresh evidence from The United States of America. *Science of the Total Environment, 754*, 142222. https://doi.org/10.1016/j.scitotenv.2020.142222

Khanna, M. (2021). COVID-19: A cloud with a silver lining for renewable energy? *Applied Economic Perspectives and Policy, 43*(1), 73–85. https://doi.org/10.1002/aepp.13102

Laherrère, J., Hall, C. A., & Bentley, R. (2022). How much oil remains for the world to produce? Comparing assessment methods, and separating fact from fiction. *Current Research in Environmental Sustainability, 4*, 100174. https://doi.org/10.1016/j.crsust.2022.100174

Lange, S., & Santarius, T. (2020). *Smart green world?: Making digitalization work for sustainability*. Routledge.

Lee, J. (2022). *The world has been using a lot more oil than we thought*. Bloomberg. Available at: https://www.bloomberg.com/opinion/articles/2022-02-13/the-world-has-been-using-a-lot-more-oil-than-we-thought

Lipiäinen, S., Sermyagina, E., Kuparinen, K., & Vakkilainen, E. (2022). Future of forest industry in carbon-neutral reality: Finnish and Swedish visions. *Energy Reports, 8*, 2588–2600. https://doi.org/10.1016/j.egyr.2022.01.191

Lugo-Morin, D. R. (2021). Global future: Low-carbon economy or high-carbon economy? *WORLD, 2*(2), 175–193. https://doi.org/10.3390/world2020012

Merlo, S., Gabarrell Durany, X., Pedroso Tonon, A., & Rossi, S. (2021). Marine microalgae contribution to sustainable development. *Water, 13*(10), 1373. https://doi.org/10.3390/w13101373

Mohsin, M., Taghizadeh-Hesary, F., Iqbal, N., & Saydaliev, H. B. (2022). The role of technological progress and renewable energy deployment in green economic growth. *Renewable Energy, 190*, 777–787. https://doi.org/10.1016/j.renene.2022.03.076

Muhammed, G., & Tekbiyik-Ersoy, N. (2020). Development of renewable energy in China, USA, and Brazil: A comparative study on renewable energy policies. *Sustainability, 12*(21), 9136. https://doi.org/10.3390/su12219136

Mustafayev, F., Kulawczuk, P., & Orobello, C. (2022). Renewable energy status in Azerbaijan: Solar and wind potentials for future development. *Energies, 15*(2), 401. https://doi.org/10.3390/en15020401

Nundy, S., Ghosh, A., Mesloub, A., Albaqawy, G. A., & Alnaim, M. M. (2021). Impact of COVID-19 pandemic on socio-economic, energy-environment and transport sector globally and sustainable development goal (SDG). *Journal of Cleaner Production, 312*, 127705. https://doi.org/10.1016/j.jclepro.2021.127705

Our World in Data. (2022). *World statistics*. Available at: https://ourworldindata.org/grapher/renewable-share-energy

Pandey, A., Pandey, P., & Tumuluru, J. S. (2022). Solar energy production in India and commonly used technologies—An overview. *Energies, 15*(2), 500. https://doi.org/10.3390/en15020500

Patle, D. S., Pandey, A., Srivastava, S., Sawarkar, A. N., & Kumar, S. (2021). Ultrasound-intensified biodiesel production from algal biomass: A review. *Environmental Chemistry Letters, 19*(1), 209–229. https://doi.org/10.1007/s10311-020-01080-z

Phour, M., Danish, M. S. S., Sabory, N. R., Ahmadi, M., & Senjyu, T. (2022). Electro-microbiology: A green approach for energy and environment sustainability. *Sustainability, 14*(17), 10676. https://doi.org/10.3390/su141710676

Pires da Mata Costa, L., Micheline Vaz de Miranda, D., Couto de Oliveira, A. C., Falcon, L., Stella Silva Pimenta, M., Guilherme Bessa, I., & Pinto, J. C. (2021). Capture and reuse of carbon dioxide (CO2) for a plastics circular economy: A review. *PRO, 9*(5), 759. https://doi.org/10.3390/pr9050759

Rausser, G., Strielkowski, W., & Korneeva, E. (2021). Sustainable tourism in the digital age: Institutional and economic implications. *Terra Economicus, 19*(4), 141–159. https://doi.org/10.18522/2073-6606-2021-19-4-141-159

Rempel, A., & Gupta, J. (2021). Fossil fuels, stranded assets and COVID-19: Imagining an inclusive & transformative recovery. *World Development, 146*, 105608. https://doi.org/10.1016/j.worlddev.2021.105608

Sadvakasova, A. K., Akmukhanova, N. R., Bolatkhan, K., Zayadan, B. K., Usserbayeva, A. A., Bauenova, M. O., & Allakhverdiev, S. I. (2019). Search for new strains of microalgae-producers of lipids from natural sources for biodiesel production. *International Journal of Hydrogen Energy, 44*(12), 5844–5853. https://doi.org/10.1016/j.ijhydene.2019.01.093

Schoeneberger, C. A., McMillan, C. A., Kurup, P., Akar, S., Margolis, R., & Masanet, E. (2020). Solar for industrial process heat: A review of technologies, analysis approaches, and potential applications in the United States. *Energy, 206*, 118083. https://doi.org/10.1016/j.energy.2020.118083

Spittler, N., Davidsdottir, B., Shafiei, E., Leaver, J., Asgeirsson, E. I., & Stefansson, H. (2020). The role of geothermal resources in sustainable power system planning in Iceland. *Renewable Energy, 153*, 1081–1090. https://doi.org/10.1016/j.renene.2020.02.046

Strielkowski, W., Civín, L., Tarkhanova, E., Tvaronavičienė, M., & Petrenko, Y. (2021). Renewable energy in the sustainable development of electrical power sector: A review. *Energies, 14*(24), 8240. https://doi.org/10.3390/en14248240

Strobel, L., Schlund, J., & Pruckner, M. (2022). Joint analysis of regional and national power system impacts of electric vehicles-a case study for Germany on the county level in 2030. *Applied Energy, 315*, 118945. https://doi.org/10.1016/j.apenergy.2022.118945

Tian, J., Yu, L., Xue, R., Zhuang, S., & Shan, Y. (2022). Global low-carbon energy transition in the post-COVID-19 era. *Applied Energy, 307*, 118205. https://doi.org/10.1016/j.apenergy.2021.118205

Valiyev, A., & Valehli, F. (2022). COVID-19 and Azerbaijan: Is the system resilient enough to withstand the perfect storm? *Problems of Post-Communism, 69*(1), 103–114. https://doi.org/10.1080/10758216.2021.1920840

Vidadili, N., Suleymanov, E., Bulut, C., & Mahmudlu, C. (2017). Transition to renewable energy and sustainable energy development in Azerbaijan. *Renewable and Sustainable Energy Reviews, 80*, 1153–1161. https://doi.org/10.1016/j.rser.2017.05.168

Zakaria, Z., Kamarudin, S. K., & Wahid, K. A. A. (2021). Fuel cells as an advanced alternative energy source for the residential sector applications in Malaysia. *International Journal of Energy Research, 45*(4), 5032–5057. https://doi.org/10.1002/er.6252

Zaky, A. S., Moirangthem, K., & Wahid, R. (2022). Biofuels: An overview. *Waste-to-Energy*, 85–144. https://doi.org/10.1007/978-3-030-91570-4_4

Zhang, Y., Zhang, X., & Feng, S. (2022). Power to gas: An option for 2060 high penetration rate of renewable energy scenario of China. *Environmental Science and Pollution Research, 29*(5), 6857–6870. https://doi.org/10.1007/s11356-021-16137-x

Chapter 25
Leadership and Global Energy Security: Sentiment Analysis of Twitter Using a Named Entity Recognition

Svitlana Kolosok, Olena Chygryn, and Svitlana Onyshchenko

Abstract Awareness of public attitudes toward the components of global energy security is essential to shaping future energy policy, implementing effective energy leadership, and ensuring sustainable development. This chapter explored the latest public opinion on energy security using data from Twitter. We applied the hashtag #energysecurity to filter 981 messages from January to March 2022. According to opinion mining in tweets, most messages were positive, not negative or neutral. Our study used neural networks for named entity recognition with machine learning based on the spaCy software library in Python to classify important energy security issues of public interest. The public was mostly concerned about gas security and less about oil and nuclear security. Many tweets have expressed worry that Russia's war on Ukraine will increase oil and gas shortages even without sanctions. Leaders seeking public support should consider energy investments in underground gas storage, energy self-sufficiency, and clean energy deployment in the near future.

Keywords Energy leadership · Twitter · Social media · Energy security

JEL Classification Q4 · L82 · N7

S. Kolosok (✉) · O. Chygryn
Sumy State University, Sumy, Ukraine
e-mail: s.kolosok@biem.sumdu.edu.ua; o.chygryn@econ.sumdu.edu.ua

S. Onyshchenko
National University "Yuri Kondratyuk Poltava Polytechnic", Poltava, Ukraine

W. Strielkowski (ed.), *Leadership, Entrepreneurship and Sustainable Development Post COVID-19*, Springer Proceedings in Business and Economics, https://doi.org/10.1007/978-3-031-28131-0_25

25.1 Introduction

Today's practical issues include the studies of leadership and energy security, which have been actively considered during the last decades. The understanding of the essence of energy security differs significantly both among theoreticians of scientific schools and among government officials of various countries of the world in the formation of energy security policy. The governments of the countries should take into account such nuances as the level of provision of energy resources, the existence of reliable and diversified energy suppliers, and dependence on monopolists in the field of energy. Also, an important aspect of national security is the possibility of using substitutes for energy resources to meet the needs of organizations, the flexibility of energy resource supply channels, and the possibility of obtaining short-term and long-term benefits from energy resource supply agreements. And the level of import dependence on energy resources is subject to special regulation and control. Countries with 100% dependence on imported gas, oil, or coal become much more vulnerable in the political arena and may suffer from the manipulation of unscrupulous suppliers of energy resources. Considering the significant shocks in prices and volumes of supply of energy resources that arose as a result of oil crises from the 1970s to the 1990s (Guliyeva et al., 2018; Didenko et al., 2020; Samusevych et al., 2021; Vasylieva et al., 2021; Kolosok et al., 2021a, b; Nyga-Łukaszewska & Napiórkowski, 2022; Yang et al., 2022), the neorealist theory of energy security policy has gained considerable popularity. Neorealists focused on exploring the possibilities of access and control over resources. According to neorealists, it is the control over resources that is the critical basis of the power of states in the international arena. And in the battle for regional dominance and leadership, international cooperation among countries is meaningless since each country cares about national interests and security first. Each country has its own energy supply and demand strategies, which affects national energy security and is a caveat to the use of a single standard energy policy for all countries without exception (Kang, 2022; Starchenko et al., 2021; Vasylieva et al., 2020; Mentel et al., 2020; Letunovska et al., 2021; Kolosok et al., 2021b; Strielkowski et al., 2021; Sedmíková et al., 2021). A classic example of global energy expansion is China's realistic approach to its energy security (Gong, 2022).

There is also an opposite point of view. According to the neoliberal theory, energy cooperation will take place in the existence of benefits for the collaborating countries. And the symmetrical interdependence of countries only strengthens their desire to trade energy resources and prevents open aggression. However, there is still no unified and universal point of view on the issue of energy security in the world community. According to the IEA, energy security means "ensuring the uninterrupted availability of energy sources at an affordable price" (Cebula & Pimonenko, 2015; Polcyn et al., 2022; Lyulyov et al., 2021a; Bilan et al., 2020a; Us et al., 2021; IEA, 2022). According to Kester, there are five approaches to energy security research. Among them are realistic and liberal political reflections; description and categorization of energy security threats; possibilities of securitization of energy security threats; the logic of thinking, discussion, and practice of energy security; as well as studies of the performativity of energy security (Lisin & Strielkowski, 2014; Bilan et al., 2020b; Lyeonov et al., 2021; George, 2020; Elnikova & Barhaq, 2020; Sahoo, 2017). However, after Russia's invasion of Ukraine, governments around the world are reassessing their energy security and leadership. The development of security problems forces countries to look for alternatives to Russian energy resources, introducing step-by-step measures to limit or prohibit their import, or even trade in such resources. Cooperation with Russian companies becomes toxic, as it can provoke sanctions for partner organizations. Many countries are refocusing on alternative markets, resources, or suppliers of energy resources. Renewable energy is developing. Such a process is not fast, but it is acceptable for some countries (Ziabina & Kovalenko, 2021; Ziabina et al., 2020; Bhowmik, 2019; Pimonenko et al., 2021; Pavlyk, 2020; El Amri et al., 2020; Us et al., 2020; Kyrychenko et al., 2021; Vasudevan & Aslan, 2022; Khalatur & Dubovych, 2022). There are still some countries in Europe that are under significant energy pressure from Russia (Balitskiy et al., 2016; Chebotareva et al., 2020). That is why it is interesting to study the latest trends and opinions in view of the radical change of energy security policy in the world in 2022, which was the purpose of this study. To identify trends in energy security and leadership, this research is based on a mixed data analysis approach and includes bibliographic analysis of scientific publications and machine analysis of tweets on energy security issues. Such analysis will describe the global behavioral tendencies in these questions.

This chapter consists of five sections. The remainder of this study is organized as follows. The second section introduces a brief literature review related to current research in energy security and leadership. The third section contains a description of the research data and methodology. The fourth section presents a description of the results and discussions. The fifth section contains conclusions and concludes this study.

25.2 Related Works

Cross-border renewable energy projects can be an effective tool for improving energy security. However, their implementation is not always possible due to significant differences between countries. In the work of Burghard et al. (2022), the results of interviews with experts from Spain, Germany, and the Netherlands, which were collected by the authors during the evaluation of the possibilities of implementing cross-border solar energy projects, are presented. They identified the negative aspects of such projects according to their research, namely, the complexity of implementation, the differences in the interests of partner countries, and the high costs of their implementation, which prevent the widespread use of such projects in the EU (Kester, 2022; Lyeonov et al., 2018). Similar conclusions were obtained by Juszczyk et al. (2022) as a result of a survey of experts from Finland and Poland regarding barriers to the spread of renewable energy. In addition to the above, experts focus on political barriers in the field of renewable energy. Polish experts emphasized the existence of a significant coal lobby in the country: in Poland, support is provided mainly to coal companies, and not to those implementing renewable energy technologies (Burghard et al., 2022; Juszczyk et al., 2022).

A similar situation with limited support for producers of renewable energy exists in Belarus. After gas disputes with Russia, the Belarusian government developed wind power to reduce its energy dependence on Russia. However, with the launch of the Astravets NPP power unit in 2020, wind power in Belarus has come to be seen as expensive, redundant, and unnecessary. Green tariffs were reduced, and wind energy quotas were introduced (Rabe et al., 2022).

But even the declarative leadership of countries aimed at decarbonizing their own energy systems does not actually mean a reduction in the consumption of "dirty" energy at all. With the implementation of strict EU ETS requirements applicable in the EEA-EFTA countries, the share of domestic fossil energy production decreased by approximately 50% in 2019 compared to 1960. At the same time, European countries experienced a simultaneous increase in fossil energy imports by approximately 40% over the same time period. Despite ambitious targets for renewable energy production in Europe, countries still rely on conventional energy consumption from fossil fuels, increasing their own energy dependence on third countries with lower environmental requirements (Novikau, 2021; Lyulyov et al., 2021b).

At the same time, there is criticism of renewable energy sources. Examining panel data for OECD countries, the scientist Cergibozan concludes that not all types of renewable resources reduce the risk of energy security. Thus, there is a positive trend toward a reduction in the risk of energy security between the total volume of renewable energy, wind energy, and hydroelectricity, while the author did not find such a trend for solar energy and biofuel (Radulescu & Sulger, 2022). In addition, Hof et al. emphasize that there is a debate about the possibility of full-scale use of bioenergy. There are concerns in the community that biomass feedstocks are not carbon neutral. And in the production of energy from biomass, there are risks of carbon leakage during its capture and storage (Cergibozan, 2022). However, many works highlight the opposite point of view – that biofuel contributes to ensuring national energy security (Hof et al., 2022; Duarah et al., 2022; Grangeia et al., 2021; Nandimandalam et al., 2022). Despite the possibilities of flexible generation of bioenergy, the deployment of bioenergy technologies is practically absent in most strategies for the development of national energy systems. Instead, national strategies focus on the development of hydropower and the construction of international integration energy networks in order to balance the unstable generation of renewable energy sources (Qiu et al., 2022).

Already existing technologies of renewable energy could ensure the transition to low-carbon energy generation and consumption and solve the issue of energy security. However, progress in the decarbonization of the energy sector remains slow. Rapid green transformations are not possible due to the existence of rigid institutions, disparate energy policies and goals, as well as power struggles and a lack of leadership for strategic changes in energy (Schipfer et al., 2022; Nwanekezie et al., 2022; Vakulenko et al., 2021).

25.3 Data and Methods

For this study, public secondary data was obtained from Twitter. De-identified text messages (tweets) were collected to study public opinion on energy security and leadership.

In general terms, collecting public opinion on energy security and leadership can be achieved through working with governments, industries, civil society, and community stakeholders to develop practical solutions to address the geopolitical, sustainable, and economic challenges posed by a changing global energy landscape. This approach offers a multidisciplinary view on sustainable development, which can contribute to achieving both energy security and environmental security in an equitable manner. The insights collected using these methods show the importance of understanding the public views of mitigating climate change to developing communications and environmental policies that are likely to be accepted and engaged. The analyses conducted for this research on public perceptions of actions for climate change mitigation allowed us to assess topics that people talked about, without having predetermined topics, contributing to identifying topics (e.g., protests, conservation, collective action) that might not have been considered in science-based reports on climate change mitigation. An innovative approach for analyzing public perceptions and behaviors is to utilize social media for understanding climate change discourse (Cabelkova et al., 2015). One of the most prevalent themes observed in regard to the people's perceptions about actions for mitigating climate change is energy – this did not vary across time points, which is consistent with earlier findings from literature (Ma et al., 2019; Cox et al., 2020). In addition, we found that the theme of energy appears to be relevant in the public perceptions of actions to mitigate climate change; it does not appear to change across time. This indicates that considering energy is the aspect that is discussed the most when it comes to taking actions to address climate change, and it is likely the aspect that is the most salient for people. For example, in the United States, the majority of Republicans support expanding the use of fossil fuel-based energy sources, whereas a majority of Democrats are still opposed.

When oil and gas companies took to Twitter in the first half of the United Nations climate meeting in Glasgow, they frequently presented themselves as part of a climate solution to climate change and talked up energy security. Social media was only a public face of oil and gas companies at the UN's Glasgow climate conference. In many ways, the messages from oil and gas companies on social media offer a window into the ways that those companies would like the public to perceive the future. Behind the scenes, the industry continues its investment in fossil fuel extraction, which is driving climate change, and its CEO has made it clear fossil fuel production is here to stay, decades into the future. Going forward, DOE will continue to lead the strategic investments to shift toward a cleaner, more domestic, more secure energy future. Extreme weather events and other impacts associated with climate change are harming the health, safety, and security of the American people, and increasing the urgency of combating climate change and speeding up the transition to a cleaner energy economy. It is, therefore, any administrations' policy to listen to the science; improve public health and safeguard our environment; provide access to clean air and clean water; limit exposure to hazardous chemicals and pesticides; hold polluters accountable, including those that disproportionately harm communities of color and lower-income communities; reduce greenhouse gas emissions; strengthen resilience to climate changes; restore and expand Our Nations national treasures and monuments; and prioritize both environmental justice and creating the high-paying, union-based jobs needed to achieve these goals. At home, any nation should address the climate crisis with an ambitious rebuilding agenda designed to reduce harmful emissions while creating good, clean-energy jobs. This analysis emphasizes that any country should put the highest priority on developing a clean-energy economy, which in turn creates good jobs. Our research therefore explains why these provisions should be included in the clean energy legislation making the case that a domestic-content tax credit would create good jobs for working people, contribute to addressing the climate crisis, and enhance the nation's economic security. The correlations with media coverage of energy and environmental topics are meant to capture media influence over public debate.

A manual advanced Twitter search with custom filters was used to form a semi-structured research base; 981 messages were filtered using the #energysecurity hashtag from January 01 to March 04, 2022. At the same time, 54 repetitions of tweets were found and deleted, that is, 927 unique messages were actually received. Subsequently, the data underwent data preprocessing, namely, punctuation marks, http links, and hashtags were removed, the lower case was set for all tweets, lemmatization and tokenization of words were performed, and stop words were removed.

According to the results of the opinion mining in tweets, unexpectedly, the majority of messages had a positive color, and not negative or neutral (see Figs. 25.1 and 25.2).

The named entity recognition method was used to classify important energy security issues that were of public interest during the research period.

For this task, neural networks with machine learning were built based on the spaCy software library in Python. This method allowed us to extract information from tweet texts about the most common organizations, countries, and parts of speech (nouns, verbs, and adjectives) mentioned in the selected tweets.

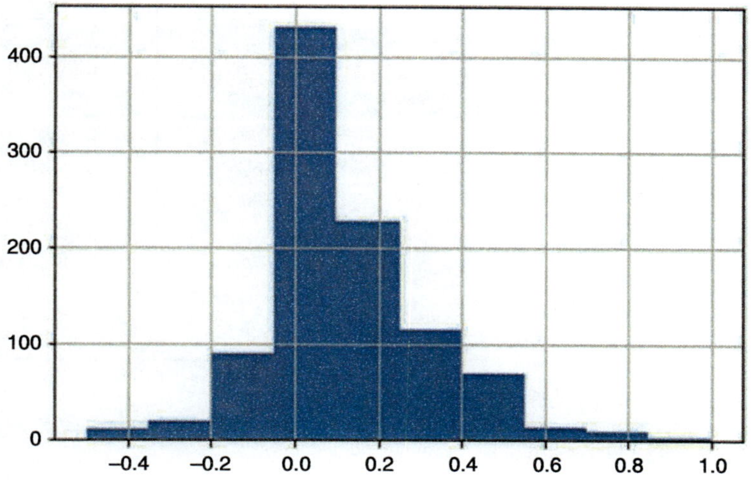

Fig. 25.1 Polarity histogram of tweets received using the hashtag "energysecurity" from January 1 to March 4, 2022. (Source: twitter.com)

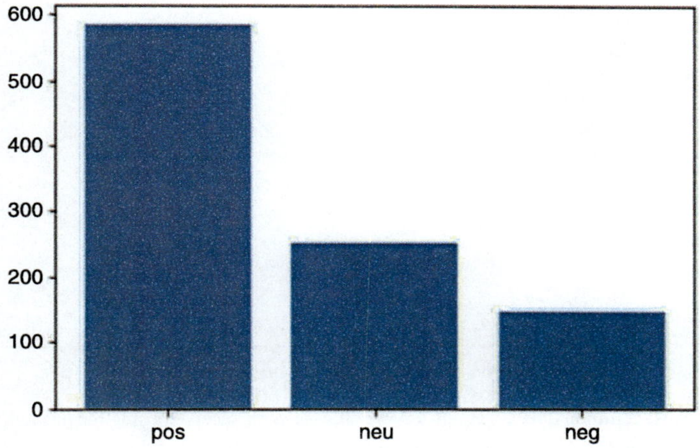

Fig. 25.2 Sentiment bar chart of tweets received using the hashtag "energysecurity" from January 1 to March 4, 2022. (Source: twitter.com)

25.4 Results and Discussion

In order to better understand the latest public opinion on energy security and leadership, we analyzed a textual dataset. Using the method of named entity recognition, the leading organizations that were most often mentioned in the messages of the sample formed from Twitter were selected (Fig. 25.3). It is noticeable that the top

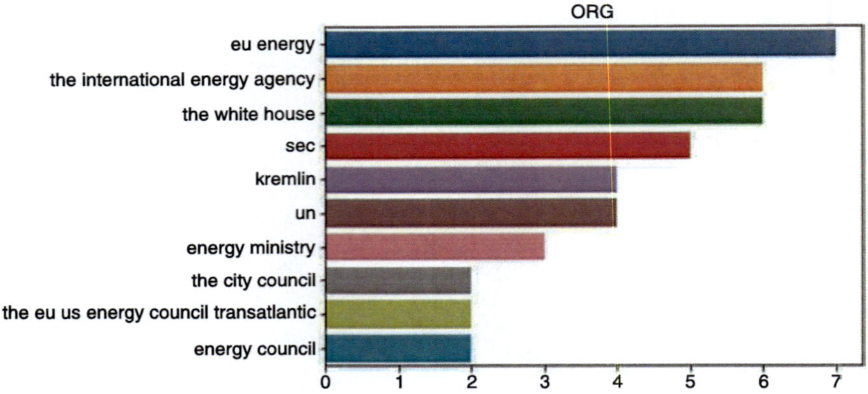

Fig. 25.3 Most common tweeted organizations (tweets received using the hashtag "energysecurity" from January 1 to March 4, 2022). (Source: twitter.com)

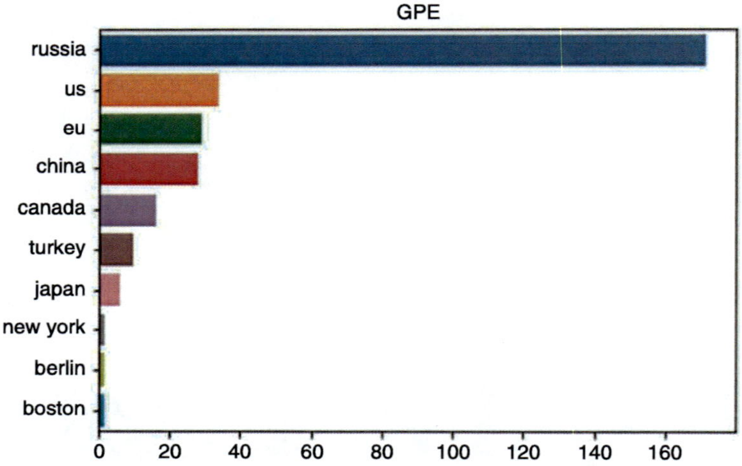

Fig. 25.4 Most common tweeted countries and cities (tweets received using the hashtag "energysecurity" from January 1 to March 4, 2022). (Source: twitter.com)

five organizations in terms of the number of mentions in Twitter messages included government structures of the EU, the USA (the White House and SEC), and Russia (the kremlin), as well as one autonomous international body (the International Energy Agency). As for the countries mentioned by the public in the context of energy security in January–March 2022, Russia should be mentioned first of all (Fig. 25.4). The state of energy security of other countries was linked to Russia's actions during this period. And significantly fewer tweets referred to the USA, the EU, China, or Canada.

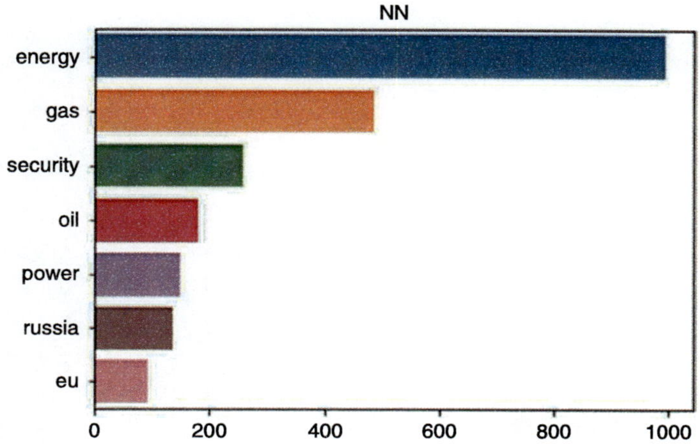

Fig. 25.5 Most common nouns (tweets received using the hashtag "energysecurity" from January 1 to March 4, 2022). (Source: twitter.com)

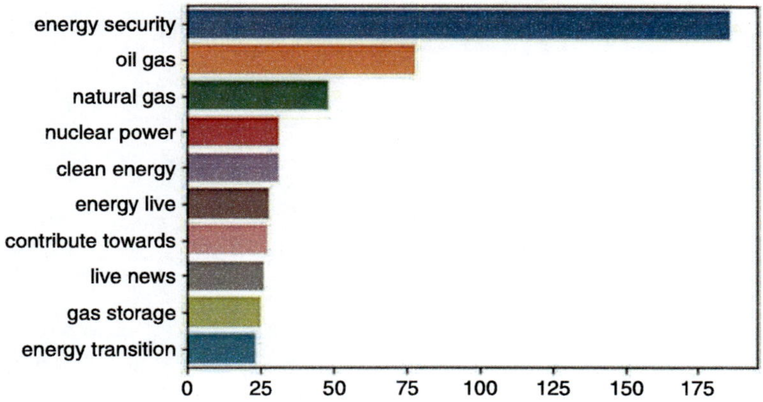

Fig. 25.6 Bigrams of tweets (after removing stop-words) received using the hashtag "energysecurity" from January 1 to March 4, 2022. (Source: twitter.com)

If we ask about different types of energy resources, according to which society was concerned about the state of national energy security, then the answer can be found among the most common nouns in tweets (Fig. 25.5). From the diagram, it is possible to understand the lack of which energy resources often worried the public. Therefore, society was most concerned about gas safety and less about oil and nuclear safety. Many tweets express concern that Russia's military actions will lead to an increase in oil and gas shortages even without the imposition of large-scale sanctions against Russia. Most of the keywords among the most common verbs in tweets are direct calls to action and leadership. The most common verbs in the tweets included "be," "ensure," "have," "take," and "reduce." At the same time, the

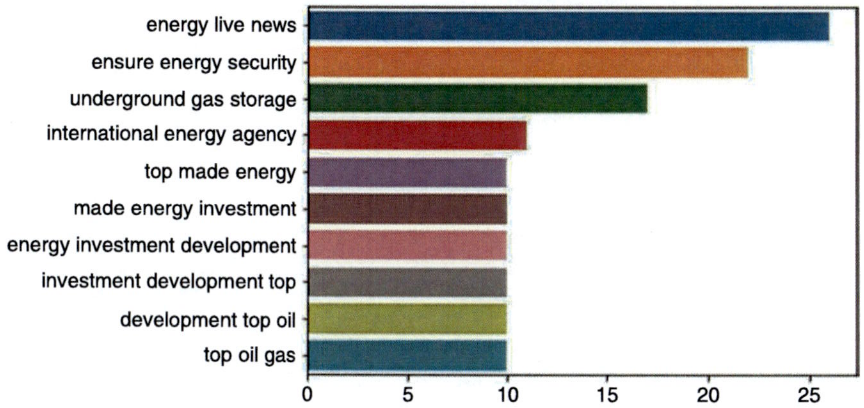

Fig. 25.7 Trigrams of tweets (after removing stop-words) received using the hashtag "energysecurity" from January 1 to March 4, 2022. (Source: twitter.com)

most common adjectives in the tweets were words like “new,” “nuclear,” “global,” “green,” “economic,” and “renewable.” During the period under review, there was a public demand for a new energy policy that could solve the energy crisis. In the spring of 2022, the tweets actively discussed the new energy policies of the EU and the USA, which involved a change in views on the conditions of natural gas supply, and the possibility of a new generation of nuclear energy and renewable energy.

From the bigrams (Fig. 25.6) and trigrams (Fig. 25.7) of the keywords of the tweets, it is possible to see that energy resources storage infrastructure is actively connected with energy security issues. The insufficiently developed energy storage infrastructure is one of the factors that increase the effects of geopolitical pressure on countries and increases the vulnerability of countries.

As a result, there is speculation on the energy resource market, and the bills of end users of energy resources have increased significantly. With the 2022 energy crisis largely linked to natural gas, calls for new underground gas storage are not surprising.

25.5 Conclusions

In 2022, humanity encountered an aggravation of energy crisis, which was associated with geopolitical pressure, sanctions, and energy blackmail. The crisis came in the unfortunate times that followed the COVID-19 pandemic and all the negative effects brought about by this catastrophe. The growing vulnerability of existing energy supply channels gives an impetus to the revision of national energy security policies. Using social platforms such as Twitter, it is possible to assess and understand public opinion on energy security issues and to shape new energy leadership policies based on the analysis of large amounts of textual data.

This work demonstrates that society was most concerned about gas safety and less about oil and nuclear safety. Many tweets expressed concern that Russia’s military actions would lead to an increase in oil and gas shortages even without the imposition of large-scale sanctions against Russia. Therefore, leaders seeking public support should soon consider energy investments in underground gas storage, increased energy self-sufficiency, and clean energy infrastructure. Further analysis should be performed to identify value characteristics in the public to change energy security vectors.

Acknowledgments This research was funded by the grants from the Ministry of Education and Science of Ukraine (grant numbers: 0122 U000788, 0120 U102002, 0123 U101920).

References

Balitskiy, S., Bilan, Y., Strielkowski, W., & Štreimikienė, D. (2016). Energy efficiency and natural gas consumption in the context of economic development in the European Union. *Renewable and Sustainable Energy Reviews, 55*, 156–168. https://doi.org/10.1016/j.rser.2015.10.053

Bhowmik, D. (2019). Decoupling CO2 emissions in Nordic countries: Panel data analysis. *SocioEconomic Challenges, 3*(2), 15–30. https://doi.org/10.21272/sec.3(2).15-30.2019

Bilan, Y., Pimonenko, T., & Starchenko, L. (2020a). Sustainable business models for innovation and success: Bibliometric analysis. *E3S Web of Conferences, 159*, 04037. https://doi.org/10.1051/e3sconf/202015904037

Bilan, Y., Srovnalikova, P., Streimikis, J., Lyeonov, S., Tiutiunyk, I., & Humenna, Y. (2020b). From shadow economy to lower carbon intensity: Theory and evidence. *International Journal of Global Environmental Issues, 19*(1–3), 196–216. https://doi.org/10.1504/ijgenvi.2020.114874

Burghard, U., Dütschke, E., Caldes, N., & Oltra, C. (2022). Cross-border concentrated solar power projects – Opportunity or dead end? A study into actor views in Europe. *Energy Policy, 163*, 112833. https://doi.org/10.1016/j.enpol.2022.112833

Cabelkova, I., Strielkowski, W., & Mirvald, M. (2015). Business influence on the mass media: A case study of 21 countries. *Transformations in Business & Economics, 14*(1), 65–75.

Cebula, J., & Pimonenko, T. (2015). Comparison financing conditions of the development biogas sector in Poland and Ukraine. *International Journal of Ecology and Development, 30*(2), 20–30.

Cergibozan, R. (2022). Renewable energy sources as a solution for energy security risk: Empirical evidence from OECD countries. *Renewable Energy, 183*, 617–626. https://doi.org/10.1016/j.renene.2021.11.056

Chebotareva, G., Strielkowski, W., & Streimikiene, D. (2020). Risk assessment in renewable energy projects: A case of Russia. *Journal of Cleaner Production, 269*, 122110. https://doi.org/10.1016/j.jclepro.2020.122110

Cox, E., Spence, E., & Pidgeon, N. (2020). Public perceptions of carbon dioxide removal in the United States and the United Kingdom. *Nature Climate Change, 10*(8), 744–749. https://doi.org/10.1038/s41558-020-0823-z

Didenko, I., Volik, K., Vasylieva, T., Lyeonov, S., & Antoniuk, N. (2020). Migration, environment, and country safety: Analysis of touchpoints. *E3S Web of Conferences, 202*, 03028. https://doi.org/10.1051/e3sconf/202020203028

Duarah, P., Haldar, D., Patel, A. K., Dong, C., Singhania, R. R., & Purkait, M. K. (2022). A review on global perspectives of sustainable development in bioenergy generation. *Bioresource Technology, 348*, 126791. https://doi.org/10.1016/j.biortech.2022.126791

El Amri, A., Boutti, R., Oulfarsi, S., Rodhain, F., & Bouzahir, B. (2020). Carbon financial markets underlying climate risk management, pricing and forecasting: Fundamental analysis. *Financial Markets, Institutions and Risks, 4*(4), 31–44. https://doi.org/10.21272/fmir.4(4).31-44.2020

Elnikova, J., & Barhaq, A. R. (2020). Transparency of responsible investment environment. *Business Ethics and Leadership, 4*(4), 68–75. https://doi.org/10.21272/bel.4(4).68-75.2020

George, B. (2020). Inclusive sustainable development in the Caribbean region: Social capital and the creation of competitive advantage in tourism networks. *Business Ethics and Leadership, 4*(3), 119–126. https://doi.org/10.21272/bel.4(3).119-126.2020

Gong, X. (2022). Energy security through a financial lens: Rethinking geopolitics, strategic investment, and governance in China's global energy expansion. *Energy Research and Social Science, 83*, 42–50. https://doi.org/10.1016/j.erss.2021.102341

Grangeia, C., Santos, L., & Lazaro, L. L. B. (2021). The Brazilian biofuel policy (Renova-bio) and its uncertainties: An assessment of technical, socioeconomic and institutional aspects. *Energy Conversion and Management: X, 13*, 100156. https://doi.org/10.1016/j.ecmx.2021.100156

Guliyeva, A., Britchenko, I., & Rzayeva, U. (2018). Global security and economic asymmetry: A comparison of developed and developing countries. *Journal of Security and Sustainability Issues, 7*(4), 707–717. https://doi.org/10.9770/jssi.2018.7.4(8)

Hof, A. F., Esmeijer, K., de Boer, H. S., Daioglou, V., Doelman, J. C., Elzen, M. G. J. D., & van Vuuren, D. P. (2022). Regional energy diversity and sovereignty in different 2°C and 1.5°C pathways. *Energy, 239*(B), 122197. https://doi.org/10.1016/j.energy.2021.122197

IEA. (2022). *Ensuring the uninterrupted availability of energy sources at an affordable price*. https://www.iea.org/areas-of-work/ensuring-energy-security

Juszczyk, O., Juszczyk, J., Juszczyk, S., & Takala, J. (2022). Barriers for renewable energy technologies diffusion: Empirical evidence from Finland and Poland. *Energies, 15*(2), 527. https://doi.org/10.3390/en15020527

Kang, H. (2022). An analysis of the relationship between energy trilemma and economic growth. *Sustainability, 14*(7), 3863. https://doi.org/10.3390/su14073863

Kester, J. (2022). The scare behind energy security: Four conceptualisations of scarcity and a never-ending search for abundance. *Journal of International Relations and Development, 25*(1), 31–53. https://doi.org/10.1057/s41268-021-00216-0

Khalatur, S., & Dubovych, O. (2022). Financial engineering of green finance as an element of environmental innovation management. *Marketing and Management of Innovations, 1*, 232–246. https://doi.org/10.21272/mmi.2022.1-17

Kolosok, S., Bilan, Y., Vasylieva, T., Wojciechowski, A., & Morawski, M. (2021a). A scoping review of renewable energy, sustainability and the environment. *Energies, 14*(15), 4490. https://doi.org/10.3390/en14154490

Kolosok, S., Vasylieva, T., & Lyeonov, S. (2021b). Machine analysis of the UK electrical energy initiatives based on the e-petitions to the UK government and parliament. *CEUR Workshop Proceedings, 2870*, 1562–1573.

Kyrychenko, K., Laznenko, D., & Reshetniak, Y. (2021). Green university as an element of forming a sustainable public health system. *Health Economics and Management Review, 2*(4), 21–26. https://doi.org/10.21272/hem.2021.4-02

Letunovska, N., Saher, L., Vasylieva, T., & Lieonov, S. (2021). Dependence of public health on energy consumption: A cross-regional analysis. *E3S Web of Conferences, 250*, 04014. https://doi.org/10.1051/e3sconf/202125004014

Lisin, E., & Strielkowski, W. (2014). Modelling new economic approaches for the wholesale energy markets in Russia and the EU. *Transformations in Business & Economics, 13*(2B), 566–580.

Lyeonov, S. V., Vasylieva, T. A., & Lyulyov, O. V. (2018). Macroeconomic stability evaluation in countries of lower-middle income economies. *Naukovyi Visnyk Natsionalnoho Hirnychoho Universytetu, 1*, 138–146. https://doi.org/10.29202/nvngu/2018-1/4

Lyeonov, S., Pimonenko, T., Chygryn, O., Reznik, O., & Gaynulina, R. (2021). Green brand as a marketing instrument: Principle, features and parameters. *International Journal of Global Energy Issues, 43*(2–3), 147–165.

Lyulyov, O., Lyeonov, S., Tiutiunyk, I., & Podgórska, J. (2021a). The impact of tax gap on macroeconomic stability: Assessment using panel VEC approach. *Journal of International Studies, 14*(1), 139–152. https://doi.org/10.14254/2071-8330.2021/14-1/10

Lyulyov, O., Pimonenko, T., Kwilinski, A., & Us, Y. (2021b). The heterogeneous effect of democracy, economic and political globalisation on renewable energy. *E3S Web of Conferences, 250*, 03006. https://doi.org/10.1051/e3sconf/202125003006

Ma, Y., Dixon, G., & Hmielowski, J. D. (2019). Psychological reactance from reading basic facts on climate change: The role of prior views and political identification. *Environmental Communication, 13*(1), 71–86. https://doi.org/10.1080/17524032.2018.1548369

Mentel, G., Vasilyeva, T., Samusevych, Y., Vysochyna, A., Karbach, R., & Streimikis, J. (2020). The evaluation of economic, environmental and energy security: Composite approach. *International Journal of Global Environmental Issues, 19*(1–3), 177–195.

Nandimandalam, H., Gude, V. G., & Marufuzzaman, M. (2022). Environmental impact assessment of biomass supported electricity generation for sustainable rural energy systems – A case study of Grenada county, Mississippi, USA. *Science of the Total Environment, 802*, 149716. https://doi.org/10.1016/j.scitotenv.2021

Novikau, A. (2021). Current challenges and prospects of wind energy in Belarus. *Renewable Energy, 182*, 1049–1059. https://doi.org/10.1016/j.renene.2021.11.011

Nwanekezie, K., Noble, B., & Poelzer, G. (2022). Strategic assessment for energy transitions: A case study of renewable energy development in Saskatchewan, Canada. *Environmental Impact Assessment Review, 92*, 106688. https://doi.org/10.1016/j.eiar.2021.106688

Nyga-łukaszewska, H., & Napiórkowski, T. M. (2022). Does energy demand security affect international competitiveness? Case of selected energy-exporting OECD countries. *Energies, 15*(6), 1991. https://doi.org/10.3390/en15061991

Pavlyk, V. (2020). Assessment of green investment impact on the energy efficiency gap of the national economy. *Financial Markets, Institutions and Risks, 4*(1), 117–123. https://doi.org/10.21272/fmir.4(1).117-123.2020

Pimonenko, T., Us, Y., Myroshnychenko, Y., Dubyna, O., & Vasylyna, T. (2021). Green financing for carbon-free growth: Role of banks marketing strategy. *Financial Markets, Institutions and Risks, 5*(3), 71–78. https://doi.org/10.21272/fmir.5(3).71-78.2021

Polcyn, J., Us, Y., Lyulyov, O., Pimonenko, T., & Kwilinski, A. (2022). Factors influencing the renewable energy consumption in selected European countries. *Energies, 15*(1), 108. https://doi.org/10.3390/en15010108

Qiu, B., Yang, C., Shao, Q., Liu, Y., & Chu, H. (2022). Recent advances on industrial solid waste catalysts for improving the quality of bio-oil from biomass catalytic cracking: A review. *Fuel, 315*, 123218. https://doi.org/10.1016/j.fuel.2022.123218

Rabe, M., Bilan, Y., Widera, K., & Vasa, L. (2022). Application of the linear programming method in the construction of a mathematical model of optimization distributed energy. *Energies, 15*(5), 1872. https://doi.org/10.3390/en15051872

Radulescu, D., & Sulger, P. (2022). Interdependencies between countries in the provision of energy. *Energy Economics, 107*, 105799. https://doi.org/10.1016/j.eneco.2021.105799

Sahoo, K. K. (2017). Transform of ethical practices in globalized economy, a special focus on Ghana. *Business Ethics and Leadership, 1*(4), 58–65. https://doi.org/10.21272/bel.1(4).58-65.2017

Samusevych, Y., Vysochyna, A., Vasylieva, T., Lyeonov, S., & Pokhylko, S. (2021). Environmental, energy and economic security: Assessment and interaction. *E3S Web of Conferences, 234*, 00012. https://doi.org/10.1051/e3sconf/202123400012

Schipfer, F., Mäki, E., Schmieder, U., Lange, N., Schildhauer, T., Hennig, C., & Thrän, D. (2022). Status of and expectations for flexible bioenergy to support resource efficiency and to accelerate the energy transition. *Renewable and Sustainable Energy Reviews, 158*, 112094. https://doi.org/10.1016/j.rser.2022.112094

Sedmíková, E., Vasylieva, T., Tiutiunyk, I., & Navickas, M. (2021). Energy consumption in assessment of shadow economy. *European Journal of Interdisciplinary Studies, 13*(2), 47–64. https://doi.org/10.24818/ejis.2021.12

Starchenko, L., Lyeonov, S., Vasylieva, T., Pimonenko, T., & Lyulyov, O. (2021). Environmental management and green brand for sustainable entrepreneurship. *E3S Web of Conferences, 234*, 00015. https://doi.org/10.1051/e3sconf/202123400015

Strielkowski, W., Civín, L., Tarkhanova, E., Tvaronavičienė, M., & Petrenko, Y. (2021). Renewable energy in the sustainable development of electrical power sector: A review. *Energies, 14*(24), 8240. https://doi.org/10.3390/en14248240

Us, Y., Pimonenko, T., Tambovceva, T., & Segers, J.-P. (2020). Green transformations in the healthcare system: The covid-19 impact. *Health Economics and Management Review, 1*(1), 48–59. https://doi.org/10.21272/hem.2020.1-04

Us, Y., Pimonenko, T., & Lyulyov, O. (2021). Energy efficiency profiles in developing the free-carbon economy: On the example of Ukraine and the V4 countries. *Polityka Energetyczna, 23*(4), 49–66. https://doi.org/10.33223/epj/127397

Vakulenko, I., Saher, L., Syhyda, L., Kolosok, S., & Yevdokymova, A. (2021). The first step in removing communication and organizational barriers to stakeholders' interaction in smart grids: A theoretical approach. *E3S Web of Conferences, 234*, 00020. https://doi.org/10.1051/e3sconf/202123400020

Vasudevan, H., & Aslan, I. (2022). Developing organizational citizenship behaviour: Organizational climate and leader-member exchange effects. *Marketing and Management of Innovations, 1*, 94–107. https://doi.org/10.21272/mmi.2022.1-07

Vasylieva, T., Machová, V., Vysochyna, A., Podgórska, J., & Samusevych, Y. (2020). Setting up architecture for environmental tax system under certain socioeconomic conditions. *Journal of International Studies, 13*(4), 273–285. https://doi.org/10.14254/2071-8330.2020/13-4/19

Vasylieva, T., Pavlyk, V., Bilan, Y., Mentel, G., & Rabe, M. (2021). Assessment of energy efficiency gaps: The case for Ukraine. *Energies, 14*(5), 1323. https://doi.org/10.3390/en14051323

Yang, B., Ding, L., Zhan, X., Tao, X., & Peng, F. (2022). Evaluation and analysis of energy security in China based on the DPSIR model. *Energy Reports, 8*, 607–615. https://doi.org/10.1016/j.egyr.2022.01.229

Ziabina, Y., & Kovalenko, Y. (2021). Regularities in the development of the theory of energy efficiency management. *SocioEconomic Challenges, 5*(1), 117–132. https://doi.org/10.21272/sec.5(1).117-132.2021

Ziabina, Y., Pimonenko, T., & Starchenko, L. (2020). Energy efficiency of National Economy: Social, economic and ecological indicators. *SocioEconomic Challenges, 4*(4), 160–174. https://doi.org/10.21272/sec.4(4).160-174.2020

Chapter 26
Leadership Toward Behavioral Change in Energy Consumption

Manuela Tvaronavičienė

Abstract The aim of this research is to study the leadership toward behavioral change in energy consumption. This research also explores how behavior change could lead to substantial greenhouse gas emission reductions across European high-income countries. The highest-impact areas of existing greenhouse gas emission reductions by behavior change are reductions in automobile trips, flying, and meat consumption – activities that are prevalent among individuals and countries with higher incomes currently, even though automobile trips are recommended less frequently in public resources or in secondary school science textbooks than are other climate actions with modest impacts, such as reducing energy use. Our results demonstrate that energy consumption behavior could be improved by various organizational factors, including leadership styles, as well as green-friendly environments. Ongoing research in many countries might inform longer-term, deeper emission reduction strategies that incorporate behavioral changes, but most longer-term strategies currently fail to capitalize on this opportunity. In general, long-run climate strategies need to explicitly address the potential for behavior change in high-impact sectors (especially neglected areas like aviation and diet, as well as more commonly studied areas like transportation and energy consumption) in developing present and future policies for formulating the leading approaches to the leading sustainable economic development of nations worldwide.

Keywords Leadership · Behavioral change · Energy consumption · Energy policy

JEL Classification D90 · O13 · 018

M. Tvaronavičienė (✉)
Department of Business Technologies and Entrepreneurship, Vilnius Gediminas Technical University, Vilnius, Lithuania

Institute of Humanities and Social Sciences, Daugavpils University, Daugavpils, Latvia
e-mail: manuela.tvaronaviciene@vilniustech.lt

W. Strielkowski (ed.), *Leadership, Entrepreneurship and Sustainable Development Post COVID-19*, Springer Proceedings in Business and Economics, https://doi.org/10.1007/978-3-031-28131-0_26

26.1 Introduction

Human behavior represents a very important factor for managing energy resources and for making continuous improvements to energy efficiency. It is crucial for the human beings to need to retrain their brains to see that they, as individuals (and teams), can contribute to their own reduction of energy use (Paone & Bacher, 2018; Harputlugil & de Wilde, 2021). There appears to be the tremendous potential in saving energy through behavior modification. This is how the potential energy savings need to be balanced with the likelihood of making change (Liobikienė & Poškus, 2019; Shevyakova et al., 2021).

What is missing represents a concentrated, structured approach for making and maintaining changes in energy behaviors. Some organizations and research institutions publish helpful case studies on behavioral change programs (Guliyeva et al., 2021; Komendantova, 2021). Another role for national governments to play in driving behavioral change could be in motivating regional and local governments to act appropriately and be centers for innovation, and to establish databases for tracking emissions and policies that mitigate emissions, or coordinate sharing of best practices (Tan et al., 2021). The highest-impact areas of existing reductions of the emissions generated by burning the fossil fuels by behavioral change are reductions in automobile trips, flying, and meat consumption activities that are prevalent among higher-income individuals and countries currently, though automobile trips are recommended less frequently in public resources or in science high school textbooks than in climate actions with modest impacts, such as reducing energy use (Vona, 2019). Overall, longer-term climate strategies should explicitly address the potential of behavior change in high-impact sectors (especially neglected areas like aviation and diet, as well as more frequently studied areas like automobile transportation and energy use) in developing present and future policies (Rochedo et al., 2021). While feedback about prior energy consumption patterns, either in absolute terms or relative to peers, drives behavior conservation, overall behavior effects vary depending on several design features (Kumar et al., 2019). Adjusting feedback frequency, for instance, does not tend to cause significant changes in energy savings, while changing feedback media does (Shen et al., 2021). Users subjected to a time-of-use tariff may also notice changes in the energy prices. The service may come in the form of an at-home energy bill report comparing use with other, similar customers. The actual energy used over a typical week usually forms the basis for the relative electricity use for unused power calculation. A large body of research shows what works for reducing household energy consumption (Sahakian et al., 2021). In that regard, the nation's healthcare system uses significant amounts of electric power. Information-based programs give customers information on energy usage, so that they can make better-informed decisions to lower energy consumption. For instance, a smartphone application could include push notifications in order to inform a consumer's energy use (Karthick & Chandrasekaran, 2021).

Utilities are already using competitions and games to interact with customers, encouraging users to set goals for cutting their own energy use in return for a small reward. For instance, Canadian electric utility BC Hydro is using multiple behavioral incentives to attract consumers into their voluntary energy-saving program, Team Power Smart, which challenges participants to lower their power use by 10% in the 1-year period participants participate. Over the 4 ½ years, its voluntary energy savings program has attracted over 91,000 households, reducing their total electrical consumption by 25.6 GWe relative to a comparable control group (Radtke, 2022). The primary rationale behind starting with the idle power reduction is that it builds a culture of energy efficiency that is either cost-effective or self-financed, and monetary savings could become the basis of future investments in energy-efficient machines. Some production leaders report installing an electronic screen in the cafeteria entryway of their enterprises for displaying energy usage and savings on a daily basis and encouraging their employees to save more energy (Mahapatra et al., 2018). In addition, other production leaders also report that an existing practice, in which printed documents showing the energy-reduction progress were taped on the blackboard, was not very interesting because the existing practices had far too much information. Trying to increase the energy efficiency, many operation leaders have also started reading up on energy audits and power usage and realized it was essential to measure actual energy usage on the equipment in order to achieve targets. When the first results from an early phase of the project showed this strategy was working well in terms of reducing energy, the operation leader gained additional confidence to proceed (Sdoukopoulos et al., 2019).

Thence, one can see that the energy costs that people are using are often not captured on their work-related budgets, so they tend to be less inclined to view their behaviors as energy consumers as costing the consumers enormous amounts of money. The right approaches and techniques to altering their energy-perceiving and assessing behavior can help them to broaden their perspectives and horizons.

26.2 Leading Approaches to Addressing Climate Change Through Behavioral Changes

In general terms, addressing climate change requires deep behavior changes, not just consumer actions, but actions as members of communities and organizations, as well as citizens with the power to affect policies (Broska, 2021; Paulikas & Paulikienė, 2022). Accelerated, equitable climate action to mitigate and adapt to the impacts of climate change is essential for sustainable development. Equitable climate action also requires recognition that higher-income countries and communities need to do more (Simpson et al., 2022). Even globally, emphasis remains on prevention, since investments and behavior changes made now will reap future benefits by reducing exposure and increasing resilience to costly future breakdowns, be they from climate change, illness, or the confluence of these or other factors (Brockhaus et al., 2021).

Given the nature and scale of socioeconomic impacts from climate change, staying put poses substantial risks. The world will therefore need to substantially ramp up and accelerate adaptation efforts so that we mitigate more direct and damaging impacts from climate change, which are either already occurring or locked in regardless of any action on decarbonization (Colgan et al., 2021). At the other extreme, delayed or restricted action would risk allowing climate change to escalate, endanger billions of lives and livelihoods, cause mass displacement, deepen political conflicts and disputes, and result in severe reductions of the global economy.

New infrastructure investments, such as the digital platforms, including those for low-carbon developments, must be built in a way that is resilient to future climate impacts, by assessing the climate risks over the life of a project (Marino et al., 2022). Some response options have carbon capture and storage capabilities, while also helping communities limit climate-related impacts (Mace et al., 2021). Economic incentive packages could help speed the transition to a climate-resilient, carbon-neutral electric grid, creating jobs in the process (Strielkowski et al., 2021). Continued innovation in technology and processes will be essential for meeting climate and other sustainable goals. Yet, just as technology innovation led us into a looming climate crisis, it could accelerate the recovery. Legislation is being passed that is a major step forward to addressing the issues of environmental justice and equity in our current climate crisis (Batres et al., 2021). The Paris agreement also seeks to enhance countries' ability to respond to climate change impacts, with adequate financial flows, new technical frameworks, and enhanced capacity building. Addressing climate change requires collective action on multiple fronts and requires both national and international collaborative business efforts (Leonavičienė & Burinskienė, 2021). The organization strategy is focused on increasing investments in economically beneficial energy efficiency, the integration of energy and environmental regulations, and establishing policies and regulations that address climate change effectively. In order to apply behavioral science to climate challenges across the planet, researchers worldwide launched some new initiatives that would identify and quantify the most influential behavior changes at the level of populations, starting with three project pilot countries (such as Mexico, India, and the USA) within transportation, energy, and food (Rutz et al., 2020). Climate action planning working groups are charged with identifying opportunities to increase the benefits of climate action programs and policies for all residents and to avoid potentially negative impacts or lost opportunities for communities of color and lower-income populations.

In general, the state climate resiliency strategy should always identify priority actions and investments that can best prepare countries and governments for the changing climate. This is why the United Nations Environmental Program (UNEP) and World Meteorological Organization (WMO) established the Intergovernmental Panel on Climate Change (IPCC) to provide policymakers with periodic scientific assessments of climate change, its impacts, and risks as well as make recommendations on adaptation and mitigation strategies. For assessment reports, experts volunteer their time to evaluate thousands of scientific papers published annually, in order to produce an integrated summary of what is known about climate change drivers, its impacts and future risks, and how adaptation and mitigation might mitigate such risks as the global warming of 1.5 degree Celsius, increase in global warming from preindustrial levels, and associated pathways to global greenhouse gas emissions (Lawal et al., 2019).

Initiative is a commitment to helping businesses and organizations reduce the climate impacts of the foods they serve, making it easier for individuals to choose businesses and organizations that are making climate-friendly choices. Behavioral policy development involves, but is also going beyond, the traditional nudges (or the smaller, often temporary changes in our environments) by exploring ways of encouraging behaviors at a population level via behaviorally informed policy choices. Voluntary efforts on an individual level are not enough by themselves to make the deep cuts needed to achieve the 1.5-degree Celsius target. Rather, households will need regulatory structures that support their behavior changes (Porter et al., 2020). Measurement and accountability, absent a recognition of the link between the warming planet and dangerous climate change, also will not lead to meaningful action.

26.3 Uncertainty in the Energy Consumption

Under conditions of uncertainty, the conventional approaches to strategic planning in the field of energy conservation and saving may prove downright hazardous. Chances are, the traditional strategy planning process is not going to do a whole lot of good. For systematically making smart strategic decisions in the face of uncertainty another approach is required – the one that escapes from that perilous binary perspective (Kurpiela & Teuteberg, 2022). No approach will eliminate the challenges of uncertainty, but this approach offers practical guidelines that lead to better-informed, confident strategic decisions. Scenario planning techniques are critical for strategizing in uncertain environments.

Such scenarios have three characteristics that make them particularly powerful tools to understand uncertainty and design strategies accordingly. They are especially helpful for developing strategies to deal with extreme events like those we have seen in the global economy lately. As the analysis behind each scenario proceeds, it is common to find a few especially potent drivers of change. Companies cannot incorporate every eventuality into their scenarios, nor should they waste much time with the least likely (Chang & Andreoni, 2020).

In the current conditions, managers should develop a series of discrete scenarios, informed by their knowledge about how the main residual uncertainties could unfold – whether, say, deregulation occurs, or competitors build new factories, or not. Such systems explicitly encourage managers to understate uncertainties to build a convincing case for their strategies. Business plans frequently extrapolate into a future trend which is obviously not sustainable (Cuypers et al., 2021). By ignoring external scenarios, and expending their energies to make modest improvements or degrades in relation to today, leaders expose themselves to abrupt changes – especially in downturns.

The pace of technological change is different, and it presents challenges to certain groups of people and societies, which we will have to recognize and address in order to avoid marginalization and political strife. With the massive digitalization brought by the COVID-19 pandemic, all aspects of human existence will be affected by the integration of digital technologies and artificial intelligence (AI) into human societies (Kesavan et al., 2022). The ways in which humans and machines are combined will be transformed – there will be many new types of human-machine symbiosis. People will become a part of the complex systems providing the emotional intelligence of humans wherever it is needed for smoother social governance (Kim et al., 2022). The uses of AI systems will expand, with greatest growth coming from systems that augment and supplement humans' capabilities and decisions. For instance, AI will help us save energy at home and on the road, by identifying exact times and temperatures that we will need, pinpointing energy sources that will be the cheapest and most effective (Gruzina et al., 2022). What could be a significant saving of energy is the reduced traffic flow of people and traffic congestion associated with it. According to energy use of information technology sectors, besides the usage of devices, data centers, networks, and computers are major sources of energy use (Razmjoo et al., 2022). With increasing usage of Big Data, energy disdain is expected to increase significantly the efficiency as well as the other forms of energy required for the systems to function, which needs better assessment.

Human attitudes toward energy, including the consciousness about the consequences, motivations, and perceived responsibility, are considered unquestioned drivers of the power systems. The tradeoffs and correlations between indoor household environments and household energy use patterns may offer insights to include human dimensions in energy policies and planning, building controls, and energy pricing. Identifying key human behavior drivers and attitudes factors of energy performance and the human-building interaction could provide relevant insights for energy policies related to strategies, such as targeted demographic segments, for improving outreach, modeling incentives, and designing the development and support programs (Ziesemer et al., 2021). This would be useful for exploring the role of human and social dimensions to increase energy policy effectiveness and efficiency gaps in the buildings sectors. Iweka et al. (2019) evaluated a number of critical issues related to the concepts, applications, and behavioral factors of building energy efficiency. The results indicated that energy-conscious users' behaviors could result in energy savings in many buildings, and nonmonetary, information-based interventions could potentially yield significant savings.

Within this context, a number of environmental and energy-related lessons from the COVID-19 pandemic challenges can also be highlighted. It is apparent that COVID-19 has exposed certain vulnerabilities of power systems. However, it has also provided opportunities for sustainable, reliable power and for resilient planning in the electricity sector (Jiang et al., 2021). The worst-case scenario offers incentives to energy producers to develop more flexible systems for renewables. Despite facing greater challenges, electricity power grids and smaller power markets are adapting to become more resilient in order to deal with future crises. Obviously, it might be impossible to recreate the challenges again for practical reasons and for security reasons (Kolomoiets et al., 2021). Some uncertainties do exist (potentially increased durability, for instance), but only at the margins. For instance, companies producing electricity – and others with businesses dependent on energy-intensive manufacturing processes – often face massive uncertainties when determining the relative costs of various fuel alternatives (Jakob, 2022).

Another example of building personal resilience in the workplace is developing and strengthening emotional acumen. Putting energy and motivation at work or having vitality is also associated with building personal resilience. Individuals can build personal resilience in their jobs by maintaining healthy work-life balance. In addition, implementing more eco-friendly waste management-related information sharing and advertising through social media platforms can be implemented in a scenario of a new lifestyle of lower energy consumption. Once such information is known, the remaining uncertainties will be limited, and the existing airlines will be able to construct a solid business case around their strategies. One can expect intelligent agents to be able to estimate likely risks and benefits from proposed treatments and procedures much better than is done now by human valuers, who, even experts, are generally poor decision-makers when faced with uncertainty.

26.4 Habitual Behaviors and Energy and Emission Reduction

Establishing habitual behaviors achieves noticeable energy and emission reduction. Achieving climate stability by meeting the Paris Agreement goal of net-zero global greenhouse gas emissions, consistent with about 85 percent emissions reductions in all sectors in the next three decades, would likely require changes from high-emitting households in the way they eat, use energy, and travel (Creutzig et al., 2022). Cities need to put forward more ambitious, easy-to-track climate targets and require better measures for greenhouse gas (GHG) emission reductions, in order to remain in line with the global ambitions for mitigating climate change. Growth targets on greenhouse gas emissions from developing countries should form part of efforts to mitigate global climate change. Reducing greenhouse gas emissions in developing countries has similar global environmental benefits to reductions in industrialized countries, although the reductions in developing countries are typically far less expensive. Reducing petroleum usage in industrialized countries would lead to lower global oil prices, which would incentivize increased usage and carbon emissions in developing countries (Wang et al., 2022). In particular, it has been estimated the USA would see its costs for meeting Kyoto targets fall by over 80% by trading with developing countries rather than by cutting emissions alone at home (Pischke et al., 2019). Moreover, by selling permits, the carbon dioxide emission reductions produce further improvements in the developed world's air quality by reducing emissions of particulates, sulfur dioxide, and nitrogen oxides. Achieving massive energy savings through behavioral changes may sound hard, even impossible, to those who are only familiar with the limited amounts of behavioral-change programs that have reduced consumption by a couple of percent on average for each household. It is important to note that almost two-thirds of the energy reduction needed to reach goal would require individuals to shift how they do things, change behavior, and eagerly adopt practices that are more efficient and healthier for the world and for themselves. Without behavioral changes, including demanding more efficient technologies and being willing to embrace a lower-carbon lifestyle, we cannot reach the emission reductions needed to achieve net-zero energy by 2050. The goal of reaching net-zero emissions is unlikely to be achieved by a prompt technical shift on its own, and beyond changes in policy and investments, it is likely to depend, in part, on reductions in the consumption of products and services that emit a large amount (Liu et al., 2022). Although behavioral changes may yield significant emission reductions, typically with individual and collective social and economic benefits, individuals may find it challenging to alter their energy consumption. Policymakers must recognize barriers to behavioral change and implement policies that facilitate individuals' ability to modify their energy usage. Adapting societal norms through raising awareness about climate change and its impacts may help to conserve energy. An important recommendation for future behavioral interventions studies is that they examine high-impact actions that have high potential for emission reductions and that they incorporate both ambitious and step-by-step behaviors for emission reductions within transportation, meat consumption, and energy usage, in order to understand which behaviors are truly effective at producing larger emission reductions (Day et al., 2019). To determine the effectiveness of historic policies for reducing carbon emissions, our research conducted energy-related greenhouse gas emission inventories of the major world's cities, incorporating inputs from various sectors, and assessed the near-term, medium-term, and long-term goal-based targets to mitigate CO_2 emissions over 2020–2050. Theoretical insights should be therefore used to study the households' preferences to cut emissions in various cities and countries. For instance, the avoidance of one round-trip flight across the Atlantic (a single-time decision made by a single individual) would produce a large quantity of emission reductions. Most of the world's largest cities can make substantial progress on reducing GHGs only if that part of their energy emissions can be better controlled. Total GHG emissions and electricity sector emissions in recent decades in the USA were significantly reduced by a shift from the use of coal to natural gas and by the transition toward a cleaner, renewable energy system, in which climate actions had already substantially contributed to the GHG emission reductions (Wilson & Staffell, 2018).

Thence, the governments' policies aimed at the changing the energy consumption behavior, as well as infrastructure developments, would have contributed to the majority of the reductions due to behavior change and need to be supported in order to restore the energy balance following the recent energy crisis in the post-COVID era.

26.5 Human Factor in the Energy Consumption

For household energy use prediction process, linear and nonlinear prediction techniques are used in all types of household energy, electric power consumption, and gas consumption. Some home energy reports and studies estimate the effect of various smart meter trials on decreasing the household electricity consumption and natural gas consumption (Rausser et al., 2018). Home energy reports, for instance, have been shown to produce savings when consumers use less air conditioning and heating or turn off appliances and lights that are not in use. Residential energy use could be further reduced by structural investments, including upgrading to more energy-efficient appliances and retrofitting existing appliances. New and existing electric plants can reduce their environmental impacts through increased generating efficiencies, installed pollution controls, and the use of cleaner power sources (Mason & Grijalva, 2019). Therefore, one can reduce environmental impacts from the electricity used in a household by buying greener electricity, as well as becoming more efficient with the power it uses. When one chooses to cut the power usage, she or he is also helping to preserve the limited natural resources that would otherwise go into fueling the plants.

Perhaps the most obvious way for cutting down energy to help the environment is through the reduced emissions from the power plants. In addition, less energy demand leads to a lower demand for the extraction of fossil fuels. Increased human consumption puts pressure on the natural processes that renew some energy resources, and it exhausts those that cannot be renewable. Social and technological innovations influence how much energy is used in human societies. Knowledge about the amounts of energy used by various processes may inform decisions on energy usage. The amount of energy used by a society, either per person or as a whole, could be therefore reduced. Efficiency, the other forms of energy needed to sustain the system, needs better assessment. This factor also illustrates the view that energy efficiency is part of individual responsibility and lists concrete examples rather than using sweeping statements. This is an impressive perspective, which emphasizes the effects on convenience and lifestyle from energy usage instead of including a rationale for energy efficiency. This factor illustrates a view among households that conservation can be motivated by environmental issues, including climate change. It also represents an uninterested, uncaring approach to domestic energy consumption; this perspective nevertheless illustrates some level of ecological awareness. This view also highlighted how, although financial savings might be a primary motivation, households did not necessarily have an understanding of their energy usage or of the impact that this might have on consumption. In terms of key factors that affected changes in overall household energy use, income level and household size were major drivers in changes in household energy consumption. As population roles shift as cities develop, increased electricity demand and new electricity products use are major reasons behind energy consumption increases as households move from rural areas to cities (Wang et al., 2020). Another kind of research suggests that the effective utilization of government facilities because of the density of the urban population would contribute to reducing household energy consumption increases (Nematchoua et al., 2021). In addition, a major driver for household energy consumption is the thermal conditions, which are essential to household energy consumption characteristics in many countries. Electricity use is a major energy use type among urban households, with different usage patterns. Today, the majority of people in the developed countries get their electricity from centrally owned electricity plants, which produce power using a variety of energy resources, such as coal, natural gas, nuclear energy, or renewable resources like water, wind, or solar power. Some distributed generation, such as distributed renewable energy, can help maintain the supply of clean, reliable electricity to customers, as well as reducing losses in electrical energy across transmission and distribution lines. Reducing one's homes energy consumption – or going "off-grid" with solar power – can be good for the environment and conserve resources (Diahovchenko et al., 2020). Changing everyday habits can help lower household electric bills, lower carbon emissions, and alleviate stress on the electric grid. By taking steps to lower energy consumption, people can help contribute to a healthier and a more prosperous world.

Successful behavioral interventions aimed at increasing energy efficiency promote conservation through the use of techniques such as feedback mechanisms that monitor households' energy use patterns, or targeted reminders that warn consumers to be particularly aware of energy usage during times of high demand. The aim is to optimize both comfort and energy savings while still allowing consumers to regulate the operational temperature according to their preferences. This may suggest that preferences and habits related to heating – such as temperature settings and use times – are harder to change than those for other energy-consuming behaviors in a house (Zhang et al., 2020). Although the disruption in demand was not deliberate nor welcomed in this case, the effect provides a quantified sense of the potential limits extreme measures could deliver under the present energy mix (such as higher household working rates or reduced consumption). Typical consumers equipped with a smart meter and an in-home display use less gas and less electricity as compared to those equipped with a traditional meter (Chalmers et al., 2019). Residents of households in engaged relationships are likely to have an aligned view of energy usage.

All in all, understanding potential improvements to energy performance in households requires more detailed data on energy consumption as potential for this sector is comprised of a lengthy list of appliance-specific and building-specific measures. Conservation involves reducing wasteful energy usage, using energy for a given purpose more effectively, making strategic choices about power sources, and cutting back on all-purpose energy consumption as the human energy needs are constantly increasing.

26.6 Conclusions

In short, governments and businesses need to use multiple behavioral incentives to attract consumers into the voluntary and desirable energy-saving programs that would ensure the leading energy-saving behavior in the post-COVID-19 uncertain epoch. Many utilities worldwide are already using competitions and games to engage with their customers, encouraging users to set goals for reducing their own energy use in return for a small reward. According to our findings, relevant literature has typically found positive relationships between perceived severity of climate change and both mitigation-related and effectiveness-related behaviors.

Notably, people's belief that energy conservation may help to mitigate climate change does not appear to relate to either curtailment- or efficiency-related behaviors among people who are not concerned at all with climate change. Without the excitement that comes with feeling anxious and acknowledging that climate change is a risk, the idea that personal or collective behavior that reduces energy use will help decrease climate change may not be perceived as being relevant at a personal level. On the other hand, stressing that only some individuals are acting in sustainable ways may, through another process, encourage the selection of sustainable energy practices, since engaging in these behaviors may reinforce one's ecological identity. Collective efficacy, operationalized as assuming many people would curb energy usage in order to decrease climate change, failed to demonstrate any statistically significant associations with the prediction of energy-saving behaviors. We found that many studies focusing on climate change specifically and self-reported energy savings behaviors document a higher willingness among those concerned about climate change to lower energy consumption. While feedback on prior energy-use patterns, either in absolute terms or relative to peers, does motivate energy conservation, the total effect on behaviors differs depending on several constructs. Adjusting feedback frequency, for instance, does not tend to cause significant changes in energy savings, while changing feedback media does. This implies that individuals might not be accurately judged as which behavioral changes are effective at reducing energy use and associated CO_2 emissions. Individuals can modify daily energy behaviors, for example, by reducing thermostat settings or the duration of showers. Energy conservation may be as simple as turning off lights or appliances when they are not needed.

There are lots of different ways one can cut back on energy usage at home, from making simple behavior changes to making major improvements in the house. There are a number of actions one can take to lower the personal energy usages' impact on the environment. Many businesses are also installing renewable energy sources and cogeneration at their buildings, which can save them money, lower environmental impacts, and give more control over energy usage. As wind and solar prices continue to drop, more and more people are buying renewables. For homes in colder regions, gas-filled windows with a low-e coating can dramatically lower heating costs. In addition to minimizing heat loss, low-e coatings on windows can also reduce heat gain, reflecting light and reducing the amount of thermal energy entering one's home. In addition, interior or exterior storm windows can reduce unnecessary heat loss. Windows are significant sources of energy loss too: they can account for up to one-fourth of the overall heating bills, and water heating is the largest contributor to the overall energy use.

Many voluntary programs for saving energy also use societal norms, encouraging participants to share their success stories as a way of maintaining motivation. Social scientists may study what factors drive energy behaviors that are sustainable versus unsustainable and explore how those factors might be addressed through energy policies and alternations to the energy system for achieving leadership toward behavioral change in energy consumption.

To summarize this research, it is clear that global climate change and ecological decline are caused in large part by human behaviors and thus could be altered if individuals engage in more consistent and sustainable energy behaviors. Long-term programs have been used to guide a drastic change in energy habits over time. This kind of exclusionary approach has managed to achieve up to eight times more participants, while providing up to three times more energy savings, than exclusionary incentives, which were tested in the same programs.

References

Batres, M., Wang, F. M., Buck, H., Kapila, R., Kosar, U., Licker, R., & Suarez, V. (2021). Environmental and climate justice and technological carbon removal. *The Electricity Journal, 34*(7), 107002. https://doi.org/10.1016/j.tej.2021.107002

Brockhaus, M., Di Gregorio, M., Djoudi, H., Moeliono, M., Pham, T. T., & Wong, G. Y. (2021). The forest frontier in the Global South: Climate change policies and the promise of development and equity. *Ambio, 50*(12), 2238–2255. https://doi.org/10.1007/s13280-021-01602-1

Broska, L. H. (2021). It's all about community: On the interplay of social capital, social needs, and environmental concern in sustainable community action. *Energy Research & Social Science, 79*, 102165. https://doi.org/10.1016/j.erss.2021.102165

Chalmers, C., Hurst, W., Mackay, M., & Fergus, P. (2019). Identifying behavioural changes for health monitoring applications using the advanced metering infrastructure. *Behaviour & Information Technology, 38*(11), 1154–1166. https://doi.org/10.1080/0144929X.2019.1574900

Chang, H. J., & Andreoni, A. (2020). Industrial policy in the 21st century. *Development and Change, 51*(2), 324–351. https://doi.org/10.1111/dech.12570

Colgan, J. D., Green, J. F., & Hale, T. N. (2021). Asset revaluation and the existential politics of climate change. *International Organization, 75*(2), 586–610. https://doi.org/10.1017/S0020818320000296

Creutzig, F., Niamir, L., Bai, X., Callaghan, M., Cullen, J., Díaz-José, J., & Ürge-Vorsatz, D. (2022). Demand-side solutions to climate change mitigation consistent with high levels of well-being. *Nature Climate Change, 12*(1), 36–46. https://doi.org/10.1038/s41558-021-01219-y

Cuypers, I. R., Hennart, J. F., Silverman, B. S., & Ertug, G. (2021). Transaction cost theory: Past progress, current challenges, and suggestions for the future. *Academy of Management Annals, 15*(1), 111–150. https://doi.org/10.5465/annals.2019.0051

Day, E., Fankhauser, S., Kingsmill, N., Costa, H., & Mavrogianni, A. (2019). Upholding labour productivity under climate change: An assessment of adaptation options. *Climate Policy, 19*(3), 367–385. https://doi.org/10.1080/14693062.2018.1517640

Diahovchenko, I., Kolcun, M., Čonka, Z., Savkiv, V., & Mykhailyshyn, R. (2020). Progress and challenges in smart grids: Distributed generation, smart metering, energy storage and smart loads. *Iranian Journal of Science and Technology, Transactions of Electrical Engineering, 44*(4), 1319–1333. https://doi.org/10.1007/s40998-020-00322-8

Guliyeva, A., Rzayeva, U., & Jafarova, N. (2021). Analysis of some indicators by means of fuzzy logic on the example of Azerbaijani energy enterprises. *E3S Web of Conferences, 250*, 02001. https://doi.org/10.1051/e3sconf/202125002001

Harputlugil, T., & de Wilde, P. (2021). The interaction between humans and buildings for energy efficiency: A critical review. *Energy Research & Social Science, 71*, 101828. https://doi.org/10.1016/j.erss.2020.101828

Iweka, O., Liu, S., Shukla, A., & Yan, D. (2019). Energy and behaviour at home: A review of intervention methods and practices. *Energy Research & Social Science, 57*, 101238. https://doi.org/10.1016/j.erss.2019.101238

Jakob, M. (2022). Globalization and climate change: State of knowledge, emerging issues, and policy implications. *Wiley Interdisciplinary Reviews: Climate Change*, e771. https://doi.org/10.1002/wcc.771

Jiang, P., Van Fan, Y., & Klemeš, J. J. (2021). Impacts of COVID-19 on energy demand and consumption: Challenges, lessons and emerging opportunities. *Applied Energy, 285*, 116441. https://doi.org/10.1016/j.apenergy.2021.116441

Karthick, T., & Chandrasekaran, K. (2021). Design of IoT based smart compact energy meter for monitoring and controlling the usage of energy and power quality issues with demand side management for a commercial building. *Sustainable Energy, Grids and Networks, 26*, 100454. https://doi.org/10.1016/j.segan.2021.100454

Kesavan, N., Seenuvasaragavan, S., Burnette, K., Heim, A. B., Patel, R. J., Johnston, J., Cao, X., Sakandar, H. A., Zhuang, Y., Oi, K. K., Zhi, Y., Diao, Y., & Strielkowski, W. (2022). Pandemic-inspired policies. *Science, 377*(6601), 22–24. https://doi.org/10.1126/science.add2046

Kim, H., Goodchild, A., & Boyle, L. N. (2022). Modeling the competing demands of carriers, building managers, and urban planners to identify balanced solutions for allocating building and parking resources. *Transportation Research Interdisciplinary Perspectives, 15*, 100656. https://doi.org/10.1016/j.trip.2022.100656

Kolomoiets, T., Tsybulnyk, N., Moroz, O., Prymachenko, V., & Khashev, V. (2021). Influence of shadow economy legalization on national security of Ukraine. *Entrepreneurship and Sustainability Issues, 8*(3), 420–437. https://doi.org/10.9770/jesi.2021.8.3(27)

Komendantova, N. (2021). Transferring awareness into action: A meta-analysis of the behavioral drivers of energy transitions in Germany, Austria, Finland, Morocco, Jordan and Iran. *Energy Research & Social Science, 71*, 101826. https://doi.org/10.1016/j.erss.2020.101826

Kumar, B., Jensen, M., Strielkowski, W., Johnston, J., Sharma, V., Dandara, C., Shehata, M., Struett, M., & Nikolaou., A., Szymanski, D., Abdul-Ghani, R., Cao, B., Varzinczak, L., Kerman, B., Arthur, P., Oda, F., Bohon, W., Ellwanger, J., Tyner, S., Jordan, E., & Murphy, C. (2019). NextGen Voices: Science-inspired sustainable behavior. *Science, 364*(6443), 822–824. https://doi.org/10.1126/science.aax8945

Kurpiela, S., & Teuteberg, F. (2022). Strategic planning of product-service systems: A systematic literature review. *Journal of Cleaner Production*, 130528. https://doi.org/10.1016/j.jclepro.2022.130528

Lawal, S., Lennard, C., & Hewitson, B. (2019). Response of southern African vegetation to climate change at 1.5 and 2.0 global warming above the pre-industrial level. *Climate Services, 16*, 100134. https://doi.org/10.1016/j.cliser.2019.100134

Leonavičienė, E., & Burinskienė, A. (2021). Research on the reflection of cultural dimensions in international business. *Insights into Regional Development, 3*(4), 101–114. https://doi.org/10.9770/IRD.2021.3.4(7)

Liobikienė, G., & Poškus, M. S. (2019). The importance of environmental knowledge for private and public sphere pro-environmental behavior: Modifying the value-belief-norm theory. *Sustainability, 11*(12), 3324. https://doi.org/10.3390/su11123324

Liu, Z., Deng, Z., He, G., Wang, H., Zhang, X., Lin, J., & Liang, X. (2022). Challenges and opportunities for carbon neutrality in China. *Nature Reviews Earth & Environment, 3*(2), 141–155. https://doi.org/10.1038/s43017-021-00244-x

Mace, M. J., Fyson, C. L., Schaeffer, M., & Hare, W. L. (2021). Large-scale carbon dioxide removal to meet the 1.5°C limit: Key governance gaps, challenges and priority responses. *Global Policy, 12*, 67–81. https://doi.org/10.1111/1758-5899.12921

Mahapatra, K., Alm, R., Hallgren, R., Bischoff, L., Tuglu, N., Kuai, L., & Umoru, I. (2018). A behavioral change-based approach to energy efficiency in a manufacturing plant. *Energy Efficiency, 11*(5), 1103–1116. https://doi.org/10.1007/s12053-017-9581-9

Marino, A., Pariso, P., & Picariello, M. (2022). Transition towards the artificial intelligence via re-engineering of digital platforms: Comparing European Member States. *Entrepreneurship and Sustainability Issues, 9*(3), 350–368. https://doi.org/10.9770/jesi.2022.9.3(21)

Mason, K., & Grijalva, S. (2019). A review of reinforcement learning for autonomous building energy management. *Computers & Electrical Engineering, 78*, 300–312. https://doi.org/10.1016/j.compeleceng.2019.07.019

Nematchoua, M. K., Nishimwe, A. M. R., & Reiter, S. (2021). Towards nearly zero-energy residential neighbourhoods in the European Union: A case study. *Renewable and Sustainable Energy Reviews, 135*, 110198. https://doi.org/10.1016/j.rser.2020.110198

Paone, A., & Bacher, J. P. (2018). The impact of building occupant behavior on energy efficiency and methods to influence it: A review of the state of the art. *Energies, 11*(4), 953. https://doi.org/10.3390/en11040953

Paulikas, J., & Paulikienė, B. (2022). Impact of the communicated information content on employee resistance to change. *Insights into Regional Development, 4*(3), 61–75. https://doi.org/10.9770/IRD.2022.4.3(4)

Pischke, E. C., Solomon, B., Wellstead, A., Acevedo, A., Eastmond, A., De Oliveira, F., & Lucon, O. (2019). From Kyoto to Paris: Measuring renewable energy policy regimes in Argentina, Brazil, Canada, Mexico and the United States. *Energy Research & Social Science, 50*, 82–91. https://doi.org/10.1016/j.erss.2018.11.010

Porter, L., Rickards, L., Verlie, B., Bosomworth, K., Moloney, S., Lay, B., & Pellow, D. (2020). Climate justice in a climate changed world. *Planning Theory & Practice, 21*(2), 293–321. https://doi.org/10.1080/14649357.2020.1748959

Radtke, J. (2022). Smart energy systems beyond the age of COVID-19: Towards a new order of monitoring, disciplining and sanctioning energy behavior? *Energy Research & Social Science, 84*, 102355. https://doi.org/10.1016/j.erss.2021.102355

Rausser, G., Strielkowski, W., & Štreimikienė, D. (2018). Smart meters and household electricity consumption: A case study in Ireland. *Energy & Environment, 29*(1), 131–146. https://doi.org/10.1177/0958305X17741385

Razmjoo, A., Gandomi, A., Mahlooji, M., Astiaso Garcia, D., Mirjalili, S., Rezvani, A., & Memon, S. (2022). An investigation of the policies and crucial sectors of smart cities based on IoT application. *Applied Sciences, 12*(5), 2672. https://doi.org/10.3390/app12052672

Rochedo, P. R., Fragkos, P., Garaffa, R., Couto, L. C., Baptista, L. B., Cunha, B. S., et al. (2021). Is green recovery enough? Analysing the impacts of post-COVID-19 economic packages. *Energies, 14*(17), 5567. https://doi.org/10.3390/en14175567

Rutz, C., Loretto, M. C., Bates, A. E., Davidson, S. C., Duarte, C. M., Jetz, W., & Cagnacci, F. (2020). COVID-19 lockdown allows researchers to quantify the effects of human activity on wildlife. *Nature Ecology & Evolution, 4*(9), 1156–1159. https://doi.org/10.1038/s41559-020-1237-z

Sahakian, M., Rau, H., Grealis, E., Godin, L., Wallenborn, G., Backhaus, J., & Fahy, F. (2021). Challenging social norms to recraft practices: A Living Lab approach to reducing household energy use in eight European countries. *Energy Research & Social Science, 72*, 101881. https://doi.org/10.1016/j.erss.2020.101881

Sdoukopoulos, E., Boile, M., Tromaras, A., & Anastasiadis, N. (2019). Energy efficiency in European ports: State-of-practice and insights on the way forward. *Sustainability, 11*(18), 4952. https://doi.org/10.3390/su11184952

Shen, M., Li, X., Lu, Y., Cui, Q., & Wei, Y. M. (2021). Personality-based normative feedback intervention for energy conservation. *Energy Economics, 104*, 105654. https://doi.org/10.1016/j.eneco.2021.105654

Shevyakova, A., Munsh, E., Arystan, M., & Petrenko, Y. (2021). Competence development for Industry 4.0: Qualification requirements and solutions. *Insights into Regional Development, 3*(1), 124–135. https://doi.org/10.9770/IRD.2021.3.1(7)

Simpson, N. P., Clarke, J., Orr, S. A., Cundill, G., Orlove, B., Fatorić, S., & Trisos, C. H. (2022). Decolonizing climate change–heritage research. *Nature Climate Change, 12*(3), 210–213. https://doi.org/10.1038/s41558-022-01279-8

Strielkowski, W., Civín, L., Tarkhanova, E., Tvaronavičienė, M., & Petrenko, Y. (2021). Renewable energy in the sustainable development of electrical power sector: A review. *Energies, 14*(24), 8240. https://doi.org/10.3390/en14248240

Tan, H., Thurbon, E., Kim, S. Y., & Mathews, J. A. (2021). Overcoming incumbent resistance to the clean energy shift: How local governments act as change agents in coal power station closures in China. *Energy Policy, 149*, 112058. https://doi.org/10.1016/j.enpol.2020.112058

Vona, F. (2019). Job losses and political acceptability of climate policies: Why the 'job-killing' argument is so persistent and how to overturn it. *Climate Policy, 19*(4), 524–532. https://doi.org/10.1080/14693062.2018.1532871

Wang, Q., Lin, J., Zhou, K., Fan, J., & Kwan, M. P. (2020). Does urbanization lead to less residential energy consumption? A comparative study of 136 countries. *Energy, 202*, 117765. https://doi.org/10.1016/j.energy.2020.117765

Wang, D., Huangfu, Y., Dong, Z., & Dong, Y. (2022). Research hotspots and evolution trends of carbon neutrality-visual analysis of bibliometrics based on CiteSpace. *Sustainability, 14*(3), 1078. https://doi.org/10.3390/su14031078

Wilson, I. A., & Staffell, I. (2018). Rapid fuel switching from coal to natural gas through effective carbon pricing. *Nature Energy, 3*(5), 365–372. https://doi.org/10.1038/s41560-018-0109-0

Zhang, Y., Bai, X., & Mills, F. P. (2020). Characterizing energy-related occupant behavior in residential buildings: Evidence from a survey in Beijing, China. *Energy and Buildings, 214*, 109823. https://doi.org/10.1016/j.enbuild.2020.109823

Ziesemer, F., Hüttel, A., & Balderjahn, I. (2021). Young people as drivers or inhibitors of the sustainability movement: The case of anti-consumption. *Journal of Consumer Policy, 44*(3), 427–453. https://doi.org/10.1007/s10603-021-09489-x

Part VI
Leadership Within the Context of Sustainable Development Goals (SDGs)

Chapter 27
Corruption as an Obstacle of Sustainable Development

Victoria Bozhenko, Anton Boyko, and Iryna Voronenko

Abstract Corruption is a leading phenomenon which is caused by a range of political, economic, institutional, social, as well as cultural factors. Corruption destroys justice and fairness, undermines vital tendering processes, damages industries, violates the business integrity, and dispossesses people of basic needs such as education, healthcare, and security. It also hampers all principles of leadership in business and economics. The main purpose of the research is to estimate the impact of corruption on reaching the Sustainable Development Goals. The relevance of this scientific problem decision is that corruption is gradually turning from a domestic problem into a global threat, and it is cross-sectoral issue. Tackling corruption is vital to achieving all of the Sustainable Development Goals, mainly SDG 16. Methodological tools of the research methods were stationary test (Levin-Lin-Chu, Hadri, Pesaran and Chin, Dickie-Fuller tests) and fully modified least squares for estimation of the parameters of panel regression. This chapter presents the results of an empirical analysis which showed that public institutions need to work toward achieving the Sustainable Development Goals by improving the institutional and regulatory framework. The results of the research can be useful for international and national anti-corruption institutions and national regulators.

Keywords Corruption · Sustainable development · Panel regression · Leadership

JEL Classification C23 · D73 · Q01

V. Bozhenko (✉) · A. Boyko
Sumy State University, Sumy, Ukraine
e-mail: v.bozhenko@uabs.sumdu.edu.ua; a.boiko@uabs.sumdu.edu.ua

I. Voronenko
National University of Life and Environmental Sciences of Ukraine, Kyiv, Ukraine
e-mail: irynav@nubip.edu.ua

W. Strielkowski (ed.), *Leadership, Entrepreneurship and Sustainable Development Post COVID-19*, Springer Proceedings in Business and Economics, https://doi.org/10.1007/978-3-031-28131-0_27

27.1 Introduction

The Sustainable Development Goals (SDG) developed and adapted by the United Nations (UN) envisage the implementation of systemic reforms by 2030, which will contribute to the qualitative transformation of the economic, social, and environmental spheres of society. However, corruption schemes and unscrupulous behavior of public administration representatives can undermine the successful implementation of all 17 Sustainable Development Goals. To monitor the level of corruption, indicators have been defined in SDG 16. Therefore, the effective anti-corruption system at the global level, which will be based on the principles of transparency, accountability, and integrity, is one of the key tasks of international organizations and national regulators.

Nowadays, corruption is not a problem of any individual countries or regions; it has become a global challenge for the international community. Kaya and Engkuchik (2021) substantiated the decrease in the intensity of bribery to public authorities in post-crisis periods of economic development. As a result of the rapid development of digital technologies, the forms of committing corruption crimes are also improving (Koudelková et al., 2015; Kuzior et al., 2022). Corruption hinders mostly the progress of transition economies, illegally or irrationally using public funds (Pilia, 2017). Resources obtained through corrupt practices often transform into economic and political pressures, thereby weakening democratic institutions and increasing corruption (Lyulyov et al., 2018; Lu et al., 2019).

Aidt (2010) empirically proved the existence of a negative correlation between corruption and sustainable development. In addition, the author substantiated that corruption increases economic instability in the country due to the eroded capital base. Therefore, the destructive consequences of corruption permeate various spheres of human life:

- Politics and government – lobbying the private interests and making favorable regulatory environment to the business activity; the imperfection of judicial control in the country (Strielkowski et al., 2021).
- Economy – tax avoidance, money laundering, illegal trade, inefficient and inappropriate use of public funds, reducing the country's investment attractiveness, and restraining the pace of innovative development (Cabelkova & Strielkowski, 2013; Nguedie, 2018; Eddassi, 2020; Zolkover & Georgiev, 2020; Samoilikova, 2020; Moskalenko et al., 2022). The corrupt environment in the country reduces the business activity of small- and medium-sized companies due to the formation of nonmarket and noncompetitive conditions for conducting business activities (Kaya, 2020).
- Society – increasing income inequality, poverty growth, limited access to essential basic needs (electricity, water, food, education, healthcare), rising violence and crime, etc. (Sardak et al., 2018; Polyakov et al., 2019; Nureev et al., 2020). The absence of democratic state processes in the country hinders the development of social capital (Mercado & Vargas-Hernández, 2019; Didenko et al., 2020).

- Ecology – irrational mining and subsoil use, depletion of forest resources due to their uncontrolled deforestation, noncompliance with the special regime of use of natural resources within the relevant territories (Bilan et al., 2020a, b; Simionescu et al., 2022).

Corruption represents one of the main threats to the macroeconomic stability and state's financial security, which requires gradual eradication and the construction of a transparent and effective system of state regulation and management (Lyeonov et al., 2018; Vasylieva et al., 2020; Rekunenko et al., 2022).

The human factor plays an important role in combating corruption (Butko et al., 2019; Fila et al., 2020). Human resource management is a constant balance between economic and ethical needs in society (Hanić & Jevtić, 2020). In order to level the role of a person at individual stages of data processing and management decision-making, it is advisable to introduce e-governance system, which contributes to reducing the number of corrupt actions by public servants (Kuzior et al., 2021; Vasilyeva et al., 2021b; Tiutiunyk et al., 2021).

Within the framework of this study, it was decided to analyze the impact of corruption on the financing of key socially necessary services – healthcare and education. Due to fraud and corruption, the world's healthcare system loses an average of about $455 billion annually out of $7.35 trillion, or 6.2% of the total. Corruption in these areas deprives a citizen of his fundamental rights to receive quality education and medicine.

Corruption in the healthcare and education system can take the form of bribery, extortion, theft, embezzlement, and undue influence (Samoilikova & Kunev, 2020; Serpeninova et al., 2020). The main types of corruption offenses in the healthcare sector are as follows:

- Nontransparent procedure for applying positions in health facilities (Smiianov et al., 2020)
- Providers of medical services can be fictitious companies that are created for money laundering and other illegal purposes
- Receipt of illegal benefits by specialists of the healthcare system to popularize the products of pharmaceutical companies (GRECO, 2020)
- Avoiding or speeding up the procedure of registration of medicinal products due to bribery of executive branch officials (UN Human Rights)
- Bribery of employees of regulatory bodies and medical professionals to provide false data about the results of clinical trials of medicinal products (UN Human Rights)
- Nontargeted and ineffective use of state funding for the healthcare system
- Nontransparent public procurement procedure, etc.

The main manifestations of corruption in the education system are as follows:

- Illegal collection of funds from parents for admission to an educational institution
- A nontransparent admission procedure for students of higher education
- Unequal conditions in the recruitment process

Thus, corruption is accompanied by a violation of ethical standards and principles of integrity, which leads to irrational or untargeted use of public funds in the field of healthcare and education.

27.2 Method and Data

A wide range of scientific methods are used to assess the presence and nature of the relationship between variables in the scientific literature: ordinary least squares (Vasilyeva et al., 2016), fully modified ordinary least square and dynamic ordinary least square (Kwilinski et al., 2022), correlation and cluster analysis (Petroye et al., 2020), panel cointegration regression (Bilan et al., 2020b; Lyulyov et al., 2021b), artificial neural network (Vasilyeva et al., 2021a, b), structural equations modeling (Brychko et al., 2021), bifurcation analysis (Kuzmenko et al., 2020), panel vector autoregressive model (Lyulyov et al., 2021a, b; Tiutiunyk et al., 2022), or the gravitational analysis (Lyeonov et al., 2020).

To test the relationship between sustainability indicators and the Corruption Perceptions Index (CPI), a model was built based on balanced panel data. Panel data combine the advantages of spatial data and time series, which allows to analyze and isolate changes at the individual level of each object. Panel data provide more informative data, greater variability, less collinearity between variables, more degrees of freedom, and greater efficiency.

Within this study, it was decided to analyze the impact of corruption on government spending on essential services: domestic general government health expenditure, % of GDP (HLT), government expenditure on education, and % of GDP (EDC). Data from The World Bank Group website and Transparency International served as the source of primary data.

In order to obtain more reliable results of assessing the impact of corruption on the financing of basic public needs, it was decided to generate panel data based on homogeneous countries according to the level of corruption in them. Thus, the average value of the index of perception of corruption during 2012–2020 was determined for each country, the minimum and maximum values of the index were calculated, and three groups of countries were distinguished based on them:

- Countries with low-level corruption ($65.1 < CPI \leq 89.4$): Australia, Austria, Belgium, Canada, Chile, Denmark, Germany, and others (26 countries)
- Countries with middle-level corruption ($40.8 < CPI \leq 65.1$): Czech Republic, Hungary, Italy, Latvia, Lithuania, Poland, Portugal, and others (36 countries)
- Countries with high-level corruption ($16.4 \leq CPI \leq 40.8$): Albania, Armenia, Egypt, Vietnam, Belarus, Bolivia, Bangladesh, Nepal, Kenya, Ukraine, and others (78 countries)

The main hypotheses of this study are as follows:

H1. Corruption affects the government spending on essential services.

H2. The level of corruption (low, medium, high) unevenly affects government spending on essential services.

Based on this, three arrays of statistical data were formed for countries with high, medium, and low levels of corruption for the period 2012–2019. All indicators have been prologarithmized, which made it possible to reduce the residuals of the model and increase their compliance with the normal distribution law.

Assessing the impact of corruption on individual indicators of sustainable development involves the following steps: testing data for stationarity, determining the presence of cointegration between the variables, estimation of parameters according to the econometric model, and checking the adequacy of the models.

27.3 Results and Discussions

The initial condition for a panel regression model is to test variables for stationarity. For panel data, it can be used the following criteria to check for unit roots: Levin-Lin-Chu, Khadri, Pesaran and Chin, and Dickey-Fuller tests. All mathematical calculations were performed in the Eviews program. Testing for the presence of a unit root in panel data involves testing the null hypothesis that the series is stationary at $p < 0.05$. The results of testing the variables for stationarity are presented in Table 27.1.

The results of estimation the statistical criteria for checking the presence of unit roots allow us to state that most of the indicators at the actual levels are nonstationary time series. To convert them into stationary variables, their first differences are determined. Among the analyzed indicators, the following variables were immediately stationary: HLT_2, HLT_3, CPI_5, and EDC_2.

Since there is integration for some indicators, it is necessary to carry out tests for cointegration between the variables – to test the assumption of long-term relationship between them. The Kao statistical test was used for verification, which involves testing the null hypothesis of the absence of cointegration relationships between variables. Since the p-value <0.05, we can assert the presence of cointegration and a long-term relationship between the variables (see Table 27.2).

Table 27.2 indicates the existence of a long-term relationship between corruption and domestic general government health expenditure for countries with low-level and middle-level corruption. In addition, there is a cointegration relationship between corruption and government expenditure on education for countries with middle-level corruption.

Taking into account the results of the previous stage on checking the cointegration relationship between variables, panel regression models were built. The following methods were used to determine the regression coefficients: there is cointegration – fully modified least squares; no cointegration – least squares. The results of the calculation of the regression parameters for three groups of countries are shown in Tables 27.3 and 27.4.

Table 27.1 Results of panel data testing for unit root

Group of countries	Indicator			Levin, Lin, and Chu test		IM, Pesaran, and Shin test	
				Statistic	Prob.	Statistic	Prob.
Low-level corruption (26 countries)	Domestic general government health expenditure (% of GDP)	CPI_1	Level	−3.965	0.000	0.665	0.747
			1st difference	−11.735	0.000	−3.812	0.000
		HLT_1	Level	−12.632	0.000	−0.759	0.224
			1st difference	−5.517	0.000	−3.170	0.022
Middle-level corruption (36 countries)		CPI_2	Level	−25.093	0.000	−5.672	0.000
			1st difference	X	X	X	X
		HLT_2	Level	−19.907	0.000	−3.481	0.002
			1st difference	X	X	X	X
High-level corruption (75 countries)		CPI_3	Level	−8.876	0.000	1.312	0.905
			1st difference	−23.419	0.000	−3.627	0.000
		HLT_3	Level	−18.513	0.000	3.232	0.001
			1st difference	X	X	X	X
Low-level corruption (22 countries)	Government expenditure on education (% of GDP)	CPI_4	Level	−7.498	0.000	−0.882	0.189
			1st difference	−10.407	0.000	−3.298	0.001
		EDC_1	Level	−5.500	0.000	−1.3378	0.084
			1st difference	−16.117	0.000	−5.531	0.000
Middle-level corruption (29 countries)		CPI_5	Level	−24.178	0.000	−5.988	0.000
			1st difference	X	X	X	X
		EDC_2	Level	−15.384	0.000	−3.325	0.000
			1st difference	X	X	X	X
High-level corruption (59 countries)		CPI_6	Level	−8.239	0.000	−0.054	0.478
			1st difference	−21.164	0.000	−7.343	0.000
		EDC_3	Level	−8.910	0.000	−0.824	0.205
			1st difference	−36.563	0.000	−10.012	0.000

Table 27.3 indicates that the level of corruption (low, middle, high) affects the financing of the healthcare system in different ways. In particular, in countries with a low and medium level of corruption, an increase in the CPI by 1 point in the long term leads to a decrease in the share of domestic general government health expenditure in GDP by 0.0439% and 0.0355%, respectively. This shows that the improvement of the anti-corruption system in countries with low and medium levels leads to

Table 27.2 Results of testing variables for cointegration

Indicator	Group of countries	t-Statistic	Prob.	Result
Domestic general government health expenditure (% of GDP)	Low-level corruption	−2.5095	0.0060	Cointegration
	Middle-level corruption	−2.0742	0.0190	Cointegration
	High-level corruption	0.1102	0.4561	No cointegration
Government expenditure on education (% of GDP)	Low-level corruption	−0.7562	0.2248	No cointegration
	Middle-level corruption	−2.8780	0.0020	Cointegration
	High-level corruption	−0.7803	0.2176	No cointegration

Table 27.3 Results of assessing the impact of corruption on domestic general government health expenditure

	CPI			
Group of countries	Coefficient	t-Statistic	Prob.	Coefficient of determination
Low-level corruption	−0.0439*	−2.4597	0.0146	0.9867
Middle-level corruption	−0.0355*	−179.4797	0.0000	0.9946
High-level corruption	0.0436*	54.9932	0.0000	0.9882

*Statistically significant

Table 27.4 Results of assessing the impact of corruption on government expenditure on education

	CPI			
Group of countries	Coefficient	t-Statistic	Prob.	Coefficient of determination
Low-level corruption	−0.0114*	−2.0604	0.0411	0.9941
Middle-level corruption	−0.0218*	−15.3380	0.0000	0.9904
High-level corruption	−0.0031	−0.2659	0.7904	0.8716

*Statistically significant

the optimization of public spending on the medical system and the involvement of private institutions in the cofinancing of medical services. At the same time, in countries with a high level of corruption, an inverse relationship has been proved: an increase in the CPI by 1 point is accompanied by an increase in the share of public spending on the healthcare system in relation to GDP by 0.0436%. The obtained results of panel regressions are adequate, as evidenced by the high level of the coefficient of determination and statistically significant regression coefficients (prob <0.05).

An increase in the corruption perception index (in fact, a decrease in the risk of corruption in the country) causes a decrease in government expenditure on education on GDP. Only for countries with high-level corruption, this pattern is not legitimate, since there is no statistically significant relationship between the analyzed variables. In confirmation of the validity of the above conclusions, we note that the constructed regression models based on panel data are adequate. The coefficient of determination for countries with low-level and middle-level corruption is 0.9941 and 0.9904, respectively, that is, the independent variables that are included in the econometric models describe the change in the performance indicator by 99.41% and 99.04%.

27.4 Conclusions

Today, corruption is probably one of the key threats to achieving the Sustainable Development Goals (SDGs). Corruption creates destructive mechanisms for stable economic development, weakens democratic institutions and the principle of the rule of law, and exacerbates the problem of uneven distribution of income and access to medical, educational, and other services. In addition, corruption destroys public trust in representatives of public administration, as well as the formation of an established opinion in society about the possibility of solving any issue through bribery or bribery. This is a worldwide concern, as corruption acts as a trigger for growing social discontent that leads to violent extremism, conflict, crime, terrorism, etc.

Our study shows that public institutions need to work toward achieving the Sustainable Development Goals by improving the institutional and regulatory framework. We now understand the role that corruption plays in undermining human development and must address its pernicious effects in full so that we can have any hope of meeting the Sustainable Development Goals (SDGs). As a custodian of the United Nations Convention Against Corruption (UNCAC), United Nations Office on Drugs and Crime (UNODC) is committed to promoting good governance, integrity, and transparency in order to assist countries to meet the SDGs. UNODC also supports countries through development and coordination of corruption-related technical assistance activities at regional, national, and international levels.

It can be seen that the United Nations Development Programme's (UNDP) approach seeks to strengthen institutions and systems that combat corruption, support implementation of the UNCAC, mitigate the risks of corruption in critical sectors, and enhance the collective actions of governments, civil society, and the private sector to combat corruption. The World Bank Group works with public and private sectors, and civil society, to support efforts to prevent corruption, enhance remedies for correcting misconduct if it occurs, and to work toward improving behaviors, norms, and standards needed to sustain anti-corruption efforts. The suite of anti-corruption initiatives expands the focus of the World Bank Group beyond developing countries to also include financial centers; engages more publicly with corruption policy than previously; uses emerging technologies to understand, confront, and prevent corruption; and incorporates insights from behavioral social sciences.

In order to reinforce and allow for a more optimized, inclusive international approach, IFAC is also committed to working together with other prominent organizations, such as the Basle Institute for Governance, Corporate Affairs, the Organization for Economic Co-operation and Development, and the World Economic Forums Gatekeepers Task Force, to better tackle and prevent corruption. To respond to this challenge, IFAC has developed the Corruption and Economic Crime Action Plan (our action plan), which provides the foundation of the accounting professions role in combating corruption and economic crime, thus furthering the United Nations Sustainable Development Goals. The day-long Climate Summit will provide the opportunity for businesses and other stakeholders to discuss effective actions and the implementation of the SDGs, including anti-corruption strategies. The 2030 Agenda has provided a huge opportunity for anti-corruption efforts, with the 16 Sustainable Development Goals, Promoting Peaceful and Inclusive Societies, Access to Justice for All, and Building Effective, Accountable, and Inclusive Institutions at All Levels. SDG Goal 16 includes Targets 16.5 (significantly reduce corruption and bribery) and 16.6 (develop effective, accountable, and transparent institutions at all levels) as well as three indicators. According to the 2030 Agenda, SDG 16 and its anti-corruption targets to reduce illicit financial flows (16.4), corruption, and bribery (16.5); develop effective, accountable, and transparent institutions (16.6); ensure responsive, inclusive, participatory, and representative decision-making (16.7); and enhance access to information (16.10) which are important not only to promoting transparency, accountability, and integrity but are essential conditions to successfully achieve all the SDGs. The implications make the implementation of anti-corruption targets under SDG 16 even more pressing than ever: not only are they important in promoting transparency, accountability, and integrity in themselves but also because they could act as the foundation to achieving successful implementation of all remaining SDGs.

In many countries, there is little institutional capacity and political will domestically to achieve the SDG 16 anti-corruption targets. In particular, health and well-being are threatened by SDG 3 unless there is an urgent response to corruption within the public health system. In particular, corruption adversely affects SDG 3 by hindering individuals access to quality healthcare services and to safe and effective medicines, and eroding systems to protect against financial risks. Several studies have provided evidence for a negative correlation between corruption and quality of public expenditure, services, and regulations. Corruption profoundly weakens the legitimacy and credibility of public institutions and shapes how individuals perceive the effectiveness of their governments and their states. Corruption flourishes in the presence of accountability, institutions, and a weak state. Corruption is a threat to more than just ordinary citizens pocketbooks; it limits economic growth, strangles investments, destroys confidence in government institutions, embezzles both public and private resources, and, worst of all, supports authoritarians and human rights abusers. These crimes delegitimize and undermine confidence in democratic institutions and governments, suppress social equality, and endanger economic growth. Corruption in public procurement undermines economic growth potential and distorts competition, impacting businesses, workers, cities, and communities.

Corruption also increases the costs of transactions, erodes fair competition, hinders foreign and domestic investment over the long term, and distorts development priorities. Regionally, corruption's presence impedes economic development, curtailing international investment and raising trade costs. Economic development is stifled as foreign direct investment is discouraged, and small businesses frequently cannot meet the startup costs required due to corruption. Money lost due to corruption is, in effect, development denied to those who are at greatest risk of being left behind. The high amounts of government and private expenditure involved (including the higher costs of governance and increased international development aid to health programs) also provide opportunities for corruption. Corruption has disproportionate impacts on poorer people and those who are more vulnerable, increasing costs and decreasing access to services, including health, education, and justice. Corruption is a significant barrier to economic and social development worldwide. As emphasized in the preamble to the UN Convention Against Corruption, corruption poses threats to stability and security in societies, weakens institutions and values of democracy and justice, and endangers sustainable development and the rule of law. The United Nations Convention Against Corruption (UNCAC), the only globally binding instrument on the fight against corruption, mobilizes actions to promote honest, transparent, and accountable government through effective anti-corruption mechanisms. Given the well-established links between corruption and lower levels of peace, justice, and equality, we have argued for some time that anti-corruption should be a core focus for the 2030 agenda of the international community on sustainable development. Corruption is also a critical topic within the UNs Sustainable Development Goals (SDGs), which seek to enhance public health, foster fairness and strong institutions, and advance sustainable human development. Corruption affects also all five pillars of sustainable development: People, Planet, Prosperity, Peace, and Partnership. Without strong anti-corruption measures, progress on other SDGs is likely to be highly limited. Most countries still have not managed to tackle corruption effectively and have not done so in almost a half century, which indicates lagging efforts by governments to tackle corruptions root causes. This constitute an enormous challenge for these countries, especially in the light of building their leadership potential and ensuring they are recognized as safe and reliable international partners and welcomed to international organizations.

Acknowledgments This research was funded by the grant from the Ministry of Education and Science of Ukraine (No. s/r 0121U100467, 0122U000783).

References

Aidt, T. S. (2010). Corruption and sustainable development. *Cambridge Working Papers in Economics*, 1061. https://doi.org/10.17863/CAM.5249

Bilan, Y., Srovnalã-KovÃi, P., Streimikis, J., Lyeonov, S., Tiutiunyk, I., & Humenna, Y. (2020a). From shadow economy to lower carbon intensity: Theory and evidence. *International Journal of Global Environmental Issues, 19*(1–3), 196–216. https://doi.org/10.1504/ijgenvi.2020.114874

Bilan, Y., Tiutiunyk, I., Lyeonov, S., & Vasylieva, T. (2020b). Shadow economy and economic development: A panel cointegration and causality analysis. *International Journal of Economic Policy in Emerging Economies, 13*(2), 173–193. https://doi.org/10.1504/IJEPEE.2020.107929

Brychko, M., Savchenko, T., Vasylieva, T., & Piotrowski, P. (2021). Illegal activities of financial intermediaries: A burden of trust crisis. *Journal of International Studies, 14*(1), 172–189. https://doi.org/10.14254/2071-8330.2021/14-1/12

Butko, M., Popelo, O., & Pishenin, I. (2019). Innovations in human resources management in Eurointegration conditions: Case for Ukrainian agro-industrial complex. *Marketing and Management of Innovations, 2*, 74–82. https://doi.org/10.21272/mmi.2019.2-07

Cabelkova, I., & Strielkowski, W. (2013). Is the level of taxation a product of culture? A cultural economics approach. *Society and Economy, 35*(4), 513–529. https://doi.org/10.1556/SocEc.2013.0007

Didenko, I., Paucz-Olszewska, J., Lyeonov, S., Ostrowska-Dankiewicz, A., & Ciekanowski, Z. (2020). Social safety and behavioral aspects of populations financial inclusion: A multicountry analysis. *Journal of International Studies, 13*(2), 347–359. https://doi.org/10.14254/2071-8330.2020/13-2/23

Eddassi, H. (2020). Fiscal regime and tax policy in resource-rich countries in the process of globalization: Literature review. *SocioEconomic Challenges, 4*(2), 67–77. https://doi.org/10.21272/sec.4(2).67-77.2020

Fila, M., Levicky, M., Mura, L., Maros, M., & Korenkova, M. (2020). Innovations for business management: Motivation and barriers. *Marketing and Management of Innovations, 4*, 266–278. https://doi.org/10.21272/mmi.2020.4-22

GRECO. (2020). *Corruption risks and useful legal references in the context of COVID-19.* Available at: https://rm.coe.int/corruption-risks-and-useful-legal-references-in-the-context-of-covid-1/16809e33e1

Hanić, A., & Jevtić, D. (2020). Human resource management between economy and ethics – Research of Serbia and Bosnia and Hercegovina. *Business Ethics and Leadership, 4*(3), 127–136. https://doi.org/10.21272/bel.4(3).127-136.2020

Kaya, H. D. (2020). Business friendliness, firm performance and owner's optimism. *Financial Markets, Institutions and Risks, 4*(3), 13–23. https://doi.org/10.21272/fmir.4(3).13-23.2020

Kaya, H. D., & Engkuchik, E. N. S. (2021). The perception of corruption among retailers in Central Asia and Eastern Europe during and after the 2008 crisis. *SocioEconomic Challenges, 5*(2), 70–80. https://doi.org/10.21272/sec.5(2).70-80.2021

Koudelková, P., Strielkowski, W., & Hejlová, D. (2015). Corruption and system change in the Czech Republic: Firm-level evidence. *Danube: Law, Economics and Social Issues Review, 6*(1), 25–46. https://doi.org/10.1515/danb-2015-0002

Kuzior, A., Mańka-Szulik, M., & Krawczyk, D. (2021). Zmiany w zarządzaniu elektronicznymi usługami publicznymi w metropolii podczas pandemii covid-19. *Polish Journal of Management Studies, 24*(2), 261–275. https://doi.org/10.17512/pjms.2021.24.2.16

Kuzior, A., Kettler, K., & Rąb, Ł. (2022). Digitalization of work and human resources processes as a way to create a sustainable and ethical organization. *Energies, 15*(1), 10.3390/en15010172.

Kuzmenko, O., Šuleř, P., Lyeonov, S., Judrupa, I., & Boiko, A. (2020). Data mining and bifurcation analysis of the risk of money laundering with the involvement of financial institutions. *Journal of International Studies, 13*(3), 332–339. https://doi.org/10.14254/2071-8330.2020/13-3/22

Kwilinski, A., Lyulyov, O., Pimonenko, T., Dzwigol, H., Abazov, R., & Pudryk, D. (2022). International migration drivers: Economic, environmental, social, and political effects. *Sustainability, 14*(11), 6413. https://doi.org/10.3390/su14116413

Lu, J., Ren, L., Qiao, J., Yao, S., Strielkowski, W., & Streimikis, J. (2019). Corporate social responsibility and corruption: Implications for the sustainable energy sector. *Sustainability, 11*(15), 4128. https://doi.org/10.3390/su11154128

Lyeonov, S. V., Vasylieva, T. A., & Lyulyov, O. V. (2018). Macroeconomic stability evaluation in countries of lower-middle income economies. *Naukovyi Visnyk Natsionalnoho Hirnychoho Universytetu, 1*, 138–146. https://doi.org/10.29202/nvngu/2018-1/4

Lyeonov, S., Żurakowska-Sawa, J., Kuzmenko, O., & Koibichuk, V. (2020). Gravitational and intellectual data analysis to assess the money laundering risk of financial institutions. *Journal of International Studies, 13*(4), 259–272. https://doi.org/10.14254/2071-8330.2020/13-4/18

Lyulyov, O., Chygryn, O., & Pimonenko, T. (2018). National brand as a marketing determinant of macroeconomic stability. *Marketing and Management of Innovations, 3*, 142–152. https://doi.org/10.21272/mmi.2018.3-12

Lyulyov, O., Lyeonov, S., Tiutiunyk, I., & Podgórska, J. (2021a). The impact of tax gap on macroeconomic stability: Assessment using panel VEC approach. *Journal of International Studies, 14*(1), 139–152. https://doi.org/10.14254/2071-8330.2021/14-1/10

Lyulyov, O., Paliienko, M., Prasol, L., Vasylieva, T., Kubatko, O., & Kubatko, V. (2021b). Determinants of shadow economy in transition countries: Economic and environmental aspects. *International Journal of Global Energy Issues, 43*(2–3), 166–182. https://doi.org/10.1504/ijgei.2021.10037735

Mercado, M. P. S. R., & Vargas-Hernández, J. G. (2019). Analysis of the determinants of social capital in organizations. *Business Ethics and Leadership, 3*(1), 124–133. https://doi.org/10.21272/bel.3(1).124-133.2019

Moskalenko, B., Lyulyov, O., & Pimonenko, T. (2022). The investment attractiveness of countries: Coupling between core dimensions. *Forum Scientiae Oeconomia, 10*(2), 153–172. https://doi.org/10.23762/FSO_VOL10_NO2_8

Nguedie, Y. H. N. (2018). Corruption, investment and economic growth in developing countries: A panel smooth transition regression approach. *SocioEconomic Challenges, 2*(1), 63–68. https://doi.org/10.21272/sec.2(1).63-68.2018

Nureev, R., Volchik, V., & Strielkowski, W. (2020). Neoliberal reforms in higher education and the import of institutions. *Social Sciences, 9*(5), 79. https://doi.org/10.3390/socsci9050079

Petroye, O., Lyulyov, O., Lytvynchuk, I., Paida, Y., & Pakhomov, V. (2020). Effects of information security and innovations on Country's image: Governance aspect. *International Journal of Safety and Security Engineering, 10*(4), 459–466. https://doi.org/10.18280/ijsse.100404

Pilia, G. (2017). Estonia and Lithuania in transition: A compared analysis of the change and its costs and benefits. *Business Ethics and Leadership, 1*(2), 12–19. https://doi.org/10.21272/bel.1(2).12-19.2017

Polyakov, M., Bilozubenko, V., Korneyev, M., & Shevchenko, G. (2019). Selection of parameters for multifactor model in the knowledge economy marketing (country level). *Innovative Marketing, 15*(1), 89–99. https://doi.org/10.21511/im.15(1).2019.08

Rekunenko, I., Zhuravka, F., Nebaba, N., Levkovych, O., & Chorna, S. (2022). Assessment and forecasting of Ukraine's financial security: Choice of alternatives. *Problems and Perspectives in Management, 20*(2), 117–134. https://doi.org/10.21511/ppm.20(2).2022.11

Samoilikova, A. (2020). Financial policy of innovation development providing: The impact formalization. *Financial Markets, Institutions and Risks, 4*(2), 5–15. https://doi.org/10.21272/fmir.4(2).5-15.2020

Samoilikova, A., & Kunev, R. (2020). The impact of health care financing on the economic growth: EU countries analysis. *Health Economics and Management Review, 1*(2), 24–32. https://doi.org/10.21272/hem.2020.2-03

Sardak, S., Korneyev, M., Dzhyndzhoian, V., Fedotova, T., & Tryfonova, O. (2018). Current trends in global demographic processes. *Problems and Perspectives in Management, 16*(1), 48–57. https://doi.org/10.21511/ppm.16(1).2018.05

Serpeninova, Y., Makarenko, I., Plastun, A., Babko, A., & Gasimova, G. (2020). Mapping of the responsible investments instruments in SDG 3 'good health and well-being' financing: EU and US experience. *Health Economics and Management Review, 1*(1), 106–115. https://doi.org/10.21272/hem.2020.1-10

Simionescu, M., Strielkowski, W., & Gavurova, B. (2022). Could quality of governance influence pollution? Evidence from the revised Environmental Kuznets Curve in Central and Eastern European countries. *Energy Reports, 8*, 809–819. https://doi.org/10.1016/j.egyr.2021.12.031

Smiianov, V. A., Vasilyeva, T. A., Chygryn, O. Y., Rubanov, P. M., & Mayboroda, T. M. (2020). Socio-economic patterns of labor market functioning in the public health: Challenges connected with COVID-19. *Wiadomosci Lekarskie, 73*(10), 2181–2187. https://doi.org/10.36740/wlek202010114

Strielkowski, W., Tarkhanova, E., Baburina, N., & Streimikis, J. (2021). Corporate social responsibility and the renewable energy development in the baltic states. *Sustainability, 13*(17), 9860. https://doi.org/10.3390/su13179860

Tiutiunyk, I., Drabek, J., Antoniuk, N., Navickas, V., & Rubanov, P. (2021). The impact of digital transformation on macroeconomic stability: Evidence from EU countries. *Journal of International Studies, 14*(3), 220–234. https://doi.org/10.14254/2071-8330.2021/14-3/14

Tiutiunyk, I. V., Zolkover, A. O., Lyeonov, S. V., & Ryabushka, L. B. (2022). The impact of economic shadowing on social development: Challenges for macroeconomic stability. *Naukovyi Visnyk Natsionalnoho Hirnychoho Universytetu, 1*, 183–191. https://doi.org/10.33271/nvngu/2022-1/183

UN Human Rights. (2022). *Special Rapporteur on the right to health.* Available at: https://www.ohchr.org/en/special-procedures/sr-health

Vasilyeva, T., Sysoyeva, L., & Vysochyna, A. (2016). Formalization of factors that are affecting stability of Ukraine banking system. *Risk Governance and Control: Financial Markets and Institutions, 6*(4), 7–11. https://doi.org/10.22495/rcgv6i4art1

Vasilyeva, T., Kuzmenko, O., Kuryłowicz, M., & Letunovska, N. (2021a). Neural network modeling of the economic and social development trajectory transformation due to quarantine restrictions during covid-19. *Economics and Sociology, 14*(2), 313–330. https://doi.org/10.14254/2071-789X.2021/14-2/17

Vasilyeva, T., Ziółko, A., Kuzmenko, O., Kapinos, A., & Humenna, Y. (2021b). Impact of digitalization and the COVID-19 pandemic on the ALM scenario: Data mining analysis for good governance. *Economics and Sociology, 14*(4), 326–354. https://doi.org/10.14254/2071-789X.2021/14-4/19

Vasylieva, T., Jurgilewicz, O., Poliakh, S., Tvaronavičienė, M., & Hydzik, P. (2020). Problems of measuring country's financial security. *Journal of International Studies, 13*(2), 329–346. https://doi.org/10.14254/2071-8330.2020/13-2/22

Zolkover, A., & Georgiev, M. (2020). Shadow investment activity as a factor of macroeconomic instability. *Financial Markets, Institutions and Risks, 4*(4), 83–90. https://doi.org/10.21272/fmir.4(4).83-90.2020

Chapter 28
Transformation of the Concept of the Business Leadership and Sustainability in the Post-COVID Era

Raisa Krayneva, Zhanargul Taskinbaikyzyh, and Tatiana Oruch

Abstract This chapter tackles the transformation of the concept of the business leadership and sustainability (especially with regard to the Sustainable Development Goals) in the post-COVID-19 period. Our study offers several key takeaways for CEOs about the business landscape after the pandemic, offering fresh perspectives on digital transformation, the future of work, transparency, and resilience. While workforce security and resilience, cost management, and organizational flexibility emerged as the highest priorities in the near term to long term, COVID-19 amplified older and introduced new business fears. The pandemic will not be the last big shock that hits businesses – and today's business leaders will likely be increasingly judged by how well they prepare for the next one.

Keywords Business leadership · Sustainable development · Transformation · post-COVID-19

JEL Classification M20 · P20 · Q01

R. Krayneva (✉)
Department of Mass Communications and Media Business, Financial University under the Government of the Russian Federation, Moscow, Russian Federation
e-mail: RKKrayneva@fa.ru

Z. Taskinbaikyzyh
Yessenov University, Aktau, Kazakhstan
e-mail: Zhanargul.taskinbaikyzy@yu.edu.kz

T. Oruch
Volga Region State University of Service, Togliatti, Russian Federation

W. Strielkowski (ed.), *Leadership, Entrepreneurship and Sustainable Development Post COVID-19*, Springer Proceedings in Business and Economics, https://doi.org/10.1007/978-3-031-28131-0_28

28.1 Introduction

As attention turns from COVID-19 back toward longer-term challenges, climate change will come back to the top of the business and social agendas, more urgently than ever. At the same time, although stopping COVID-19 and stabilizing economies and communities are top priorities, even the United Nations pointed to the 2030 Agenda as the natural organizing principle once the world is able to shift global attention toward rebuilding (Linner & Wibeck, 2021). The vulnerabilities exposed by the COVID-19 crisis could serve as an underscore for why ecological issues were becoming the world's highest policy priorities prior to the COVID-19 catastrophe (Orgad & Hegde, 2021). Left unchecked, global ecological emergencies like climate change and the loss of biodiversity may result in social and economic harms much greater than the damage caused by the coronavirus crisis. COVID-19 is creating a humanitarian crisis on global scale, affecting hundreds of thousands of lives. Due to long-lasting effects on the ongoing pandemic on societal and economic well-being, the current COVID-19 pandemic has created global challenges to the sustainability of healthcare provision and businesses (Kesavan et al., 2022). Although the COVID-19 pandemic has brought an enormous shock for organizational sustainability, it is worth noting that it might provide businesses with an excellent opportunity to shift toward real business social responsibility and ethical decision-making (Kakderi et al., 2021). COVID-19 has also proven a true test for the values and purposes of corporations. From small to big companies, businesses successfully dealing with the COVID-19 pandemic and ensuing downturn often have switched to business models conducive both to survival in the short-term and resilience and long-term growth. The reality of how companies are dealing with the crisis – and positioning themselves to recover – tells a much different story, one that is pivoting to a business model that is conducive to short-term survival, and also long-term resilience and growth. Companies are now recalibrating their business models and making strategic adjustments to a new consumer paradigm (Ozdemir et al., 2021).

In today's globalized world, companies must do a good job at what they do already, but they must also constantly look for new models for growth in order to find sustained success. Successful companies will reimagine the role of operations within their enterprises, creating new value by being much more responsive to their ultimate customers – including, but not limited to, speeding up product and customer experience innovation, mass customization, improving environmental sustainability, and managing a more connected, agile ecosystem (Pfeiffer, 2021). Corporate leaders need to take quick decisions and take swift actions, to support the operations of the enterprise in order to meet the needs of customers and communities as well as to protect and sustain workers. Companies must understand what customers will value, post-COVID-19, and design new use cases and tailored experiences around these insights (Butt, 2021). Just as so much risk and financial modeling had to be rebuilt in the wake of the financial crisis in 2008, data use and analytics need to be recalibrated to account for post-COVID-19 realities. Future research could focus on testing the effects of different strategies, such as diversifying products, maintaining aggressive innovations in service provision, and opening up to collaborations with market leaders about sustainability in an ongoing pandemic. Many businesses have reinvented themselves to meet the new needs created by the pandemic – even in traditionally slow-moving sectors like healthcare and education – and much will need to be done to meet the multifaceted challenges the remainder of the decade presents (Strielkowski et al., 2022).

If companies are to sustain balanced growth in an ongoing pandemic era, in which challenges and opportunities coexist, they need to fully integrate internal and external resources and count on digital transformation for survival, growth, and renewal (Su et al., 2022). As city leaders struggle to keep their citizens and communities safe while COVID-19 reverberates around the globe, the ideas about sustainable development and meeting Sustainable Development Goals (SDGs) by 2030 might appear to be unimportant in light of their present demands. Good leadership is critical, and perhaps there is one glimmer of hope, namely, the COVID-19 situation represents an opportunity for leaders to build greater cohesion and team innovation in the face of adversity. It is not hard to imagine who would do better in the more agile business landscape after COVID-19, where value creation is shared and strategic partnerships count for more. Yet the current crisis makes clear that the usual way of doing things is inadequate for this challenge (Koroma et al., 2022).

In the recent years, business has shown a greater appreciation for its collective dependence on a prosperous society and planet but is still light years from the size and speed of partnerships needed to build a new economic model where individual firms can win and lose, but only when the entire community and planet earth are prosperous (Peters et al., 2022). Business needs to make partnerships central to what it does at every turn, not only around particular issues (e.g., sourcing palm oil) but about systemic changes as well (e.g., the global food system). Topics cover the knowledge areas crisis management, resilience, corporate resilience, the impact of digitalization on resilience, sustainable business practices, business implications of COVID-19, and leadership in COVID-19.

28.2 COVID-19 as a Global Disruption

In general terms, the idea of tipping points in technological adoption or in the digital disruption is nothing new, but all available evidence indicates that the COVID-19 crisis is a tipping point, and it is the point that requires even greater changes as economic and human conditions change. The rapid global spread of COVID-19 has rapidly surpassed other recent pandemics in both scale and scope. In addition to the lethal human cost and destruction of millions of lives brought by the coronavirus disease, economic harm is already substantial and widespread. While all are facing unprecedented challenges, women are bearing the burden of COVID-19's economic and social impact. Women who are poor and marginalized are at even higher risk for transmission and death from COVID-19, loss of livelihood, and increased violence

(Thompson, 2022). While children seem largely insulated from this pandemic's direct mortality impacts, the indirect effects resulting from stressed health systems and the destruction of vital health services, such as immunizations and prenatal care, could lead to devastating increases in children's deaths. As the risk of violence against children has increased as a consequence of the COVID-19 pandemic, child protective services have been weakened, due in part to measures implemented to contain the spread of the virus (Bhatia et al., 2021).

New HIV infections among younger children have decreased by half over the past decade; however, the service interruptions caused by this pandemic may undo these gains. More than 94 million children are at risk for missing the measles vaccine because of suspended measles campaigns in 26 countries (as of November 2020) as part of efforts to contain COVID-19, which is likely to worsen ongoing measles outbreaks. Schools for over 168 million children worldwide were closed entirely for nearly a full year because of this pandemic lockdown, as of March 2020. At least one-third of the world's schoolchildren – 463 million children worldwide – were not able to access distance learning as COVID-19 closed down their schools. With more than 290 million students globally withdrawn from school because of COVID-19, the United Nations' demand for online offerings, curricula, and platforms is likely to accelerate (Feldman et al., 2021).

Subnational governments, regions, and municipalities are on the frontline in managing the crisis and rebuilding, and they are facing asymmetric impacts on health, economy, society, and finance of COVID-19 across countries, as well as across regions and subregions. While alignment with a common strategy and strong leadership has long been markers of success in a crisis or transition, the magnitude of the role played by technologies to differentiate the crisis in COVID-19 is dramatic (Doll et al., 2021). Customer needs can change drastically in crises like these, typically moving away from rationality toward emotion, and companies have an important role in navigating that change. For instance, some companies responded to the COVID-19 crisis by mandating a sweeping increase in inventories, fearing they would be running low on needed supplies. Such decisions must carefully account for unexpected consequences and disruptions. Many companies had instituted at least temporary solutions to address the myriad of new demands placed upon their companies, and far faster than they thought was feasible prior to the COVID-19 pandemic (Dewick et al., 2021).

The case of COVID-19 illustrates how many companies are completely unaware of their supply chains vulnerabilities to global disruptions, optimizing for efficiency rather than for sustainability. These are the faces behind the headlines, those who are the people who are being hit hardest by the economic impacts of the pandemic. Their plight is only going to get worse if policies do not deliberately include measures that provide economic relief and that purposefully focus on small and medium businesses and their livelihood security.

28.3 Social and Ecological Policy Trade-Offs in Post-COVID Era

Given the fact that the world's economic governance is entering a post-COVID-19 era alongside digital transformation, this chapter attempts to examine the mechanisms underlying digital economies effects on economic development as well as to examine the strategies for development in a post-COVID-19 era.

It can be stated that technology innovation could be a key driver for the achievement of SDGs in the post-COVID-19 period (Korneeva et al., 2021). The need for such studies has also been highlighted by many researchers who call for a study of policies that will allow the achievement of SDGs in a post-COVID-19 world (Sharma et al., 2021). To the best of our knowledge, this study is one of the few to examine the impact of technology innovations on SDGs and employment in the post-COVID-19 era. As COVID-19 prevalence does not lead to any change in socio-ecological impacts of innovations, and COVID-19's character has only amplified these impacts, consideration of data sets of many countries for prescribing policy frameworks after COVID-19 is therefore logical. It could be said that the COVID-19 incidence has supplemented the ecological benefits of technology innovations but has not brought about any major transformations in the ecological impacts of innovations (Ratten et al., 2021). We suggest that, given the expected socio-ecological trade-offs in a COVID-19-driven policy-making process, policy reorientation is needed. Whether an innovation will be an ecological panacea or will open Pandora's Boxes of social imbalances is emerging as the main issue of the post-COVID-19 period. Responding to the COVID-19 crisis requires policies that are mutually supportive and integrated between economic, social, and environmental objectives in order to tackle the deeper-seated nature of the COVID-19 crisis (Lauriola et al., 2021). As in SDGs, effective mechanisms of policy coordination across different levels are needed to consult, coordinate, negotiate, and adjudicate policy decisions in order to address the many dimensions of this pandemic and to deal with its economic, social, and environmental impacts. Building political commitment and leadership needs to be done at the highest level. The interconnected nature of health, economic and social crises caused by the COVID-19 crisis, together with other global challenges, such as climate change, suggests that issues of policy coherence will be more challenging, necessitating alignment between policy agendas (Bongardt & Torres, 2022). Beyond the development cooperation dimension, applying a policy coherence lens, in both the pre- and post-assessment phases, is crucial for dealing with the multidimensional and transboundary impacts from the COVID-19 crisis. Accordingly, when formulating socioeconomic development policies in a post-pandemic context, researchers and policy-makers need to pay attention both to the short-term recovery and the longer-term development. Containing the COVID-19 pandemic and avoiding further negative socioeconomic impacts require rapid and effective interventions, not only by continuing prevention, screening, and treatment measures but also by effectively spreading the vaccines around the world. The existence of highly infectious diseases that become pandemics, like COVID-19, poses severe health and economic challenges due to various global social and environmental transformations that have occurred due to economic development. Model social choices consist of trade-offs between limits to the black death and levels of economic activity. Most global pandemics significantly reduced world economic output and increase unemployment. The COVID-19 pandemic has hampered global markets, slowing progress on Economic Growth (SDG 8) and Production and Consumption (SDG 6) for 80 percent and 70 percent of world countries, respectively (Zhao et al., 2022).

International and national organizations that are involved in formulating strategies to transform after COVID are faced with fiscal and energy constraints in the aftermath of COVID-19, when the governments are obliged to look for solutions that minimize trade-offs, foster synergies, and tackle the challenges of addressing the multidimensional challenges of development to promote systematic SDGs (Barquet et al., 2022). Identifying the proper economic policies to address the COVID-19 outbreak and the spread among humans has proven to be exceptionally difficult, as major measures aimed at containing infection levels and disease mortality rates involve a large economic cost in terms of lost levels of revenue and economic output. EU member states have been prepared to adopt measures to counteract COVID-19, which entail significant economic costs, as well as creating risks of an economic recession and severe unemployment (Howarth & Quaglia, 2021). In addition, many young people, in Europe and beyond, weighed up this trade-off in the course of the pandemic, as the communications strategies around this trade-off affected their preferences of policies to guide a recovery in economic and social activities, and, finally, their intentions to follow the recommended behaviors. This shows the effectiveness of the counter-COVID-19 policies.

28.4 Designing Post-Pandemic Comprehensive Policies

Based on the available data and comprehensive analysis of the literature, this chapter highlights ongoing challenges facing SDGs, determines the impact of the COVID-19 pandemic on SDG progress, and presents a systematic framework to facilitate SDGs progress in the post-pandemic era. The ongoing pandemic has affected the social, economic, and ecological systems around the world, slowing and disrupting the progress made toward achieving the Sustainable Development Goals (SDGs). The COVID-19 pandemic is also posing an enormous challenge for achieving the SDGs. Shulla et al. (2021) have identified a severe impact of ongoing COVID-19 pandemic on countries' capacity to achieve SDGs. The 2020 SDG report has shown that countries with better SDG scores responded to the COVID-19 pandemic most effectively. The SDGs have promoted systemic designs that enable governments and private partnerships to respond in a more holistic, connected way to the pandemic and beyond. Well-designed domestic policy frameworks and partnerships are essential for embedding resilience into patterns of consumption and production (Forman et al., 2022). Gaps between the SDG vision and practical capacity hamper efforts at national level the effective implementation of SDGs within countries depends on their inclusion within the country's socioeconomic development plans. Achieving specific SDGs enhances living conditions and brings stability to the economy's demand side by fostering sustainable development, which is manifested through improved labor market conditions, reductions in poverty, and improvements in health and education. Following the endorsement of the 17 United Nations Sustainable Development Goals in 2015, a number of empirical studies (Pires et al., 2021; Leal Filho et al., 2022) suggest policies should be adopted to meet these different SDGs. The need for such studies has also been highlighted by Horisch (2021), who called for a policy study that would allow for achieving SDGs during the post-COVID-19 timeframe. After introducing the COVID-19 pandemic, all HEIs research outputs must be designed in an objective manner, consistent with UN SDGs and university visions, in order to address the challenges of the community's reality. The challenges are because many communities will become poorer due to increased inequalities caused by this pandemic, and in order to solve it, higher education institutions (HEIs) must start considering interventions that could mitigate such impacts, creating an interplay between their curriculums and real-life social problems using the UN SDGs as project frames within our various communities (Useh, 2021).

During the COVID-19 pandemic, the need is to examine the effects of e-learning in education sectors and to propose appropriate policy implications to address the Sustainable Development Goal 4, which is quality education, in the United Nations Sustainable Development Goals. Addressing this requires comprehensive research into the impacts of the pandemic on sustainable development (Wang et al., 2022). We anticipate that our study will advance knowledge about underlying challenges and will contribute to policy interventions that mitigate the adverse impacts of the ongoing pandemic on SDGs in the developing countries. For achieving the 2030 Agenda, it is necessary to thoroughly understand the impacts of the pandemic on the SDGs, so that policy-makers can design interventions to mitigate negative impacts. The recovery plan must focus on rebuilding activities and programs disrupted by the ongoing pandemic, as well as developing new ones that will speed up progress in different SDGs, building on past experiences. Overall, one could conclude that, if people do not adhere to a road map for SDG achievement, then poverty effects affected by the pandemic will be prevalent for a decade, at the very least. The pandemic has had multiple impacts, but with regard to SDG Target 2, impacts can be assessed using two methods. In addition, Sustainable Development Goals 15, 13, and 16 are the most lightly affected, while Goal 14 has had net positive impacts from the pandemic. Clearly, SDGs 1 and 8 are the most impacted goals, meaning more than half the targets of these goals are heavily affected by the pandemic. The effect of COVID-19 on progress on SDGs COVID-19 has led to reduced SDG advancement that needs to be reverted as soon as possible for ensuring the recovery of business and economics.

The COVID-19 pandemic has limited progress toward achieving SDGs, which is evident from the marked decrease of the Global SDG Index scores in 2020. The impact of the ongoing pandemic on sustainable development goals (SDGs) in the world should be assessed and monitored for the short-term (within 1 year), mid-term (within next 5 years), and long-term (by 2030).

Overall, building upon the Millennium Development Goals that were established and aimed to end extreme poverty, protect the planet, and promote prosperity for all by 2030. In fact, SDGs are multifaceted and represent significant improvements compared with other measures of development, including Human Development Index, Gender Equity Index, Livability Index, Governance Index, and Index of Sustainable Economic well-being. One of the ways in which one can guarantee the concomitant achievement of UN SDGs and development of communities is by having the graduate projects after COVID-19 directed specifically toward the resolution of the UN SDGs in order to have positive societal impacts on our communities.

28.5 Conclusions

Overall, it is quite obvious that the recent COVID-19 pandemic has exposed systemic weaknesses in our global economy and societies. The World Health Organization (WHO) declared novel coronavirus COVID-19 as a global pandemic and a public health emergency in March 2020 and since then the global spread of COVID-19 has had multiple impacts on the sustainability of the world's production and consumption of a variety of commodities. Amid the coronavirus pandemic, the global supply chains and manufacturing networks are going through an extremely distressed phase. These global supply chain and manufacturing systems are mostly impacted by widespread spread of the pandemic. The outbreak of COVID-19 is impacting global and domestic production systems and trade at large scales. The impact of COVID outbreak on the trends of production and sustainable consumption will be worth noting; hence, significant changes are likely to be observed over the coming months and years. As mentioned above, post-COVID-19 timeframe will prompt organizations to transform their production and supply chain systems more sustainably. A circular economy serves as a key for creating sustainable supply chains after COVID-19. The implementation of an industrial strategy that seeks both higher-quality jobs and ecological sustainability within the agriculture sector would be especially critical, aligning with both the post-COVID-19 jobs needs and Green Deal objectives. The integration of public and corporate policies with technical innovations could alleviate negative impacts of economic activities on the SDGs. In particular, the outbreak of COVID-19 has substantially improved environmental resilience in organizations, although it has reduced consumer economies and raised challenges in managing the workforce in industry. In addition, an effective response to the COVID-19 crisis requires enhanced cooperation among scientists and decision-makers.

Many countries have established governance mechanisms for the coordination of scientific advice across the various ministries involved in efforts to contain, delay, and mitigate the effects of COVID-19, helping to move beyond the policy silos of pandemic response, as well as providing guidance for how reconstruction efforts could promote a more inclusive, resilient, and resilient economy that could underpin efforts at the Sustainable Development Goals. The briefings by OECD and other organizations seek to provide a road map to policy coordination. Effective engagement of key stakeholders, making sure that different voices are heard across the community, will be crucial for ensuring inclusive and sustainable recovery from the COVID-19 crisis. Reflecting on the unprecedented mobilization and impacts of the COVID-19 response may inspire new ways of thinking and can help humanity to take a leap forward to effect change. A sustained recovery requires cohesive policies, which account for cross-border impacts, in order to avoid prolonged economic decline, worsening the global impacts of the COVID-19 crisis, and undoing the gains made in SDGs. Reconstruction requires interconnected, integrated approaches for implementing policies and designing reconstruction, in which countries are successful in transforming into green economies and building resilience against future shocks. The primary lesson that we may have learned from the pandemic is that preparedness for disasters, at the local and global levels, is best built upon the resiliency and buffers that need to be built into systems, including, in our case, the food systems, but that is not going to happen without appropriate investments. Investing in green systems and in clean energy will be difficult, since before COVID-19, governments were estimated to be trillions of dollars shy of being able to meet SDGs. In the light of COVID-19, it is difficult to see how the economic costs, in isolation, or risk of economic downturn, could still be used as a valid argument against action on environmental issues or a transition toward resilience. Sustainable and healthy diets for all in the course of the COVID-19 outbreak appear to be far more dependent on social and economic conditions than technical aspects of food production and processing. Greater attention is also being paid to such complications within policy areas beyond epidemiology and public health, because health crises such as COVID-19 carry wide-ranging consequences for individuals and societies as a whole. The global community will need years, perhaps decades, to evaluate the full scope of COVID-19 and the implications COVID-19 has on our societies, including its effects on inequalities, health, and the well-being of citizens. Although it is controversial in some ways, the SDGs provide guidance and encouragement for mitigating negative consequences of the pandemic, and they offer a framework within which needed leadership and accountability can be implemented. Each SDG and objective needs to be evaluated with regard to the degree of post-COVID-19 priority, the degree to which it promotes development, not merely growth, and its capacity to build resilience against global shocks and stresses. As the objective of the majority of governments is to support a strategic dimension to policy-making in the field, we believe that the immediate and longer-term impacts of the pandemic on food systems and food governance must be considered and, thus, the credibility of proposed transformations and recommendations that will be provided. It suggested we reflect on what activities we wish to see suspended for good and what we wish to reestablish: what entirely new activities or habits we wish to cultivate following the COVID-19 pandemic and how workers or entrepreneurs who have been deprived by reshaped economies may be helped to move to different, more sustainable, or more sustainable roles or activities. Changes in attitudes referred to the possibility that the public will become more supportive of environmental policies as a result of COVID-19. Recently, the surveys of organizations from a range of industries demonstrated that senior leaders said that they expected the fallout of the COVID-19 pandemic to dramatically change how they do business in the next 5 years, and nearly as many said that the crisis would have an ongoing effect on the needs of their customers. All in all, it becomes clear that the world faces an unprecedented, multidimensional crisis, one that requires coherent policy responses. Thence, appropriate institutional and operational policies are needed to address production losses and to improve the pattern of consumption, both of which will boost the economy even more.

References

Barquet, K., Järnberg, L., Alva, I. L., & Weitz, N. (2022). Exploring mechanisms for systemic thinking in decision-making through three country applications of SDG synergies. *Sustainability Science, 17*(4), 1557–1572. https://doi.org/10.1007/s11625-021-01045-3

Bhatia, A., Fabbri, C., Cerna-Turoff, I., Turner, E., Lokot, M., Warria, A., & Devries, K. (2021). Violence against children during the COVID-19 pandemic. *Bulletin of the World Health Organization, 99*(10), 730. https://doi.org/10.2471/BLT.20.283051

Bongardt, A., & Torres, F. (2022). The European green deal: More than an exit strategy to the pandemic crisis, a building block of a sustainable European economic model. *JCMS: Journal of Common Market Studies, 60*(1), 170–185. https://doi.org/10.1111/jcms.13264

Butt, A. S. (2021). Supply chains and COVID-19: Impacts, countermeasures and post-COVID-19 era. *The International Journal of Logistics Management, ahead-of-print*, ahead-of-print. https://doi.org/10.1108/IJLM-02-2021-0114

Dewick, P., Hofstetter, J. S., & Schröder, P. (2021). From panic to dispassionate rationality-Organizational responses in procurement after the initial COVID-19 pandemic peak. *IEEE Engineering Management Review, 49*(2), 45–56. https://doi.org/10.1109/EMR.2021.3079630

Doll, K., Ragan, M., Calnin, G., Mason, S., & House, K. (2021). Adapting and enduring: Lessons learned from international school educators during COVID-19. *Journal of Research in International Education, 20*(2), 114–133. https://doi.org/10.1177/14752409211034399

Feldman, A. G., O'Leary, S. T., & Danziger-Isakov, L. (2021). The risk of resurgence in vaccine-preventable infections due to coronavirus disease 2019-related gaps in immunization. *Clinical Infectious Diseases, 73*(10), 1920–1923. https://doi.org/10.1093/cid/ciab127

Forman, R., Azzopardi-Muscat, N., Kirkby, V., Lessof, S., Nathan, N. L., Pastorino, G., & Mossialos, E. (2022). Drawing light from the pandemic: Rethinking strategies for health policy and beyond. *Health Policy, 126*(1), 1–6. https://doi.org/10.1016/j.healthpol.2021.12.001

Horisch, J. (2021). The relation of COVID-19 to the UN sustainable development goals: Implications for sustainability accounting, management and policy research. *Sustainability Accounting, Management and Policy Journal, 12*(5), 877–888. https://doi.org/10.1108/SAMPJ-08-2020-0277

Howarth, D., & Quaglia, L. (2021). Failing forward in Economic and Monetary Union: Explaining weak Eurozone financial support mechanisms. *Journal of European Public Policy, 28*(10), 1555–1572. https://doi.org/10.1080/13501763.2021.1954060

Kakderi, C., Komninos, N., Panori, A., & Oikonomaki, E. (2021). Next City: Learning from cities during COVID-19 to tackle climate change. *Sustainability, 13*(6), 3158. https://doi.org/10.3390/su13063158

Kesavan, N., Seenuvasaragavan, S., Burnette, K., Heim, A. B., Patel, R. J., Johnston, J., Cao, X., Sakandar, H. A., Zhuang, Y., Oi, K. K., Zhi, Y., Diao, Y., & Strielkowski, W. (2022). Pandemic-inspired policies. *Science, 377*(6601), 22–24. https://doi.org/10.1126/science.add2046

Korneeva, E., Olinder, N., & Strielkowski, W. (2021). Consumer attitudes to the smart home technologies and the Internet of Things (IoT). *Energies, 14*(23), 7913. https://doi.org/10.3390/en14237913

Koroma, B., Macarthy, J. M., Conteh, A., & Osuteye, E. (2022). Sustaining community-based responses to the COVID-19 pandemic: Learning platforms at work in Freetown, Sierra Leone. *Urbanisation*, 24557471221115030. https://doi.org/10.1177/24557471221115030

Lauriola, P., Martín-Olmedo, P., Leonardi, G. S., Bouland, C., Verheij, R., Dückers, M. L., & Zeka, A. (2021). On the importance of primary and community healthcare in relation to global health and environmental threats: Lessons from the COVID-19 crisis. *BMJ Global Health, 6*(3), e004111. https://doi.org/10.1136/bmjgh-2020-004111

Leal Filho, W., Vidal, D. G., Chen, C., Petrova, M., Dinis, M. A. P., Yang, P., & Neiva, S. (2022). An assessment of requirements in investments, new technologies, and infrastructures to achieve the SDGs. *Environmental Sciences Europe, 34*(1), 1–17. https://doi.org/10.1186/s12302-022-00629-9

Linner, B. O., & Wibeck, V. (2021). Drivers of sustainability transformations: Leverage points, contexts and conjunctures. *Sustainability Science, 16*(3), 889–900. https://doi.org/10.1007/s11625-021-00957-4

Orgad, S., & Hegde, R. (2021). Crisis-ready responsible selves: National productions of Covid-19 pandemic. *International Journal of Cultural Studies, 25*(3–4), 287–208. https://doi.org/10.1177/13678779211066328

Ozdemir, O., Dogru, T., Kizildag, M., Mody, M., & Suess, C. (2021). Quantifying the economic impact of COVID-19 on the US hotel industry: Examination of hotel segments and operational structures. *Tourism Management Perspectives, 39*, 100864. https://doi.org/10.1016/j.tmp.2021.100864

Peters, M. A., Rizvi, F., McCulloch, G., Gibbs, P., Gorur, R., Hong, M., & Misiaszek, L. (2022). Reimagining the new pedagogical possibilities for universities post-COVID-19: An EPAT Collective Project. *Educational Philosophy and Theory, 54*(6), 717–760. https://doi.org/10.1080/00131857.2020.1777655

Pfeiffer, S. (2021). The greater transformation: Digitalization and the transformative power of distributive forces in digital capitalism. *International Critical Thought, 11*(4), 535–552. https://doi.org/10.1080/21598282.2021.2005656

Pires, A. P., Soto, C. R., & Scarano, F. R. (2021). Strategies to reach global sustainability should take better account of ecosystem services. *Ecosystem Services, 49*, 101292. https://doi.org/10.1016/j.ecoser.2021.101292

Ratten, V., da Silva Braga, V. L., & da Encarnação Marques, C. S. (2021). Sport entrepreneurship and value co-creation in times of crisis: The COVID-19 pandemic. *Journal of Business Research, 133*, 265–274. https://doi.org/10.1016/j.jbusres.2021.05.001

Sharma, H. B., Vanapalli, K. R., Samal, B., Cheela, V. S., Dubey, B. K., & Bhattacharya, J. (2021). Circular economy approach in solid waste management system to achieve UN-SDGs: Solutions for post-COVID recovery. *Science of The Total Environment, 800*, 149605. https://doi.org/10.1016/j.scitotenv.2021.149605

Shulla, K., Voigt, B. F., Cibian, S., Scandone, G., Martinez, E., Nelkovski, F., & Salehi, P. (2021). Effects of COVID-19 on the sustainable development goals (SDGs). *Discover Sustainability, 2*(1), 1–19. https://doi.org/10.1007/s43621-021-00026-x

Strielkowski, W., Firsova, I., Azarova, S., & Shatskaya, I. (2022). Novel Insights in the leadership in business and economics: A post-coronavirus update. *Economies, 10*(2), 48. https://doi.org/10.3390/economies10020048

Su, R., Obrenovic, B., Du, J., Godinic, D., & Khudaykulov, A. (2022). COVID-19 pandemic implications for corporate sustainability and society: A literature review. *International*

Journal of Environmental Research and Public Health, 19(3), 1592. https://doi.org/10.3390/ijerph19031592

Thompson, R. J. (2022). Changing realities for women and work: The impact of COVID-19 and prospects for the post-pandemic workplace. *Merits, 2*(3), 164–169. https://doi.org/10.3390/merits2030012

Useh, U. (2021). Sustainable development goals as a framework for postgraduate future research following COVID-19 pandemic: A new norm for developing countries. *Higher Education for the Future, 8*(1), 123–132. https://doi.org/10.1177/2347631120972064

Wang, X. Y., Li, G., Malik, S., & Anwar, A. (2022). Impact of COVID-19 on achieving the goal of sustainable development: E-learning and educational productivity. *Economic Research-Ekonomska Istraživanja, 35*(1), 1950–1966. https://doi.org/10.1080/1331677X.2021.1927789

Zhao, W., Yin, C., Hua, T., Meadows, M. E., Li, Y., Liu, Y., & Fu, B. (2022). Achieving the Sustainable Development Goals in the post-pandemic era. *Humanities and Social Sciences Communications, 9*(1), 1–7. https://doi.org/10.1057/s41599-022-01283-5

Chapter 29
Personnel Development as a Leading Strategy of Adult Learning in the Framework of Achieving the Sustainable Development Goal 4 "Quality Education"

Yuriy Petrushenko, Kateryna Onopriienko, and Zhanna Dovhan

Abstract Personnel development is becoming increasingly important as one of the areas of adult education in achieving the Sustainable Development Goal 4 "Quality Education." As with any economic process, staff training and development must be valued in terms of benefits and costs. This chapter hypothesizes the effectiveness of the contribution to education in the context of the COVID-19 pandemic. Based on the results of the analysis of financial statements and calculations of enterprises, an assessment of the effectiveness of the personnel training system was made, and recommendations were given for its improvement in the context of achieving the Sustainable Development Goal 4. A correlation-regression method was used, which provided additional evidence to confirm the proposed hypothesis. The relationship between the average wage and the share of adult participation in education has been established. Data from UNESCO, Eurostat, as well as financial documents of enterprises and statistics of the European Union countries on adult learning, were used as an information base.

Keywords Personnel development · Education · SDG · Quality · Leadership

JEL Classification I25 · J24 · L21

Y. Petrushenko · K. Onopriienko (✉)
Sumy State University, Sumy, Ukraine
e-mail: y.petrushenko@biem.sumdu.edu.ua; k.onoprienko@uabs.sumdu.edu.ua

Z. Dovhan
West Ukrainian National University, Ternopil, Ukraine

Eastern Europe in the public health insurance company Barmer, Wuppertal, Germany

W. Strielkowski (ed.), *Leadership, Entrepreneurship and Sustainable Development Post COVID-19*, Springer Proceedings in Business and Economics, https://doi.org/10.1007/978-3-031-28131-0_29

29.1 Introduction

Nowadays, adult education is an extremely relevant issue within the framework of the Sustainable Development Goals, or rather goal number 4. The principles of adult education have been actively formulated in the last few decades. By choosing education independently, adults can pursue it independently.

The economic crisis during and after the COVID-19 pandemic and the slowdown in economic growth that Europe faced, caused a high level of unemployment among young people and demanded from European politicians extensive investments in the development of employee competencies (Vorontsova et al., 2018). Today, human resources are an important component for its development. Since the company's employees organize and carry out the production process, as a result, the development of the company's human resources is a component of the increase in the efficiency of the organization, business, and the country in general. At the same time, it is absolutely obvious that high-quality and effective work of the company's personnel cannot be achieved without proper education throughout life (Guliyeva et al., 2021).

The United Nations General Assembly Resolution "Transforming our world: A 2030 Agenda for Sustainable Development" adopted in 2015 outlined an action plan to move the world onto a new development trajectory. All UN member states have joined the global process of ensuring sustainable development in each European country each global goal is considered taking into account the specifics of the state system (Artyukhov et al. 2021a, b, c).

The goal of the program remains the same – to make the planet a better place to live. However, the approach has changed. The UN realized that instead of focusing on solving problems in individual countries, it is much better to achieve sustainable development throughout the world, which would become the foundation for the continuous improvement of life on the planet. Moreover, one of the most important in this process is the development of personnel at the enterprise.

Although states are not legally bound to achieve the Sustainable Development Goals, the opportunities offered by the implementation of the program are attractive for doing so. For example, opening markets worth more than $12 trillion and creating more than 380 million jobs by 2030 (Elfert, 2019).

However, taking into account the fact that if the goals are achieved, our lives will improve, and this is not only a matter for government bodies but also for all of us. Everyone can contribute, and business in particular, because this is an opportunity to open new markets and develop in different directions. In the scientific literature, the issue of adult education within the framework of the goal of sustainable development 4 was investigated by a number of scientists (Khoshnava et al., 2019; Artyukhov et al., 2022; Volk et al., 2021; Novikov, 2021).

VOSviewer software and the Scopus database were also used for bibliography research, which allowed to see the number of scientific works on the keywords "adult education" and "sustainable development goals," as well as how much they were combined. Therefore, we can see that the concept of the goals of sustainable development intersects with the concept of human, training, and education (Fig. 29.1).

The next point on which we focused our attention was the number of scientific works dedicated to personnel development. Therefore, analyzing the Scopus database, we see that the greatest growth of scientific works in this topic has been observed since the 2000s, that is, with the appearance of IT and its development. This forced the business to invest more and more in the education of its personnel (Fig. 29.2).

Let us consider what aspects the global goal for sustainable development No. 4 consists of. Quality education: ensure inclusive and equitable quality education and promote lifelong learning opportunities for all. It includes such aspects as: by 2030, substantially increase the number of youth and adults who have relevant skills, including technical and vocational skills, for employment, decent jobs and entrepreneurship (Shkarlet et al., 2019). To ensure that all youth and a substantial proportion of adults, both men and women, achieve literacy and numeracy. So we see that not only adult education, but also the issue of personnel improvement is gaining new importance (Čábelková et al., 2020).

A key role in the implementation of the goal of sustainable development No. 4 is the management and financing of the continuing education system. Lifelong learning refers to "all lifelong learning that serves to improve knowledge, personal, civic, social or employment-related perspectives." Let us consider some examples of financing education during life in European countries.

By 2022, Finland has seen an increase in direct costs for personnel training for 11 years in a row. It costs an average of $1096 per employee to train and upskill them (Serpeninova et al., 2020). As an integral part of Finnish adult liberal

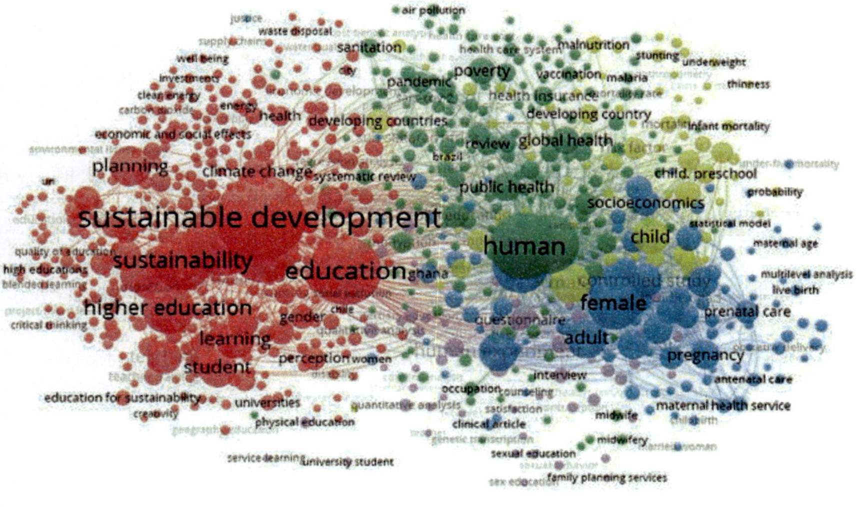

Fig. 29.1 Network map between the keywords of articles published on sustainable development. (Source: Own results using VOSviewer software)

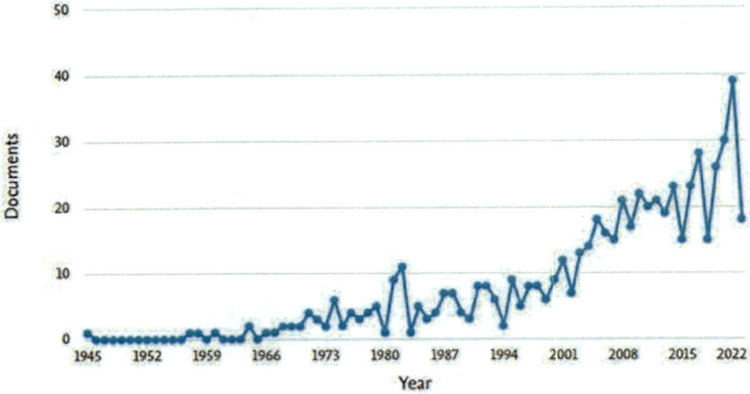

Fig. 29.2 The trend of the number of scientific works by the two keywords "personnel development" and "sustainable development goals". (Source: Own results based on Scopus database)

education, adult education centers provide general and vocational education services according to regional needs. The functioning of these centers is provided by 53% from the state budget and 28% from the municipal budget, and only 15% consists of funds paid by students or employees themselves (Polyakov et al., 2019; Strielkowski et al., 2020).

In France, in accordance with the current legislation, mandatory deductions from the wage fund of enterprises in the amount of 0.25% (for enterprises with fewer than 10 employees) and 1.5% (for enterprises with more than 10 employees) are concentrated on personnel training (Lyeonov et al., 2021). If the enterprise consists of one person, no deductions are made. The actual expenditure of French enterprises on personnel development is on average 3% of the wage fund of enterprises. The company has no right to deny an employee training (Degtjarjova et al., 2018).

In Germany, government organizations anticipate a shortage of skilled workers due to the country's demographic situation. The federal government supports further education through an education bonus of up to €500 per year. Adults with professional qualifications can apply for an increase in the grant to 650 euros and 80 euros per month for books (Voronkova et al., 2019).

In Sweden, the adult education is largely financed by the state, regions, and municipalities. The Swedish National Agency for Higher Vocational Education makes the final decisions on the distribution of public funding and the list of programs for which adult students can qualify for financial assistance. Most adult education providers charge tuition, but with some programs, students may qualify for financial aid (Polyakov et al., 2020).

29.2 Prerequisites for the Sustainable Personnel Development

Personnel development is the process of improvement and mastering of new high-quality professional, business skills and abilities, personal competencies as a result of lifelong learning and self-development of the individual. Employee development takes place in the conditions of a harmonious combination of personal, intellectual, moral, emotional, and psychophysiological qualities, the quality combination of which ensures a high level of effectiveness and efficiency and increases the overall level of competitiveness of the enterprise (Meresa, 2019; Mercado & Vargas-Hernández, 2019). Personnel development is a continuous process that aims to ensure the comprehensive development of employees and increase the overall efficiency of the enterprise.

Personnel development is provided by measures for industrial adaptation, personnel evaluation for the purpose of certification, professional career planning of workers and specialists, and stimulation of personnel development (Babchynska, 2021; Yarovenko et al., 2021).

The process of personnel development has many aspects, which is evidenced by numerous points of view and different methods of managing such programs. Such conditions include professional growth, continuing education in the workplace, staff development, updating human resource management methods, talent management, and other combinations of these aspects (Kuzior & Kwilinski, 2022).

With the accelerating pace of technological change, and in the aftermath of the COVID-19 pandemic, the skills required to carry out work activities have become obsolete at a much faster rate. In modern sectors of the economy, updating competencies to maintain competitiveness and increase labor productivity indicators becomes necessary every 12–18 months (Sardak et al., 2018).

Employees and employers are objectively interested in increasing various components of human capital. General human capital includes knowledge and skills that can be used in various activities and are easily transferred from one employer to another without any significant loss of value (Kotenko et al., 2021). Recently, adult education has been given increasing importance due to the very effective updating of competencies that require situational response. Without considering long-term investment in employees, it is challenging for an organization to maintain their competitiveness and meet their goals.

According to human capital theory, on-the-job training is considered the most flexible form of skills development. It is aimed at the accumulation of specific human capital and most quickly increases labor productivity due to the short period of training and great practical orientation (Kuzior et al., 2022b).

Training during the working day, combined with self-education, supports the development of two types of human capital at once – general and specific, while the fact that self-education is initiated directly by the employee becomes essential (Boeren, 2019; Mujtaba & Senathip, 2020).

Adults learn more effectively when the acquisition of new knowledge is necessary to solve specific problems or life needs (Skrynnyk & Vasilyeva, 2020). One of the most important incentives in this context is considered to be the lack of skills felt by the employee or his employer to perform current labor operations.

The procedure for the implementation of learning and development of the personnel of the enterprise includes the following:

- Assessment of the need for personnel
- Formation of a personnel development plan
- Conducting staff training and knowledge testing based on its results
- Analysis of the effectiveness of the application of knowledge by trained employees and documentation of learning outcomes
- Assessment of the effectiveness of training (Podchalymova et al., 2021; Didenko et al., 2020a)

In the course of the study, we conducted a survey on the topic of the development of the company's personnel. 97 employees of the international company Porsche Ukraine aged 23–56 were surveyed. In terms of the age structure of respondents, it can be seen that the age category ranging from 23 to 35 years (60% of respondents) was predominant in the company. The second largest sample of respondents is the age category from 36 to 45 years at 30%. The rest of the respondents (10%) were above 45 years of age.

When asked about the application of the acquired knowledge, the employees identified the problem. Often the acquired knowledge is applicable in one situation, but unacceptable for another. Consequently, the content of training does not always coincide with the content of professional activity. Revealing the answer to the question about the criteria for assessing the effectiveness of training, people focused on the fact that, in principle, the enterprise does not have a procedure for assessing effectiveness. Therefore, there are no criteria.

From the survey, we can conclude that people under 40 are more motivated to lifelong education (Liuta et al., 2021). Also, employees intend to improve their skills and undergo additional training only if they are provided with financial support. At the same time, almost 35% of respondents are ready to invest their own money in their further education.

An analysis of the data in the context of the amount of money that employees of various professional categories are expected to spend on their training showed that the higher the status of an employee, the greater the amount he can independently invest in the passage of educational programs.

On the one hand, this is due to the preferred forms of training: the higher the professional level an employee occupies, the more complex and, therefore, financially costly programs he chooses. In addition, the higher the position of an employee in the professional hierarchy, the more heterogeneous skills he needs in connection with his position, and the more money he is ready to spend on his own.

On the other hand, the amount of money that workers could spend on additional education is proportional to their average monthly salary.

At this point, we stopped our attention. A sample was made from European countries according to two indicators: the average salary and the percentage of participation in adult education. The following result was found – the higher the average salary, the more people in the country take part in adult education (Fig. 29.3).

It was also found that in the period after the COVID-19 pandemic, that is, 2021 and early 2022, the trend of adult participation in education went up. This is due to two main factors – the transition of education to online and, in this regard, teaching adults digital skills.

However, it is worth noting that it is difficult to assess in monetary terms whether enough is being invested in adult education and staff development, because it is not just the money that is invested that is important. Another clear indicator of interest in learning and development is time.

When it comes to the organization of the staff development process, the time spent by the organization is a very valuable resource that the company could use for many other tasks, such as providing services/selling products or attracting new customers. In addition, the employer's labor costs are not characterized by the cost of labor but by the amount of money the employee costs the organization (Brimah et al., 2020; Plys, 2021).

That is, labor costs should include not only costs that are directly related to a person but also costs aimed at increasing his ability to work, increasing productivity with the help of professional qualifications, work motivation, training, and development (Hanić & Jevtić, 2020). For example, the company Porsche Ukraine spent an average of 45.1 h per employee in 2021 to help him develop the necessary competencies in the context of the new strategy. The researched enterprise Porsche Ukraine prefers teachers and trainers under outsourcing conditions. All courses, such as language, digital literacy, and psychological courses, are conducted by external companies.

Paying for the services of an external teacher usually results in significant savings compared to paying under an organization contract. However, it should be

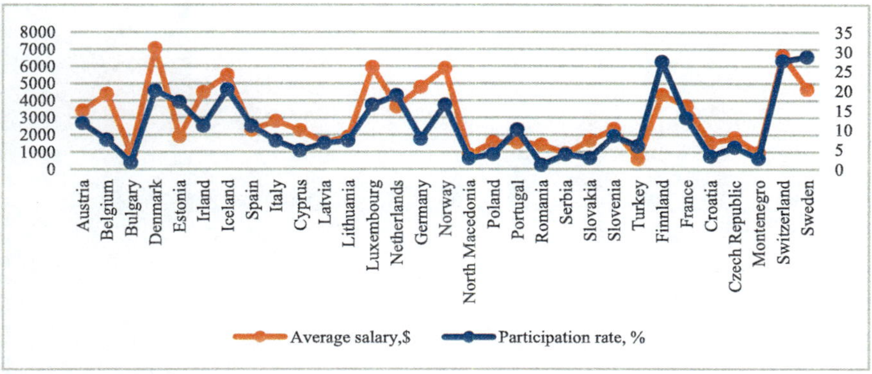

Fig. 29.3 Variation of the average salary and the share of people's participation in adult education by country. (Source: Own results)

taken into account that an external teacher may require a separate fee for the development of the educational event itself as well as for the creation of materials and slides for it. It should also not be forgotten that the guarantee of the quality of the educational event in this case is based only on the authority of the teacher, and in the previous case, the guarantor is the provider company (Didenko et al., 2021; Tolokonnikova & Dashko, 2022; Didenko et al., 2020b).

In the course of the research, we hypothesized the existence of a direct relationship between the indicators of net income and the share of expenditure on adult education at the enterprise. Eleven enterprises were selected for evaluation. It was established that the highest share of expenditures on personnel development is 3%, and the lowest in the studied enterprises is 1.85% (Fig. 29.4).

Using the correlation-regression method, the dependence between these indicators was established, and the Pearson correlation coefficient was $R = 0.56$; this indicator shows that there is a direct average relationship. As a result of the analysis, the following equation was obtained: $y = 0.0018x + 2.67$. From it, we can conclude that with an increase in investment in the development of people by 1%, the company will receive 2.67 thousand dollars in profit. Having calculated the coefficient of determination $R^2 = R^2 \times 100$, we found that 31% of the profit depends on investing in personnel development.

A different situation is observed between macro indicators. For the study, a hypothesis was put forward about the relationship between the indicators of the average salary and the percentage of participation of European countries in adult education. In this correlation, the Pearson correlation coefficient was $R = 0.76$. So the higher is the average salary, the more people engage in their own development. The resulting equation is $y = 32.87x + 3510.2$, which means that with an increase in the average salary, the employee will spend 32% more on education. Having

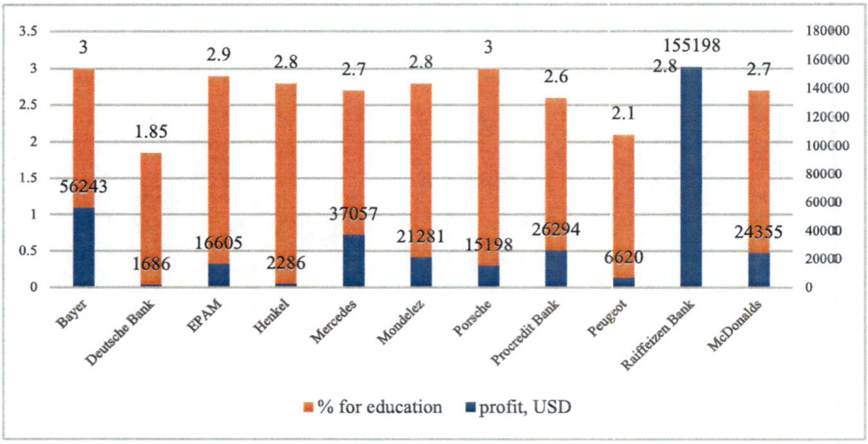

Fig. 29.4 Variation of income and expenses for adult education by enterprises. (Source: Own results)

calculated the coefficient of determination $R^2 = R^2 \times 100$, we found that the average salary depends on the involvement of staff in adult education by 49%, which shows the feasibility of adult education for the development of the enterprise.

29.3 Discussion of Results

The COVID-19 pandemic has affected almost all areas of life. Moreover, it has led to a global failure in the functioning of educational systems around the world. According to UNESCO estimates, about 78% of students did not attend universities, about 56% of enterprises switched to online work due to the pandemic, and as a result – online education (Neuwirth et al., 2021; Smiianov et al., 2020).

It becomes clear that COVID-19 has already impacted on graduate recruitment into the workforce, reducing the number of graduates being hired this year by large firms and small- and medium-sized enterprise. McKinsey recently called COVID-19 the "biggest destroyer of jobs, ever" noting 195 million jobs were lost in Q2 of 2020 alone, with two service industries (accommodation/food; wholesale/retail) accounting for 40% of at risk jobs with 80% of customer service/sales jobs at risk (Zhao & Watterston, 2021). This has had a disproportionate impact on minorities and marginalized communities, increasing existing social cleavages.

Teaching and learning have moved mostly online, and the use of remote video conferencing has become commonplace. After the COVID-19 pandemic, not only businesses but also educational organizations are experiencing significant difficulties in organizing the educational process. The organization of distance learning during this period was brought to the fore and became the main approach that ensures the development of personnel at the enterprise (Samusevych et al., 2021). These changes also affected adult education programs, much of which is based on face-to-face classes.

With the passage of time, it was found many advantages that online education provides. These are as follows:

- Activity: since an adult is fully aware of the importance of the educational process, distance education can help unlock potential and open new opportunities to gain valuable knowledge in any part of the world.
- Openness: each adult student can independently choose the form of obtaining knowledge that is necessary for him. It can be either listening to or viewing lecture material or reading educational literature (Klyuka, 2022).
- Flexibility: which is expressed in the possibility of studying educational materials at a time convenient for an adult student.
- Accessibility: every student who has free access to the Internet has the opportunity to get an education regardless of their place of residence or location.
- Modularity: the construction of the educational process is aimed at obtaining only the knowledge that the student needs; the study of educational material is

structured according to the type of modules, which greatly facilitates the learning process (Dutta et al., 2020).

Analysis of the impact of COVID-19 on adult education in a global dimension shows that the distance form of education remains relevant for most education providers to this day. Therefore, online provision is a key strategy in response to COVID-19 (Volk et al., 2022).

During the analysis of online adult education, we identified two trends. The first is the growing popularity of online education. This form of education will become part of the further development of adult education. In order to ensure the effectiveness of distance education, teachers must master the latest teaching methods, including online teaching.

The next trend is the growing demand for shorter and cheaper educational courses. Considering the fact that the professional market is changing at a rapid pace, adult education should provide such courses that would be short but would precisely satisfy the needs of the consumer (Volk et al., 2022), because technologies develop very quickly and knowledge can become outdated in 6 months.

According to the forecasts of the analytical company Oxford Economics, by 2030 about 20 million factory workers in the world will change jobs. More than 76 million people will lose their jobs by 2023, according to the 2018 World Economic Forum (Bilan et al., 2022). Thus, providers of adult education already need to provide an opportunity for continuous education, which will become a necessary condition for the competitiveness of a specialist (Lockee, 2021; Antonyuk et al., 2021).

Distance education of adults after the COVID-19 pandemic has become a powerful factor of socialization and a means of achieving a certain level of education, general culture, and professional competence.

However, few people say that the implementation of online education requires digital education skills, that is, the ability to work with a computer, in various educational platforms, with various circular tools (Vorontsova et al., 2021). In the world, one of the key problems in adult education remains digital literacy. The ability of all strata of the population to use modern devices: computers and smartphones, especially among the older generation, continues to be important. There is an urgent need for adult education, which would teach the population to use digital tools correctly.

In today's conditions, it is necessary to understand that digital literacy already includes a much wider set of basic skills than 10 years ago (Kaya, 2020). In general, digital education has a very positive effect on the general level of preparation of adults for unpredictable changes in the format of work. Today, important areas of digital education include digital hygiene, personal data protection, ethical behavior in the digital environment, and the basics of handling computer equipment (Kuzior et al., 2022a).

Understanding and being able to recognize phishing links, filter spam, leave passwords, and personal data in forms on Internet sites are the basis of a citizen's digital security.

29.4 Conclusions

Thus, personnel development is one of the important points within the framework of the Sustainable Development Goal No. 4. The analysis showed that not only the company but also the employee is ready to invest in his development, if in the future he will understand what financial and social prospects it can open for him.

A direct relationship was established between indicators of net income on investments in adult education at the enterprise and between indicators of average salary and the share of participation of European countries in adult education. The proposed hypotheses were confirmed as a result of the calculation using the correlation-regression method.

Under the influence of the COVID-19 pandemic, companies have dramatically switched to online work and training, which has created a pressing challenge in training employees online, as well as teaching them digital skills. Therefore, we see this as a further strategy for personnel development and adult education in general, because the time in which we live today can be characterized as a period of change and progress, especially in the fields of development of modern information technology and development of related services based on these technologies.

Acknowledgments The research was funded by a grant from Ministry of Education and Science of Ukraine, "Convergence of economic and educational transformations in the digital society: modeling the impact on regional and national security" (state registration number: 0121 U 109553) and "Reforming the lifelong learning system in Ukraine to prevent labor emigration: a cooperative model of institutional partnership" (state registration number: 0120U102001).

References

Antonyuk, N., Plikus, I., & Jammal, M. (2021). Human capital quality assurance under the conditions of digital business transformation and COVID-19 impact. *Health Economics and Management Review, 2*(3), 39–47. https://doi.org/10.21272/hem.2021.3-04

Artyukhov, A., Dluhopolskyi, O., Vasylieva, T., Lyeonov, S., Dluhopolska, T., & Tsikh, H. (2021a). *Local (university) rankings and quality of education: Identification of publication activity indicators*. Paper presented at the 2021 11th International Conference on Advanced Computer Information Technologies, ACIT 2021 – Proceedings, 246–249. https://doi.org/10.1109/ACIT52158.2021.9548380

Artyukhov, A. Y., Vasylieva, T. A., & Lyeonov, S. V. (2021b). An integrated method for evaluating the quality of education and university performance. *Naukovyi Visnyk Natsionalnoho Hirnychoho Universytetu, 3*, 148–154. https://doi.org/10.33271/nvngu/2021-3/148

Artyukhov, A., Volk, I., Vasylieva, T., & Lyeonov, S. (2021c). The role of the university in achieving SDGs 4 and 7: A Ukrainian case. *E3S Web of Conferences, 250*, 04006. https://doi.org/10.1051/e3sconf/202125004006

Artyukhov, A. E., Volk, I. I., & Vasylieva, T. A. (2022). Agile methodology in higher education quality assurance system for SDGs 4, 8 and 9 achievement: National experience. *CEUR Workshop Proceedings, 3085*, 81–94.

Babchynska, O. I. (2021). Tools for forming a personnel development system of an innovatively active enterprise. *Bulletin Of The Khmelnytskyi National University, 3*, 169–173.

Bilan, S., Šuleř, P., Skrynnyk, O., Krajňáková, E., & Vasilyeva, T. (2022). Systematic bibliometric review of artificial intelligence technology in organizational management, development, change and culture. *Business: Theory and Practice, 23*(1), 1–13. https://doi.org/10.3846/btp.2022.13204

Boeren, E. (2019). Understanding Sustainable Development Goal (SDG) 4 on "quality education" from micro, meso and macro perspectives. *International Review of Education, 65*, 277–294. https://doi.org/10.1007/s11159-019-09772-7

Brimah, B. A., Olanipekun, W. D., Bamidele, A. G., & Ibrahim, M. (2020). Knowledge management and its effects on financial performance: Evidence from Dangote Flour Mills, Ilorin. *Financial Markets, Institutions and Risks, 4*(2), 34–42. https://doi.org/10.21272/fmir.4(2).34-42.2020

Čábelková, I., Blaginin, V., Strielkowski, W., & Platitsyn, A. (2020). Sustainable education: Altering the opinions on renewable energy sources. *E3S Web of Conferences, 208*, 09001. https://doi.org/10.1051/e3sconf/202020809001

Degtjarjova, I., Lapina, I., & Freidenfelds, D. (2018). Student as stakeholder: "voice of customer" in higher education quality development. *Marketing and Management of Innovations, 2*, 388–398. https://doi.org/10.21272/mmi.2018.2-30

Didenko, I., Paucz-Olszewska, J., Lyeonov, S., Ostrowska-Dankiewicz, A., & Ciekanowski, Z. (2020a). Social safety and behavioral aspects of populations financial inclusion: A multicountry analysis. *Journal of International Studies, 13*(2), 347–359. https://doi.org/10.14254/2071-8330.2020/13-2/23

Didenko, I., Volik, K., Vasylieva, T., Lyeonov, S., & Antoniuk, N. (2020b). Migration, environment, and country safety: Analysis of touchpoints. *E3S Web of Conferences, 202*, 03028. https://doi.org/10.1051/e3sconf/202020203028

Didenko, I., Volik, K., Vasylieva, T., Lyeonov, S., & Antoniuk, N. (2021). Environmental migration and country security: Theoretical analysis and empirical research. *E3S Web of Conferences, 234*, 00010. https://doi.org/10.1051/e3sconf/202123400010

Dutta, P., Dutta, U., Hasan, S., Sarkar, S., & Sana, T. (2020). Educate, empower and protect people through timely authentic information: Explore how Bangladeshi newspapers response to COVID-19 pandemic. *SocioEconomic Challenges, 4*(3), 93–103. https://doi.org/10.21272/sec.4(3).93-103.2020

Elfert, M. (2019). Lifelong learning in Sustainable Development Goal 4: What does it mean for UNESCO's rights-based approach to adult learning and education? *International Review of Education, 65*, 537–556. https://doi.org/10.1007/s11159-019-09788-z

Guliyeva, A., Aliyeva, T., Rzayeva, U., & Huseynova, R. (2021). Lifelong education as a challenge of digital era. *Turkish Journal of Computer and Mathematics Education (TURCOMAT), 12*(6), 1282–1289. https://doi.org/10.17762/turcomat.v12i6.2468

Hanić, A., & Jevtić, D. (2020). Human resource management between economy and ethics – Research of Serbia and Bosnia and Hercegovina. *Business Ethics and Leadership, 4*(3), 127–136. https://doi.org/10.21272/bel.4(3).127-136.2020

Kaya, H. D. (2020). Business friendliness, firm performance and owner's optimism. *Financial Markets, Institutions and Risks, 4*(3), 13–23. https://doi.org/10.21272/fmir.4(3).13-23.2020

Khoshnava, S. M., Rostami, R., Zin, R. M., Štreimikienė, D., Yousefpour, A., Strielkowski, W., & Mardani, A. (2019). Aligning the criteria of green economy (GE) and sustainable development goals (SDGs) to implement sustainable development. *Sustainability, 11*(17), 4615. https://doi.org/10.3390/su11174615

Klyuka, A. I. (2022). Staff motivation as a factor in improving the efficiency of the education system: Qualifying work for obtaining a bachelor's degree in the specialty 053 "Psychology".

Kotenko, S., Kobushko, I., Heiets, I., & Rusanov, O. (2021). KPI model impact on employee motivation and competitiveness of private healthcare facilities. *Health Economics and Management Review, 2*, 31–42. https://doi.org/10.21272/hem.2021.2-04

Kuzior, A., & Kwilinski, A. (2022). Cognitive technologies and artificial intelligence in social perception. *Management Systems in Production Engineering, 30*(2), 109–115. https://doi.org/10.2478/mspe-2022-0014

Kuzior, A., Kettler, K., & Rąb, Ł. (2022a). Digitalization of work and human resources processes as a way to create a sustainable and ethical organization. *Energies, 15*(1), 172. https://doi.org/10.3390/en15010172

Kuzior, A., Kettler, K., & Rabą, Ł. (2022b). Great Resignation—Ethical, cultural, relational, and personal dimensions of generation Y and Z employees' engagement. *Sustainability, 14*(11), 6764. https://doi.org/10.3390/su14116764

Liuta, O., Lieonov, S., Artyukhov, A., Sushko-Bezdenezhnykh, M., & Dluhopolskyi, O. (2021). Student survey as a tool for quality assurance in higher education: The case of Ukrainian university. *Naukovyi Visnyk Natsionalnoho Hirnychoho Universytetu, 2021*(4), 158–164. https://doi.org/10.33271/nvngu/2021-4/158

Lockee, B. B. (2021). Online education in the post-Covid era. *Nature Electronics, 4*(1), 5–6. https://doi.org/10.1038/s41928-020-00534-0

Lyeonov, S., Vasilyeva, T., Bilan, Y., & Bagmet, K. (2021). Convergence of the institutional quality of the social sector: The path to inclusive growth. *International Journal of Trade and Global Markets, 14*(3), 272–291. https://doi.org/10.1504/IJTGM.2021.115712

Mercado, M. P. S. R., & Vargas-Hernández, J. G. (2019). Analysis of the determinants of social capital in organizations. *Business Ethics and Leadership, 3*(1), 124–133. https://doi.org/10.21272/bel.3(1).124-133.2019

Meresa, M. (2019). The Effect of Strategic Management Practices on the institutional Performance; the case of Dedebit credit and saving institution in Eastern Tigray. *SocioEconomic Challenges, 3*(3), 80–97. https://doi.org/10.21272/sec.3(3).80-97.2019

Mujtaba, B. G., & Senathip, T. (2020). Workplace mobbing and the role of human resources management. *Business Ethics and Leadership, 4*(1), 17–34. https://doi.org/10.21272/bel.4(1).17-34.2020

Neuwirth, L. S., Jović, S., & Mukherji, B. R. (2021). Reimagining higher education during and post-Covid-19: Challenges and opportunities. *Journal of Adult and Continuing Education, 27*(2), 141–156. https://doi.org/10.1177/1477971420947738

Novikov, V. (2021). Intercept of financial, economic and educational transformations: Bibliometric analysis. *Financial Markets, Institutions and Risks, 5*(2), 120–129. https://doi.org/10.21272/fmir.5(2).120-129.2021

Plys, K. S. (2021). Continuous education as a tool for management and development of intellectual capital of the Enterprise. *Bulletins of Higher Educational Institutions. Series: Economy, Finance And Production Management, 1*(47), 71–79.

Podchalymova, G. N., Ilyina, I. V., & Belova, S. N. (2021). The strategy of development of the adult continuing education system as a mechanism for ensuring the quality of education. *Pedagogical Education and Science, 1*, 63–70.

Polyakov, M., Bilozubenko, V., Korneyev, M., & Shevchenko, G. (2019). Selection of parameters for multifactor model in the knowledge economy marketing (country level). *Innovative Marketing, 15*(1), 89–99. https://doi.org/10.21511/im.15(1).2019.08

Polyakov, M., Bilozubenko, V., Korneyev, M., & Nebaba, N. (2020). Analysis of key university leadership factors based on their international rankings (QS world university rankings and times higher education). *Problems and Perspectives in Management, 18*(4), 142–152. https://doi.org/10.21511/ppm.18(4).2020.13

Samusevych, Y. V., Novikov, V. V., Artyukhov, A. Y., & Vasylieva, T. A. (2021). Convergence trends in the "economy – education – digitalization – national security" chain. *Naukovyi Visnyk Natsionalnoho Hirnychoho Universytetu, 6*, 177–183. https://doi.org/10.33271/NVNGU/2021-6/177

Sardak, S., Korneyev, M., Dzhyndzhoian, V., Fedotova, T., & Tryfonova, O. (2018). Current trends in global demographic processes. *Problems and Perspectives in Management, 16*(1), 48–57. https://doi.org/10.21511/ppm.16(1).2018.05

Serpeninova, Y., Makarenko, I., Plastun, A., Babko, A., & Gasimova, G. (2020). Mapping of the responsible investments instruments in SDG 3 'good health and well-being' financing: EU and US experience. *Health Economics and Management Review, 1*(1), 106–115. https://doi.org/10.21272/hem.2020.1-10

Shkarlet, S., Kholiavko, N., & Dubyna, M. (2019). Information economy: Management of Educational, innovation, and research determinants. *Marketing and Management of Innovations, 3*, 126–141. https://doi.org/10.21272/mmi.2019.3-10

Skrynnyk, O., & Vasilyeva, T. (2020). Comparison of open learning forms in organizational education. *CEUR Workshop Proceedings, 2732*, 1314–1328.

Smiianov, V. A., Vasilyeva, T. A., Chygryn, O. Y., Rubanov, P. M., & Mayboroda, T. M. (2020). Socio-economic patterns of labor market functioning in the public health: Challenges connected with covid-19. *Wiadomosci Lekarskie, 73*(10), 2181–2187. https://doi.org/10.36740/wlek202010114

Strielkowski, W., Kiseleva, L. S., & Sinyova, A. Y. (2020). Trends in international educational migration: A case of Finland. *Integration of Education, 24*(1), 32–49. https://doi.org/10.15507/1991-9468.098.024.202001.032-049

Tolokonnikova, A. V., & Dashko, I. M. (2022). Organization and objectives of staff training at the enterprise. *Modern Youth in the World of Information Technologies: Materials, 23*(4), 12–22.

Volk, I., Artyukhov, A., Vasylieva, T., & Hordiienko, V. (2021). *Quality of education and socio-economic development: Decision making approach for assessing system performance*. Paper presented at the 2021 IEEE 8th International Conference on Problems of Infocommunications, Science and Technology, PIC S and T 2021 – Proceedings, 256–260. https://doi.org/10.1109/PICST54195.2021.9772207

Volk, I., Artyukhov, A., & Lyeonov, S. (2022). *Modeling of information system for blended education quality assurance and socio-economic impact*. Paper presented at the Proceedings – 16th International Conference on Advanced Trends in Radioelectronics, Telecommunications and Computer Engineering, TCSET 2022, 590–593. https://doi.org/10.1109/TCSET55632.2022.9766905

Voronkova, O., Hordei, O., Barusman, A. R. P., & Ghani, E. K. (2019). Social integration as a direction for humanization of economic relations and improvement of social welfare. *SocioEconomic Challenges, 3*(4), 52–62. https://doi.org/10.21272/sec.3(4).52-62.2

Vorontsova, A. S., Lieonov, S. V., Vasylieva, T. A., & Artiukhov, A. Y. (2018). Innovations in the financing of lifelong learning system: Expenditure optimization model. *Marketing and Management of Innovations, 2*, 218–231. https://doi.org/10.21272/mmi.2018.2-18

Vorontsova, A., Vasylieva, T., Lyeonov, S., Artyukhov, A., & Mayboroda, T. (2021). *Education expenditures as a factor in bridging the gap at the level of digitalization*. Paper presented at the 11th International Conference on Advanced Computer Information Technologies, ACIT 2021 Proceedings, 242–245. https://doi.org/10.1109/ACIT52158.2021.9548338

Yarovenko, H., Bilan, Y., Lyeonov, S., & Mentel, G. (2021). Methodology for assessing the risk associated with information and knowledge loss management. *Journal of Business Economics and Management, 22*(2), 369–387. https://doi.org/10.3846/jbem.2021.13925

Zhao, Y., & Watterston, J. (2021). The changes we need: Education post Covid-19. *Journal of Educational Change, 22*(1), 3–12. https://doi.org/10.1007/s10833-021-09417-3

Chapter 30
Managing Green Innovations in European Union: Should We Expect Updates in the Concepts of Leadership, Entrepreneurship, and Sustainable Development?

Anargul Belgibayeva, Liudmyla Saher, and Ihor Vakulenko

Abstract The energy crisis, which is unfolding in the world against the background of the military conflict in Ukraine, is affecting all sectors of the world economy, not limited to the energy sector. Due to the shortage of energy resources, there are redistributions of state and local budgets and company budgets. In many cases, such redistribution can lead to delays, reductions, or cancellations of investments in strategic projects. Many such projects are innovative. Accordingly, the current situation may harm the innovative development of the economy, individual industries, or companies. Due to the complex logistics links inherent in today's economy, one innovative company's difficulties can threaten the entire logistics chain. Innovations aimed at achieving sustainable development goals are among the most vulnerable in the current situation. Green innovation can be most negatively affected by global adverse processes.

This study analyzes the expected impact of the energy crisis on the logistics chains of innovation products in the European Union. It identifies vulnerabilities in innovation policy that may be significantly affected by economic difficulties due to the energy crisis.

However, scientists also see the crisis as an opportunity to develop individual industries or companies. That is why the study, along with the negative conse-

A. Belgibayeva
Sh. Ualikhanov Kokshetau University, Kokshetau, Kazakhstan

L. Saher (✉) · I. Vakulenko
Sumy State University, Sumy, Ukraine

The Institute of Economic Research of the Slovak Academy of Sciences, Bratislava, Slovak Republic
e-mail: l.sager@kmm.sumdu.edu.ua; vakulenko@ms.sumdu.edu.ua

W. Strielkowski (ed.), *Leadership, Entrepreneurship and Sustainable Development Post COVID-19*, Springer Proceedings in Business and Economics, https://doi.org/10.1007/978-3-031-28131-0_30

quences of the crisis, identifies promising areas that can benefit from a generally unfavorable economic situation.

Assessing the impact of the deteriorating economic situation of European countries due to the Russian-Ukrainian war, which causes, among other things, the global energy crisis, on the prospects of innovative development, updates in the concepts of leadership, entrepreneurship, and sustainable development were carried out using economic and mathematical methods of data analysis obtained from World Data Bank, Eurostat, and other open access sources providing relevant data on innovation. The results of the quantitative analysis were obtained using IBM SPSS Statistics software.

Keywords Green economy · Green innovations · Innovation policy · Sustainable Development Goals · European Union · Leadership · Entrepreneurship

JEL Classification O32 · Q28 · Q48

30.1 Introduction

After the beginning of Russia's military aggression against Ukraine which began in February 2022, the world's economic and political situation, particularly in the European Union (EU), changed significantly. The most significant factors that affected the deterioration of business conditions and social well-being were (a) disruption of logistics chains as a result of hostilities in Europe; (b) sanctions imposed against Russia by the member states of the European Union; and (c) Russia's use of energy as a weapon to put pressure on the European Union.

The most noticeable effect of Russian aggression against Ukraine on the countries of the European Union was the increase in the cost of energy resources. This led to a rise in the prices of goods and services and an increase in public expenditures of the European Union member states and the EU as a whole.

It is worth noting that the situation observed now is predictable. Its reasons lie in the fact that for several decades, the countries of the European Union have been building relations with Russia as a critical supplier of energy resources. During this time, Russia has become an irreplaceable supplier for some EU countries. For some, it is the main one, so much so that supplies cannot replace its share from other countries in the short term.

Russia was preparing the conditions for blackmailing the EU through the supply (or its termination) of energy resources in advance. Under sanctions since the occupation of Crimea and the creation of the conflict in Donbas, Russia used its dominant role as an energy supplier to the EU to launch Nord Stream 2. This pipeline was supposed to increase the EU's energy dependence on Russia (Balitskiy et al., 2016; Strielkowski et al., 2021).

The energy issue lies on the surface. It attracts the attention of politicians and the public. However, only a component of more profound consequences will be visible shortly. An important question that needs to be answered is the expected impact of the energy crisis and the disruption of a significant number of logistics chains on European business. How strong will this blow be for the business, how will it react to changes in the external environment, and whether it will be able to implement innovative projects? The latter is significant because, as is known, the crisis not only has negative consequences and worsens the situation. Many scientists and entrepreneurs regard the crisis as a time of opportunity that should be used for business development because competitors may lose market positions or the ability to compete effectively for a while. Innovation occupies a prominent place in these processes. However, what to do in the situation that has developed with system innovations? Projects involve many participants such as technology developers, component manufacturers, service organizations, and logistics companies. Will companies be able to keep existing projects and implement them? Is it possible to expect a reduction in innovative activity? This crisis will be a window of opportunity or dark times for European businesses as well as climate change issue (Lu & Strielkowski, 2022). These questions must be answered based on the actions taken by companies and the authorities of the EU countries (including the governing bodies of the EU) to support economic development and social stability in the region.

It is essential to investigate whether governments and companies will be able to support innovative activities aimed at sustainable development. It largely depends on whether the priorities will be changed from the long-term trend of supporting sustainable development technologies to the need to minimize the negative consequences of the energy crisis when the price is the rejection of projects aimed at achieving the goals of sustainable development. Any changes in European development financial institutions (DFIs) will affect the prospects of innovative activities. Similarly, the reduction of investment opportunities for enterprises and organizations will also affect the number of innovative projects that will be financially supported. Taking into account changes in the investment of innovative projects by enterprises and organizations and DFIs is important in their relationship because the efficiency of investment increases when they act together (Clò et al., 2022).

The energy crisis, which entails heavy economic consequences for the European Union countries, exposes the existing problems in the policy of the European Union and individual Member States regarding the support of the innovation and sustainable development (Simionescu et al., 2020).

The analysis of the steps already taken by the government and businesses shows how viable the innovative development model built in Europe is. In addition, based on such an analysis of the situation, it is possible to identify those sectors of the economy that can develop in the established economic conditions and profit from the case, which is generally characterized as complex and negative.

30.2 Materials and Methods

The assessment of the impact of the deterioration of the economic situation of European countries as a result of the Russian-Ukrainian war, which causes, among other things, the global energy crisis, on the prospects of innovative and sustainable development was carried out using the analysis of scientific literature and economic and mathematical methods of data analysis from open sources, in particular, obtained from the World Data Bank, Eurostat, and other sources providing relevant statistics (Table 30.5). The results of the quantitative analysis of the interdependencies between the indicators key to the issue under investigation and the cost of energy resources as a catalyst for economic changes were obtained using the IBM SPSS Statistics software.

30.3 Results and Discussions

Investment is a key factor in increasing the efficiency of natural resource use (Chygryn & Krasniak, 2015; Gruzina et al., 2021; Lehmann et al., 2022) and large-scale application of the principles of additive economics (Melnyk et al., 2022).

To solve specific problems by investing in projects aimed at ensuring sustainable development, it is advisable to broadly consider the forms of attracting financing (Keliuotytė-Staniulėnienė & Daunaravičiūtė, 2021). The study (Lukšić et al., 2022) says that "environmental, social, governance/sustainability-linked bonds and debt-for-climate swap investments as innovative financial instruments that hold promise in leveraging additional finance to support the sustainability goals." Any investment mechanism should be implemented according to the model of innovative sustainable development (Kozmenko et al., 2010) or the national innovation system, which is based on the relationship between science, education, and innovation (Khanin et al., 2019).

It is worth it because research and innovation investments depend on natural resources' availability. Countries rich in natural resources spend less on developing innovative technologies than those with limited access to natural resources and have to pay a high price for them (Kamguia et al., 2022). Such a situation today should affect the behavior of the governments of the countries of the European Union because, in the conditions of the increase in the cost of energy resources and their shortage, it is necessary to find ways out of the situation to maintain the appropriate level of social and economic development (Vasylieva et al., 2020). However, two aspects are of concern. First, it takes time to develop and implement innovative technologies. Sometimes this time is stretched over the years. Second, today it is not clear which technology funds will be invested. The priority area for investments will be green energy, which can rid the countries of the European Union of their dependence on Russia. Or, perhaps, priority investment areas will be projects to diversify the supply of fossil energy resources. That is, the current situation will only increase investment in traditional energy. The last option is highly undesirable for sustainable development. Another factor adds pessimism to the rapid future achievement of the desired indicators of sustainable development goals. We are talking about the instability of access to energy resources. According to Shinwari et al. (2022) and Karaoulanis (2018), investment in renewable energy and innovative technologies is key to sustainable development. However, the volume of such investments is lower if there is no constant access to energy resources or the amount of available energy resources changes unpredictably. We can observe exactly such a situation in the European energy market. It is unknown what number of resources will be available to the countries of the European Union shortly. The reason for this is the uncertainty regarding the volume of Russian energy supplies to Europe, on the one hand, and the unknown volume and price of energy resources that will be supplied to the European market by non-Russian suppliers. Accordingly, in such conditions, it can be predicted that optimistic forecasts regarding investment in advanced technologies, in particular in renewable energy, will not come true.

Such a situation is negative not only for energy (Samusevych et al., 2021a, b). Other innovations aimed at sustainable development are also under attack. And this, in turn, harms the development of a circular economy (Platon et al., 2022; Mishenin et al., 2020; Polyakov et al., 2021) as on both at the global level and at the level of an individual enterprise (Kuzior et al., 2022a, b).

The prospect of a decrease in investment in large-scale and global innovation projects causes concern. Governments are more inclined to invest in global and risky projects than businesses (Orlando et al., 2022). This is due to the availability of the necessary resources and the need to ensure economic, energy, and environmental security (Štreimikienė et al., 2022). An essential factor that restrains the investment of private companies in innovative projects is the obsolescence of innovations. When innovation activity is too intense, new technologies are created very often. Too often for the market, because it makes no sense to invest in innovative technology, spending significant money on its development, if it will become obsolete faster than the maximum effect from its use will be obtained. Competitors who introduce newer technology will be at an advantage in this case. Therefore, today's situation, where we assume a reduction in investment in innovative projects and technologies, can help to spread already existing technologies. Paradoxically, this can positively affect the results of innovative activities of enterprises. A study by (Ye et al., 2022) examines this issue in detail. Also, reducing process innovations, particularly in renewable energy, can positively affect the implementation of system innovations (Li & Shang, 2022). Such trends may lead to a change in the company's policy on innovation leadership. After all, what will come to the fore is not the ability to quickly bring innovative products or technologies to the market that are not of fundamental importance but the ability to support previously implemented system innovation solutions. This point of view is entirely consistent with the scientific results obtained by (Ye et al., 2022) and research (Starchenko et al., 2021a, b), which asserts the priority for entrepreneurs of the model of achieving long-term sustainable development, instead of short-term profit maximization. This approach corresponds to the concept of environmental responsibility of firms (Taliento & Netti, 2020; Serpeninova et al., 2020); although the number of innovative products and technologies waiting to enter the market is decreasing, the long-term environmental impact can be significant. Accordingly, the strategy of companies claiming leadership in their markets may become different compared to the one we have observed in recent years.

However, when determining the prospects for investment and innovative technologies in the conditions that have developed in the economy of the European Union as a result of Russian aggression against Ukraine, it is essential to take into account the impact of state subsidies on the costs of small- and medium-sized enterprises for design and research work, the result of which should be innovations. At the same time, it is worth considering not only the effectiveness of state support in financing (Kozmenko & Vasyl'yeva, 2008), which private companies master. It is worth considering whether such government support encourages the company to invest its funds in continuing research and design work on innovative products and technologies. The study's results (Kiman & Jongmin, 2022) show a linear effect of government subsidies on research and development investment for small- and medium-sized enterprises. The conclusions of this study indicate that although state financial support has a significant impact on innovative activity, it does not fully solve the problem of stimulating research and development costs in the production sector. Accordingly, under the existing conditions, we can state that the reduction of state support for innovative activities of private companies will lead to a quantitative reduction of innovations that have entered the market. It will also harm policies to address regional inequality (Singh, 2018). At the same time, according to the reasoning above, we can assume that this will be an additional factor that will strengthen the market positions of enterprises that have recently implemented system innovation projects. However, at the level of the state or the European Union, this will lead to aggravation of the functional links between the level of competitiveness and the country's innovative capacity (Samoilikova, 2020).

However, the ability to take advantage of new trends in the market to benefit one's company depends not only on global factors but also on top management. Research findings (Zhao & Yuan, 2022) suggest that top management characteristics affect the intensity of R&D investment. The effectiveness of innovations largely depends on this. At the same time, it is essential to pay attention to the factors that most significantly influence the behavior of the company's top managers, namely, that equity incentives play a more significant role than salary incentives. Considering the role of management as a factor of successful activity in challenging economic conditions, it is advisable to pay attention to ensuring conditions for ease of starting a business (Kaya, 2020). This will support small businesses and contribute to the growth or preservation of the necessary level of economic activity, including the production of innovative products and the development of green entrepreneurship (Starchenko et al., 2021a, b; Lyeonov et al., 2021). This will act as a compensatory mechanism against the background of the expected reduction in investment in innovation by medium and large businesses. In addition, some industries were affected by COVID-19 and suffered losses even before the deterioration of the social and economic situation in the EU due to the large-scale Russian invasion of Ukraine in February 2022, in particular, the tourism industry (Kuzior et al., 2021).

Underinvestment in energy innovation leads to increased time to market for advanced innovation technologies (Han et al., 2022). This restrains the development of renewable energy and smart energy grids as a progressive direction of energy sector development (Kolosok et al., 2021). Also, inconsistency in energy policy will affect climate policy outcomes (Bhowmik, 2019; El Amri et al., 2020), considering the relationship between energy consumption and carbon dioxide emissions (Kuzior et al., 2022a, b). This leads to the deterioration of the quality of the environment and causes a chain reaction, where the resources needed for further development migrate to countries or territories with better living conditions (Didenko et al., 2020, 2021; Letunovska et al., 2021), to which we should also add the institutional component and transparency of the business environment (Bilan et al., 2020).

It is necessary to get clarity on the issues of the future policy on overcoming the energy crisis and the measures that will be taken. It cannot be expected that private companies will increase or at least continue to invest in innovation projects at the current (as of the end of 2021 – beginning of 2022) level, because uncertainty leads to a reduction in investment in innovation (Kyaw, 2022). As a compensatory mechanism for reducing investments in innovative activities by private companies, the intensification of the efforts of international institutions can be considered. However, they have limited potential (Bhandari & Bhattarai, 2017).

The cooperation of enterprises from different industries allows for an increase in the efficiency of the implementation of innovative products and technologies (Hyde & Fu, 2022) and the implementation of principles of managing sustainable enterprise development (Kasych & Vochozka, 2017; Samusevych et al., 2021a, b; Vysochyna et al., 2020). In addition, creating clusters, for example, medical engineering clusters, can have a positive effect (Zaloha et al., 2020). There is a wide field of opportunities for companies in various sectors. After all, the Russian-Ukrainian war affected logistical connections, which caused damage to a significant number of enterprises and their market position and ability to compete in the market.

To understand the impact of the energy crisis on the prospects of economic development of the EU, innovative activity, and the scale of economic changes, it is advisable to carry out a series of simple calculations to establish a correlation between the critical issues considered in this work. In particular, it is crucial to determine how the cost of energy resources (natural gas and oil) affects gross national income (Table 30.1).

The calculation of Pearson's pairwise correlation coefficient shows a strong relationship between gross national income and the price of natural gas. The same conclusion is valid for the relationship between gross national income and oil price because the coefficient's value ranges from 0.7 to 0.9. Moreover, the established relationship is linear.

It is also worth noting that the cost of energy resources significantly affects investments in clean energy (Table 30.2), which should replace fossil fuels that are increasing in price (Table 30.2 and 30.3). However, we can now assume that this dependence will not work in connection with the diversification of the supply of fossil fuels in the EU.

At the same time, it is worth noting that the price of natural gas is a stronger incentive for investing in renewable energy generation than changes in oil prices.

A similar conclusion can be drawn regarding the impact of oil and natural gas price fluctuations on the volume of investments in development and research in renewable energy (Table. 30.4).

Table 30.1 The relationship between gross national income and the price of energy resources

Correlations				
		Gross national income (US dollars per capita)	Europe Brent Spot Price FOB (dollars per barrel)	Gas prices for household consumers (US dollars per million metric British thermal unit)
Gross national income (US dollars per capita)	Pearson correlation	1	−.090	.037
	Sig. (two-tailed)		.722	.886
	N	18	18	18
Europe Brent Spot Price FOB (dollars per barrel)	Pearson correlation	−.090	1	.746**
	Sig. (two-tailed)	**.722**		.000
	N	18	18	18
Gas prices for household consumers (US dollars per million metric British thermal unit)	Pearson correlation	.037	.746**	1
	Sig. (two-tailed)	**.886**	.000	
	N	18	18	18

**Correlation is significant at the 0.01 level (two-tailed)

Table 30.2 Impact of oil and natural gas prices on clean energy investments

Correlations				
		New investment in clean energy in Europe (billion US dollars)	Europe Brent Spot Price FOB (dollars per barrel)	Gas prices for household consumers (US dollars per million metric British thermal unit)
New investment in clean energy in Europe (billion US dollars)	Pearson correlation	1	.628*	.566*
	Sig. (two-tailed)		.012	.028
	N	15	15	15
Europe Brent Spot Price FOB (dollars per barrel)	Pearson correlation	**.628***	1	.916**
	Sig. (two-tailed)	.012		.000
	N	15	15	15
Gas prices for household consumers (US dollars per million metric British thermal unit)	Pearson correlation	**.566***	.916**	1
	Sig. (two-tailed)	.028	.000	
	N	15	15	15

*Correlation is significant at the 0.05 level (two-tailed)
**Correlation is significant at the 0.01 level (two-tailed)

Table 30.3 Impact of oil and natural gas prices on investments in renewable energy production capacity

Correlations				
		Global renewable energy capacity investment (billion US dollars)	Europe Brent Spot Price FOB (dollars per barrel)	Gas prices for household consumers (US dollars per million metric British thermal unit)
Global renewable energy capacity investment (billion US dollars)	Pearson correlation	1	.221	.068
	Sig. (two-tailed)		.411	.801
	N	16	16	16
Europe Brent Spot Price FOB (dollars per barrel)	Pearson correlation	.221	1	.891**
	Sig. (two-tailed)	**.411**		.000
	N	16	16	16
Gas prices for household consumers (US dollars per million metric British thermal unit)	Pearson correlation	.068	.891**	1
	Sig. (two-tailed)	**.801**	.000	
	N	16	16	16

**Correlation is significant at the 0.01 level (two-tailed)

Table 30.4 Impact of oil and natural gas prices on investments in development and research in the field of renewable energy

Correlations				
		Renewable energy R&D (billion US dollars)	Europe Brent Spot Price FOB (dollars per barrel)	Gas prices for household consumers (US dollars per million metric British thermal unit)
Renewable energy R&D (billion US dollars)	Pearson correlation	1	.144	−.043
	Sig. (two-tailed)		.595	.876
	N	16	16	16
Europe Brent Spot Price FOB (dollars per barrel)	Pearson correlation	.144	1	.891**
	Sig. (two-tailed)	**.595**		.000
	N	16	16	16
Gas prices for household consumers (US dollars per million metric British thermal unit)	Pearson correlation	−.043	.891**	1
	Sig. (two-tailed)	**.876**	.000	
	N	16	16	16

**Correlation is significant at the 0.01 level (two-tailed)

30.4 Conclusions

The economic crisis, a consequence of Russia's invasion of Ukraine, caused many adverse effects that will affect the development of the EU economy in the future. Along with the breakdown of logistics links that have affected competitive markets and forced companies to change their competitive strategies, the energy crisis has created some challenges. As a central sector of the economy, the rise in the cost of energy resources has triggered a chain reaction. Investment models of companies are subject to change. The reduction of private investment volumes will lead to a change in priorities in innovation policy. Instead of process innovations, system innovations will become more effective. In such conditions, companies that conducted large-scale innovative projects before the crisis will have a competitive advantage because they will be able to receive a more significant economic benefit due to the weakening of the effect of innovation obsolescence.

Despite the observed trend in the key economic sector – energy – to intensify investment activities in the development and implementation of renewable energy technologies in response to the increase in the price of energy resources, it may not be valid shortly if the investment resources of governments and companies are directed to the diversification of the supply of fossil energy resources.

Acknowledgments The research was funded by the grant from Ministry of Education and Science of Ukraine, 0122 U000780, "Cognitive model of innovations' commercialization in the conditions of Industry 4.0: intellectual capital protection, marketing and communications"; 0122 U000769 "Transfer of green innovations in the energy sector of Ukraine: a multiplicative stochastic model of the transition to a carbon-neutral economy."

Appendix

Table 30.5 Output dataset

	Gross national income (US dollars per capita)	New investment in clean energy in Europe (billion US dollars)	Global renewable energy capacity investment (billion US dollars)	Renewable energy R&D (billion US dollars)	Renewable energy investment (billion US dollars)	Europe Brent Spot Price FOB (dollars per barrel)	Gas prices for household consumers (US dollars per million metric British thermal unit)
2004	25,854,569	30,100	40,100	3600	23,300	38,260	3755
2005	26,813,959	38,900	60,400	3900	31,600	54,570	5915
2006	29,170,877	53,100	88,200	4400	40,700	65,160	8213
2007	31,059,795	75,200	120,200	5000	64,500	72,440	8143
2008	32,251,130	88,300	155,700	6100	79,100	96,940	13,138
2009	31,849,415	91,300	146,500	8700	76,500	61,740	8855
2010	32,962,420	123,100	213,000	8600	112,200	79,610	8223
2011	34,461,166	137,800	264,700	9100	131,700	111,260	10,597
2012	34,892,986	98,400	239,900	8900	91,100	111,630	11,981
2013	36,146,071	70,200	211,700	9200	57,700	108,560	11,189
2014	37,091,426	78,500	265,100	8800	68,700	98,970	10,463
2015	38,111,426	73,800	300,300	8500	61,100	52,320	7305
2016	40,610,718	77,700	280,000	9400	71,500	43,640	4353
2017	42,739,576	58,400	315,000	12,000	49,100	54,130	5744
2018	44,751,378	74,500	280,200	13,300	60,800	71,340	7916
2019	46,102,108	–	–	13,400	58,400	64,300	4455
2020	44,756,438	–	–	–	–	41,960	3178
2021	48,454,877	–	–	–	–	70,860	15,907

Source: (International Energy Agency, 2022; European Environment Agency, 2022; Eurostat, 2022; World Bank, 2022)

References

Balitskiy, S., Bilan, Y., Strielkowski, W., & Štreimikienė, D. (2016). Energy efficiency and natural gas consumption in the context of economic development in the European Union. *Renewable and Sustainable Energy Reviews, 55*, 156–168. https://doi.org/10.1016/j.rser.2015.10.053

Bhandari, M. P., & Bhattarai, K. (2017). Institutional architecture for sustainable development (SD): A case study from Bangladesh, India, Nepal, and Pakistan. *SocioEconomic Challenges, 1*(3), 6–21. https://doi.org/10.21272/sec.1(3).6-21.2017

Bhowmik, D. (2019). Decoupling CO2 emissions in Nordic countries: Panel data analysis. *SocioEconomic Challenges, 3*(2), 15–30. https://doi.org/10.21272/sec.3(2).15-30.2019

Bilan, Y., Srovnalã-KovÃi, P., Streimikis, J., Lyeonov, S., Tiutiunyk, I., & Humenna, Y. (2020). From shadow economy to lower carbon intensity: Theory and evidence. *International Journal of Global Environmental Issues, 19*(1–3), 196–216. https://doi.org/10.1504/ijgenvi.2020.114874

Chygryn, O., & Krasniak, V. (2015). Theoretical and applied aspects of the development of environmental investment in Ukraine. *Marketing and Management of Innovations, 3*, 226–234. https://doi.org/10.21272/mmi.2015.3-20

Clò, S., Frigerio, M., & Vandone, D. (2022). Financial support to innovation: The role of European development financial institutions. *Research Policy, 51*(10), 104566. https://doi.org/10.1016/j.respol.2022.104566

Didenko, I., Volik, K., Vasylieva, T., Lyeonov, S., & Antoniuk, N. (2020). Migration, environment, and country safety: Analysis of touchpoints. *E3S Web of Conferences, 202*, 03028. https://doi.org/10.1051/e3sconf/202020203028

Didenko, I., Volik, K., Vasylieva, T., Lyeonov, S., & Antoniuk, N. (2021). Environmental migration and country security: Theoretical analysis and empirical research. *E3S Web of Conferences, 234*, 00010. https://doi.org/10.1051/e3sconf/202123400010

El Amri, A., Boutti, R., Oulfarsi, S., Rodhain, F., & Bouzahir, B. (2020). Carbon financial markets underlying climate risk management, pricing and forecasting: Fundamental analysis. *Financial Markets, Institutions and Risks, 4*(4), 31–44. https://doi.org/10.21272/fmir.4(4).31-44.2020

Gruzina, Y., Firsova, I., & Strielkowski, W. (2021). Dynamics of human capital development in economic development cycles. *Economies, 9*(2), 67. https://doi.org/10.3390/economies9020067

Han, J., Guo, J., Cai, X., Lv, C., & Lev, B. (2022). An analysis on strategy evolution of research & development in cooperative innovation network of new energy vehicle within policy transition period. *Omega (United Kingdom), 112*, 102686. https://doi.org/10.1016/j.omega.2022.102686

Hyde, D., & Fu, E. (2022). Cross-technology innovation trends and evidence with patent and funding data. *World Patent Information, 70*, 102129. https://doi.org/10.1016/j.wpi.2022.102129

Kamguia, B., Keneck-Massil, J., Nvuh-Njoya, Y., & Tadadjeu, S. (2022). Natural resources and innovation: Is the R&D sector cursed too? *Resources Policy, 77*, 102725. https://doi.org/10.1016/j.resourpol.2022.102725

Karaoulanis, A. (2018). Strategic transformation and innovation towards Blue Ocean creation in a changing corporate reality. *Business Ethics and Leadership, 2*(2), 49–55. https://doi.org/10.21272/bel.2(2).49-55.2018

Kasych, A., & Vochozka, M. (2017). Theoretical and methodical principles of managing enterprise sustainable development. *Marketing and Management of Innovations, 2*, 298–305. https://doi.org/10.21272/mmi.2017.2-28

Kaya, H. D. (2020). Business friendliness, firm performance and owner's optimism. *Financial Markets, Institutions and Risks, 4*(3), 13–23. https://doi.org/10.21272/fmir.4(3).13-23.2020

Keliuotytė-Staniulėnienė, G., & Daunaravičiūtė, K. (2021). The global green bond market in the face of the COVID-19 pandemic. *Financial Markets, Institutions and Risks, 5*(1), 50–60. https://doi.org/10.21272/fmir.5(1).50-60.2021

Khanin, I., Shevchenko, G., Bilozubenko, V., & Korneyev, M. (2019). A cognitive model for managing the national innovation system parameters based on international comparisons (the case of the EU countries). *Problems and Perspectives in Management, 17*(4), 153–162. https://doi.org/10.21511/ppm.17(4).2019.13

Kiman, K., & Jongmin, Y. (2022). Linear or nonlinear? Investigation an affect of public subsidies on SMEs R&D investment. *Journal of the Knowledge Economy, 13*(3), 2519–2546. https://doi.org/10.1007/s13132-021-00823-9

Kolosok, S., Bilan, Y., Vasylieva, T., Wojciechowski, A., & Morawski, M. (2021). A scoping review of renewable energy, sustainability and the environment. *Energies, 14*(15), 4490. https://doi.org/10.3390/en14154490

Kozmenko, S., & Vasyl'yeva, T. (2008). Specialized innovative investment banks in Ukraine. *Banks and Bank Systems, 3*(1), 48–56. https://doi.org/10.21511/bbs.3(1).2008.01

Kozmenko, S., Vasil'yeva, T., & Leonov, S. (2010). The structuring of components of the net domestic product according to the innovation criterion. *Innovative Marketing, 6*(4), 30–41. https://doi.org/10.21511/im.6(4).2010.01

Kuzior, A., Lyulyov, O., Pimonenko, T., Kwilinski, A., & Krawczyk, D. (2021). Post-industrial tourism as a driver of sustainable development. *Sustainability (Switzerland), 13*(15), 8145. https://doi.org/10.3390/su13158145

Kuzior, A., Arefieva, O., Poberezhna, Z., & Ihumentsev, O. (2022a). The mechanism of forming the strategic potential of an enterprise in a circular economy. *Sustainability (Switzerland), 14*(6), 3258. https://doi.org/10.3390/su14063258

Kuzior, A., Postrzednik-Lotko, K. A., & Postrzednik, S. (2022b). Limiting of carbon dioxide emissions through rational management of pro-ecological activities in the context of CSR assumptions. *Energies, 15*(5), 1825. https://doi.org/10.3390/en15051825

Kyaw, K. (2022). Effect of policy uncertainty on environmental innovation. *Journal of Cleaner Production, 363*, 132645. https://doi.org/10.1016/j.jclepro.2022.132645

Lehmann, C., Cruz-Jesus, F., Oliveira, T., & Damásio, B. (2022). Leveraging the circular economy: Investment and innovation as drivers. *Journal of Cleaner Production, 360*, 132146. https://doi.org/10.1016/j.jclepro.2022.132146

Letunovska, N., Lyuolyov, O., Pimonenko, T., & Aleksandrov, V. (2021). Environmental management and social marketing: A bibliometric analysis. *E3S Web of Conferences, 234*, 00008. https://doi.org/10.1051/e3sconf/202123400008

Li, D., & Shang, C. (2022). When does environmental innovation crowd out process innovation? A dynamic analysis. *Managerial and Decision Economics, 43*(6), 2275–2283. https://doi.org/10.1002/mde.3524

Lu, J., & Strielkowski, W. (2022). Russia: Could cryptocurrency mining exploit unsold fuel? *Nature, 605*(7909), 225–225. https://doi.org/10.1038/d41586-022-01267-0

Lukšić, I., Bošković, B., Novikova, A., & Vrbensky, R. (2022). Innovative financing of the sustainable development goals in the countries of the Western Balkans. *Energy, Sustainability and Society, 12*(1), 1–19. https://doi.org/10.1186/s13705-022-00340-w

Lyeonov, S., Pimonenko, T., Chygryn, O., Reznik, O., & Gaynulina, R. (2021). Green brand as a marketing instrument: Principle, features and parameters. *International Journal of Global Energy Issues, 43*(2–3), 147–165. https://doi.org/10.1504/ijgei.2021.115141

Melnyk, L., Matsenko, O., Kubatko, O., Korneyev, M., & Tulyakov, O. (2022). Additive economy and new horizons of innovative business development. *Problems and Perspectives in Management, 20*(2), 175–185. https://doi.org/10.21511/ppm.20(2).2022.15

Mishenin, Y., Klisinski, J., Yarova, I., & Rak, A. (2020). Ensuring healthy environment: Mechanisms of cluster structures development in the field of waste management. *Health Economics and Management Review, 1*(2), 78–90. https://doi.org/10.21272/hem.2020.2-09

Navickas, V., Petroke, I., Baciuliene, V., & Vasylieva, T. (2021). The impact of the sharing economy as an ecosystem on the tourism sector. *Journal of Tourism and Services, 12*(22), 66–88. https://doi.org/10.29036/jots.v12i22.241

Orlando, B., Ballestra, L. V., Scuotto, V., Pironti, M., & Giudice, M. D. (2022). The impact of R&D investments on eco-innovation: A cross-cultural perspective of green technology management. *IEEE Transactions on Engineering Management, 69*(5), 2275–2284. https://doi.org/10.1109/TEM.2020.3005525

Platon, V., Pavelescu, F. M., Antonescu, D., Frone, S., Constantinescu, A., & Popa, F. (2022). Innovation and recycling—Drivers of circular economy in EU. *Frontiers in Environmental Science, 10*, 902651. https://doi.org/10.3389/fenvs.2022.902651

Polyakov, M., Khanin, I., Bilozubenko, V., Korneyev, M., & Shevchenko, G. (2021). Factors of uneven progress of the European Union countries towards a circular economy. *Problems and Perspectives in Management, 19*(3), 332–344. https://doi.org/10.21511/ppm.19(3).2021.27

Samoilikova, A. (2020). Financial policy of innovation development providing: The impact formalization. *Financial Markets, Institutions and Risks, 4*(2), 5–15. https://doi.org/10.21272/fmir.4(2).5-15.2020

Samusevych, Y., Maroušek, J., Kuzmenko, O., Streimikis, J., & Vysochyna, A. (2021a). Environmental taxes in ensuring national security: A structural optimization model. *Journal of International Studies, 14*(2), 292–312. https://doi.org/10.14254/2071-8330.2021/14-2/19

Samusevych, Y., Vysochyna, A., Vasylieva, T., Lyeonov, S., & Pokhylko, S. (2021b). Environmental, energy and economic security: Assessment and interaction. *E3S Web of Conferences, 234*, 00012. https://doi.org/10.1051/e3sconf/202123400012

Serpeninova, Y., Makarenko, I., Plastun, A., Babko, A., & Gasimova, G. (2020). Mapping of the responsible investments instruments in sdg 3 'good health and well-being' financing: EU and US experience. *Health Economics and Management Review, 1*(1), 106–115. https://doi.org/10.21272/hem.2020.1-10

Shinwari, R., Yangjie, W., Payab, A. H., Kubiczek, J., & Dördüncü, H. (2022). What drives investment in renewable energy resources? Evaluating the role of natural resources volatility and economic performance for China. *Resources Policy, 77*, 102712. https://doi.org/10.1016/j.resourpol.2022.102712

Simionescu, M., Strielkowski, W., & Tvaronavičienė, M. (2020). Renewable energy in final energy consumption and income in the EU-28 countries. *Energies, 13*(9), 2280. https://doi.org/10.3390/en13092280

Singh, S. N. (2018). Regional disparity and sustainable development in NorthEastern states of India: A policy perspective. *SocioEconomic Challenges, 2*(2), 41–48. https://doi.org/10.21272/sec.2(2).41-48.2018

Starchenko, L., Lyeonov, S., Vasylieva, T., Pimonenko, T., & Lyulyov, O. (2021a). Environmental management and green brand for sustainable entrepreneurship. *Paper Presented at the E3S Web of Conferences, 234*. https://doi.org/10.1051/e3sconf/202123400015.

Starchenko, L. V., Samusevych, Y., & Demchuk, K. (2021b). Social and eco-friendly enterpreneurship: The keys to sustainability. *Business Ethics and Leadership, 5*(1), 118–126. https://doi.org/10.21272/bel.5(1).118-126.2021

Štreimikienė, D., Samusevych, Y., Bilan, Y., Vysochyna, A., & Sergi, B. S. (2022). Multiplexing efficiency of environmental taxes in ensuring environmental, energy, and economic security. *Environmental Science and Pollution Research, 29*(5), 7917–7935. https://doi.org/10.1007/s11356-021-16239-6

Strielkowski, W., Sherstobitova, A., Rovny, P., & Evteeva, T. (2021). Increasing energy efficiency and modernization of energy systems in Russia: A review. *Energies, 14*(11), 3164. https://doi.org/10.3390/en14113164

Taliento, M., & Netti, A. (2020). Corporate social/environmental responsibility and value creation: Reflections on a modern business management paradigm. *Business Ethics and Leadership, 4*(4), 123–131. https://doi.org/10.21272/bel.4(4).123-131.2020

Vasylieva, T., Machová, V., Vysochyna, A., Podgórska, J., & Samusevych, Y. (2020). Setting up architecture for environmental tax system under certain socioeconomic conditions. *Journal of International Studies, 13*(4), 273–285. https://doi.org/10.14254/2071-8330.2020/13-4/19

Vysochyna, A., Samusevych, Y., & Starchenko, L. (2020). Convergence trends of environmental taxation in European countries. *E3S Web of Conferences, 202*, 03031. https://doi.org/10.1051/e3sconf/202020203031

Yarovenko, H., Bilan, Y., Lyeonov, S., & Mentel, G. (2021). Methodology for assessing the risk associated with information and knowledge loss management. *Journal of Business Economics and Management, 22*(2), 369–387. https://doi.org/10.3846/jbem.2021.13925

Ye, F., Paulson, N., & Khanna, M. (2022). Are renewable energy policies effective to promote technological change? The role of induced technological risk. *Journal of Environmental Economics and Management, 114*, 102665. https://doi.org/10.1016/j.jeem.2022.102665

Zaloha, O., Vakulenko, I., & Jasnikowski, A. (2020). Prerequisites and possibilities of creating a medical engineering cluster as the part of regional territorial marketing. *Health Economics and Management Review, 1*(2), 51–60. https://doi.org/10.21272/hem.2020.2-06

Zhao, M., & Yuan, F. (2022). The top management team and enterprise innovation: An empirical study from growth enterprise market listed companies in China. *Managerial and Decision Economics, 43*(6), 2066–2082. https://doi.org/10.1002/mde.3509

Chapter 31
Commitment to Sustainability: Innovations, Drivers, and Changemakers in the Times of Digitalization in the Post-COVID Era

Maria Skivko, Kirill Gerasimov, and Elena Morozova

Abstract This chapter emphasizes the necessity for changes in modern businesses in the post-COVID era and focuses on possibilities of sustainable commitment in the time of digitalization that was further deepened by the pandemic. Furthermore, it explains the importance of creating the positive environmental impact and implementing innovative ideas in the production process. In particular, this chapter defines the idea of sustainable innovations, analyzes key drivers for sustainable changes, and explains the significance of educating new changemakers. The concluding idea stresses the importance of cooperation of innovations, drivers, as well as changemakers for the transition to sustainability.

Keywords Changemakers · Digitalization · Sustainable innovations · Leadership

JEL Classification L1 · L2 · O3

31.1 Introduction

Today's discussion on climate change is more than just a public speech: the global community needs a concrete plan to provide sustainable development and lower negative environmental impact. By introducing the idea of sustainability in the Brundtland Commission report in 1987 (The Brundtland Report, 1987), World Commission on Environment and Development declared the official necessity of changes at all levels. These changes should approach international organizations,

M. Skivko (✉) · K. Gerasimov · E. Morozova
Samara National Research University, Samara, Russia
e-mail: gerasimov.kb@ssau.ru

W. Strielkowski (ed.), *Leadership, Entrepreneurship and Sustainable Development Post COVID-19*, Springer Proceedings in Business and Economics, https://doi.org/10.1007/978-3-031-28131-0_31

national governments, big corporations, small and medium companies, and civil society (Hermundsdottir & Aspelund, 2022; Jia et al., 2022; Morales et al., 2022).

In particular, businesses meet challenging conditions for sustainable development that includes creating and implementing innovations at different stages of the production process, solving environmental issues, and searching for more sustainable solutions and new ideas. Such contributions are significant for global economy, ecological and social inputs, business development, as well as for the company's reputation and public image. Globalization; different economic, political, and social conditions; technological progress; digitalization resulting in transformation of consumer preferences; and consumer values force modern companies to follow and adapt various sustainable principles. Global and local environmental problems also affect the situation in the global market.

Furthermore, the role of changemakers in sustainable transformation of businesses is crucial: by educating, stimulating, and supporting changemakers it is possible not only to solve several sustainable issues but bring the company to new qualitative level. The transition to sustainability through active and well-educated business leaders opens the positive perspectives for businesses, national governments, and citizens.

Sustainable design continues to advance in response to emerging technologies and insights. Nowadays, all aspects of life are connected to the principles of sustainability, and the pandemic even reinforced that trend (Pelikánová et al., 2021). For example, architects and designers are thinking about the future, and creating buildings that are built around a broad conception of sustainability, a conception that encompasses much more than just improved energy efficiency (Abdelaal & Guo, 2021). In fact, the drive toward sustainability is already starting to change the competitive landscape, forcing companies to rethink their products, technologies, processes, and business models (Korneeva et al., 2021a). Instead of waiting for market changes that will create incentives to pursue sustainability practices, companies are creating these changes that enable new forms of corporate sustainability (Van Zanten & van Tulder, 2021). Governments in many developing countries are becoming environmentally conscious and encouraging companies to implement sustainable products and processes, particularly for those at the bottom of the pyramid. Ongoing changes to companies operating environments, such as the digitization and the advent of new business models, means sustainable innovation is becoming the primary goal for companies of all types and sizes (Korneeva et al., 2021b). After all, innovation, both conventional and sustainable, involves developing new products, services, or processes. Sustainability innovation involves making deliberate changes in the products, services, or processes of the enterprise in order to produce lasting societal and ecological benefits, as well as generating financial profits for the enterprise. In addition, the sustainability-driven innovation involves improving the operations and processes of the firm in order to be more efficient, with a focus on drastically reducing costs and waste (Linde et al., 2021). Sustainability-driven innovation is disruptive, as it may lead to better business models, better processes, streamlining resource flows, reducing waste and costs, and creating new segments entirely, making it more difficult for businesses to protect the status quo. As companies get better at doing so, the experience will take them into the last phase of sustainable innovation, when the effects of the new product or process spread beyond the boundaries of one market. Sustainable innovation means companies are looking at ways in which they can maintain ongoing innovation/improvement to grow the business, gain competitive advantages, and grow their market share (Akter & Haque, 2022). The proper corporate structure can help to ensure innovation is a sustainable practice. In future cycles of the innovation process, systems thinking could further be used to help identify patterns of disconnected efforts within a sustainable innovation ecosystem and attempt to bring them together in order to consolidate their effects and improve progress toward sustainability goals.

All countries worldwide must promote appropriate approaches, such as deploying technology and all of the innovations and services it brings, so as to enable developing and developed countries alike to enter an equitable model of engagement between all stakeholders, at the private and public, as well as at international levels, for all stakeholders at individual, private, public, as well as global. It becomes apparent now that digital technologies may produce innovations for social sustainability, which enable equal opportunities for everyone's involvement in societies and markets, diminishing various kinds of gaps (e.g., cultural, educational, technological) between developed and developing countries, providing products that are built upon users' preferences and satisfaction levels, and creating potential benefits to workers health conditions (Biancone et al., 2021). Sustainability can fuel innovation by introducing new design constraints that shape the way that critical resources – energy, carbon, water, materials, and waste – are used in products and processes. For significant social and environmental innovation to occur, companies need to change their planning practices, moving away from constraints-based to needs-based planning (Strielkowski et al., 2016). Where customer preferences change, demanding more environmentally sound services and products, companies need to come up with new solutions and innovations that address those changing customer demands. The choices made around these new design constraints can very well define if the firm is going the disruptive path versus the sustaining one when it comes to innovation (Carayannis et al., 2015). For instance, if areas for potential collaboration arise during a product design stage, the firm may have opportunities to collaborate in terms of innovations in order to develop products that are a lasting and end-to-end proposition. Organizations that develop larger, strategic plans around sustainability not only drive innovation within organizations that are larger enterprises – including transformations to critical processes – but may also affect what their customers want and how their suppliers work (Strielkowski et al., 2021). To achieve this new frontier, leading organizations are taking a deeper look at the internals of their operations and throughout the supply chain, assessing where they stand, prioritizing initiatives, and then crafting a broad strategy for sustainability that promotes innovation in products and processes in order to meet their goals (Grabs & Carodenuto, 2021). Central to building a sustainable supply chain are operational innovations that drive greater energy efficiency and lower the firm's reliance on fossil fuels. The examples of sustainable innovations mentioned earlier (i.e., biofuels made from coffee waste, smartphones with modular designs, and similar things) illustrate how companies can participate in these types of sustainable innovations.

Therefore, innovation management might need an updated model for different characteristics driving sustainable innovation. Companies engaged in sustainable innovation are thinking in long-term terms, aligning themselves with consumers' demands regarding equitable labor conditions, ecologically sound processes and products, and improvements to communities. These companies realize that consumers assess the extent to which the organization is ethical and sustainable, and they base their decisions on whether to support a firm based on those assessments (Velter et al., 2020). Few innovations, be it compliance with regulations or creating new lines of products, can advance in today's world if companies do not build coalitions with other businesses, NGOs, and governments. In my work in sustainable development, one can witness inadequacies in the use of corporate social responsibility, charities, and social businesses to tackle the underlying causes of serious development challenges, like poverty. This is particularly the case when we talk about the climate crisis, the biggest global challenge that we have to address in the coming years and one that will affect our ability to achieve all 17 Sustainable Development Goals (SDGs) defined by the United Nations.

This chapter explores the ways how businesses can contribute sustainable practices and perform the drivers for transition. Moreover, this chapter analyzes the meaning of sustainable innovations for modern businesses and identifies the challenges on its implementations. Furthermore, this chapter defines key drivers for sustainable development and explains some perspectives to apply changes and innovations. Finally, this chapter emphasizes the necessity to educate modern sustainable leaders to facilitate changes and support the transition to sustainability.

31.2 On the Way to Sustainable Innovations

Innovations applied to sustainable development signify the possibility of better meeting human needs and values by the process of changes. It includes cost-effective strategies, ways to deal with uncertainty, and search for an optimal solution to fulfill the audience expectations. Implementation of innovations has to consider current economic problems and priorities, changing social values, consumer preferences and attitudes, and potential environmental impact of new products or technologies. Additionally, flexibility, creativity, and multidisciplinary approach of business leaders can foster innovations implementation.

One of the core tasks in implementing innovations for business today is to find the ways to decrease the costs and increase the possibilities for introducing changes. Therefore, there is a strong need for appropriate skills and competencies and strategic and systematic approaches in order to consider suitable capacities for different business sizes and reduce risks by significant improvements. Furthermore, it is crucial to review market rules and regulations in order to better implement innovations and more efficient stimulate the consumer interest.

Sustainable innovations can include technologies to reduce energy or water use, waste or emissions production, and air pollution; to apply efficient resource use; and to construct value chain or manage risks (McKinsey, 2021). Among the most efficient and demandable innovations, it is possible to denote, for instance, the use of sustainable energy by the public transportation system, the use of private electric vehicles, cheap energy storage solutions, plastic recycling technologies, and the use of energy-efficient LED lighting. Usually, sustainable innovations include different ideas, concepts, and practices that represent the input to the environmental, economic, and social pillars of sustainable development.

Furthermore, sustainable innovations can relate social changes aiming to improve well-being and welfare of individual and social groups (Loorbach et al., 2017). Various innovations based on activities and ideas with social purpose as affordable housing or medical support, education training, or employment opportunities lead to upgrading the quality of life, both of producers and consumers.

Sustainable innovations are profitable for businesses by achieving economic performance through environmental-friendly solutions and technologies. From local administration to international organizations, different institutions offer benefits to businesses (such as subsidies, tax breaks, grants) from the usage of environmental-friendly solutions. Moreover, these innovations respond to consumer demands for sustainable products and services. With the growth of sustainable development discussion, a new type of consumer that appreciates positive environmental input from businesses determines new, more sustainable and more conscious, consumer attitudes. Additionally, sustainable innovations represent economic development thinking that combines new technologies and innovative processes, environmental protection, production of sustainable products with delivery of social impact, and serving social goals.

Deloitte's innovation branch, Doblin, proposes ten types of sustainable innovations to implement (Doblin, 2021). This practical model allows identifying the potential for business models, products, or marketing. First, this model focuses on the ways how to improve a business model. For example, it concerns a profit model and the opportunities to earn money more sustainably, partnership and networking by creating value or sharing technologies, talents, or resources; existing tangible and intangible company assets as tools, apps, or knowledge and training programs to generate revenue; and the ways to boost efficiency and reduce waste through applying new techniques, equipment, or software.

Second, the model suggests the possibilities to improve the entire product, its quality or characteristics, performance, or idea. It relates the work on product performance by reducing emissions or waste and the product system and its relations to other products. Finally, the model describes the principles to improve the customer experience. It concerns customer service practices to enhance the product system and delight customers; the distribution channels and delivery systems to upgrade customer experience; positive brand performance by providing transparency and creating value; and customer engagement through meaningful and memorable interaction.

31.3 Drivers for Sustainable Changes

It is necessary to investigate the key drivers leading to actual changes among different social, economic, or environmental transformations that occur due to sustainable innovations. Drivers signify those resources or conditions necessary for implementation, development, and business growth. As some scholars mention in their work, drivers can concern business solutions, stakeholders' demands, or current sustainability needs (Littlewood et al., 2018). Drivers for sustainable changes facilitate such issues as general cost reduction, implementation of environmental-friendly or low-carbon technologies, increasing energy efficiency, reducing fossil fuel use, and lowering negative environmental impact. Additionally, drivers assist in enhancing company image and corporate reputation, build trust, decrease costs, increase employee satisfaction, manage risks, and improve operations (Lopes et al., 2022; Salam & Ali, 2020).

As Saeed and Kersten denote, there are primary and secondary drivers, and their influence on business differ due to certain factors (Saeed & Kersten, 2019). The former are different shareholders, suppliers, employees, or financial institutions that directly influence business processes; the latter indirectly affects through reputation, media image, or communities.

Moreover, the business can meet as external (society, markets, governmental regulations) as well as internal (existing resources, corporate culture, company's mission) drivers that lead to strategic decisions and innovative solutions. National and international regulations, standards and certifications, and governmental rules as external drivers can foster or prohibit changes in any business structure. As external social drivers, mass media, NGOs, and communities can affect business by raising consumer awareness and attracting attention to certain sustainable or unsustainable businesses. The external pressure of markets, from stakeholders, competitors, and suppliers, leads the market competition and contributes to consumer demands by facilitating the core changes for sustainable development. The internal drivers comprise business leaders, corporate, and organizational strategy and influence the resource operations (Lozano, 2015). Furthermore, organizational and corporate cultures motivate employees to follow sustainable initiatives and empower to shape the corporate mission and improve the public image.

It is possible to focus precisely on business, sustainability, and key stakeholders' drivers that influence the path to sustainable development. Business drivers include the responses to the climate change demands (physical, institutional, market-based) and opportunities to act, test, and experiment with new business solutions and strategies. For instance, it can concern the changes in the supply chain (Foerstl et al., 2015), process- and product-related transformations, and marketing (Tollin & Christensen, 2019). Moreover, business drivers encourage better choosing sustainable suppliers or challenging suppliers with sustainable regulations to implement environmental laws in the production process. Additionally, these drivers work on reducing waste production and managing better resources, water, and energy use.

Business models that incorporate and adopt sustainable solutions obtain greater chances at the market; these models also can include suitable organizational culture, legal regulations, and coherence between strategy and business model to achieve sustainability (Rauter et al., 2017). Some scholars emphasize the difference in adopting and implementing sustainable innovations in large corporations and small and medium companies (Grigorescu et al., 2017). As more significant polluters, big corporations have more considerable responsibility and probably larger negative environmental impact; at the same time, they also have more opportunities (economic, technological) to provide changes. On the contrary, small and medium companies, having less financial or technical support, are more flexible to faster implement innovations and incorporate changes in the production process.

Sustainability drivers stimulate managers to develop and include some social mission into business processes; it can also include the employees' motivation for a more sustainable lifestyle or a more sustainable corporate culture (Hristov et al., 2022). Some scholars denote that openness for innovations and the ability to deal with ambiguity and creative problem-solving stimulate sustainable thinking and maintain more sustainable organizational practices in the company (Kajzer Mitchell & Walinga, 2017). Moreover, technical, managerial, and marketing skills developed by business leaders also facilitate the implementation of sustainable practices.

Stakeholders represent government, community, investors, customers, suppliers, and employees' demands for sustainable actions. Accordingly, they stimulate business management to follow certain governmental acts and institutional requirements that regulate and define sustainable practices and strategies.

Furthermore, some scholars emphasize the influence of digital and information technologies on implementing of sustainable solutions and define modern drivers for the supply chain (Luthra et al., 2020). In order to create a long-term sustainable relationship between business and stakeholders, it is necessary to organize an efficient collaboration among stakeholders and production. Moreover, it is crucial to maintain proper management support and developed infrastructure to operate existing facilities (Gadekar et al., 2022; Jaiswal & Zane, 2022). Technological advancements and expert knowledge can extend the workforce knowledge and support managerial decisions with actual innovations.

Consumers and consumer demands can play the role of drivers, too. According to The Economist Intelligence Unit, commissioned by WWF, the popularity of consumer online search for sustainable products has increased by 71% globally since 2016 (WWF, 2021). More often, consumers expect from the business a clear position and certain sustainable actions to lower environmental impact and confirm the significant role of the business in the sustainable discussion (The Guardian, 2013; Campaign, 2016). Furthermore, the trend for conscious consumption and sustainable brands expands globally and pretends to stay longer (Insights, 2019).

It is crucial to mention that the range of changes and the speed of innovations implementation caused by any type of drivers differ from one business to another. The reasons are such factors as company size, business sector, national and cultural contexts, institutional regulations, and other economic, political, and social characteristics.

31.4 Educating Changemakers as Modern Sustainable Leaders

Today, it is crucial to educate a new generation of changemakers who will face environmental problems and continue searching for sustainable solutions (Mohamed et al., 2022; Vosen et al., 2022). Many scientists and analysts predict that future generations will experience more often the effects of climate change and the problem of unemployment (Friends of the Earth Policy, 2021). The time of uncertainty, environmental crisis, technological progress, and various social, cultural, and political changes demand for new skills and knowledge to be prepared for the job market. Society will react to those changes by new job positions, as, for instance, green jobs that will require specific skills and experience that differ from the traditional education system.

The creation of new jobs, especially within the sustainable perspective, will provide benefits for the population with job positions as well as for the planet with some sustainable solutions and businesses (Changemakers, 2021). Additionally, it will strengthen the development of social entrepreneurship projects that require those creative and innovative approaches and leadership skills to foster sustainable initiatives. In particular, there is a need for social entrepreneurship leaders who can simultaneously lead the profitable business and generate the positive environmental input.

Educating young generations and supporting youth leaders for sustainable development represent today's crucial global tasks for modern education (Armstrong & Gough, 2019; Ki-Moon, 2016). It is necessary to define and promote the importance of youth leadership and create the possibilities to develop it. Moreover, it is significant to advance and deepen leadership skills already at the level of school education. Additionally, it is essential to motivate various institutions worldwide to establish and support special education programs for young participants (Walker, 2017; Kensler, 2012). Future generations as agents of change become strategic players in applying and promoting sustainable changes, so they need current support, access to knowledge, and successful world practices.

At the same time, it is necessary to educate modern business leaders of any age and experience; reconfiguring and advancing the knowledge of sustainable business strategies, environmental-friendly solutions, and best worldwide practices of sustainable products and technologies can help to reestablish existing production processes and accomplish global sustainable goals. The possibilities of professional development and training as a part of lifelong learning can be implemented into the entire business development. In addition, it is crucial for modern business leaders to clearly identify the benefits of sustainable education and implementation of the best sustainable practices at the local environment.

Business leaders are critical to express sustainable attitudes and to perform sustainable solutions. To prepare and motivate business leaders is another global task for providing sustainable changes. For significant changes and in the time of uncertainty and turbulence, there is a strong need for sustainability-minded leaders

(Rimanoczy, 2014; Cole & Snider, 2019). Creativity, innovative approach, and sustainable way of thinking can help find a proper solution for turbulence management (Ansell et al., 2022; Whewell et al., 2022). Some scholars denote that educational programs for sustainable business leaders should focus on understanding of sustainability, creating a personal connection, and empowerment to act for sustainability (Haney et al., 2020). Additionally, building a community of business leaders will extra stimulate the process of changemaking.

However, business leaders can meet some challenges while deepening their sustainability competencies (Crews, 2010). First, it is needed to fulfill the expectations of stakeholders by integrating their interests and needs. Second, it is important to include sustainable attitudes in the organizational culture, company's mission, or core values. Third, the holistic thinking of business leaders has to combine the most successful cases and methods of the past and the critical thinking, creativity, and openness for novelties of the modern time. Fourth, organizational learning must be up-to-date and embrace the best management practices, motivate individuals to follow sustainable attitudes, and foster sustainable culture. Those practices include talent management, training initiatives, mentoring, and educational programs for stimulating leadership development. Finally, measurement and reporting and monitoring systems evaluate the level of business changes.

31.5 Conclusions

This chapter presented the perspective on the commitment to sustainability and the arguments of educating sustainable leaders. In particular, this chapter focused on the ideas of sustainable innovations, drivers for sustainable changes, and changemakers as significant sustainable leaders in the post-COVID era. On the one hand, there is a demand from the global community, stakeholders, international organizations, and national governments to provide sustainable changes and fulfill the global requirements for sustainable development. On the other hand, consumers today prioritize more often sustainable brands and products, pay more attention to the sustainable commitment of a company, and change consumer attitudes for more sustainable and conscious. Therefore, from both perspectives, there is a need for some changes and innovations to incorporate.

The pandemic restrictions and the time of uncertainty defined new perspectives for business development; in this way, more sustainable practices and more efficient solutions can help businesses to survive and overcome modern challenges. The global sustainable agenda can stimulate innovative ideas and support those businesses aiming to join the sustainable discussion on the local or global levels. Environmental drivers demand for innovative solutions in the production process; economic drivers stimulate to recalculate the costs and search for costs optimization through sustainable practices. Social drivers aim to promote well-being and welfare by providing possibilities for individuals and social groups to produce and consume more sustainable products and services.

The coronavirus outbreak has permanently altered our experiences – as customers, employees, citizens, and people. As a consequence, people are changing their attitudes and behaviors. There is no doubt that the COVID-19 pandemic has changed a lot of aspects of our lives, particularly related to ISIS-related fields like information behaviors, business models, cybersecurity, as well as the data privacy. The COVID-19 pandemic has created disruptions in the short term while also spurring longer-term changes to how the world lives and does business in the new normal. Organizations made major changes in response to post-COVID-19 – there is no going back now, just the way forward. The COVID-19 pandemic has precipitated a shift toward digital with the majority of business companies and state institutions saying that they were able to execute initiatives they would previously face resistance and red tape of all sorts.

As COVID-19 is now becoming endemic across most of the world, people should be shifting their focus toward the more sustainable and inclusive growth. No doubt, COVID-19 is challenging the status quo and traditional thinking in all industries, but acknowledging some of the trends mentioned above, and taking prompt actions, will help individuals and businesses to get a leg up in this disruptive world. While COVID-19 has clearly affected workplaces and organizational dynamics in myriad ways, there is reason to hope that after the pandemic, there is renewed, deeper acknowledgment – one that will drive organizational policy-making – that we need individuals capable of breaking down traditional discipline silos. Because COVID-19, a new coronavirus, shares aspects of complexity, uncertainty, and value judgments that characterize big challenges, it will take unified, sustained contributions across disciplines globally to address a pandemic like this one. There are areas where responsible research on the information system may yield knowledge that could aid the fight against an ongoing pandemic, the newest of the great challenges that mankind is facing.

When it comes to the implications for further research, in the discussion above, we suggested how conducting responsible and balanced research could help tackle the COVID-19 pandemic, a new grand challenge, together with other Sustainable Development Goals. One possibility for the academics and scholars alike would be to contribute to this ongoing endeavor is by building upon the collective knowledge generated while continuing the fight against the COVID-19 pandemic and responding rapidly to developing new research capabilities to address grand challenges by conducting responsible information systems research. This might and most probably would inspire the key insights for leading executives about the business landscape after the pandemic, offering fresh perspectives on digital transformation, the future of work, transparency, and resilience. Of course, each of these perspectives brings its own important sets of lights that should illuminate the way that we move forward in a post-COVID-19 world. We showed this by analyzing the articles exploring the impact of the pandemic in areas including health care, operations, and sustainable, inclusive growth showing how industries were shaped by this experience and how leaders are looking forward. The imperative now for leaders is to evaluate the effect that each human capital trend will have on the operations and strategic goals of their organizations, determine what requires immediate action, and evaluate to what extent the human resources trends changed the strategic goals and plans prior to COVID-19. There may be a post-pandemic recovery on the horizon, but that is likely to hinge on business leaders' ability to react to productivity and growth jolts caused by COVID-19. The challenges to reigniting the world economy after an economic downturn caused by the coronavirus are, of course, different. More generally, the pace and depth of the COVID-19 crisis have shown that the central principle underlying the global economy prioritizing short-term economic growth and efficiency at the expense of longer-term sustainability can come at enormous social costs. The vulnerabilities exposed by the pandemic can serve as an underscore for why ecological issues were becoming the world's highest policy priorities prior to the COVID-19 outbreak. A new canvassing of experts on technology, communications, and social change finds many expected similar impacts from the COVID-19 pandemic and all its negative effects.

The World Health Organization (WHO) declared the COVID-19 outbreak, a new strain of coronavirus, to be a global pandemic on March 11, 2020. The coronavirus outbreak has been affecting our personal and professional lives ever since. For the past 2 years, COVID-19 pandemic has led to unprecedented levels of business and public health concerns and is likely to change how businesses function in the future. The COVID-19 pandemic also has some implications for businesses advertising, marketing, and advertising practices. This also includes practical best practices suggestions to help businesses understand their obligations and strengthen regulatory compliance programs to minimize risks associated with advertising, marketing, and other promotional activities following the COVID-19 pandemic. In addition, it embeds the exploration of the role of environmental health in strengthening pandemic resilience and COVID-19 and the transition to low-carbon energy supply. It appears that in the post-COVID-19 era, companies must consistently deliver new disruptive value for customers, not just one time. After the COVID-19 pandemic is over, there will be an increased need for innovation and to get out from under the comfort of doing business as usual. Most of the business organizations are likely to continue increasing the use of contingent workers in order to retain greater workforce flexibility in the post-COVID-19 period, and they will also look at adopting other labor models that they saw in the midst of the pandemic, such as talent sharing and proper compensation for work. However, even now it looks like the majority of employees are likely to work remotely at least some of the time post-COVID-19, up from just a small share in the pre-pandemic period (which was subjected to complex administrational arrangements, labor laws, and other rules and regulations). In addition, it appears that the coronavirus pandemic helped to boost the development of artificial intelligence-based tools and solutions that are likely not even to stay but to be promoted even further for more applications related to dealing with customers and employees. Expectations are quite high at the moment, as the majority of CEOs would probably tend to use AI-based customer engagement tools to accomplish this, with chatbots handling up to two-thirds of customer service traffic during post-COVID-19 at some organizations.

All in all, educating young leaders within the sustainable agenda should become the integral part of all the levels of modern education. In the same way, lifelong practices can offer the ways for professional development and necessary skills training in accordance with modern challenges and social, economic, and environmental problems.

Furthermore, it is essential to add that innovations will not work properly without the protection and support of changemakers with their ability to create, explain, and distribute innovative approach. Additionally, changemakers would need some drivers from business sector, government, or consumers to focus on some problematic areas for changes or some efficient solutions and practices. Finally, drivers without innovations will not achieve the final goal to motivate, encourage, stimulate, and engage the society. Therefore, only productive cooperation and networking, interinfluence, and collaborative approach between innovative ideas, driving power, and encouraging leaders can bring a positive input and long-expected results to the sustainable development discussion and transition.

References

Abdelaal, F., & Guo, B. H. (2021). Knowledge, attitude and practice of green building design and assessment: New Zealand case. *Building and Environment, 201*, 107960. https://doi.org/10.1016/j.buildenv.2021.107960

Akter, J. S., & Haque, S. M. (2022). Innovation management: Is big data necessarily better data? *Management of Sustainable Development, 14*(2), 27–33. https://doi.org/10.54989/msd-2022-0013

Ansell, C., Sørensen, E., & Torfing, J. (2022). Co-creation for sustainability: SDG 17 as a lever of change. *Emerald Open Research, 4*(3), 3. https://doi.org/10.35241/emeraldopenres.1114920.1

Armstrong, P., & Gough, A. (2019). Developing and motivating young leaders for sustainability. In D. B. Zandvliet (Ed.), *Culture and environment: Weaving new connections* (pp. 163–187). Brill. https://doi.org/10.1163/9789004396685_010

Biancone, P., Secinaro, S., Marseglia, R., & Calandra, D. (2021). E-health for the future. Managerial perspectives using a multiple case study approach. *Technovation, 102406*. https://doi.org/10.1016/j.technovation.2021.102406

Campaign. (2016). *Brands, not government, must drive social change, say consumers in study*. https://www.campaignlive.co.uk/article/brands-not-government-drive-social-change-say-consumers-study/1380448

Carayannis, E. G., Sindakis, S., & Walter, C. (2015). Business model innovation as lever of organizational sustainability. *The Journal of Technology Transfer, 40*(1), 85–104. https://doi.org/10.1007/s10961-013-9330-y

Changemakers. (2021). *What are "green jobs" and why do they matter?* https://medium.com/change-maker/what-are-green-jobs-and-why-do-they-matter-fbaaab141e47

Cole, R., & Snider, B. (2019). Managing in turbulent times: The impact of sustainability in management education on current and future business leaders. *Journal of Cleaner Production, 210*, 1622–1634. https://doi.org/10.1016/j.jclepro.2018.11.135

Crews, D. E. (2010). Strategies for implementing sustainability: Five leadership challenges. *SAM Advanced Management Journal, 75*(2), 15–21.

Doblin. (2021). *Ten types of innovations*. https://doblin.com/ten-types

Foerstl, K., Azadegan, A., Leppelt, T., & Hartmann, E. (2015). Drivers of supplier sustainability: Moving beyond compliance to commitment. *Journal of Supply Chain Management, 51*(1), 67–92. https://doi.org/10.1111/jscm.12067

Friends of the Earth Policy. (2021). *An emergency plan on green jobs for young people*. https://policy.friendsoftheearth.uk/sites/default/files/documents/2021-03/EMERGENCY_PLAN_GREEN_JOBS_FEB_2021.pdf

Gadekar, R., Sarkar, B., & Gadekar, A. (2022). Investigating the relationship among industry 4.0 drivers, adoption, risks reduction, and sustainable organizational performance in manufacturing industries: An empirical study. *Sustainable Production and Consumption, 31*, 670–692. https://doi.org/10.1016/j.spc.2022.03.010

Grabs, J., & Carodenuto, S. L. (2021). Traders as sustainability governance actors in global food supply chains: A research agenda. *Business Strategy and the Environment, 30*(2), 1314–1332. https://doi.org/10.1002/bse.2686

Grigorescu, A., Maer-Matei, M. M., Mocanu, C., & Zamfir, A. -M. (2017). Circular economy and decision models among European SMEs. *Sustainability (Switzerland), 9*(9). https://doi.org/10.3390/su9091507

Haney, A. B., Pope, J., & Arden, Z. (2020). Making it personal: Developing sustainability leaders in business. *Organization & Environment, 33*(2), 155–174. https://doi.org/10.1177/10860266188062

Hermundsdottir, F., & Aspelund, A. (2022). Competitive sustainable manufacturing-sustainability strategies, environmental and social innovations, and their effects on firm performance. *Journal of Cleaner Production, 370*, 133474. https://doi.org/10.1016/j.jclepro.2022.133474

Hristov, I., Camilli, R., & Cristofaro, M. (2022). The key value drivers of the company's sustainable development strategy: An integrated framework. In Academy of management proceedings 2022, 1, 14945 : Academy of Management.

Insights. (2019). *A 'natural' rise in sustainability around the world*. https://www.nielsen.com/eu/en/insights/article/2019/a-natural-rise-in-sustainability-around-the-world/

Jaiswal, M., & Zane, L. J. (2022). Drivers of sustainable new technology diffusion in national markets: The case of electric vehicles. *Thunderbird International Business Review, 64*(1), 25–38. https://doi.org/10.1002/tie.22243

Jia, H. M., Wang, Y., Tian, Q., & Fan, D. (2022). Marketing innovations and sustainable development for the future ecosystem. *Journal of Cleaner Production, 133295*. https://doi.org/10.1016/j.jclepro.2022.133295

Kajzer Mitchell, I., & Walinga, J. (2017). The creative imperative: The role of creativity, creative problem solving and insight as key drivers for sustainability. *Journal of Cleaner Production, 140*, 1872–1884. https://doi.org/10.1016/j.jclepro.2016.09.162

Kensler, L. A. (2012). Ecology, democracy, and green schools: An integrated framework. *Journal of School Leadership, 22*(4), 789–814. https://doi.org/10.1177/105268461202200406

Ki-moon, B. (2016). Sustainability—Engaging future generations now. *Lancet, 387*(10036), 2356–2358. https://doi.org/10.1016/S0140-6736(16)30271-9

Korneeva, E., Skornichenko, N., & Oruch, T. (2021a). Small business and its place in promoting sustainable development. *E3S Web of Conferences, 250*, 06007. https://doi.org/10.1051/e3sconf/202125006007

Korneeva, E., Olinder, N., & Strielkowski, W. (2021b). Consumer attitudes to the smart home technologies and the Internet of Things (IoT). *Energies, 14*(23), 7913. https://doi.org/10.3390/en14237913

Linde, L., Sjödin, D., Parida, V., & Wincent, J. (2021). Dynamic capabilities for ecosystem orchestration a capability-based framework for smart city innovation initiatives. *Technological Forecasting and Social Change, 166*, 120614. https://doi.org/10.1016/j.techfore.2021.120614

Littlewood, D., Decelis, R., Hillenbrand, C., & Holt, D. (2018). Examining the drivers and outcomes of corporate commitment to climate change action in European high emitting industry. *Business Strategy and the Environment, 27*(8), 1437–1449. https://doi.org/10.1002/bse.2194

Loorbach, D., Frantzeskaki, N., & Avelino, F. (2017). Sustainability transitions research: Transforming science and practice for societal change. *Annual Review of Environment and Resources, 42*(1), 599–626. https://doi.org/10.1146/annurev-environ-102014-021340

Lopes, J. M., Gomes, S., Pacheco, R., Monteiro, E., & Santos, C. (2022). Drivers of sustainable innovation strategies for increased competition among companies. *Sustainability, 14*(9), 5471. https://doi.org/10.3390/su14095471

Lozano, R. (2015). A holistic perspective on corporate sustainability drivers. *Corporate Social Responsibility and Environmental Management, 22*(1), 32–44. https://doi.org/10.1002/csr.1325

Luthra, S., Kumar, A., Zavadskas, E. K., Mangla, E. K., & Garza-Reyes, J.-A. (2020). Industry 4.0 as an enabler of sustainability diffusion in supply chain: An analysis of influential strength of drivers in an emerging economy. *International Journal of Production Research, 58*(5), 1505–1521. https://doi.org/10.1080/00207543.2019.1660828

McKinsey. (2021). *The next normal arrives: Trends that will define 2021—And beyond*. https://www.mckinsey.com/featured-insights/leadership/the-next-normal-arrives-trends-that-will-define-2021-and-beyond

Mohamed, B. H., Disli, M., Al-Sada, M., Bin, S., & Koç, M. (2022). Investigation on human development needs, challenges, and drivers for transition to sustainable development: The case of Qatar. *Sustainability, 14*, 3705. https://doi.org/10.3390/su14063705

Morales, P., Flikkema, M., Castaldi, C., & de Man, A. P. (2022). The effectiveness of appropriation mechanisms for sustainable innovations from small and medium-sized enterprises. *Journal of Cleaner Production, 133921*. https://doi.org/10.1016/j.jclepro.2022.133921

Pelikánová, R. M., Němečková, T., & MacGregor, R. K. (2021). CSR statements in international and Czech luxury fashion industry at the onset and during the COVID-19 pandemic-slowing down the fast fashion business? *Sustainability, 13*(7), 3715. https://doi.org/10.3390/su13073715

Rauter, R., Baumgartner, R. J., & Jonker, J. (2017). Going one's own way: Drivers in developing business models for sustainability. *Journal of Cleaner Production, 140*, 144–154. https://doi.org/10.1016/j.jclepro.2015.04.104

Rimanoczy, I. (2014). A matter of being: Developing sustainability-minded leaders. *Journal of Management for Global Sustainability, 2*(1), 95–122. https://doi.org/10.13185/JM2014.02104

Saeed, M., & Kersten, W. (2019). Drivers of sustainable supply chain management: Identification and classification. *Sustainability (Switzerland), 11*(4). https://doi.org/10.3390/su11041137

Salam, M. A., & Ali, M. (2020). Building reputation through sustainable supplier selection: The case of an emerging economy. *European Journal of Management and Business Economics, 29*(3), 315–332. https://doi.org/10.1108/EJMBE-12-2019-0217

Strielkowski, W., Lisin, E., & Tvaronavičiene, M. (2016). Towards energy security: Sustainable development of electrical energy storage. *Journal of Security and Sustainability Issues, 6*(2), 235–244. https://doi.org/10.9770/jssi.2016.6.2(4)

Strielkowski, W., Guliyeva, A., Rzayeva, U., Korneeva, E., & Sherstobitova, A. (2021). Mathematical modeling of intellectual capital and business efficiency of small and medium enterprises. *Mathematics, 9*(18), 2305. https://doi.org/10.3390/math9182305

The Brundtland Report. (1987). *Our common future*. UN. https://sustainabledevelopment.un.org/content/documents/5987our-common-future.pdf

The Guardian. (2013). *How can business drive social change?* https://www.theguardian.com/sustainable-business/how-can-business-drive-social-change

Tollin, K., & Christensen, L. (2019). Sustainability marketing commitment: Empirical insights about its drivers at the corporate and functional level of marketing. *Journal of Business Ethics, 156*(4), 1165–1185.

Van Zanten, J. A., & van Tulder, R. (2021). Analyzing companies' interactions with the sustainable development goals through network analysis: Four corporate sustainability imperatives. *Business Strategy and the Environment, 30*(5), 2396–2420. https://doi.org/10.1002/bse.2753

Velter, M. G. E., Bitzer, V., Bocken, N. M. P., & Kemp, R. (2020). Sustainable business model innovation: The role of boundary work for multi-stakeholder alignment. *Journal of Cleaner Production, 247*, 119497. https://doi.org/10.1016/j.jclepro.2019.119497

Vosen, E., Wilhelm, S., & Habisch, A. (2022). Empowering women as key changemakers: Why female-driven social innovation matters. In *The global impact of social innovation* (pp. 217–234). Springer. https://doi.org/10.1007/978-3-031-03849-5_18

Walker, C. (2017). Tomorrow's leaders and Today's agents of change? Children, sustainability education and environmental governance. *Sustainability Education and Environmental Governance Children & Society, 31*(1), 72–83. https://doi.org/10.1111/chso.12192

Whewell, E., Caldwell, H., Frydenberg, M., & Andone, D. (2022). Changemakers as digital makers: Connecting and co-creating. *Education and Information Technologies*, 1–23.

WWF. (2021). *The eco-wakening*. Available at: https://explore.panda.org/eco-wakening

Chapter 32
Europe's Energy Innovation: Global Leadership Scenarios and Prospects for Achieving Sustainable Development Goals

Ihor Vakulenko, Liudmyla Saher, and Oleh Skorba

Abstract The strategic plans for developing the European Union's energy sector did not take into account the military aggression and the war waged by the Russian Federation in Ukraine that started in February 2022. Energy resources are used as a means of political struggle. The world has faced a new global energy crisis. There are several scenarios for overcoming the crisis. This study focuses on the role of energy innovation in implementing baseline crisis scenarios. It assesses the prospects for achieving SDGs and the long-term goals of the European Union's energy policy. Renewable Energy and Smart Grids are considered critical components to achieving sustainable development in each scenario. Each scenario is determined based on actual energy policy measures taken by governments and intergovernmental institutions to overcome the crisis and the opinions of experts and leading scientists using up-to-date statistics from The International Energy Agency, the World Data Bank, and Eurostat. Quantitative data analysis methods with SPSS Software were applied to analyze the impact of the business environment on the energy sector. The results of such an analysis are valuable not only for understanding the prospects for the development of the energy sector. They provide a partial answer to the question of achieving the goals of the climate policy and improving the quality of the environment and, accordingly, a deeper understanding of the value caused by the war waged by the Russian Federation of the global energy crisis in Europe and the whole world.

I. Vakulenko · O. Skorba
Sumy State University, Sumy, Ukraine
e-mail: vakulenko@ms.sumdu.edu.ua; o.skorba@biem.sumdu.edu.ua

L. Saher
Sumy State University, Sumy, Ukraine

Institute of Economic Research of the Slovak Academy of Sciences, v.v.i., Bratislava, Slovak Republic
e-mail: l.sager@kmm.sumdu.edu.ua

W. Strielkowski (ed.), *Leadership, Entrepreneurship and Sustainable Development Post COVID-19*, Springer Proceedings in Business and Economics, https://doi.org/10.1007/978-3-031-28131-0_32

Keywords Sustainable development · Green economy · Renewable energy · Smart grids · Energy crisis · Competitiveness · Global leadership

JEL Classification O32 · Q28 · Q48

32.1 Introduction

The global community, both in the European Union and worldwide, has developed a plan to achieve the Sustainable Development Goals to combat climate change, which poses a global and imminent threat to humanity, and to make rational use of available resources to maintain a balance between current needs and the needs of future generations (Khoshnava et al., 2019). It can be seen that the goals of sustainable development cover all crucial areas of human life. However, it is worth recognizing that one of the most important directions for achieving sustainable development on a global scale is energy (Kwatra et al., 2020). Every branch of the economy and every aspect of social life are imbued with energy, and people in the developed Western economies are used to cheap and affordable electricity, oil, and gas (Strielkowski, 2017). Despite the predictions that were made in the twentieth century regarding the reduction of energy consumption in the future, which for us is modernity, this did not happen.

On the contrary, we can observe a steady trend toward energy consumption growth. This is explained by the fact that the forecast for the reduction of energy consumption was based on the assumption that improvements in engineering and technology will lead to a specific decrease in energy consumption per appliance (Simionescu et al., 2022). This assumption was correct. However, it is worth noting that another factor affecting energy consumption was not considered in the forecasts. We are talking about the invention of new devices and tools, which nothing could be known in the second half of the twentieth century when similar predictions were made. In any case, it can be stated that the forecast of a decrease in energy consumption over time is currently not valid because from 1965 to 2021, the total energy consumption in the world increased by 381% (Energy Data Explorer, 2022). At the same time, energy consumption growth indicates energy's growing importance and new requirements for the energy system. The classic energy system, formed in the twentieth century, is based on the widespread use of fossil fuels. In growing energy consumption conditions, more fossil fuels are needed to meet energy needs. However, such a situation is unacceptable because it increases the volume of greenhouse gases, which negatively affects the climate and reduces the quality of the population's living environment (Brożyna et al., 2020).

To solve this problem, scientists, government, and businesses have been working side by side to create new environmentally friendly technologies for the past decades. The result of this cooperation was the spread of renewable energy, the share of which has increased 38 times over the past 20 years. However, introducing green technologies in the energy sector encountered two problems from the beginning. The first problem is technological imperfection. Technologies that are used cannot effectively replace traditional (fossil) fuels.

The second problem is the insufficient economic efficiency of green energy projects. As a result, renewable energy projects did not arouse the interest of investors for a long time. This restrained the development of renewable energy. Both problems are gradually being solved. However, the current state is far from ideal, although it is significantly better than a few years ago. International policy, embodied in numerous programs and plans for the development of renewable energy and the introduction of energy-efficient technologies, in particular, the creation of smart grids, significantly impacted progress in this matter. However, these plans worked in stable macroeconomic and political conditions, where processes are developing under the forecasts made. After Russia launched a war against Ukraine, the largest in Europe since World War II, the situation changed dramatically.

The leadership of the Russian Federation considers energy resources as a mechanism of political influence on states whose energy systems and economies depend on the energy resources imported from Russia. The rise in the cost of oil and gas over the past few years is the result of political rather than economic processes (even though the impacts of the recent COVID-19 pandemic with its shutdowns and restrictions can also be blamed). During the spring and summer of 2022, the problem of their shortage was added to the increase in the cost of energy carriers. This is partly a consequence of the sanctions policy pursued by the countries of the world, particularly the European Union, against the Russian Federation. However, this is partly the result of the policy of the Russian Federation, which is trying to create an artificial deficit of energy resources in the European Union to incline it to political decisions beneficial to itself (Lisin et al., 2014).

Regardless of the reasons that led to the shortage of energy resources and the increase in their cost, this issue needs to be resolved. There are two obvious solutions to the problem. Both are highly complex and require significant investment and time to implement. The first option is diversification of the supply of energy resources to replace those supplied by Russia. The second option is to increase energy production from renewable sources. Based on these options, we will further describe options for exiting the energy crisis for the countries of the European Union. However, in spite of the way out of the energy crisis, it can already be stated that the direction of development of the energy sector of the European Union must undergo significant changes, and the whole concept of its energy policy should be revised considerably, even if the economic costs would raise high.

As it has been already mentioned earlier, implementing measures to overcome the energy crisis requires significant resources. First of all, money and time. Accordingly, the time needed to solve the energy problem can be shortened by increasing investments in energy. However, the required investment is unplanned. This is a significant problem. After all, the funds that should be directed to minimize or eliminate the consequences of the energy crisis should be withdrawn from other directions or projects.

At the same time, the negative impact of the situation on the energy market is felt by all market participants: governments and authorities, enterprises and organizations, and households.

Additional costs incurred by governments, local authorities, international organizations, and energy sector enterprises can significantly affect the implementation of current plans and the strategy's performance for achieving sustainable development goals, both in the energy sector and in other industries closely related to it. In this context, the question arises about the future of renewable energy, which significantly depends on which option to overcome the energy crisis will be implemented.

32.2 Materials and Methods

In this research, an analysis of scientific publications on energy, sustainable development, and innovation was carried out to carry out the research. Statistical data processing methods were used to establish relationships between the studied phenomena. In particular, the SPSS software was employed by the authors in order to process an array of statistical data from open sources.

32.3 Results and Discussions

Sustainable development of society is one of the main tasks that the governments of the European Union countries are trying to solve. Achieving this goal is connected with forming a favorable environment for conducting socially and environmentally responsible business. Entrepreneurship is the key to economic development. That is why, in the conditions of crisis phenomena in the economy and society, it must be supported by providing the requirements (Starchenko et al., 2021a, b; Taliento & Netti, 2020). A study by He (2019) discusses the connection between entrepreneurship and trade and improving the quality of the natural environment. There are few similar studies. Perhaps that is why the development of entrepreneurship and the creation of conditions for increasing trade turnover are generally not considered a tool for achieving sustainable development goals. It is impossible to ignore the results of such studies. Still, it is necessary to remember that the positive impact of the business environment on the environment is possible only if certain conditions are met, in particular, compliance with the rules of social and environmental responsibility.

The European Union's energy sector mainly depends on Russian energy resources. Such a situation was considered by many politicians, public figures, and scientists as threatening for a long time. However, neither the governing bodies of the EU nor the governments of the EU member states made efforts to change this situation (Stępień & Miciuła, 2016; Balitskiy et al., 2016).

How negative will be the consequences of the EU's past refusal to diversify the supply of energy resources will become apparent shortly.

It remains to be stated that the energy issue for the European Union countries is as acute as never before. The problem must be solved shortly. Currently, the European community is at a turning point, when the current energy development trends may be significantly changed (Bilotskiy & Danylova, 2014; Kuzior et al., 2021b; Polcyn et al., 2022). However, how it will be resolved is debatable.

In general, several possible scenarios can be identified. The first scenario is the most likely. This can be argued by looking at the political decisions made by the leadership of the European Union. This scenario consists in diversifying the supply of fossil energy resources. Russian oil and gas should be replaced by oil and gas from other countries. At first glance, the solution seems logical and effective in the short term. But analyzing the problems that must be solved to replace such a large amount of Russian energy resources, we can conclude that this solution is not so good in the long run. The change in the flow of energy resources requires the construction of new and modernization (increasing the throughput) of the existing oil and gas transport infrastructure. First, construction and modernization require huge investments. Second, which is very dangerous, having invested a significant amount of money and oil and gas infrastructure, investors will seek to get the maximum return. The projects must pay off in full and bring significant profit. This cannot happen if renewable energy is actively developed simultaneously. After all, in such a case, it will be necessary to decommission the built and modernized oil and gas infrastructure before it fully completes its service life. Considering that the service life of such infrastructure is 50 or more years, renewable energy prospects become vague. Another signal that such a scenario of exiting the energy crisis will be implemented is the recent decision of the EU governing bodies to recognize natural gas as a green fuel.

The second scenario, which seems less likely, is an increase in investments in renewable and smart energy (Serpeninova et al., 2020; Sedmíková et al., 2021). This option of exiting the crisis is more profitable from the point of view of achieving the goals of sustainable development (Kolosok, 2021a ; Matvieieva, 2022) and avoiding energy gaps (Vasylieva et al., 2021) and overcoming energy poverty (Didenko et al., 2021). However, previous experience with renewable energy development shows that this scenario will take time to materialize.

The opportunities for the development of intelligent networks have been growing in recent years, and the forecast for further acceleration of their implementation is exceptionally optimistic (Rossi et al., 2022; Kolosok et al., 2021a, b).

Large-scale and long-term renewable energy projects cannot be stopped. This will harm the achievement of climate policy goals because the relationship between renewable energy and CO2 emission reduction is more pronounced in the long run. Still, in the short run, it is insignificant (Us, 2021; Kuang et al., 2022). Financial instruments play an important role in these stimulating enterprises to reduce emissions (El Amri et al., 2020, 2021; Starchenko et al., 2021a, b).

There are opportunities to improve energy efficiency, particularly in transport (Ovaere & Proost, 2022; Kuzior, 2021). Obviously, the more efforts governments make to reduce harmful emissions and implement various measures to combat climate change, the greater the result will be. However, everything is ambiguous. The experience of energy efficiency policies of individual countries varies considerably, even with similar measures to combat climate change (Bhowmik, 2019). This leads to the conclusion that there is no universal concept for achieving climate policy goals that will be equally effective for different countries.

Renewable energy consumption and technological innovation are significant contributors to economic growth (Deng et al., 2022). Indicators of production and consumption of renewable energy and diversity of energy production sources are important components of building an energy-efficient economic system (Ziabina et al., 2020; Salihaj & Pryimenko, 2017; Shkarupa & Chasnyk, 2014) and circular economy (Polyakov et al., 2021). An important place in this system is occupied by enterprises that implement energy-efficient and environmentally safe technologies (Sotnyk & Kulyk, 2015; Kuzior, 2021a), improving the quality environment (Letunovska et al., 2021).

Significant investments were also made in shaping the population's worldview regarding the need for economic and rational use of natural resources for sustainable development to ensure proper conditions and quality of the living environment for future generations. Environmental university education was an influential and essential component of forming an innovative climate (Kyrychenko et al., 2021). However, the efforts can be partially offset by overcoming the energy crisis by diversifying the supply and promoting oil and gas development as the primary energy resource.

We can talk about the third scenario, which can be implemented. It is a combination of the first two scenarios, that is, diversification of fossil fuel supply in the short term with a transition to renewable energy in the medium and long term. However, there are reasons to believe this will not happen, for the reasons described above.

All scenarios increase the energy security of the EU and EU member states (Samusevych et al., 2021a, b; Didenko et al., 2020; Mentel et al., 2020; Simionescu et al., 2020). At the same time, a significant role in this issue for an extended period was attributed to environmental taxation as an effective regulatory tool (Samusevych et al., 2021a, b; Vysochyna et al., 2020; Štreimikienė et al., 2022). However, in the case of implementation of the first scenario, the role of this regulator should be revised.

The answer to the question of achieving sustainable development goals and implementing programs aimed at this depends on which of the scenarios for overcoming the energy crisis will be chosen. Judging by the actual steps already taken, it can be assumed that the speed of solving the problem has been selected as the main priority. European politicians are trying to avoid destabilizing the socioeconomic situation in their countries and return to the status quo, that is, to a situation where energy resources are available in the required amount and at an acceptable price.

Thus, one can lean toward the version that the first described scenarios will be realistic. But what prompts the choice of this particular scenario, which essentially goes against the plans to achieve sustainable development in the European Union?

We have already guessed the reasons for this above. The new oil and gas infrastructure must be profitable and used for a long time. Is the choice of the first scenario (diversification of fossil fuel supply) forced? Answering this question, it is advisable to analyze the dependence of energy consumption on the prices of the main types of energy resources, oil, and natural gas. The results of the correlation analysis are shown in Table 32.1. The data in Table 32.1 show no significant dependence of the towers on the price of oil and natural gas and their consumption. This indicates that, despite the increase in the cost of energy carriers, companies and households do not reduce their consumption. We can assume that there are two reasons for this. The first is that the price is still within the range consumers are willing to pay. The second, which is more important for our research, is that it is impossible to abandon the consumption of these energy resources (only in small amounts) without negative economic consequences. The second reason is predominant in the EU's intentions to diversify its oil and gas supply rather than abandoning them in the short term in favor of renewable energy.

Thus, it is clear that the fundamental motive in choosing a way out of the energy crisis will be to preserve the volume of energy consumption to avoid economic collapse. It is necessary to ask whether renewable energy can be a tool for replacing fossil fuels in the short term. A partial answer to this question is given in Table 32.2, which shows that when the cost of traditional energy resources (oil and natural gas) increases, the production of renewable energy increases. Thus, it can be argued that replacing conventional energy with renewable energy occurred more intensively during the rise in the price of oil and gas on the energy market.

However, despite the realism of the scenario of replacing fossil fuels with renewable ones, especially in the current conditions, there remains one more factor that has a significant impact – time. To assess the real possibilities of replacing oil and gas with renewable energy, it is worth looking at the volumes of renewable energy production and their growth dynamics (Table 32.3). It is clear from the data in the table that renewable energy is not a quick substitute for traditional fuels.

Table 32.1 The relationship between the cost of energy resources and total energy consumption

Correlations

		Final energy consumption	Europe Brent Spot Price FOB	Gas prices for household consumers
Final energy consumption	Pearson correlation	1	−.277	−.198
	Sig. (two-tailed)		.318	.480
	N	15	15	15
Europe Brent Spot Price FOB	Pearson correlation	−.277	1	.916**
	Sig. (two-tailed)	.318		.000
	N	15	15	15
Gas prices for household consumers	Pearson correlation	−.198	.916**	1
	Sig. (two-tailed)	.480	.000	
	N	15	15	15

**Correlation is significant at the 0.01 level (two-tailed)

Table 32.2 The relationship between renewable energy production and the cost of fossil fuels (oil and natural gas)

Correlations				
		Global capacity in renewable power	Europe Brent Spot Price FOB	Gas prices for household consumers
Global capacity in renewable power	Pearson correlation	1	−.063	−.141
	Sig. (two-tailed)		.825	.616
	N	15	15	15
Europe Brent Spot Price FOB	Pearson correlation	−.063	1	.916**
	Sig. (two-tailed)	.825		.000
	N	15	15	15
Gas prices for household consumers	Pearson correlation	−.141	.916**	1
	Sig. (two-tailed)	.616	.000	
	N	15	15	15

**Correlation is significant at the 0.01 level (two-tailed)

Table 32.3 Statistical data on energy consumption, energy generation from renewable sources, and the dynamics of oil and natural gas prices over 15 years

	Final energy consumption (million tons of oil equivalent)	Global capacity in renewable power (GW)	Europe Brent Spot Price FOB (dollars per barrel)	Gas prices for household consumers (US dollars per million metric British thermal unit)
2004	1189.6	227	38.260	3755
2005	1193.8	245	54.570	5915
2006	1196.9	281	65.160	8213
2007	1177.4	311	72.440	8143
2008	1185.0	354	96.940	13.138
2009	1118.9	414	61.740	8855
2010	1167.1	481	79.610	8223
2011	1114.7	564	111.260	10.597
2012	1116.3	658	111.630	11.981
2013	1115.7	749	108.560	11.189
2014	1067.9	854	98.970	10.463
2015	1090.3	982	52.320	7305
2016	1110.4	1120	43.640	4353
2017	1122.6	1279	54.130	5744
2018	1123.9	1443	71.340	7916

Source: (Eurostat, 2022; World Bank, 2022; European Environment Agency, 2022; International Energy Agency, 2022)

Table Accordingly, in conditions of limited time, restore the necessary energy supply volumes. The scenario of replacing fossil fuels with renewables is ineffective.

However, to achieve the goals of sustainable development, the third scenario is also acceptable, which provides for the short-term replacement of fossil energy resources from Russia with fossil energy resources from other countries. This option is intermediate, however, as mentioned above, unlikely. After all, new suppliers enter the EU energy market under long-term contracts. At the same time, significant funds are invested in the infrastructure necessary for the transportation of energy resources. It can be assumed that under such conditions, a strong lobby of the oil and gas industry will operate, which will harm the plans for deploying renewable energy and, as a result, the achievement of the goals of sustainable development.

32.4 Conclusions

Overall, it becomes quite clear that the war waged by the Russian Federation in Ukraine, at the doorstep of the European Union, has already caused serious economic as well as humanitarian damages. However, in our opinion, the energy crisis has not yet reached its full peak. The use of energy resources as a tool of political struggle forces the EU countries to look for an alternative to the Russian energy resources which might be cumbersome and expensive. In our research, we have researched three scenarios for resolving and exiting the energy crisis that can be implemented by the EU for solving this problem. The first scenario is the diversification of the supply of fossil energy resources to replace Russia in the European energy market. The second scenario is the development of renewable energy with the same goal. The third scenario is a combination of the previous two, which is to develop renewable energy, but first, diversify the supply of fossil fuels to pass the peak of the crisis. It is advisable to use the second or third scenarios to achieve sustainable development goals. However, the analysis of the conditions in which the EU countries have to solve the energy issue requires the conclusion that the EU will implement the first scenario, which is harmful to sustainable development and, in particular, to achieve the goals of the EU's climate policy. Accordingly, the energy crisis in Europe shortly may lead to a serious aggravation of environmental problems, primarily related to climate change. All of this should be taken into consideration by the current European and world leaders who have to operate in the present conditions which are worsened by the aftermath of the COVID-19 crisis that has also impacted the world economies with adverse and negative effects

Acknowledgments The research was funded by a grant from Ministry of Education and Science of Ukraine, 0122 U000769, "Transfer of green innovations in the energy sector of Ukraine: a multiplicative stochastic model of the transition to a carbon-neutral economy."

References

Balitskiy, S., Bilan, Y., Strielkowski, W., & Štreimikienė, D. (2016). Energy efficiency and natural gas consumption in the context of economic development in the European Union. *Renewable and Sustainable Energy Reviews, 55*, 156–168. https://doi.org/10.1016/j.rser.2015.10.053

Bhowmik, D. (2019). Decoupling CO2 emissions in Nordic countries: Panel data analysis. *SocioEconomic Challenges, 3*(2), 15–30. https://doi.org/10.21272/sec.3(2).15-30.2019

Bilotskiy, S., & Danylova, N. (2014). Economic expansion of the European renewable energy market in case of European Union law. *Marketing and Management of Innovations, 4*, 136–145. https://doi.org/10.21272/mmi.2014.4-16

Brożyna, J., Strielkowski, W., Fomina, A., & Nikitina, N. (2020). Renewable energy and EU 2020 target for energy efficiency in The Czech Republic and Slovakia. *Energies, 13*(4), 965. https://doi.org/10.3390/en13040965

Deng, W., Akram, R., & Mirza, N. (2022). Economic performance and natural resources: Evaluating the role of economic risk. *Resources Policy, 78*, 102840. https://doi.org/10.1016/j.resourpol.2022.102840

Didenko, I., Volik, K., Vasylieva, T., Lyeonov, S., & Antoniuk, N. (2020). Migration, environment, and country safety: Analysis of touchpoints. *E3S Web of Conferences, 202*, 03028. https://doi.org/10.1051/e3sconf/202020203028

Didenko, I., Volik, K., Vasylieva, T., Lyeonov, S., & Antoniuk, N. (2021). Environmental migration and country security: Theoretical analysis and empirical research. *E3S Web of Conferences, 234*, 00010. https://doi.org/10.1051/e3sconf/202123400010

El Amri, A., Boutti, R., Oulfarsi, S., Rodhain, F., & Bouzahir, B. (2020). Carbon financial markets underlying climate risk management, pricing and forecasting: Fundamental analysis. *Financial Markets, Institutions and Risks, 4*(4), 31–44. https://doi.org/10.21272/fmir.4(4).31-44.2020

El Amri, A., Oulfarsi, S., Boutti, R., Sahib Eddine, A., & Hmioui, A. (2021). Carbon financial markets underlying climate change mitigation, pricing and challenges: Technical analysis. *Financial Markets, Institutions and Risks, 5*(1), 5–17. https://doi.org/10.21272/fmir.5(1).5-17.2021

Energy Data Explorer. (2022). Available at: https://ourworldindata.org/explorers/energy?tab=table&facet=none&country=USA~GBR~CHN~OWID_WRL~IND~BRA~ZAF&Total+or+Breakdown=Total&Energy+or+Electricity=Primary+energy&Metric=Annual+consumption

He, S. (2019). The impact of trade on environmental quality: A business ethics perspective and evidence from China. *Business Ethics and Leadership, 3*(4), 43–48. https://doi.org/10.21272/bel.3(4).43-48.2019

Khoshnava, S. M., Rostami, R., Zin, R. M., Štreimikienė, D., Yousefpour, A., Strielkowski, W., & Mardani, A. (2019). Aligning the criteria of green economy (GE) and sustainable development goals (SDGs) to implement sustainable development. *Sustainability, 11*(17), 4615. https://doi.org/10.3390/su11174615

Kolosok, S., Bilan, Y., Vasylieva, T., Wojciechowski, A., & Morawski, M. (2021a). A scoping review of renewable energy, sustainability and the environment. *Energies, 14*(15). https://doi.org/10.3390/en14154490

Kolosok, S., Vasylieva, T., & Lyeonov, S. (2021b). Machine analysis of the UK electrical energy initiatives based on the e-petitions to the UK government and parliament. *CEUR Workshop Proceedings, 2870*, 1562–1573.

Kuang, H., Akmal, Z., & Li, F. (2022). Measuring the effects of green technology innovations and renewable energy investment for reducing carbon emissions in China. *Renewable Energy, 197*, 1–10. https://doi.org/10.1016/j.renene.2022.06.091

Kuzior, A., & Staszek, M. (2021). Energy management in the railway industry: A case study of rail freight carrier in Poland. *Energies, 14*(21), 6875. https://doi.org/10.3390/en14216875

Kuzior, A., Kwilinski, A., & Hroznyi, I. (2021a). The factorial-reflexive approach to diagnosing the executors' and contractors' attitude to achieving the objectives by energy supplying companies. *Energies, 14*(9), 2572. https://doi.org/10.3390/en14092572

Kuzior, A., Lobanova, A., & Kalashnikova, L. (2021b). Green energy in Ukraine: State, public demands, and trends. *Energies, 14*(22), 7745. https://doi.org/10.3390/en14227745

Kwatra, S., Kumar, A., & Sharma, P. (2020). A critical review of studies related to construction and computation of sustainable development indices. *Ecological Indicators, 112*, 106061. https://doi.org/10.1016/j.ecolind.2019.106061

Kyrychenko, K., Laznenko, D., & Reshetniak, Y. (2021). Green university as an element of forming a sustainable public health system. *Health Economics and Management Review, 2*(4), 21–26. https://doi.org/10.21272/hem.2021.4-02

Letunovska, N., Saher, L., Vasylieva, T., & Lieonov, S. (2021). Dependence of public health on energy consumption: A cross-regional analysis. *E3S Web of Conferences, 250*, 04014. https://doi.org/10.1051/e3sconf/202125004014

Lisin, E., Lebedev, I., Sukhareva, E., & Komarov, I. (2014). Analysis of scenario of structural and technological modernization of the power industry in the context of competitive electricity markets. *International Economics Letters, 3*(3), 105–114. https://doi.org/10.24984/iel.2014.3.3.3

Matvieieva, Y. (2022). Modelling and forecasting energy efficiency impact on the human health. *Health Economics and Management Review, 2*, 78–85. https://doi.org/10.21272/hem.2022.2-09

Mentel, G., Vasilyeva, T., Samusevych, Y., Vysochyna, A., Karbach, R., & Streimikis, J. (2020). The evaluation of economic, environmental and energy security: Composite approach. *International Journal of Global Environmental Issues, 19*(1–3), 177–195. https://doi.org/10.1504/ijgenvi.2020.10037581

Ovaere, M., & Proost, S. (2022). Cost-effective reduction of fossil energy use in the European transport sector: An assessment of the fit for 55 package. *Energy Policy, 168*, 113085. https://doi.org/10.1016/j.enpol.2022.113085

Polcyn, J., Us, Y., Lyulyov, O., Pimonenko, T., & Kwilinski, A. (2022). Factors influencing the renewable energy consumption in selected European countries. *Energies, 15*(1), 010108. https://doi.org/10.3390/en15010108

Polyakov, M., Khanin, I., Bilozubenko, V., Korneyev, M., & Shevchenko, G. (2021). Factors of uneven progress of the European union countries towards a circular economy. *Problems and Perspectives in Management, 19*(3), 332–344. https://doi.org/10.21511/ppm.19(3).2021.27

Rossi, J., Srivastava, A., Hoang, T. T., Tran, Q. T., & Warneryd, M. (2022). Pathways for the development of future intelligent distribution grids. *Energy Policy, 169*, 113140. https://doi.org/10.1016/j.enpol.2022.113140

Salihaj, T., & Pryimenko, S. (2017). Modification of the international energy agency model (the IEA model of short-term energy security) for assessing the energy security of Ukraine *SocioEconomic Challenges, 1*(4), 95–103. https://doi.org/10.21272sec.1(4).95-103.2017.

Samusevych, Y., Maroušek, J., Kuzmenko, O., Streimikis, J., & Vysochyna, A. (2021a). Environmental taxes in ensuring national security: A structural optimization model. *Journal of International Studies, 14*(2), 292–312. https://doi.org/10.14254/2071-8330.2021/14-2/19

Samusevych, Y., Vysochyna, A., Vasylieva, T., Lyeonov, S., & Pokhylko, S. (2021b). Environmental, energy and economic security: Assessment and interaction. *E3S Web of Conferences, 234*, 00012. https://doi.org/10.1051/e3sconf/202123400012

Sedmíková, E., Vasylieva, T., Tiutiunyk, I., & Navickas, M. (2021). Energy consumption in assessment of shadow economy. *European Journal of Interdisciplinary Studies, 13*(2), 47–64. https://doi.org/10.24818/ejis.2021.12

Serpeninova, Y., Makarenko, I., Plastun, A., Babko, A., & Gasimova, G. (2020). Mapping of the responsible investments instruments in SDG 3 'good health and well-being' financing: EU and US experience. *Health Economics and Management Review, 1*(1), 106–115. https://doi.org/10.21272/hem.2020.1-10

Shkarupa, O., & Chasnyk, O. (2014). Sustainable energy management in a "green" economy – The case of Ukraine. *Marketing and Management of Innovations, 4*, 192–199. https://doi.org/10.21272/mmi.2014.4-22

Simionescu, M., Strielkowski, W., & Tvaronavičienė, M. (2020). Renewable energy in final energy consumption and income in the EU-28 countries. *Energies, 13*(9), 2280. https://doi.org/10.3390/en13092280

Simionescu, M., Strielkowski, W., & Gavurova, B. (2022). Could quality of governance influence pollution? Evidence from the revised environmental Kuznets curve in central and eastern European countries. *Energy Reports, 8*, 809–819. https://doi.org/10.1016/j.egyr.2021.12.031

Sotnyk, I., & Kulyk, L. (2015). Efficient energy management: Theoretical basis of the financial activity of energy service companies. *Marketing and Management of Innovations, 3*, 212–225. https://doi.org/10.21272/mmi.2015.3-19

Starchenko, L., Lyeonov, S., Vasylieva, T., Pimonenko, T., & Lyulyov, O. (2021a). Environmental management and green brand for sustainable entrepreneurship. *E3S Web of Conferences, 234*, 00015. https://doi.org/10.1051/e3sconf/202123400015

Starchenko, L. V., Samusevych, Y., & Demchuk, K. (2021b). Social and eco-friendly entrepreneurship: The keys to sustainability. *Business Ethics and Leadership, 5*(1), 118–126. https://doi.org/10.21272/bel.5(1).118-126.2021

Stępień, P., & Miciuła, I. (2016). Liberalization of the polish energy market and the EU commitments. *Czech Journal of Social Sciences, Business and Economics, 5*(2), 25–33. https://doi.org/10.24984/cjssbe.2016.5.2.3

Štreimikienė, D., Samusevych, Y., Bilan, Y., Vysochyna, A., & Sergi, B. S. (2022). Multiplexing efficiency of environmental taxes in ensuring environmental, energy, and economic security. *Environmental Science and Pollution Research, 29*(5), 7917–7935. https://doi.org/10.1007/s11356-021-16239-6

Strielkowski, W. (2017). Social and economic implications for the smart grids of the future. *Economics & Sociology, 10*(1), 310–318. https://doi.org/10.14254/2071-789X.2017/10-1/22

Taliento, M., & Netti, A. (2020). Corporate social/environmental responsibility and value creation: Reflections on a modern business management paradigm. *Business Ethics and Leadership, 4*(4), 123–131. https://doi.org/10.21272/bel.4(4).123-131.2020

Us, Y., Pimonenko, T., & Lyulyov, O. (2021). Energy efficiency profiles in developing the free-carbon economy: On the example of Ukraine and the V4 countries. *Polityka Energetyczna, 23*(4), 49–66. https://doi.org/10.33223/epj/127397

Vasylieva, T., Pavlyk, V., Bilan, Y., Mentel, G., & Rabe, M. (2021). Assessment of energy efficiency gaps: The case for Ukraine. *Energies, 14*(5). https://doi.org/10.3390/en14051323

Vysochyna, A., Samusevych, Y., & Starchenko, L. (2020). Convergence trends of environmental taxation in European countries. *E3S Web of Conferences, 202*, 03031. https://doi.org/10.1051/e3sconf/202020203031

Ziabina, Y., Pimonenko, T., & Starchenko, L. (2020). Energy efficiency of national economy: Social, economic and ecological indicators. *SocioEconomic Challenges, 4*(4), 160–174. https://doi.org/10.21272/sec.4(4).160-174.2020

Conclusions

Wadim Strielkowski

The proceedings volume provided an account of the pandemic-inspired policies for leadership, entrepreneurship, as well as sustainable development set within the context of the challenges posed by the recent war in Ukraine.

The impacts of COVID-19 are not limited to human health, because challenges related to managing global trade, geopolitics, or sustainable development are also present. There is a high level of uncertainty associated with the COVID-19 pandemic: it could have serious impacts on the world with regard to the quality of life; sustainable development in politics, environment, and economy; as well as on the world economy. Not only the economic losses but also social parameters such as changes in psychological sustainability are affected by the pandemic. In a way, COVID-19 has changed the way humans live on a global scale, by impacting the context of social, economic, and ecological aspects. To rescue economies and the mental health of citizens from the COVID-19 pandemic, many countries developed strategies, but years of struggles are needed to restore economic stability. A post-pandemic recovery may emerge, but this is likely to hinge on business leaders' ability to cope with the shocks in productivity and growth caused by the coronavirus crisis.

It becomes obvious that leadership, entrepreneurship, and sustainable development were all deeply affected by the COVID-19 pandemic. What seemed impossible some 2–3 years ago has now become a reality. The whole concept of leadership and leaders in the post-COVID-19 world has also undergone profound transformation. The recent pandemic caused a digital surge in the ways business or education are perceived or conducted, or the energy sources are handled. Thanks to it, frontier technologies such as metaverse, fintech, or cryptocurrencies became more ubiquitous and widespread. Hence, there exists a pressing need for eliciting recent updates

W. Strielkowski
Prague Institute for Qualification Enhancement, Prague, Czech Republic
e-mail: strielkowski@prizk.cz

W. Strielkowski (ed.), *Leadership, Entrepreneurship and Sustainable Development Post COVID-19*, Springer Proceedings in Business and Economics, https://doi.org/10.1007/978-3-031-28131-0

in the concepts of leadership, entrepreneurship, and sustainable development that would bring new insights into this topic and define the place and the role of leadership within the context of the European Union's Green Deal and the United Nations' Sustainable Development Goals (SGDs).

To conclude our volume of proceedings entitled *Leadership, Entrepreneurship and Sustainable Development Post COVID-19* which is based on the results and outcomes of the 8th International Leadership Conference 2022 "Novel Insights in the Leadership in Business and Economics After the COVID-19 Pandemic" (NILBEC2022) held in Prague, Czech Republic, from June 24 to June 25, 2022, by the Prague Institute for Qualification Enhancement, I would like to ask myself as well as our potential readers one important question, namely: *What are the positive changes for the leadership in business and economics based on lessons learned during the pandemic?*.

The world we live in keeps adapting and changing, so should we. The main results and outcomes of the 8th NILBEC2022 PRIZK International Conference might speak for themselves and provide us with plenty of food for thought.

Best wishes,
Wadim Strielkowski (Prague Institute for Qualification Enhancement)